D1084724

Praise for *Oil, Power, and War*

"Beautifully written and marvelously translated, *Oil, Power, and War* provides a detailed history of oil's impact on economic and technical advances—and, in turn, their impact on oil—over the past century. Extending its narrative through the events of early 2018, it offers a profound new understanding of oil's role in war and peace, growth and stagnation; and it casts new light on the foundations of national power and the challenge that lies ahead. A terrific education and an engrossing read."

—Dennis Meadows, coauthor of *The Limits to Growth*

"The definitive history of the rise and eventual fall of oil, brilliantly told. Auzanneau illuminates the history of our time driven by cheap oil and the persistent search for more at all costs. Insightful, authoritative, and essential reading. A dazzling and wise book."

—David Orr, author of *Dangerous Years*;
Paul Sears Distinguished Professor of Environmental
Studies and Politics, Oberlin College

"Matthew Auzanneau's ambitious new history of oil is a must read for anyone intrigued by the instrumental role of energy in the ebb and flow of modern civilization. This is a richly documented and beautifully written book, which tells a story that has not been fully told—until now. Auzanneau masterfully reveals the vast extent to which the arteries of today's politics, economics, and culture have been indelibly shaped by the rise—and decline—of the world's most abundant fossil fuel. In years to come, historians will refer back to Auzanneau's work as a definitive guide to the real role of oil in some of the most pivotal events in world history."

—Nafeez Ahmed, editor of INSURGE intelligence;
visiting research fellow at the Global Sustainability
Institute at Anglia Ruskin University

"Auzanneau's *Oil, Power, and War* is a fascinating and excellent book. It sets out in detail the extraordinary story of oil's discovery, production, pricing, and control, and throws light on the fears, misapprehensions, power plays, and conflicts that our addiction to this cheap and flexible form of energy has engendered. Auzanneau is particularly good at explaining the importance of oil in the sustenance of modern society, and therefore why the coming constraint to the global oil supply—caused by the current resource-limited plateau (and soon decline) in the global production of conventional oil—is likely to be so difficult. Hopefully lessons learned from our past mistakes, laid out so well in this book, can help guide us through the oil challenges that lie ahead."

—R. W. Bentley, editor of *The Oil Age*;
author of *Introduction to Peak Oil*

"Matthieu Auzanneau's book is a must for anyone who wants to understand the modern world. Our consumer society is based on cheap energy. Thus if you want to know the sources of the world's current wealth and how our economy is likely to evolve in the future, you must study the history of world oil since 1859. This book tells that story more fully, fairly, accurately, and entertainingly than any other to date. Indeed, previous accounts of the history of oil are now effectively obsolete."

—Jean Laherrère, petroleum engineer;
president of ASPO France

"An absolutely great book, and a nearly unbelievable summary of the history of oil. But this is not just the story of oil, it is also the story of humankind during the past two centuries or so, and it shows how almost everything that happened during those centuries links back to oil. Auzanneau presents a treasure trove of information. Did you know that Mussolini was lured into his disastrous Ethiopian campaign by hopes of finding oil there? Did you know that the British won the Battle of Britain partly because the fuel of their Spitfires had a higher octane number than that of the German Messerschmitts? Did you know that the Marshall Plan to rebuild the European economies was based on the idea of replacing Europe's dependence on American oil with a dependence on US-controlled Middle East oil? There's all this and much more in *Oil, Power, and War*, and the story of oil and humankind is not yet concluded. In the future it will be mostly about getting rid of oil before oil gets rid of us."

—Ugo Bardi, author of *Extracted*

OIL POWER AND WAR

OIL POWER AND WAR

A DARK HISTORY

MATTHIEU AUZANNEAU

Translated by John F. Reynolds
Foreword by Richard Heinberg

Post Carbon Institute
Corvallis, Oregon

Chelsea Green Publishing
White River Junction, Vermont
London, UK

Commissioning Editor: Shaun Chamberlin
Editor: Joni Praded
Project Manager: Alexander Bullett
Copy Editor: Deborah Heimann
Proofreader: Angela Boyle
Indexer: Ruth Satterlee
Designer: Melissa Jacobson

Printed in the United States of America.
First printing October, 2018.
10 9 8 7 6 5 4 3 2 1 18 19 20 21 22

Our Commitment to Green Publishing

Chelsea Green sees publishing as a tool for cultural change and ecological stewardship. We strive to align our book manufacturing practices with our editorial mission and to reduce the impact of our business enterprise in the environment. We print our books and catalogs on chlorine-free recycled paper, using vegetable-based inks whenever possible. This book may cost slightly more because it was printed on paper that contains recycled fiber, and we hope you'll agree that it's worth it. Chelsea Green is a member of the Green Press Initiative (www.greenpressinitiative.org), a nonprofit coalition of publishers, manufacturers, and authors working to protect the world's endangered forests and conserve natural resources. *Oil, Power, and War* was printed on paper supplied by Thomson-Shore that contains at least 30% postconsumer recycled fiber.

Library of Congress Cataloging-in-Publication Data

Names: Auzanneau, Matthieu, author. | Reynolds, John F., translator. | Heinberg, Richard, author of foreword.
Title: Oil, power, and war : a dark history / Matthieu Auzanneau ; translated by John F. Reynolds ; foreword by Richard Heinberg.
Other titles: Or noir. English
Description: White River Junction, Vermont : Chelsea Green Publishing, 2018. | Translation of: Or noir : la grande histoire du petrole. | Includes bibliographical references and index.
Identifiers: LCCN 2018012052 | ISBN 9781603587433 (hc) | ISBN 9781603587440 (ebook)
Subjects: LCSH: Petroleum industry and trade—History. | Petroleum industry and trade—Political aspects.
Classification: LCC HD9560.5 .A75813 2018 | DDC 338.2/728—dc23
LC record available at https://lccn.loc.gov/2018012052

Chelsea Green Publishing
85 North Main Street, Suite 120
White River Junction, VT 05001
(802) 295-6300
www.chelseagreen.com

Contents

PART TWO
SPRING
1945–1970

PART THREE
SUMMER
1970–1998

PART FOUR
AUTUMN
1998–20??

Foreword

Come and listen to my story 'bout a man named Jed
A poor mountaineer barely kep' his fam'ly fed
And then one day he was shootin' at some food
And up through the ground come a-bubblin' crude.
Oil, that is. Black gold. Texas tea.

Well the first thing you know old Jed's a millionaire.
The kinfolk said, "Jed, move away from there."
They said, "Californy is the place you ought to be,"
So they loaded up the truck and they moved to Beverly.
Hills, that is. Swimming pools. Movie stars.

—Paul Henning

Perhaps the most instantly recallable verse on the subject of petroleum, the theme-song lyric to the hit 1960s television series *The Beverly Hillbillies* tells a tale of sudden wealth. It is a perfect touchstone for the real story of humanity's experience with liquid hydrocarbons.

In the real story, riches consisted both of the billions accumulated by the great magnates of the petroleum industry—including John D. Rockefeller, J. Paul Getty, H. L. Hunt, and Charles and David Koch—and also the quickly growing economic output of industrial civilization once it came to be fueled by oil. This novel source of energy spawned entire new industries—notably the automotive, aviation, and plastics industries—while revolutionizing existing ones (agriculture, forestry, fishing, shipping, manufacturing, lubricants, chemicals, paints, dyes, cosmetics, road paving, and pharmaceuticals). It propelled humanity into an age of mobility and rising expectations.

But sudden acquisition of wealth is just the initial theme in both narratives—that of *The Beverly Hillbillies* and that of the modern industrial world. The saga of Jed Clampett and his family is a comedy in which city slickers try to siphon off some of the Clampetts' fortune. While Jed, Granny, Ellie May, and Jethro always manage to get the better of the various grifters and hangers-on they encounter, we suspect that their affluence may be transitory and that the final episode may see the Clampetts return to shooting squirrels in order to fill Granny's soup pot.

Similarly, the real story of oil is of fortunes lost, betrayal, war, espionage, and intrigue. In the end, inevitably, the story of oil is a story of depletion. Petroleum is a nonrenewable resource, a precious substance that took tens of millions of years to form and that is gone in a comparative instant as we extract and burn it. For many decades, oil-hungry explorers, using ever-improving technology, have been searching for ever-deteriorating prospects as the low-hanging fruits of planet Earth's primordial oil bounty gradually dwindle. Oil wells have been shut in, oil fields exhausted, and oil companies bankrupted by the simple, inexorable reality of depletion.

It is impossible to understand the political and economic history of the past 150 years without taking account of a central character in the drama—oil, the magical wealth-generating substance, a product of ancient sunlight and tens of millions of years of slow geological processes, whose tragic fate is to be dug up and combusted once and for all, leaving renewed poverty in its wake. With *Oil, Power, and War*, Matthieu Auzanneau has produced what I believe is the new definitive work on oil and its historic significance, supplanting even Daniel Yergin's renowned *The Prize*, for reasons I'll describe below.

The importance of oil's role in shaping the modern world cannot be overstated. Prior to the advent of fossil fuels, firewood was humanity's main fuel. But forests could be cut to the last tree (many were), and wood was bulky. Coal offered some economic advantages over wood. But it was oil—liquid and therefore easier to transport; more energy-dense; and simpler to store—that turbocharged the modern industrial age following the development of the first commercial wells around the year 1860.

John D. Rockefeller's cutthroat, monopolist business model shaped the early industry, which was devoted mostly to the production of kerosene for lamp oil (gasoline was then considered a waste product and often discarded into streams or rivers). But roughly forty years later, when Henry Ford developed the automobile assembly line, demand for black gold was suddenly as explosive as gasoline itself.

Speaking of explosions, the role of petroleum in the two World Wars and the armament industry in general deserves not just a footnote in history books but serious and detailed treatment—such as it receives in this worthy volume. Herein we learn how Imperial Japan and Nazi Germany literally ran out of gas while the Allies rode to victory in planes, ships, and tanks burning refined US crude. Berlin could be cut off from supplies in Baku or North Africa, and Tokyo's tanker route from Borneo could be blockaded—but no one could interrupt the American war machine's access to Texas tea.

In the pages that follow, we learn about the origin of the decades-long US alliance with Saudi Arabia, the development of OPEC, the triumph of the petrodollar, and the reasons for both the Algerian independence movement and the Iranian Revolution of 1979. Auzanneau traces the postwar growth of the global economy and the development of consumerism, globalization, and car culture. He recounts how the population explosion and the Green Revolution in agriculture reshaped demographics and politics globally—and explains why both depended on petroleum. We learn why Nixon cut the US dollar's tether to the gold standard just a year after US oil production started to decline, and how the American economy began to rely increasingly on debt. The story of oil takes ever more fascinating turns—with the fall of the Soviet Union after its oil production hit a snag; with soaring petroleum prices in 2008 coinciding with the onset of the global financial crisis; and with wars in Iraq, Syria, and Yemen erupting as global conventional oil output flatlined.

As I alluded to above, comparisons will inevitably be drawn between *Oil, Power, and War* and Daniel Yergin's Pulitzer-winning *The Prize*, published in 1990. It may be helpful therefore to point out four of the most significant ways this work differs from Yergin's celebrated tour de force.

The most obvious difference between the two books is simply one of time frame. *The Prize*'s narrative stops in the 1980s, while *Oil, Power, and War* also covers the following critical decades, which encompass the dissolution of the Soviet Union, the first Gulf War, 9/11, the US invasions of Afghanistan and Iraq, the global financial crisis of 2008, and major shifts within the petroleum industry as it relies ever less on conventional crude and ever more on unconventional resources such as bitumen (Canada's oil sands), tight oil (also called shale oil), and deepwater oil.

Second, and more importantly, to a greater extent than Daniel Yergin, Matthieu Auzanneau displays a keen understanding of the central role of energy in society and the economy, and therefore of oil's pivotal significance in the unprecedented economic growth that occurred during the twentieth century. All manifestations of human power, whether economic, military, or political, are physically grounded in energy. Power, after all, is energy over time (one watt equals one joule per second). Therefore a recounting of political, economic, and military history—even one that concerns itself with the history of the oil industry—will fail to successfully trace the sources, routes, and consequences of power if it is not based on a sound understanding of how energy works. In order to appreciate oil's role in recent history, we must begin by understanding it as a concentrated, cheap, and portable store of energy. Yergin understands that oil is a valuable commodity, but *The Prize* never quite manages to explain why it is valuable, or why it is so closely linked with military, economic, and political power. Because Auzanneau begins his introduction with an explanation of oil's energetic qualities, the reader is far better prepared to understand the historic human power plays centered on this remarkable substance.

A third distinction lies in point of view. Yergin has been criticized for telling the story of oil from the perspective of the winners—the major oil companies, the oil barons, and the Anglo-American elites who have shaped global economic and political realities for the past century and more. Auzanneau brings an outsider's perspective, one that is far more critical of, for example, US political interference in Iran in the early 1950s. While Yergin repeats the usual explanation for the 1970s oil crises, Auzanneau digs deeper and shows why failing US oil production provided a motive for American policy makers to quietly convince their Arab client states to hike prices so as to enable US-based oil companies to earn higher profits. Yergin acts essentially as a cheerleader for the oil industry; Auzanneau is a journalist who is aware of the enormous ecological and social consequences of our dependency on petroleum.

Finally, unlike Yergin and other historians of the oil industry, Auzanneau frames his tale of petroleum as a life cycle, with germination followed by spring, summer, and autumn. There is a beginning and a flourishing, but there is also an end. This framing is extremely helpful, given the fact that the world is no longer in the spring or summer of the oil era. We take petroleum for granted, but it's time to start imagining a world, and daily life, without it.

Taken together, these distinctions indeed make *Oil, Power, and War* the definitive work on the history of oil—no small achievement, but a judgment well earned.

Over the past decade, worrisome signs of global oil depletion have been obscured by the unabashed enthusiasm of energy analysts regarding growing production in the United States from low-porosity source rocks. Termed "light tight oil," this new resource has been unleashed through application of the technologies of hydrofracturing (fracking) and horizontal drilling. Total US liquid fuels production has now surpassed its previous peak in 1970, and well-regarded agencies such as the Energy Information Administration are forecasting continued tight oil abundance through mid-century.

Auzanneau titles his discussion of this phenomenon (in chapter 30), "Nonconventional Petroleum to the Rescue?"—and frames it as a question for good reason: Skeptics of tight oil hyperoptimism point out that most production so far has been unprofitable. The industry has managed to stay in the game only due to low interest rates (most companies are heavily in debt) and investor hype. Since source rocks lack permeability, individual oil wells deplete very quickly—with production in each well declining on the order of 70 percent to 90 percent in the first three years. That means that relentless, expensive drilling is needed in order to release the oil that's there. Thus the tight oil industry can be profitable only if oil prices are very high—high enough, perhaps, to hobble the economy—and if drilling is concentrated in the small core areas within each of the productive regions. But these "sweet spots" are being exhausted rapidly. Further, with tight oil the energy returned on the energy invested in drilling and completion is far less than was the case with American petroleum in its heyday.

It takes energy to fell a tree, drill an oil well, or manufacture a solar panel. We depend on the energy payback from those activities to run society. In the miraculous years of the late twentieth century, oil delivered an averaged 50:1 energy payback. It was this, more than anything else, that made rapid economic growth possible, especially for the nations that were home to the world's largest oil reserves and extraction companies. As the world relies ever less on conventional oil and ever more on tight oil, bitumen, and deepwater oil, the overall energy payback of the oil industry is declining rapidly. And this erosion of energy return is reflected in higher overall levels of debt in the oil industry and lower overall financial profitability.

Meanwhile the industry is spending ever less on exploration—for two reasons. First, there is less money available for that purpose, due to declining financial profitability; second, there seems comparatively little oil left to be found: Recent years have seen new oil discoveries dwindle to the lowest level since the 1940s. The world is not about to run out of oil. But the industry that drove society in the twentieth century to the heights of human economic and technological progress is failing in the twenty-first century.

Today some analysts speak of "peak oil demand." The assumption behind the phrase is that electric cars will soon reduce our need for oil, even as abundance of supply is assured by fracking. But the world is still highly dependent on crude oil. We have installed increasing numbers of solar panels and wind turbines, but the transition to renewables is going far too slowly either to avert catastrophic climate change or to fully replace petroleum before depletion forces an economic crisis. While we may soon see more electric cars on the road, trucking, shipping, and aviation will be much harder to electrify. We haven't really learned yet how to make the industrial world work without oil. The simple reality is that the best days of the oil business, and the oil-fueled industrial way of life, are behind us. And we are not ready for what comes next.

How could a story so essential to our understanding of the present and recent past be so poorly understood by such huge swathes of the general public? *Oil, Power, and War* helps enormously by offering us a sweeping yet also detailed view of how we got to this juncture. Where we go from here, as always, is up to us.

Richard Heinberg
Senior Fellow, Post Carbon Institute
April 2, 2018

Introduction

Progress has long been considered a given: a guaranteed occurrence in human societies, a wheel that once set in motion continues to spin, aided only by human intelligence and innovation. But what really sparks or tempers progress?

The answer is energy potential—a physical reality measured by its capacity to change the nature of other things around it, to alter the order of the world, or to strengthen it. Each time that we put something in motion, each time that the state of something changes in one way or another, a flow of energy is in play. The economy—the framework around which our industrial society is ordered—and all of the technical progress it mobilizes are no exceptions.

In other words, energy is the ultimate universal currency. As Georges Bataille wrote in 1949, "Essentially wealth is energy: Energy is the basis and the end of production."[1] Without adequate energy sources, ingenuity would be rendered impotent, its fruits out of reach. Progress would not progress.

Today, fossil fuels provide four-fifths of the energy we use. Nothing has changed on this front since the early days of the steam locomotive—apart from the ever-increasing amount of energy used, which has been compounded many times over to keep the economic machinery churning. Since the end of the Second World War and today more than ever, oil remains, among fossil fuel sources, the principal and most precious fuel of what I call "technical humanity." It is omnipotent, versatile, polymorphic, ubiquitous.

Yet energy is ambivalent about what it sets in motion. It also has limits. And it is clear to anyone who actually looks that material limits define what is possible and what is impossible. Science fiction writer Frank Herbert put it this way: "Energy absorbs the structure of things, and builds with these structures."[2] This obvious fact, however, is often ignored. Since the dawn of technological progress, the human spirit has fooled itself into believing that we are progress's empowering force. But we are merely driving the course of progress. The tank of its real empowering force is still largely fueled by oil.

This book explores the paths followed so far by our primary source of energy, as well as the manner in which these paths have determined many dynamics and balances of power among us. For more than a century and a half, black gold has remained the most secure source of wealth. The industry that extracts it from beneath Earth's surface has experienced revenues as much as ten times higher than any other industry's. Economic and military power structures, as well as many critical aspects of our way of life, have been metabolized by the energy derived from oil, and shaped by forms specific to it.

The destiny of the strongest industrial nations is pegged to oil, starting with the United States, where more wells have been drilled than anywhere else on Earth. Absorbed by a nearly universal appetite for the American way of life, the world's population has become, in spite of our misfortunes, the largest and most affluent in history. Now where are we headed?

It was on 9/11 that I began to ask deeper questions about the link between power and energy. Watching events unfold from France, where I was working as a journalist, I found myself saying, "Just wait and see, Saddam will end up taking the punishment." Afterward, it was extremely hard for me to believe that Saudis were allowed to board a plane and leave the United States immediately after the attacks, without having to answer any questions from US authorities. With great astonishment I learned that the Bush and bin Laden families had actual direct connections. And I ended up wondering, what were the odds that these kinds of connections would exist?

Like many who grew up during the end of the great American century, I had long been immersed in American culture. I had an insatiable appetite for its cinema, its music, its literature, and more generally a sustained fascination with its creativity, humor, and vitality. I was dazzled by its intimacy with each innovation that the new technological world offered to me, a child of an "old Europe" so often tired of itself. That's why the tragic catastrophe of the 2003 invasion of Iraq was a personal affront to my idea of America, the one I cherish so much.

And then I wondered if the Bush phenomenon constituted a singularity in American history, or if, on the contrary, it revealed a *filum*, a logical historical continuity between successive incarnations of US power and the energy sources essential for the expression of this unprecedented power.

Oil, Power, and War is the result of my inquiry.

I have chosen to divide the story that follows into seasons. Oil's power took root in the United States a century and a half ago, entering its springtime around 1945, a period of time often recounted with nostalgia. That power passed through its summer solstice shortly before the decisive oil crisis of 1973, and is likely to extinguish somewhere in the course of the twenty-first century—giving birth to a new era, which will either be better or worse than this one. It's up to us, today, to decide.

Part One of the book, "Germination," recounts the earliest history of oil, before explaining how the Oil Age was, so to speak, born naturally—from the ground up—in the United States. We learn how the source of incomparable profit that black gold constitutes gave rise to many crucial aspects of modern capitalism as well as some of the fundamental structures of Wall Street. We look at how the availability of oil ushered in machinery that required ever more machinery and ever more oil, how the power of oil changed our notions of labor and shaped the outcome of the First World War, and how Iraqi oil was already then among the spoils of war shared by the victors. We learn how Big Oil forged close links with Nazi industry, how the knots of the Second World War were tightened around access to crude for Germany and Japan, and finally how those knots were eventually undone thanks to the then unequaled superiority of the energy power Mother Nature entrusted (by chance) to the United States of America.

It was this same energy power that presided over the arrival of spring for the oil era (Part Two) and created the golden age of the American Empire. It was an age during which the streams of power flowed directly from oil reserves controlled by the "princes" of the capitalist world establishment—be they the earliest (the Rockefeller dynasty) or those preparing their advent (the Bush clan, in particular)—all the while relying on a few secret structures that unite the ultimate networks of industry, finance, and intelligence. And it was an age when the waves of power stirred by oil often formed the prevailing matrix of political power, became the source of its dubious financing, and opened the floodgates to the great upheaval that defines the modern world: unprecedented population explosion, transformed lifestyles, and greedy aspirations to push any limits constraining human passions, for better and for worse.

The summer solstice (Part Three) was reached with the relentless decline of conventional oil across the United States, beginning in 1970. This ecological event, which revealed the existence of certain "limits to growth," caused a cascade of economic and political turbulences, starting with the oil shock of 1973—a shock that some in Washington saw coming from afar and even encouraged, in order to let American hegemony mutate and perpetuate itself. That crisis led the American empire (again, naturally) to move toward the fabulous sources of black gold in the Persian Gulf, employing its tools of policy and power—tools that were both financial and military, and often covert and terrifying. It was in this summer of the Oil Age that a Faustian bargain was made around what one of the founders of OPEC came to call "the devil's excrement": Oil provided, to those who could play, a pinnacle of overindulgence; others, as we will see, paid the price.

The spiral accelerated faster with autumn (Part Four), which began at the turn of the millennium, when it finally became obvious that oil is not an inexhaustible fountain of youth. Nor is it innocuous—whether concerning economic growth, climate change, terrorism, or war. We explore the connections between the jostling to secure oil assets in a world whose easily accessible resources are dwindling, the occupation of Iraq and the endless wars in the Middle East, the financial crisis of 2008, and other defining features of recent decades, such as the appearance of the latest fix to human oil addiction: shale oil. More than any others, those years make starkly clear why overcoming the free reign that oil has exercised on humanity is an urgent objective. An objective that remains decidedly out of reach at present.

So, now, we can expect winter to come. How will we prepare to face the harsh next season that will soon be upon us?

Energy from petroleum arrived at a key moment in the development of our species. Today, the very thing that drives technical humanity has also become its nemesis. Risks multiply as the combustion of fossil fuels generates greenhouse gases, which in turn upset the planet's climate. This danger, for the time being unresolved, masks another that is more direct and perhaps more immediate. Once consumed in our various machines, energy dissipates and disintegrates irretrievably in the form of heat, becoming almost useless. In this way our limited reserves of fossil energy are being consumed with ever-increasing greed. The history of those reserves is one of a series of depletions offset by the discovery of new stocks. But, lacking accessible reserves able to compensate for the natural decline of the large number of fields discovered in the last century, we risk a drastic disruption in supply—perhaps before 2030, even as early as around 2020. In any case, much too quickly for an industrial humanity that has sprouted from fields of crude to learn to live in peace with a perpetual lack of crude.

The perils of our dependency on fossil fuels have been understood since the beginning of the Industrial Revolution, though they've been consistently ignored.[3] And so, the abundance of energy offered to us in the form of oil has been able to transform the world by itself. The pages ahead tell the story of how that happened, explore the conditions that allowed this abundance to precipitate turbulence, and revisit the history of societies lifted to excess and the lives of people caught up in the storm. *Oil, Power, and War* is also the history of people and institutions—corporations, nations—who fight to stay in the eye of the cyclone, believing it possible to impose their own direction on the vortex, or to transgress its limits.

PART ONE

GERMINATION

. . . TO 1945

ONE

A Seed Is Planted

A little temple stands on a dry and desolate patch of land one hour's hike from the Caspian Sea, on the outskirts of Baku. Its construction is so simple that upon its discovery one puzzles at who might have built it, exactly where Asia meets Europe to the east of the Caucasus Mountains. Clearly a place of ritual, it exudes a primitive strength.

Placed in the center of a dusty courtyard in what was probably a caravansary—a fortified rest stop on the Silk Road—the structure is a cube of limestone with thick walls, opened by arches on each side and topped with a four-sided dome that lends it an archaic appearance. A uniquely distinctive, fine piece of forged iron is attached to one of the dome faces: It's the trident-like trishula of the Hindu god Shiva, a symbol of creation, perpetuation, and destruction, pointing to the sky more than 3,000 kilometers from the Indus Valley.

Four short chimneys extend upward from the walls of the temple. Below the dome stands a large fireplace; a few meters to the side of the temple is a sixth source of incandescence, an eternal flame, which had once glowed from the center of a circular pit. Worshipped for centuries, it burned methane and naphtha, which seeped spontaneously from the porous rock beneath the foundation.

In its present form, the temple seems to have been built in the seventeenth century by a small community of merchants from India. But the origins of the Ateshgah, the Fire Temple of Baku, are much older—forgotten in the fog of ancient times. The Zoroaster faithful have long maintained that the flame has been revered since the deluge. Zoroastrianism—one of the oldest monotheistic religions, perhaps the oldest—was the major religion of Persia's Achaemenid Empire, which included up to fifty million people (perhaps one-third of the human population during that era), from the sixth to the fourth century BC. One of the principal tenets of the Zoroaster doctrine is a belief in free will. Early Zoroastrians took to heart, maybe for the first time in recorded history, the notion that humans can choose between good and evil. Fire was considered an agent of purity, capable of exposing these opposing forces—a belief perpetuated by Judaism, Islamism, Christianity, and across the gnostic spectrum.[1]

The presence of naturally occurring fire along the Caspian Sea must have had special significance for the Zoroaster faithful, who still today practice their religion in Zoroastrian fire temples. Evidence of fire worshipers attracted to the area can be traced back to the Middle Ages. As far back as the seventeenth century, Christian travelers sighted Zoroastrians and Hindu yogis at the Ateshgah, located in a place they called Surakhani, which seems to mean "red fountain." The German naturalist Engelbert Kaempfer described seeing seven eternal fire pits during his visits in 1683.

Two centuries later, in 1883, the Ateshgah was abandoned. The rush for black gold in the capital city of Baku was so intense that the air had become fetid, unbreathable. Exploitation of hydrocarbon resources had begun to transform the human experience. Greedy men hearing wild tales of fortune were attracted to Baku. Some said that, offshore from the distant Muslim city, the waters of the Caspian Sea can catch fire when enormous bubbles of gas capable of overturning boats rise to the surface. These fires erupting from the sea have, moreover, the fascinating property of being consumed while emitting an imperceptible heat. The oil kingpin of Baku was Ludvig Nobel, brother of Alfred Nobel, the inventor of dynamite and creator of the famous prize. One day the wealthy Swedish engineer crossed, unscathed, through one of these walls of fire in his steam-powered yacht.[2]

Shortly before the First World War, the region of Baku became, for a time, the main source of black gold for all of humanity. The wells drilled around its natural outcrops of naphtha produced more than half of the world's crude oil. On a 1919 stamp printed by the ephemeral Republic of Azerbaijan, subsequently conquered by the Red Army and assimilated into the Soviet Union, the Baku Fire Temple appears. Behind it, five strange pyramids—oil derricks.

After a century of intense exploitation, the oil reserves dried up; the eternal fire of Ateshgah stopped burning in 1969.[3] Today, the flame is rekindled for tourists using a pipeline that transports gas tapped further and further out to sea. Decade after decade, both Soviet and Western corporations have had to move their rigs, to drill ever deeper wells more distant from the shores of the closed Caspian Sea. Since 2011, despite recent titanic efforts, the production of offshore platforms is slowing in Azerbaijan, maybe irreversibly so.[4]

Around the temple of fire a few rusty pumps remain, often immobile, frozen in time. Between them snakes the highway to the airport.

A Massive Stock of Vital Energy, Transformed

Oil began to form on Earth more than a billion years ago. Its arrival coincided with the explosion of complex cell life in the oceans. In all their forms, from the heaviest (bitumens) to the lightest (natural gas) to conventional liquid oil, hydrocarbons are fossils: They originate from decomposed organisms, compressed and heated in the depths of sedimentary layers throughout geological eras, all around the planet. Unlike coal, which is formed from decayed plants (wood, leaves, seeds, and the like), hydrocarbons come from tiny marine organisms deposited on the seabed. These fossil energies are solar energy metabolized by photosynthesis and then stored for eons in the Earth's crust.

More than half of the oil exploited today formed one hundred million to two hundred million years ago during the Jurassic and Cretaceous—the time of the dinosaurs.* During these periods, tectonic forces gradually raised the oceans more than 200 meters above their current levels. The continents were invaded by water. The shores hosted environments favorable to the development of phytoplankton, then to the accumulation of its debris in the sediment of shallow, calm, warm seas, and lagoons. The phenomenon was particularly widespread along the banks of the Tethys, a great ocean of this era, which separated the paleocontinents from Laurasia to the north (which would form North America, Europe, and Asia) and Gondwana to the south (South America, Africa, India, Antarctica, Australia). It is there, on the shores of the ancient Tethys, that most of the giant oil deposits were discovered, spanning from the Middle East to Mexico in a path traveling through North Africa. Random movements of the tectonic plates over millions of years created the borders that unite and divide the nations of the Oil Age.

The concentration of carbon dioxide in the atmosphere was much higher than it is today, as was the average atmospheric temperature (which likely reached 20°C [68°F], compared with 14.6°C [58.28°F] in 2012). During these intensely hot and humid periods, it would probably have been possible to swim at the South Pole. The microscopic life-forms that became petroleum accumulated in the sediment because, contrary to the fate usually reserved for dead organisms, they lacked the oxygen needed for decomposition. Hundreds of millions of tons of plankton, asphyxiated, subsequently transformed into petroleum. According to a hypothesis formulated by several paleoclimatologists, it is possible that the current greenhouse gas effect could lead to such anoxic conditions.

---

* Sixty-five million years ago, the impact of the asteroid responsible for the dinosaurs' extinction contributed to the formation of the Cantarell oil field, off the coast of Mexico—one of the largest in the world, and among those that is quickly depleting today.

An irony of history? If we continue on the current path, humankind could release enough carbon dioxide into the atmosphere to raise its temperature more than 6°C (10.8°F) by the next century. By consuming fossil fuels, we have initiated a phenomenon capable of bringing temperatures back to the level of one hundred million years ago, when a fantastic accumulation of marine-life debris spurred the formation of petroleum.

In just a century and a half of industrial development, humankind has pumped nearly half of the Earth's conventional crude oil, which geological evolution took tens of millions of years to produce. In so doing, it has altered the conditions for the development of life on Earth at an unprecedented pace—and probably irreversibly, on the timescale of human societies.

A Sea Soup Pressure-Cooked over Millions of Years

In order for oil to appear, a very particular concurrence of geological phenomena is necessary. The base material is kerogen, the solid organic substance that forms when anaerobic bacteria slowly decomposes plankton and other organic matter within very fine marine clay sediments. Sediment containing kerogen, which generally represents less than 5 percent of its mass, is then covered by new sedimentary layers: They are gradually buried at depths of between 2 and 10 kilometers, sometimes more. The Earth's crust then acts as an oven: As it is buried and slowly compressed, the kerogen is cooked at temperatures ranging from 50°C to 300°C (122°F to 572°F), depending on the depth. Kerogen gradually "cracks": The large organic molecules that constitute it are reduced to smaller molecules of hydrocarbons—so called because they combine hydrogen and carbon atoms. What is called crude oil is actually a mixture of different types of oils, or various hydrocarbon molecules. This mixture varies greatly from one deposit to another and, to a lesser extent, within the same deposit. The purity of the mixture is frequently impaired by the presence of sulfur from volcanic activity. The more violent and intense the thermal cracking of kerogen, the smaller the hydrocarbon molecules become. Oils are formed, either heavy and viscous or fluid and light. When there are fewer than five carbon atoms per molecule, natural gas is produced. Rock that contains kerogen is called bedrock. Hydrocarbons that form there are pushed toward the surface by strong pressure from below—rising above the more dense groundwater and seeping out through permeable rocks. But in order for a field of oil or gas to form, somewhere above the bedrock there must be an impermeable layer of rock capable of sealing the hydrocarbons under a subterranean fold known as an *anticline*, or under a fault trap. Below this barrier, oil and natural gas concentrate in the interstices of a porous and highly permeable rock (usually sandstone or limestone) that petroleum geologists refer to as a reservoir. If the impermeable layer, usually consisting of clay or salt, is absent, the hydrocarbons slowly reach the surface. There, the lightest evaporate and the heaviest degrade to form bitumen.[5]

The simplest hydrocarbons, such as methane, are among the very first molecules spontaneously formed in the interstellar vacuum of supernovas, thanks to gravity. Under high pressure, in terrestrial rock soup, a phenomenal diversity of mixtures of naphthenic, formic, or asphaltic hydrocarbons emerges: essentially alkanes (such as methane, ethane, propane, butane, pentane, octane, and pentacontane), capable of supplying an almost infinite variety of saturated, unsaturated, cyclic, aromatic, alkene, and alkyne molecules; or even simple and complex alcohols, such as methylene, ethylene, benzene, butadiene, propylene, glycerol, acetone, toluene, polyamide, phenol, polyurethane, and the like. By simple distillation followed by fractionation; by cracking, oxidation, hydrogenation, reforming, or visbreaking; by assembling monomers into polymeric macromolecules, an incredibly diverse and precious array of products can be offered to technical humanity, forming the crux of the industrial era's pivotal advances and mass consumption.

Build, Coat, Lubricate, Heal, Burn: Since the Dawn of Civilization

In *The Epic of Gilgamesh*, one of the earliest known literary works, the protagonist, who resembles the Bible's Noah, uses 18,000 liters of bitumen to make his ark watertight.[6] It is written in Genesis that, after the flood, Noah's descendants had access to bitumen to erect the Tower of Babel: "They said one to another, 'Come! Let us make bricks and bake them in the fire!' The brick served as stones and the bitumen was used for mortar. They said, 'Come! Let us build a city and a tower whose summit penetrates the heavens! Let us make a name and not be scattered all over the world.'"[7] The myth of God's destruction of the Tower of Babel originated somewhere around the ruins of Babylon, a cradle of civilizations situated in the heart of present day Iraq, one of the largest oil producers. The Greek historian Herodotus attests to the use of "hot bitumen" to bind the bricks that were used to build the ancient capital of Mesopotamia: "At eight days from the city of Babylon is the town of Is, situated on a small river of the same name, which flows into the Euphrates.

A great quantity of bitumen was extracted from this river and used to seal the walls of Babylon."[8] The Bible also relates how the mother of Moses "coated with asphalt and pitch" the basket of papyrus in which she concealed her child among the reeds of the Nile.[9]

Wherever it flourished, and long before it asserted itself as indispensable to the extraction of all other raw materials, men had discovered all sorts of uses for petroleum—a word derived from the medieval Latin *petroleum*, meaning "rock oil." Akkadian tablets from 2200 BC refer to *naptu*—the source of the Arabic name for oil, *naft*, and the Greek one, *naphtha*. Ancient Egyptians seem to have used oil for the conservation of their mummies (the Arabic word *mumia* means bitumen).[10*] The black substance outlining the eyes on Mesopotamian funerary statues was made of bitumen as well. From springs of black gold in what is now Los Angeles, the Yokuts people of California collected bitumen to caulk their canoes, and the women used it like starch to stiffen their clothes.[11]

All around the world, the medicinal uses of "rock oil" are as old as they are numerous. On the island of Sumatra, oil has always been used as a compress to treat rheumatism. In the first century BC, in his *Natural History*, the Roman scientist Pliny the Elder evoked the therapeutic virtues of petroleum, listing naphtha deposits exploited in Mesopotamia, Judea, and Syria. In the thirteenth century *Book of Marvels*, the Venetian merchant traveler Marco Polo relates that, on Armenia's northern frontier, bordering Georgia, "there is a spring that gushes oil at such a rate that one hundred ships could load at once. It is not good to eat, but it burns well and is good for salving scabies on men and animals alike and for treating itching and mange on camels. Men come from far off to fetch this oil, and in all the lands around they burn no other oil than this."[12] In China, bitumen was rubbed on patients' bodies to soothe ulcers and cure ringworm, scabies, and arrow wounds. It also was used in alchemy. A seventeenth-century Chinese text recommends ingesting oil to regrow teeth and hair.[13] Rembrandt's "The Night Watch" is actually a daytime scene: This masterpiece owes its nickname to a poorly dried coat of bitumen extracted from Judea.

Small wells flowing with petroleum are mentioned in two-thousand-year-old chronicles of the Middle Kingdom. During the centuries following, wells of 300 to 1,000 meters deep were dug in China, where buckets were lowered to the bottom. These were salt mines and the bitumen was a secondary product.[14] When the Muslims conquered Mesopotamia and Persia in the year 640 AD,

* Buffon reports that the ambassador of Persia offered "mummy balm or mumia" to King Louis XIV who kept it in his cabinet inside "two boxes of gold." The French naturalist noted: "This balm was only bitumen, and had merit only in the spirit of those who offered it" (Georges-Louis Leclerc de Buffon, natural history, *Œuvres Complètes de Buffon*, vol III, Rapet and Cie, Paris, 1818, p. 204).

they found hundreds of bitumen and naphtha quarries. Bitumen's importance was so great that as early as the ninth century, in order to control the quarries, the Abbasid caliphate appointed nafta walis, governors of oil. To one of them, a disillusioned friend one day addressed these verses: "You, where is your modesty? / As if you'd been given the throne itself! / If by guarding the stinking wells / You have gained such aloofness / How would you behave if instead / You were guarding amber and musk?"[15]

The Chinese and the Burmese were soon able to light up, using torches made of bamboo rods filled with bitumen. The Chinese and the Romans used oil to lubricate the axles of their chariots. About 1070 AD, Chinese intellectual Shen Gua described a method of making ink from the burning residue of rock oil, pronounced "shiyou" in Chinese.[16] As early as the dawn of the Middle Ages, Baghdad's roads were frequently covered with asphalt. (It was not until 1838 that a Parisian street was covered with asphalt for the first time).[17] The *Secretum Secretorum*, a work attributed to Rhazes, a ninth-century Persian scholar who lived in Baghdad, confirms the likely common use of oil lamps. This book, which for Western readers became one of the most influential works of the Middle Ages, describes several methods of distilling lamp oil (*naft abyad*, or white oil) using an alembic, ten centuries before oil lamps illuminated the birth of industry in the West.[18]

From the first steps of civilization, of agriculture, and of maritime trade in Mesopotamia, the drive to control hydrocarbons—like the need to secure water access—was one of the major causes of war, because bitumen was necessary for waterproofing irrigation canals and boats.[19]Among all the different uses for oil, its military function became the most prominent, from China to Europe, long before the industrial era began. In 578 AD, Emperor Wu of the Northern Zhou dynasty used bitumen to set fire to the Turks' battle gear. The grease burned intensely, even in contact with water, and saved the city of Jinquan from the attack.[20]

Greek fire, an incendiary weapon capable of setting the sea aflame, repelled numerous naval and land offensives during the Arabs' first siege of Constantinople, from 674 to 678 AD. An ancestor of napalm, Greek fire was launched using clay hand grenades, catapults, or flame-throwing siphons. It was manufactured in Constantinople by a special corps of closely watched workmen and masters, and only a regiment of specialized soldiers, the siphonarios, could use it. The secret of Greek fire was considered essential for the preservation of Byzantium's precarious power. Emperor Constantine VII Porphyrogenitus, whose reign lasted from 913 to 959, thus warned his heir: "You must above all things be cautious and pay close attention to the liquid fire which is launched by means of tubes; and if anyone should ask you how we make it, you must reply that this fire was shown and revealed by an angel to the great and holy first Christian emperor Constantine."

One commentator made it clear: "By this message and by the angel himself, he was enjoined . . . to prepare this fire only for Christians, only in the imperial city, and never elsewhere."[21] In 1204, during the fourth crusade, Constantinople was sacked by Frankish crusaders led by the Christian Venetian fleet, and the secret of Greek fire spread throughout the Latin world.[22]

The Byzantines, however, were not the only people in the East who knew how to use oil to make war. As early as the middle of the ninth century, the caliphate of Baghdad instituted a regiment of incendiary soldiers, the *naffatun*. In 1168, when Amalric I, king of Jerusalem, besieged Cairo, the Fatimid vizier of Egypt ordered the evacuation of the city before burning it with "20,000 pots of naphtha and 10,000 lightning bombs." The fire burned for 54 days.[23] John de Joinville, who accompanied Saint Louis in the seventh crusade, wrote in his memoirs that the "Greek" fire used by the Saracens "seemed like a dragon flying through the air."[24]

Toward Industry

Whether used for lubrication, for lighting, as a remedy, or as a weapon, oil subsequently never ceased to be exploited in a fairly rudimentary and limited way, from the Chinese province of Shannxi to Bavaria, throughout Burma, Baku, Baghdad, and Mosul, in Romania, Galicia, Sicily, and the Po Valley. In China, it was used to lubricate the workings of proto-industrial machinery: mechanical hammers driven by water mills, for example.[25] In France, in 1734, tests were carried out on the distillation of crude oil collected on the site of a very old natural geyser, Pechelbronn's *fontaine de poix*, or fountain of pitch, in northern Alsace. Eleven years later, King Louis XV authorized the sale of "fats, oils and other goods" extracted from the Pechelbronn "asphalt mine."[26] In Baku, once scooped or sponged up with coarse fabrics, oil was transferred to goatskin bags that were loaded onto the backs of camels, and transported far away in caravans. At the beginning of the nineteenth century, this city on the banks of the Caspian Sea had more than one hundred bitumen quarries. From the 1830s on, in Baku, Galicia, and elsewhere in Europe, small refinery workshops multiplied, supplying paraffin, vaseline, oils, and solvents.

Little by little, the first industrial uses of petroleum appeared. The new machines demanded all kinds of lubricants, and oil refining offered an almost unlimited range: from thick fats for locomotives to the lightest oils for watches.*

* It took a long time before these fine oils seriously competed with vegetable oils, when fractional distillation became commonplace.

Collected at the edge of the Dead Sea, the "bitumen of Judea," which had the property of hardening when exposed to light, was the secret ingredient of the daguerreotype, invented by Nicéphore Niépce around 1826. He had the idea of using it to coat the tin plates that became the precursor to modern photographic film. Paraffin from petroleum was well suited to making candles, and could be used as a coating to preserve meat, offering an alternative to tallow and other animal fats. Most important of all, a lightweight oil that burned with a soft, strong light appeared in drugstores in major European cities. They called it "lamp oil."

Since the earliest days of the industrial revolution, the mysterious question of oil's origin has divided the best scientists. The illustrious Russian intellectual Mikhail Lomonosov was considered the first to have formulated, in 1757, the idea that petroleum had biological origins. Many great nineteenth-century scientists espoused the theory that hydrocarbons had mineral, or abiotic, origins, believing they were generated in the very heart of the terrestrial mantle. Among them were the French chemists Berthelot and Gay-Lussac, the German von Humboldt, and the Russian Dmitri Mendeleev, the father of the periodic table of elements. Now discredited, the abiotic hypothesis persisted, especially in the Soviet Union, where it became the official theory in the aftermath of the Second World War, when fear of scarcity drove the Kremlin to launch immense prospecting efforts. This abiotic theory continues today with the recurrent myth of an inexhaustible oil supply, perpetually reconstituting itself in the depths of the Earth.

Samuel Kier's "Magic Liquid"

Today, no trace remains of the first oil rush that exploited and exhausted the supply found in the beautiful, wooded hills of Pennsylvania. In the northeastern United States, south of the great Lake Erie, many have forgotten that their now sleepy valleys, located midway between Pittsburgh and Cleveland on an 80 kilometer strip at the foot of the Appalachian Mountains, were for a long time called the oil region, a fiery melting pot of the American oil industry. Since the late 2000s, however, Pennsylvania has been the epicenter of what some call a fossil fuel renaissance, which may prove to be one of the ultimate oil rushes: this time, in pursuit of shale gas and oil.

Samuel Kier was an enterprising man full of candor and instinct. With his good-humored air and frank, almost childish looks, he succeeded in seizing many opportunities offered to the pioneers of American capitalism in the mid-nineteenth century. He was born in Pennsylvania in 1813, to a family of Scottish and Irish immigrants who worked in small salt mines. In 1838, Samuel Kier founded a river company that transported coal from Pittsburgh to Philadelphia. His associate in that endeavor was James Buchanan, who became the fifteenth American president—Abraham Lincoln's predecessor to the White House. Kier also invested in several foundries, but it was the family salt mines that ensured his fortune. It was during the 1830s that the salt-mining technique used in China for more than one thousand years arrived in Europe and then in the United States. The West copied and, importantly, added steam power when mimicking the Chinese extraction process. In 1847, Kier and his father launched new drilling operations in Tarentum, Pennsylvania, near the Allegheny River. But the brine extracted from these new wells was polluted by a blackish, foul-smelling liquid. The salt miners did not know what to make of this oil, and often just poured it into the river.

In 1848, an apothecary sold Samuel Kier's wife an "American medicinal oil," a "rock oil" from Kentucky, to treat her tuberculosis. When Samuel realized that it was the same oil that flowed from his salt mines, he decided to bottle his own. He hired traveling merchants to crisscross the region on gaudy wagons, selling a quarter of a liter of crude oil for 50 cents. These phony doctors presented Kier's Petroleum, extracted from "four hundred feet below the earth's surface," as a panacea that could be ingested or applied as an ointment, able to heal (according to an advertisement that Kier printed on a fake bank note) liver disease, bronchitis, gout, and even blindness. Another advertisement extolled the miracle oil in this little quatrain: "The healthful balm, from nature's secret spring / The bloom of health and life to man will bring / As from her depths the magic liquid flows / To calm our sufferings and assuage our woes."[27]

Despite the fact that his rock oil cost nothing to extract, Kier failed to make a satisfactory profit. He sold his carts, and in 1849 had the idea of sending a sample of his oil to a Philadelphia chemist named James Booth. Booth recommended that it be distilled to obtain a precious solvent for preparing latex. In 1850, Kier set up the first petroleum distillation plant to the west of the Atlantic, in Pittsburgh. Through trial and error, he succeeded in producing lamp oil, which was quite nauseating and emitted a lot of smoke. He gradually improved the process, and in 1851 began to sell his "carbon oil" to coal miners in Pittsburgh, along with lamps that he manufactured. Kier earned considerable wealth but never obtained a patent. James Booth wrote to him later: "We missed it by letting this thing slip."[28]

During the 1850s, following Samuel Kier's lead, dozens of distillation workshops were set up in Pittsburgh, New York, and Boston. They produced tens of thousands of liters of lamp oil every day. In a patent filed in 1854 in

New York by Abraham Gesner, an opportunistic Canadian entrepreneur, white oil—so called because of its transparency, and known throughout Islam since the early Middle Ages—earned its English name, kerosene. Meanwhile, the seeds of the petroleum industry continued to grow in Europe. In 1857, in Ploiesti, Romania, the first major oil-refining plant was opened. The following year, a thousand lanterns illuminated the streets of Bucharest.

1859: "Colonel" Drake Sparks the First Black Gold Rush

1859 is considered year zero for the petroleum industry. Indeed, late in the afternoon on Saturday, August 27 of that year, on the banks of a small Pennsylvania river, after months of effort and many tribulations, a fake American Army colonel named Edwin Drake managed to raise crude oil from a drilling depth of 21 meters. The following day, Drake received a previously drafted order to stop drilling. It was issued by the banker who gave him the false, reputation-boosting title of "colonel." Short of money, James Townsend had abandoned all hope of success.

Drake's operation, undertaken with the help of salt-mining specialist William "Uncle Billy" Smith, was located on an island in the middle of a small tributary of the Allegheny River known for its oily deposits: Oil Creek. The well was nicknamed "Drake's Folly" by local loggers who were skeptical of its potential. A small wooden derrick supported a percussion drill; the idea was simply to ram a heavy piece of metal into the ground, powered by a wood-fired steam motor. (Rotary boreholes existed in Europe for water wells but were only later used for oil wells).

As an occasional train conductor and jack of all trades, Edwin Drake was on the cutting edge of modernity, well aware of the era's latest technology. He was hired by a visionary, polyglot entrepreneur, George Bissell, himself advised by Benjamin Silliman Jr., a chemist from prestigious Yale University, who confirmed what other chemists had already determined before him: that it was possible to distill crude oil to extract a large number of precious products.

"Colonel" Drake, with his top hat, his long beard, and his solemn, determined demeanor, is still heralded as the great pioneer, the authentic father of petroleum exploitation, the first to have succeeded in drilling a well and extracting oil from it. However, Russian and Azeri historians dispute this claim, citing another 21-meter-deep well drilled in Baku in 1846, thirteen years before Drake's. Launched at the behest of Vasiliy Semyonov, an advisor to the council of the Central Administrative Committee of the territory of

Transcaucasia, that well was drilled under the direction of Major Alekseyev, an officer to the tsar, who supervised the many oil pits on the shores of the Caspian Sea.[29] Major Alekseyev used a percussion drill, driven not by steam but by the force of eight men who pulled on a cable suspended 10 meters high with a tripod.[30] During the same time period, around 1850, Samuel Kier was already operating his wells and refining oil in Pittsburgh. In 1858, in Ontario, Canada (one year before Drake's drilling), James Miller Williams, who ran a bitumen quarry, drilled a well in search of water during a drought. Oil bubbled up.[31] Perhaps the black gold industry, long dominated by Americans, was reluctant to attribute its paternity to the Russians, a Canadian, or a "medicine man" associated with a US president who was unable to prevent the coming Civil War (and who is still considered, along with George W. Bush and Donald Trump, to be one of the most incompetent figures ever to occupy the White House).*

The exploitation of shale gas also reaches far back in time, particularly in the Appalachians, where it is almost as old as the coal industry but infinitely more modest. As early as 1825, in the village of Fredonia, near Lake Erie, natural gas was captured from shallow crevices and conveyed by a wooden pipe to illuminate a few streets. In 1857, an entrepreneur named Preston Barmore drilled a well about 30 meters deep in search of shale gas. Failing to obtain a satisfactory flow, he sent gunpowder down to the bottom of his well and made it explode with a metal bar heated to white: His fracturing technique proved to be modestly effective, a century and a half before the current boom in shale gas and oil.[32]

Perhaps his was not the first oil well of the industrial era, but Edwin Drake's drilling certainly initiated the first oil rush. This formative event occurred just ten years after the great California Gold Rush of 1849. Mark Twain, the period's most iconic American writer, saw it as a crucial moment in American history, "the watershed event that sanctified a new money worship and debased the country's founding ideals."[33] The news that it was possible to extract "rock oil" in large quantities attracted a flood of prospectors to the Oil Creek Valley in the weeks that followed. The derricks began to grow like mushrooms in the rain.

---

* The case of Samuel Kier was downplayed by the celebrated hagiographer of oilmen, Daniel Yergin, an American historian closely linked to the black gold industry. Yergin is the author of the 1991 book *The Prize*. This colossal and valuable history of oil, by far the most cited in this domain, is also a textbook case of a story told exclusively from the winner's perspective. Yet it was nevertheless by seeing the drilling rigs depicted on an ad boasting about the virtues of Samuel Kier's "rock oil" that George Bissell imagined his project, and it was in Tarentum, where Kier had long been collecting his crude oil, that Edwin Drake went to observe drilling technology. (See John A. Harper, "Yo-Ho-Ho and a Bottle of Unrefined Complex Liquid Hydrocarbons," *Pennsylvania Geology* 26, no. 1, 1995, searchable at oil150.com.)

Oil Saves the Whales and Births an Industry Synergized with War

There was no source of lubricant and lighting abundant enough to meet new and ever-increasing needs. At the dawn of the US oil industry, whale oil was the most sought-after source for candles and lamps, street lights, and the lubrication of all sorts of mechanisms. Spermaceti in particular, a fine oil extracted from the head of the sperm whale, was considered the "oil of the kings." It garnered the highest price of any oil, and whalers—from the French of Le Havre to the Americans of New Bedford—scoured the globe for it. It was for this oil that the *Pequod's* sailors chased Moby Dick, the white sperm whale, in Herman Melville's 1851 novel. A century later, after industrial civilization had been launched, Starbuck, Captain Ahab's wise first mate, declared in the movie adaptation of *Moby Dick*: "It is our task in life to kill whales, to furnish oil for the lamps of the world. If we perform that task well and faithfully, we do a service to mankind that pleases Almighty God."[34]

The emergence of the petroleum industry undoubtedly saved the sperm whales, seals, elephant seals, and other marine mammals hunted for their fats. The whalers had been forced to travel farther and farther toward the poles, in search of an ever-decreasing number of animals. In the United States, the whaling fleet reached its peak in 1846 and then began shrinking, about the same time that the production of whale and sperm whale oil declined.[35] The Pennsylvania oil rush accelerated the changeover from whale oil to petroleum. With luck and endless toil, whalers could extract up to 2,000 liters of spermaceti from the enormous skull of a sperm whale; meanwhile, 3,000 liters of crude oil rose daily from Edwin Drake's well. A drawing published in 1861 in *Vanity Fair* magazine depicted "a grand ball given by the whales in honor of the discovery of the oil wells in Pennsylvania."[36]

The American Civil War caused tremendous upheaval. It allowed the American petroleum industry to profit from rapid distribution. The US population was only thirty million at that time. The new industries of the north prospered with Appalachian coal. The southern slave states, which seceded after the election of abolitionist president Abraham Lincoln in November 1860, hoped that the European powers would wind up on their side due to heavy dependence on "king cotton," an industry built upon the work of black slaves. This was not the case, and it was an even more powerful king that triumphed over the Confederate army: the "king coal" of the northern states.

From April 1861 to May 1865, the Civil War greatly stimulated the production of kerosene, allowing it to advance unchallenged by competing products.

Troops requisitioned, immobilized, or destroyed most of the whaling fleet. The war interrupted deliveries of turpentine (or pitch), oily pulp from pines and other resinous trees from the southern states. At that time, the essence of turpentine was used, like oil, to clean the wounds of the injured, and as a cheap and odorous illuminant called "camphrene" in the United States. Turpentine also produced a substance similar to the rosin used by violinists, which was used to lubricate the wheels and other parts of cars.[37] Rather cheap, this vegetable fat was probably, before oil arrived, the form of lubricant that offered the best value for the price, better than tallow—the fat of sheep or beef—or marine mammals.[38]

The American Civil War was the first mechanized conflict. It primed the pump for amassing the fortunes of those who refined lubricants and solvents destined for armament factories, railways, artillery equipment, or the wheels of the first armored warships. In February 1865, at the Battle of Wilmington, North Carolina, in which the last deep-water port controlled by the Confederates fell, the northern army was able to rain down a hundred shells per minute.

With the war barely over, many were already aware of the transformation that this black liquid was causing. An American journalist described it: "From Maine to California it lights our dwellings, lubricates our machinery, and is indispensable in numerous departments of arts, manufactures and domestic life. To be deprived of it now would be setting us back a whole cycle of civilization. To doubt the increased sphere of its usefulness would be to lack faith in the progress of the world."[39] As early as 1865 Congressman James Garfield, future twentieth president of the United States, had already stated, "Oil, not cotton, is King now in the world of commerce."[40]

First Wild Tremors

It often happens that oil is discovered in inhospitable territories, far from settled territory, and so the new "cycle" of civilization began in desolate places. Photographs taken at that time in the Oil Creek region, where a year after Edwin Drake's success seventy-five wells already had been drilled, suggest a universe of anarchy and filth.[41] The razed forests gave way to wooden derricks often erected so close together that they were almost touching, on bare hillsides and shapeless terrain, through which muddy water flowed. Emissions often made the air suffocating. Distillation residues that no one knew what do with (for example, gasoline, unnecessary before the invention of the combustion engine) soaked the soil and ended up in rivers by the millions of liters. Drivers of steam ships sailing the Cuyahoga River from the oil region to Cleveland avoided throwing hot coal overboard for fear of the water catching fire.[42]

These were dangerous places: Drake's well was destroyed by fire a few weeks after its opening, in autumn 1859. In 1861, just one week after the Civil War began, the first "gusher" (a well in which the pressure is so strong that crude oil spouts out like a gigantic fountain) ignited and killed nineteen people. Fortune closely followed death: Within three days after the fire was extinguished, the gusher ejected an unbelievable quantity of crude oil: three thousand barrels per day (the 159-liter wooden barrels in Pennsylvania and beyond set the standard unit of measure for crude oil production still in use today).[43] During the years of frenzied civil war, so many fires occurred in the wells and rudimentary refineries around Oil Creek that producers posted signs warning: "Smokers Will Be Shot."[44]

With the advent of the new oil discoveries, towns began to pop up almost overnight—Oil City, Oleopolis, Pithole, and more—attracting crooks, prostitutes, and liquor salesmen. Then when the wells dried up, these became ghost towns almost overnight. The many Civil War veterans who came to seek their fortune in these cities were not necessarily out of their element: "The whole place smells like a corps of soldiers when they have the diarrhoea," noted a visitor to Pithole.[45] The town existed for a little over a year, from the success of the first boreholes in January 1865, to the exhaustion of the principal wells in January 1866. Between these two dates, Pithole had time to accumulate up to twenty thousand residents, two banks, two telegraph offices, a newspaper, and more than fifty hotels (including an unknown number of brothels). Once the prospectors had gone, the surrounding plots, which had once changed hands several times a week for more than $1 million, were sold for less than $5.

Even more than the gold market, the petroleum market was subjected from its beginning to an abrupt succession of overproduction and shortage cycles, devastating for producers and, in turn, for their customers. Edwin Drake himself was ruined after making some risky investments. When in September of 1861 an explosively prolific new gusher, dubbed the Empire Well, entered production, the price of a barrel of oil fell to only 10 cents. This was cheaper than water, although shippers continued to charge $3.00 per barrel to transport it. During the last months of the Civil War, the price of a barrel reached $12.00, and then fell to less than $2.50.

Against these formidable odds, a young accountant, as sharp as he was ambitious, quickly honed his method and hitched his harness to the wild, emerging power of oil.

TWO

John D. Rockefeller, the Power of Petroleum, and the Spiral of Expansion

Was John D. Rockefeller Sr. a virtuoso of capitalism? Or a lone genius, like those who populate the arts? On the contrary, it seems that he was, among a thousand other cogs in the wheel, simply the one who proved to be the best-suited to harness and propel the immense force already in motion. When he arrived in the oil regions during the feverish years of miraculous profits generated by the Civil War, the man who was about to become the richest of all time appeared, instantly and in every respect, to be the perfect person in the perfect place to raise from underground the fabulous fortune that had lain dormant for millions of years, the germ of the true new world.

The creator of the oil company that is still, today, one of the richest private companies on the planet was a natural tamer of the emerging beast. Born into a German family who had arrived in the United States in the early eighteenth century, he was fully convinced that his Christian salvation was somehow responsible for the fortunes wrought by his labor. He flawlessly embodied the ideal of the pro-capitalist Protestant that sociologist Max Weber described a few years later: He believed his professional career path had been inspired by God himself.[1] But certain peculiar stigmas that Rockefeller carried with him from childhood also helped shape him into the superman of capitalism.

John Davison Rockefeller was born on July 8, 1839, in northern New York State: a remote rural setting, nevertheless quickly disrupted by the arrival of the steam engine and the bank. His mother, a devout Baptist abandoned by her husband, instilled in him a strong sense of morality. His authoritarian, charismatic father—a big drinker, a sharp shooter, a ventriloquist—was not satisfied with working among the humble pioneers of American industry. Shortly after his marriage, Bill Rockefeller ran one of the myriad small sawmills that built the United States, supplying lumber for buildings and coal mines and fueling many steam engines. But his whimsical, brutal, and secretive temperament drove him to remain, throughout his life, a fierce individualist attracted by the lawless, troubled world of the "wild frontier," where he led a double life. Nicknamed "Devil Bill" by the townsfolk, he was charged with rape and fled to become a charlatan doctor, selling miraculous remedies to gullible peasants under the pseudonym "Dr. Levingston." No one knows if the fake doctor sold vials of Kier's Petroleum or any other form of petroleum that was then commonly found in the pharmacopoeia of itinerant healers in the northeastern United States.[2]

Although he did not say much about Bill Rockefeller, John D. Rockefeller idolized his lying, always-absent father. "He taught me the principles and methods of business," he confided.[3] Bill Rockefeller once explained to a neighbor the approach he took to educating his sons: "I trade with the boys and skin 'em and I just beat 'em every time I can. I want to make 'em sharp."[4] John, his eldest son, became the sharpest of them all.

From a "Little Side Issue" to Rockefeller Hegemony

John Rockefeller was not a prospector who was any happier or more skilled than other prospectors of his day. He was a wholesaler. Just twenty years old in 1859 when he established his own small trading house in Cleveland, this accountant with hollow cheeks and narrow lips had long known how to conceal his impatient desire to get rich. Cleveland, extending along the shore of Lake Erie, not far from the mines and quarries of the Appalachian Mountains, was one of the main nodes of industrialization in the United States. With his partner Maurice Clark, twelve years his senior, Rockefeller built a small warehouse from which he negotiated the purchase and sale of all kinds of high-demand products for the developing and expanding region. "Grain, fish, water, lime, plaster, coarse fine solar and dairy salt," the company's first advertisement proclaimed.[5]

Located way up in the north of the young nation, Cleveland had the advantage of being far enough away from the approaching war front but close enough to take advantage of it. In 1863, the year of the northern army's victory at Gettysburg, John Rockefeller's nascent fortune was already ample enough to allow him to both make his first investments in the railways and respond favorably to a proposal for a new type of business. Samuel Andrews, a penniless, self-taught chemist, asked Rockefeller and Maurice Clark to invest

$4,000 in the construction of what then appeared to the two partners as "a little side issue," an oil-refining workshop.[6] Andrews, like Clark, an Englishman of humble origins, was a jack of all trades who had learned the traditional techniques of distilling animal fat and had applied that knowledge to refining oil extracted from bituminous shale (also called oil shale), already exploited at the time in Pennsylvania.[7] Rockefeller later recounted that he had merely relied on the enthusiasm of Clark and Andrews.[8]

The intermediary position of his trading establishment shielded Rockefeller from the brutal hazards of oil extraction, and placed him in an ideal position to take advantage of it. He undoubtedly possessed a profound intuitive understanding of the physical constraints inherent in the interplay of inert masses and human masses. Unlike many of his competitors, he didn't build his refinery in the remote forests of the oil regions. He bought land in the nearby countryside of Cleveland, on the tributary of a river that flows into Lake Erie, and near which, a few weeks later, on November 3, 1863, the first rail line connecting the city to New York was opened by a locomotive sporting Union colors.[9]

The Civil War greatly benefited Cleveland's wholesalers, who celebrated both the closure of access to the Mississippi and other trade routes in the southern United States and the high inflation of commodity prices. The price of a barrel of oil, in particular, was soaring. With little effort, crude oil producers enjoyed brilliant profits, and the refiners' profits were still more magnificent. The latter's initial investments were ridiculously modest: a hearth, some recuperation barrels and, eventually, a still. It was not even necessary to buy extra fuel; the oil itself was perfectly sufficient to run the operation. Like many of his contemporaries, the pious Rockefeller perceived a clear sign of divine providence in the discovery of Pennsylvania's oil fields: "These vast stores of wealth were the . . . bountiful gifts of the great Creator," he declared ecstatically while taking his first steps into the budding industry.[10]

Rockefeller was far from being the only businessman attracted by the manna. Refineries proliferated quickly in Pittsburgh, as well as in New York, Philadelphia, and Baltimore. By the end of 1866, Cleveland alone had no fewer than fifty fairly important distilleries; the smell they exuded above the city was so foul that it spoiled beer and milk.[11] The United States, however immense, was still only the humble backyard of the European economy. Two-thirds of the kerosene produced in Cleveland was exported to Europe.[12] It was shipped through New York, where John then dispatched his younger brother William to invest the oil profits on Wall Street, very near the docks where the barrels arrived from Pennsylvania. No other industry was so destined to become international.

Rockefeller expanded his oil business with astonishing confidence and boldness. He constantly borrowed huge sums but always reimbursed them with interest. At twenty-five, he built Cleveland's largest refinery, capable of processing five hundred barrels per day. Rockefeller calculated quickly and accurately.[13] His wealth was immediately absorbed and metabolized by financial institutions in which he acquired shares very early in his career: In 1868, only twenty-nine years old, he sat on the board of a bank for the first time, the Ohio National Bank. Many other such appointments followed.[14]

With so many hazards and risks associated with oil fields, investing relentlessly in refineries rather than oil wells made it possible to position oneself at the true heart of the business. Rockefeller understood that the flow of black gold could be controlled not by extracting the oil, but by marketing it and reducing transport costs as much as possible. In the spring of 1868, he entered into a secret agreement with railway magnate Jay Gould to finance the first-ever large, short-distance pipeline system serving the Oil Creek wells. In return, he enjoyed a discount of no less than 75 percent on the cost of transporting his oil on the Gould railway. With oil, Rockefeller later explained, the railways had access to "a large, regular volume of business, such as had not hitherto been given to the roads in question" (not even with coal, which was consumed near its extraction site by the iron and steel industry).[15] Railway owners became Rockefeller's constant allies in his ascension toward the quasi-monopoly of oil sales in the United States.

Economic Darwinism: The Creation of Standard Oil and the Cleveland Massacre

The Standard Oil Company was registered in the state of Ohio on January 10, 1870; this forefather of ExxonMobil would soon become the leading privately owned company on the planet, although it didn't possess a single oil well. With $1 million in capital, it already controlled 10 percent of American refineries. It had a barrel factory, a fleet of special cars, warehouses, and docks. Its name, Standard Oil, highlighted the uniform quality of its oil, at a time when poor quality refineries were frequently responsible for fatal explosions. John Rockefeller was the company's president and majority shareholder. He owned 26.67 percent of the initial capital. He always retained more than a quarter of the shares. With the shares of his brother William and those held through the former company absorbed by Standard Oil, the Rockefellers controlled more than 40 percent of the company. The other two senior partners in Standard Oil were Samuel Andrews (responsible for the proper operation of the refineries) and Henry Flagler, an investor nine years older than John Rockefeller who had built his initial fortune by selling grain to the Union army during the Civil War. At age thirty, Rockefeller openly pursued hegemonic goals. He was not afraid to tell a businessman in Cleveland that "the Standard Oil Company will some day refine all the oil and make all the barrels."[16]

In the early 1870s, the oil industry was on the brink of apoplexy. Its bounty had exceeded even the wildest of dreams. It had spoiled the refiners too much; many "were disappointed if they did not make one hundred percent profit in a year—sometimes in six months," Rockefeller lamented.[17] Speculation was so intense at the end of the Civil War that the existing refineries were capable of distilling three times the amount of crude oil generated by the region's wells.[18]

It was then that John Rockefeller decided to push, with all his power and all conceivable means, on the accelerator of an economic Darwinism. Judging that the time was right to wipe out the "ruinous competition" between refiners, the future champion of industrial capitalism decided to work toward replacing the competition with a cartel dominated and controlled by Standard Oil, in order that it alone (or nearly alone) might enjoy the extraordinary profits that, by its nature, the oil trade realized.[19]

Rockefeller had in no way invented the principle of a cartel, nor was he the originator of discounts on transport, which allowed him to build his own quasi-monopoly. In the 1870s, many secret agreements emerged in the United States, aimed at controlling prices and production levels in activities as diverse as the trade in salt, grains, beef, rope, sugar, and whiskey. In the petroleum industry, those same oil-region producers—who soon nicknamed Standard Oil "the octopus," and depicted John Rockefeller as the kind of chilling shadow one invokes to frighten children—were members of the Oil Creek Association, which also plotted to limit extractions in order to increase prices.

At the end of 1871, Rockefeller refined his offensive. As kerosene prices fell further and further, Standard Oil secretly bought New York's largest wholesaler and managed to raise enough funds from banks to increase its capital from $1 million to $3.5 million on January 1, 1872. Rockefeller had a devastating plan in the works: a conspiracy known as the Cleveland Massacre, which dealt the competition a deadly blow and propelled Standard Oil on the road to monopoly.

But the boss of Standard Oil was not the originator of this plot. Through an intermediary who came to meet him on November 30, 1871, at a New York hotel, Tom Scott, president of the Pennsylvania Railroad, proposed a cryptic alliance between his company (then the most powerful in the world), some refiners (most notably Standard Oil), and the two other powerful American rail networks: Cornelius Vanderbilt's New York Central and Jay Gould's Erie Railroad. The agreement was ratified under a deliberately misleading legal shell called the South Improvement Company (SIC). It is considered to be one of the first examples of a financial holding.[20] The SIC was a double cartel that put an end to the price war among refiners as well as the no less ferocious war railway companies were waging to dominate oil transport. Forty-five percent of the petroleum transports ended up going to the Pennsylvania Railroad; the Erie and New York Central each got 27.5 percent. In exchange for the arrangement, Standard Oil and other refiners received significant discounts (around 40 percent) on the normal price for the transport of each barrel, and also benefited from another 40 percent discount for the transport of each barrel belonging to the refiners who were excluded from the scheme.

Although not the initiator, Rockefeller enthusiastically adopted the plan, and Standard Oil, by far the largest refining company involved in the SIC, played the key role of arbiter. Exclaimed Rockefeller, "Indeed, the project grows on me."[21] Although he realized that the SIC could be a failure, he saw no need for ethical concern: "It was right before me and my God," he declared years later.[22]

On February 26, 1872, Oil Creek's producers were astounded to learn that oil transportation tariffs had just doubled for everybody—or almost everybody. The plot was quickly exposed. Standard Oil wagons were attacked, burned, or emptied of their contents. Rockefeller had to temporarily dismiss 90 percent of his workers. Unlike Gould, Vanderbilt, and Scott, the other conspirators, he was still a complete unknown to the general public. The press began calling him the "Mephistopheles of Cleveland."

The plan backfired: In the face of public scorn, the railway companies retreated. Rockefeller was the last to back down. On April 8, he acknowledged that the SIC contracts had been nullified. Since February, however, Standard Oil had acquired twenty-two of its thirty-six refining competitors from Cleveland. One of them explained: "There was a pressure brought to bear upon my mind . . . that if we did not sell out we would be crushed out."[23]

New Fiefdoms, New Vassals: Standard Oil Reaches Unstoppable Critical Mass

By the age of thirty-one, Rockefeller headed the world's largest refining company, with nearly a quarter of the existing capacity in the United States.

One month after the SIC was dissolved, he set up a new, official cartel with Pittsburgh's principal refiners. This became the "Pittsburgh Plan." Meanwhile, during the summer, he caught wind of a counteroffensive launched by the oil region producers—the same ones being favorably portrayed in the press as champions of free competition, victims of Standard Oil. Once again, as during the Civil War, these producers had agreed to halt their drills in order to raise prices. In reality, Rockefeller was perfectly satisfied with the initiative, which he hoped would help solve the problem of overproduction. Free and undistorted competition probably had fewer supporters in the oil region than in any other young industry riddled with hidden agreements. Nature was too generous, oil too abundant, its exploitation too inexpensive, and new discoveries came too fast. Competition is the natural path of capitalism only when physical capital, a factor of production, is both rare and expensive. Rarity is the most common economic situation. But oil was an exception, at least during the ascendant phase of its exploitation's history.

Although it became a major driver of economic development, oil remained an anomaly for economic science. An anomaly that, strangely, was hardly ever recognized: According to most economists, what is precious is rare and expensive; like water, black gold proved to exist in fantastic quantities, while still being terribly precious.

All of the agreements between the American oil pioneers derailed quickly. The small refineries integrated into the Pittsburgh Plan cheated and exceeded their quotas. Some speculators bought stills for the explicit purpose of blackmailing Standard Oil, only to be bought out.[24] In the oil region, it was worse: Producers failed to limit their extractions. They attacked each other, setting fire to wells and destroying steam pumps. Early in 1873, Rockefeller decided, almost literally, they could all go to hell: "They don't want to be saved. They want to go on and serve the devil and keep on in their wicked ways," he said of the refiners.[25]As for alliances between producers and refiners, they were only a "rope of sand."[26] He was now ready to pursue his hegemonic project to the end, and for the sole benefit of his enterprise.

Beyond his uncompromising determination and his powerful abilities, Rockefeller benefited greatly from the advantageous nature of the oil market to carry out his grandiose project. As he well understood, exploiting black gold did not require big investments (a few pipes, pumps, stills, wagons) compared to the impressive profit margins it provided. In addition, petroleum allowed discreet upward or downward selling-price fluctuations, since a can of kerosene was cheap to produce and affordable for the millions of Americans and Europeans for whom it had become a necessity. At the end of the Cleveland Massacre, Rockefeller possessed a critical mass of capital. He controlled enough refineries to sell more kerosene, more cheaply than anyone else, which brought other refiners to their knees. By increasing its prices in cities where the competition had already been absorbed, Standard Oil was able to lower its oil prices in other locations for long enough to suffocate the local competition. In 1874, Standard Oil absorbed Charles Pratt & Company, its main competitor in New York, and Lockhart, Frew & Company, Pittsburgh's main refiner (and the first in the United States to export). Their executives were generally pleased about joining Standard Oil, which seemed capable of offering them perpetual wealth and stability. From that point on, the company's momentum was unstoppable and led Standard Oil to control, by 1877, nine-tenths of the refining industry of the United States, from whence it dominated the Earth's oil supply, already accounting for nearly three-quarters of its sales with overseas business. Standard Oil and its future descendants, in particular Exxon, remained the biggest refiners on the planet, the primary providers of petroleum products, now essential for greasing the wheels of the world economy.

The rise of Standard Oil was all the more unstoppable because Rockefeller was a perfectionist who practiced scrupulous management. After observing a worker sealing a can of kerosene with forty drops of solder, he congratulated himself for saving "several hundred thousand dollars" by ordering that the operation be completed with thirty-nine drops.[27]Above all, Rockefeller showed a merciless tactical pragmatism, mingled with paranoia (inherited from his childhood, no doubt). When he absorbed his competitors, he insisted that they retain their company names and conceal their membership in Standard Oil. In August 1878, when Samuel Andrews—the associate who fifteen years earlier had convinced him to get involved with petroleum and benefit from Andrews's knowledge of refining—was overwhelmed by the company's aggressive changes, Rockefeller sent a messenger to buy out Andrews's shares; he railed against the "ignorant Englishman" who was upset by the unlimited ambition of Standard Oil's patron.[28*] Rockefeller had the skill to win over his most tenacious enemies and make them vassals. In the mid-1870s, two of them became his principal associates: Henry H. Rogers, the refiner's representative from New York who convinced Tom Scott to abandon the SIC; and John D. Archbold, an independent refiner who referred to Rockefeller in the press as a "great anaconda" during the Cleveland Massacre.[29]

* With a large portion of the $1 million dollars from the payoff, Sam Andrews built a five-story, one-hundred-room house and was the subject of ridicule for suggesting that Queen Victoria visit.

"It Was Forced Upon Us": Marx, Rockefeller, Convergences

While other industrial sectors have seen companies that crush the competition grow, none of those companies have endured like Standard Oil and its subsidiaries. For Rockefeller, this state of affairs was due to something innate, physical, or even (to use a term he would have related to) providential. "It was forced upon us," he said of the initiative that had led to his company's hegemony. "We had to do it in self-defense. The oil business was in confusion and daily growing worse. Someone had to make a stand." Concentration of industry is a fundamental and inevitable economic phenomenon, he said in a historic interview: "This movement was the origin of the whole system of economic administration. It has revolutionized the way of doing business all over the world. The time was ripe for it. It had to come, though all we saw at the moment was the need to save ourselves from wasteful conditions." And he concluded: "The day of combination is here to stay. Individualism is gone, never to return."[30] Speaking to students in 1902, to describe yet again the organic necessity for the death of free competition, he (or his speechwriter) chose a bucolic metaphor: "The American Beauty Rose can be produced in the splendor and fragrance . . . only by sacrificing the early buds which grow up around it."[31]

A striking convergence with Marxist authors exists around this vision of capitalism's inevitable evolution toward cartels or monopolies. In *Capital*, Karl Marx writes that the development of credit and financial markets is "a new and terrible weapon in the battle for competition and is finally transformed into an enormous social mechanism for the centralization of capitals."[32] This junction between the great champion and the great scourge of capitalist industry is confirmed notably through the work of Austrian economist Joseph Schumpeter.[33] At the end of the 1930's crisis, this "creative destruction" theorist recognized the relevance of Marx's historical analysis, while advocating in favor of capitalist monopolies, which he considered to be valuable "fortresses" in times of recession. (Schumpeter, it turns out, was the favorite Harvard economics professor of David Rockefeller, grandchild of Standard Oil's founder and eventually the "pope" of Wall Street banks after the Second World War.)[34]

John D. Rockefeller did not discover oil and was not a pioneer in refining oil. Nor was he the first to have exported petroleum products to Europe. He did not dream up the cartels that were organized around cheap transport and that allowed the competition to be crushed. However, from the office of his first warehouse in Cleveland, he saw the immense wave of oil fortunes approaching. A wave whose power was born from the commercial qualities peculiar to it: abundant, cheap, instantly indispensable. Rockefeller paddled on that wave with force and chose the exact right moment to surf it. Once he began riding that wave, he left thousands of others with similar aspirations behind. In spite of his immense pride in the empire he had built, Rockefeller recognized that he was only a simple agent (though certainly a zealous, efficient one, the greatest of all) of a necessary, providential phenomenon, which was far greater than him. This humility was probably more a matter of realism than of faith.

Great Fear, Immense Hopes

At the end of the nineteenth century, light from coal gas was still a privilege of the richest quarters of the wealthiest cities. As it spread through the Old Continent, oil had such a revolutionary impact on daily life that it seemed to trouble the collective consciousness of the bourgeoisie. Until oil's advent, abundant light had been a clear sign of distinction within society. During the Paris Commune, the revolutionary insurrection of the Parisian people in the spring of 1871, accidental fires caused by the first oil lamps were attributed to the *pétroleuses*—women who long remained, in the eyes of French high society, the epitome of the enraged revolutionary. But the *pétroleuse* was a myth. During the "Bloody Week" of repression that marked the end the Commune, some claimed to have seen Communardes (working-class and socialist women of Paris) set fire to many buildings using oil lamps. Yet when the Commune was raided, not one among the thousands of women arrested was condemned as an arsonist. Justice quickly concluded that the few Communardes accused of arson had been unjustifiably charged.[35] Still, the myth of the *pétroleuse*, sparked by reactionary publications, persisted for decades.

But in every layer of society, the technical expansion amplified by oil's arrival was met with exalted enthusiasm. To the bourgeoisie, it offered countless opportunities for enrichment; for the young working class, or in its name, it germinated even stronger dreams of liberation. Debates between revolutionaries were filled with questions about the role of the machine. Precocious and short-lived labor movements such as the Luddites in early-nineteenth-century England directed their anger at the looms, new machines they accused of stealing the bread of the workers. However, nobody, supposedly, can stop progress. In spite of the gloomy suffering inflicted in mines and workshops, as competition spread throughout the Industrial Revolution, the machine itself was increasingly considered a source of hope. Already, the powerful new energy source was changing eating habits, which it helped to make more sophisticated. (In Paris, for instance, the "trains from the sea" coming from Le Havre made river fish less popular). In 1880, Frenchman Paul Lafargue, son-in-law of Karl Marx, concluded his manifesto *Le Droit à la Paresse* (*The Right to Be Lazy*) by describing machines as "[t]he breath of fire, with limbs of steel . . . inexhaustible . . . the savior of humanity, the god who shall redeem man from the sordid arts and from working for hire, the god who shall give him leisure and liberty."[36]

The US Oil Industry Is Born without Science and against Scientists

The free flow of oil changed the world by itself. It was a cycle of involution in which individual genius counted for nothing and social Darwinism (or "memetics") accomplished everything, a cycle irreducible to the scale of individuals, even among the most brilliant and powerful in the world.* There is no Thomas Edison or Marie Curie in this story: For over a quarter of a century, until the advent of the automobile and, later, the rise of petrochemistry, the oil phenomenon had little need for scientists to express itself and upset society. When the first drill rigs were built in Pennsylvania, the techniques used for drilling were already very old, designed initially to drill for water. The craft of refining was older still, and purely empirical. The distillation of fats, oils, and alcohols had been mastered for centuries, as had the distillation of "naft" by the Arabs. Remember, too, that in 1863, in Cleveland, Sam Andrews, the man who attracted Rockefeller to the petroleum business and was later repudiated by him, acquired his know-how in the animal-fat refining workshop where he had worked.[37] For ages, we have known how to make tallow with the fat from cows, sheep, and goats. This common lubricant can also be used to make candles and soap, or to waterproof animal skins. The first petroleum products that irreversibly replaced tallow were derived through the same production techniques. Rudimentary petroleum grease for locomotive axles was initially extracted without even using distillation; rather crude oil was filtered through the ashes of animal bones—a method traditionally used for making vegetable and animal oils.[38]

Rockefeller prided himself on never having felt the need to seek out scientific knowledge.[39] During its first quarter century, the American petroleum

* Memetics examines developments in culture through an extended Darwinian approach. It is based on the concept of "memes" to study social cultures, by analogy with the concept of "gene" in genetics.

industry did not employ any full-time scientists.[40] (A project lasting a few months was entrusted to a German chemist, Herman Frasch, in the mid 1870s; it did wonders for improving the paraffin production process for chewing gums and candles.)[41] In fact, American oilmen had a frank aversion to scientists, who they referred to as troublemakers. Often incompletely distilled, oil was responsible for very frequent, often appalling, accidents. For example, in 1869, in New Orleans, fifty people were burned alive in a single fire. "There are no words strong enough to stigmatize the criminals who sell kerosene," wrote chemist Charles Frederick Chandler in 1870 in the very first issue of *American Chemist* magazine, in the aftermath of a kerosene lamp explosion in Brooklyn that killed five people, including four children.[42] Chandler, a professor at Columbia University, discovered that commercial kerosene often contained too much of the C_5 to C_8 hydrocarbons. These light hydrocarbons, not yet known as "gasoline," were industrial waste at that time, by-products that were of no use until the combustion engine arrived. Despite their danger, some refineries did not exclude them, due either to incompetence or the lure of gains: They increased the volume of commercialized kerosene, and thus the profits.[43] Thanks to the high quality of its refining practices, Standard Oil was able to rise above the jungle of competition.

The "Miracle," Its Properties, Its Functions

"The whole process seems a miracle," said Rockefeller one day. "What a blessing the oil has been to mankind!"[44] The sudden availability of miraculous quantities of petroleum sustained the Industrial Revolution during this defining period. Oil, from its heaviest to its lightest forms, had physical properties capable of providing a staggering number of services, without the intervention of scientific progress.

Incandescence came first. Lantern oil was the first cheap, efficient, abundant light source ever available to mankind. Thanks to the merchant marine and steam trains, its arrival in the most isolated markets (along with sugar, tea, and coffee) increased labor productivity and intensified lifestyles. Lamp oil made it possible to work longer in barns, workshops, mines, offices, libraries, and laboratories.

Petroleum also was a good source of heat. Fuel tanks quickly became commonplace in the cellars of houses and other buildings. In 1900, in the United States, before demand for fuel exploded and despite widespread use of lamp oil, 70 percent of the oil produced was used to generate heat (electricity remained a rarity).[45] Heavy fuel was also very quickly adopted in a large number of industrial processes (starting with the refining of oil itself). In particular, it replaced crushed coal as the ideal fuel to create the flame of about 1,500°C (2,732°F) needed to heat the mixture of limestone and clay from which cement is made. Oil's abundance thus facilitated a massive and increasingly systematic use of the new king of building materials: concrete. Thanks to this agglomeration of cement and sand (or gravel), it became easier to erect towns and to extend roads where, for lack of stone, it previously had been very difficult or impossible. Stone was often a luxury product. Concrete, its new ersatz, much stronger but also more energy hungry than brick (fired with coal), became omnipresent. In the United States, heavy fuel oil was often used as a complement to coal in Pennsylvania's blast furnaces. It replaced coal in boilers and steam engines more and more frequently, because it was a denser form of energy. As for kerosene, it became the fuel of soldering torches. To assemble the steel plates and beams of ships or bridges, workers used "hot riveting," plunging their rivets in a cauldron filled with kerosene.

The fluid property of black gold was no less important. Unlike coal, which must be sought underground, crude oil often springs out from its own pressure (at least when oil fields are in their early prime). Moreover, like any liquid, oil flows from tanks to reservoirs thanks to the power of gravity, without having to be broken up and then continually carried along like coal. The distillation of crude oil made it possible, for the first time in history, to produce an infinite range of greases, oils, solvents, and detergents in abundance. Poured drop by drop or slathered on the biggest steel beams, lubricants produced from an abundance of oil reduced problems that had caused prior setbacks, particularly with the steam engine. Various liquids derived from basic petroleum also found their way to a multitude of applications in the domain of paints, dyes and inks, cleaning supplies, pharmaceuticals, and cosmetics.* From the outset, Rockefeller and his competitors sold Vaseline (or "petroleum jelly") as well as benzine, an excellent solvent.[46] The refinement of crude oil also became the most economical production source for another valuable viscous liquid, a hydrophilic and slightly sweet lubricant: glycerin. Glycerin, originating from petroleum (rather than from vegetable and animal fats, as before) gradually became an ingredient in recipes for a very large number of soaps, syrups, moisturizing creams, suppositories, and toothpastes.† Combined with

---

* In 1885, the Genevan pharmacist Charles Hahn started selling his famous capillary lotion, Pétrole Hahn, with debatable effectiveness.

† In 1871, French chemist Charles Friedel obtained for the first time a full synthesis of glycerol from propylene.

nitrates, it also accelerated the production of nitroglycerin (for dynamite), as petroleum refineries were able to supply unprecedented amounts of this minor by-product for its production. At the beginning of the First World War, in one of the earliest industrial uses of petrochemicals, Royal Dutch Shell was the first to replace coal with petroleum to produce toluene, a very effective solvent and, moreover, a basic molecule of TNT (trinitrotoluene).[47]

The natural plasticity of liquid and solid hydrocarbons when heated was also exploited almost half a century before the invention of PVC plastic. In 1874, Standard Oil created specialized subsidiaries dedicated solely to the emerging bitumen and asphalt market for roads, as well as the already flourishing market for paraffin wax, used to make chewing gum, hair gel, and candles and to waterproof clothing and wrap candies, meat, ammunition, and explosives.[48]

Lighting, heating, lubricating, cleaning, protecting, or even flattening: For all these precious uses, oil was never the result of a scientific or technical revolution. Its natural properties were enough to offer many advantages in quality and, above all, quantity over products historically derived from vegetables and animals (such as tallow, lard, pitch, lanolin, turpentine oil, whale oil, and beeswax). Two uses for oil took longer to emerge. They were technically more complex, more cumbersome to implement. But it was these uses that created the greatest impact, despite the original reticence of the future mass market: farmers and motorists.

As early as the 1860s, Rockefeller was among the refiners who attempted, with considerably different levels of commercial success, to sell their residual sulfuric acid, discovered twenty years earlier to be a useful ingredient in fertilizer (thanks in particular to the work of German chemist Justus von Liebig).[49] This was the petroleum industry's first foray into agricultural fertilizers, the first step in a series of advances that ultimately had huge demographic consequences.

The automobile was an age-old dream, rooted in ideas that went back two centuries before the necessary technical and energy conditions were available for its realization. The Dutch scientist Christiaan Huygens was among those who, in the seventeenth century, invented an internal combustion engine before adequate fuel came along. (Huygens devised the idea of driving pumps with gunpowder; Father Jean de Hautefeuille envisioned a similar solution for the gardens of the Château de Versailles.) As early as 1668, the Flemish Jesuit Ferdinand Verbiest, on a mission in Beijing, designed what appeared to be the first prototype of a land vehicle driven by a steam jet. Two centuries later, in 1859, the same year as Edwin Drake's drilling success, the Franco-Belgian engineer Jean-Joseph Étienne Lenoir filed the patent for the first operational internal combustion, two-stroke engine. It ran on the same coal gas that made it possible to illuminate London, Paris, and an increasing number of Europe's wealthy cities. But coal gas is poor in calorific value and can only support weak levels of mechanical power. The very nature of petroleum, an extraordinarily dense and malleable fuel, arrived as a "miracle" to invoke combustion's latent potential, as it had lighting's. In 1870, Austrian inventor Siegfried Marcus developed the first petroleum gasoline engine for a pushcart. It was an alternative to steam cars, which were rare and far too heavy. In 1876, Germans Gottlieb Daimler, Nikolaus Otto, and Wilhelm Maybach began to develop a four-stroke engine. And ten years later, in 1886, their compatriot Karl Friedrich Benz filed the patent for the first gasoline-powered automobile. It took two more decades, until the dawn of the twentieth century, before the rise of technology and the availability of raw materials gave birth to the automobile era.

Energy and Raw Materials: The Steps of a Grand Staircase

Inventions are not the decisive factors in a society's technical evolution: The availability of raw materials and the physical force capable of implementing the inventions at scale are more important. Long before the Industrial Revolution, human ingenuity knew how to design sophisticated, complex, and precise mechanisms for watchmaking. But the main material of the watchmakers of the sixteenth and seventeenth centuries was brass, a soft metal alloy that can be easily worked cold and that, precisely for this reason, did not offer the necessary resistance needed to fabricate the large powerful machines of the Industrial Revolution, which were often much less complex than a clock. Fabricating those machines required colossal energy sources to produce iron and steel in large quantities.[50]

Often, invention is not even the first requirement for technical evolution. Much like the genetic evolution of organisms, the memetic evolution of societies spontaneously begins in search of optimal forms, generation after generation, just as the bodies of fish evolve toward certain hydrodynamic curves. Optimal forms do not preexist "in the heaven of ideas," they belong to the physical world. They are inherent, waiting to eventually be discovered and expressed. Many technical transformations seem to be spontaneously activated by the materials and physical forces that are brought to bear, not attributed to an inventor or even seen as true inventions. Their true source: the properties of nature. Examples include water mills and even windmills (rarer and more delicate), which, since antiquity, have been invented and reinvented in various parts of the world. This was particularly the case when massive oil extraction began, when the expansion of complex societies demanded the pursuit of more abundant and efficient sources of energy, materials, lubricants, and solvents, sources that were cheaper than coal, wood, stone, fats, and organic alcohols.

Hydrocarbons have by no means taken the place of coal, nor has coal replaced wood, in order to constantly increase the "driving power of fire."* Overall, the consumption of all three of humankind's primary energy sources has never ceased to grow, right up to today. These combustibles are the stepping stones of technology, acting like the increasingly wide steps of a grand staircase onto which more and more raw materials are added. With the addition of petroleum, the staircase sprang into a vertiginous spiral: The sources of energy and raw materials multiplied their effects as they intertwined.

Long before the start of the Industrial Revolution in England, people began to mine coal, not because it was a better fuel but because there was a chronic lack of wood. From the fourteenth century on, English forests were increasingly consumed by the growing population and by shipbuilding. Like the France of the Bourbons, the England of the Stuarts had a strong need for charcoal to fuel its chimneys and stoves. Some 80 hectares of forests were required to supply the charcoal absorbed in one year by a single foundry.[51] The construction of each ship demanded hundreds, even thousands, of mature trees. In France, Jean-Baptiste Colbert had ordered that many forests be replanted in order to avoid scarcity. The English had to import wood from the Baltic, and even from the forests of New England, on the other side of the Atlantic.

The volumetric and weight energy density of coal is certainly greater than that of wood and charcoal, but its extraction demands more effort. It is more difficult (in other words, more energetically expensive) to dig a mine than to cut down a forest; moreover, coal is rarer than wood and therefore requires specific transport networks.[52] The presence of numerous navigable waterways near the first coal mines on the island of Great Britain greatly facilitated this industry's early development; Northumberland and Wales, where the largest British mines were located, were very near the sea.[53] Although fed by coal, the Industrial Revolution was actually born of traditional renewable energy derived from solar energy: the force of the wind, the water cycle, wood, and manpower. These were the energies that made it possible to open the first

* The "driving [or motive] power of fire" is taken from the title of a book published in 1824 by French engineer Sadi Carnot, the father of thermodynamics, when he was twenty-four years old: *Réflexions sur la Puissance Motrice du Feu et sur les Machines Propres à Développer cette Puissance.*

quarries and coal mines in Great Britain, in the north and the east of France, Belgium, the Ruhr, and the Appalachians. At the end of the sixteenth century, two centuries before James Watt refined the steam engine, the Flemish engineer Simon Stevin gave the tiny Netherlands an unparalleled industrial foundation when he brought radical mechanical improvements to the energy efficiency of Dutch mills. The steam engine appeared in the picture early on, but for a long time it played a limited, minor role, only widely utilized later, after the miners' muscles had conquered the first easily accessible layers of coal. It was precisely for this purpose that the steam engine had been conceived: to take over. From its very first use in the industry (by Englishman Thomas Savery in 1698, the "Miners' Friend"), the steam engine was mainly used to pump groundwater from the bottom of the mine, so that the shovels and picks could get at the coal.[54] Coal-powered steam engines made it possible to dig deeper and faster for coal and other minerals.

Signs of the Industrial Revolution were not obvious to anyone living in England until the 1830s, when coal mines attained enough production to allow industrial development to fuel itself and become fully operational: Coal was sufficiently abundant to provide the energy steam engines needed to replace manpower, including, increasingly, at the bottom of the mines.[55] Fossil energy had begun to take the place of renewable energy and to multiply its impact. Harder to exploit than wood, coal provided, in return, more energy, whether it be to melt metals, make glass, generate steam, or extract more coal and carry it to the surface. By the mid nineteenth century, coal had replaced wood charcoal in the blast and glass furnaces. The Chrystal Palace and the London sewer system became plausible.* But once the kerosene lamp solved the problem of darkness at the bottom of the mines, miners and their picks were constantly on the job, and ever larger quantities of lumber were needed to support the tunnels dug very deep in "subterranean forests" of coal strata. By overlapping each other, the sources of energy and raw materials allowed us to dig deeper and farther, to constantly push the specter of depletion further into the future.

Historically, the flow of materials has obeyed physical necessities. Metal foundries absorbed much larger quantities of fuel than of other raw materials. This explains why, in England in particular, metallurgical factories were built not near metal mines but near coal mines.[56] When the petroleum industry was born, the United States had more horsepower than the United Kingdom.[57] The development of the oil regions was greatly facilitated by an abundance of timber and coal previously exploited from the Appalachians, so much so that in

* The Chrystal Palace was an industrial marvel of cast iron and glass at the Great Exhibition of the Works of Industry of All Nations in London in 1851.

retrospect it seems logical that the black gold industry would have blossomed there rather than in the treeless plains (in Mesopotamia, on the banks of the Caspian), where abundant fields of crude oil, with their wells and pits, had been discovered long ago. Before oil prospectors arrived, woodcutting was the only activity in the village closest to Oil Creek, Titusville, which was destined to be abandoned once the surrounding hills were clearcut.[58] Edwin Drake's steam engine was fueled by charcoal. Oil barrels, derricks, reservoirs, and even the first pipelines were made of wood. Without coal, locomotives, or steamboats, it would have been impossible to bring the black gold to market.

Oil and the Energy-Complexity Spiral

The coextensive combination of energies and raw materials could intensify radically thanks to petroleum, even before the advent of the internal combustion engine. The availability of steam power in the United Kingdom and the United States more than quintupled between 1850 and 1875, on the eve of and a short time after the spectacular arrival of the petroleum industry.[59] The numerous lubricants extracted from crude oil were indispensable; without them, steam engines would have been incapable of realizing their full potential. Between 1865 and 1900, American production of oil and grease extracted from petroleum rose from thirty-five thousand to nearly five million barrels.[60]

The last quarter of the nineteenth century saw large quantities of petroleum products become available at the same time that the technical projects that would have been impossible without them were launched. Such a concomitance is certainly not the result of luck: Black gold and its numerous precious functions did more than any other new product to meet the ever more demanding physical constraints imposed by industrialists' grandiose designs. Without petroleum lubricants, the coal used by steam engines would have been incapable of accomplishing the wonders made possible by oil. Begun in 1881, the Panama Canal project took thirty-three years to complete, primarily due to its level of technical difficulty. Compared to the construction of the Suez Canal, completed in 1869 largely thanks to the manual labor of thousands of Egyptians forced to work without pay, the excavation of the Panama isthmus demanded much more energy and machinery—in particular many mechanical steam shovels, which required immense quantities of high-quality grease, in addition to coal. When the Panama Canal opened in 1914, the first oil-fired bulldozers began to multiply; they ran on heavy fuel, a superior energy source that soon made all sorts of other exploits possible.

The reopening of Andalusia's Rio Tinto mines, too, was made possible by petroleum. Since the ancient times of the Phoenicians and the civilization of Tartessos, the mines of the Rio Tinto ("Red River," named for the iron dissolved in its waters) have brought fortune and power to those who controlled them. They were among the principal mines exploited by the Romans, who extracted, in particular, the silver from which a great number of denarii, the Imperial silver coins, were produced.[61] Trajan, the emperor at the height of the Roman Empire, was born near these mines. They fell dormant for more than a millennium after the fall of Rome, most likely because the slaves had reached the end of all accessible ledges. The Visigoths and Moors subsequently exploited them only superficially. The crown of Spain was no more successful, preferring, for the next two centuries, to exhaust the fabulous veins of the New World that had been left intact by the native civilizations, whether in Potosi in the Andes or in Guanajuato in Mexico.[62] (The Portuguese did the same in Brazil: In the early nineteenth century, the Portuguese Minas Gerais virtually ceased to give gold because the surrounding forests, which provided the essential fuel for the mining, had been devastated).[63]

Beginning in 1873, the Rio Tinto Company was able to breathe new life into the Andalusian mines, building docks and opening railway lines in a Spain that was almost devoid of industry. This consortium, led by the British, dug water reservoirs and transported thousands of tons of coal, cement, steel, grease, oil, asphalt, excavators, pumps, and cranes, and hundreds of workers. Numerous ancient artifacts were discovered during the excavation of the new Rio Tinto quarries, including two stone heads of Ashtaroth, the bellicose Phoenician goddess, treasurer of hell in the Judeo-Christian tradition.[64] As early as 1877, the Rio Tinto Company became the world's largest producer of copper, only four years after its creation.[65]

Graham Bell had filed his patent for the telephone in Washington the previous year, and in 1878 in New York, inventor Thomas Edison founded the Edison Electric Light Company, ancestor of the industrial giant General Electric. The need for copper never ceased to grow, and the industry never ran out of electrical cables, causing an irrepressible rise in demand for thermal energy sources. A chain reaction began to unfold, with an ever-increasing demand for the ways in which oil burns, lubricates, and washes. One consequence among many others in this multilayered chain reaction: First in the United States before 1900, then in almost all the rest of the world until the oil shock of 1973, oil ranked third on the roster of most-used energy sources for electric turbines, just behind coal and dams. And in every isolated outpost where modern technology advanced, fuel inevitably was relied upon to run the generators.[66] Thanks to the diverse properties of petroleum, progress was able to drive further and deeper into the Earth, to enlarge the canals, to raise the dikes, to double, to quadruple the railway lines, and even to collect the sand needed for concrete, which was then processed by oil heat. As for the Rio Tinto Company, soon controlled by the Rothschild family, it sent its ever more powerful excavation machines to the very ends of the world. Today, it remains one of the biggest mining companies in existence, despite the exhaustion and closure in 2001 of its ultimate Andalusian mine.

The birth of the petroleum industry was made possible by the industrial tools that preceded it; likewise, its expansion helped meet increasingly extensive industrial needs. There is a "spiral" between the complexity of a society's technology and the amount of energy that technology requires, said American anthropologist Joseph A. Tainter.[67] Because it is useful to solve problems, this complexity increases with time. And in order to increase (commensurate with an increase in technical mechanisms and industrial organizations, offices, laws, controls), this complexity needs essentially only one thing in return: more energy. Joseph Tainter observed that "energy and complexity tend to be intertwined, they either go up together or down together . . . : you cannot have complexity without energy, and if you have energy, you will have complexity."[68]

The Idea of an End Is Born: 1865, the Coal Question, and Entropy

Energy use and complexity inevitably rise alongside efficiency: This axiom of social evolution was foreseen as early as 1865 by one of the founders of economic science. In his book *The Coal Question*, the Englishman Stanley Jevons observed the industry's first steps toward efficiency and tried to draw some lessons from them. His early, penetrating analysis revealed an apparent paradox, which is in fact a fundamental consequence of technical progress: Jevons remarked that, far from diminishing, the consumption of coal increased as steam-driven machinery improved. "It is wholly a confusion of ideas to suppose that the economical use of fuel is equivalent to a diminishing consumption. The very contrary is the truth," wrote Jevons. This paradox is now called the "rebound effect" of energy consumption. It is easily explained. When steam locomotives become less carbon-intensive, they are also able, thanks to this economy, to travel farther distances and to pull heavier trains. A more economical machine is more efficient: it can accomplish more work, and thus help to make more machines—leading to an increase, rather than a decrease in overall energy consumption. Technical progress liberates access to more energy, and this energy can meet more needs (including finding new sources of energy). More energy yields more complexity, which in turn calls for more energy, and so on: The spiral begins. Generally, people only learn how to accelerate its rotation when attempting to solve problems.

The Jevons paradox uncorks the corollary that most preoccupies the author of *The Coal Question*—as indicated by the book's subtitle: *An Inquiry Concerning the Progress of the Nation, and the Probable Exhaustion of Our Coal Mines*. Stanley Jevons wrote: "In the increasing depth and difficulty of coal mining we shall meet that vague, but inevitable boundary that will stop our progress."[69] Then he issued a warning: If, due to (and not in spite of) technical progress, demand for coal was bound to increase, the day would come when Britain's coal reserves would be depleted and would no longer satisfy demand. As we shall see, the English economist's concerns about the future of the mines of his country proved to be fully justified.

Capable of solving innumerable problems, the energy-complexity spiral itself tends toward its own great problem, its absolute physical limit. However, the abundance of Earth's resources long allowed us to ignore, if not dismiss, the "question" of exhaustion that Stanley Jevons raised about British coal during the dawn of the oil industry. In 1865, German physicist Rudolf Clausius gave the concept mathematical expression when describing entropy, based on the second law of thermodynamics discovered in 1824 by the engineer Sadi Carnot. That law states that during any physical phenomenon, energy irreversibly dissipates in the form of heat, passing from an exploitable ordered state to a disordered state, no longer viable for human use. *Entropy* refers to the amount of energy that cannot be transformed into work. Its characteristic increase over time thus measures the inevitably growing disorder of the world. Energy is the price of the cohesion of every system: The increase of the order of a system necessarily leads to an increase of the disorder of the world that surrounds it. Once dissipated by a machine, the energy is lost.

THREE

Sharing the World Market: The First Attempts, at the Cradle of Royal Dutch Shell

When Robert Nobel, the eldest brother of the illustrious Swedish industrialists, first landed in Baku in March 1873, the exploitation of the immense oil resources in this city on the banks of the Caspian Sea was stagnant. Isolated and remote, the small peninsula of oil-producing land lacked timber to make barrels and derricks. Extraction remained rudimentary and production was low. Oil was often poorly refined. Its transport to Moscow, via the port of Astrakhan and then along the Volga, was as slow as it was restricted. Tsar Alexander II had decided, however, to take the brakes off the main obstacle to oil development around the city. Until then, the empire's corrupt administration had rotated concession owners every four years in order to better skim their profits. In March 1872, at the same time as the Cleveland Massacre in the United States, St. Petersburg authorized the auction of some 1,300 hectares of oil fields and triggered intense speculation that marked the beginning of the Russian oil boom.[1]

The Ingenious Nobel Brothers Give Rise to Baku, the "Black City"

Upon his arrival on the shores of the Caspian, Robert Nobel immediately recognized the potential of Baku's black gold. The Nobel family—made famous when the youngest living brother, Alfred, invented dynamite in 1867—was already firmly established in Russia. Ludvig, the middle child, was one of the main suppliers of the Tsar's army. His foundries and mechanical factories in

St. Petersburg manufactured steam engines, cars, gun carriages, shells, mines, and rifles. Robert had come to Baku to buy hardwood to make rifle butts. But he decided instead to invest the 25,000 rubles that his brother Ludvig had entrusted to him in the purchase of an oil concession and a small refinery previously owned by a steamship captain. He created the Nobel Brothers Petroleum Company.[2] Branobel (the Russian contraction of "Nobel brothers") soon displayed a logo in the form of a shield depicting the Ateshgah, the Fire Temple of Baku. A decade later it became, for a few brief years, the world's largest petroleum producer.

An accomplished chemist, Robert Nobel quickly discovered how to treat his kerosene with caustic soda to stop it from releasing the usual copious smoke.[3] The Branobel company brought in a professional chemist from Sweden, making it the first petroleum company ever to employ a full-time scientist.[4] The first barrels of the Nobel brothers' oil reached St. Petersburg in October 1876. Ludvig Nobel joined his brother Robert in Baku the same year, realizing that the petroleum trade was even more profitable and promising than that of arms manufacturing. In Paris, Alfred Nobel contributed by negotiating various financial arrangements, particularly a loan from Crédit Lyonnais that appears to be the first loan pledged on future oil production—a practice that became common.[5]

Ludvig and Robert Nobel introduced a series of technical innovations in the refining processes that were gradually imitated everywhere, including the United States. Beginning in 1877, Ludvig Nobel set out to overcome the geographical barriers that prevented the large-scale commercialization of Baku oil. Until then, each barrel had to be stowed under the chassis of special carts, which ten thousand men hauled every day over the 10 kilometers separating the wells from the refineries.[6] In Glasgow, Ludvig Nobel built 5-inch diameter tubes through which oil flowed thanks to a 27-horsepower steam pump. The pipes were sent to Baku for installation. The cart drivers tried to sabotage the project, but Branobel erected eight watchtowers and hired Cossacks who shot trespassers on sight.[7]

In January 1878, Ludvig Nobel ordered the very first tanker, delivered that same year by a Swedish shipyard. Brilliantly designed, the *Zoroaster* was a 184-foot (54-meter) vessel capable of carrying 242 tons of kerosene in twenty-one different compartments, to ensure structural stability. The *Zoroaster* and its successors were designed to face the tempestuous Caspian Sea and then be able to navigate the Volga and its canals to reach St. Petersburg and the Baltic Sea. In 1881, during its loading along the docks of Baku, a sister ship of the *Zoroaster* had exploded, killing half of its crew and causing one of the first oil spills in history. Despite this, oil tankers replaced the use of barrels and drums. Never patented, Branobel's pivotal innovation was soon imitated. Eight years later, the first ship specifically dedicated to the transport of oil crossed the Atlantic. The black gold market could now become global. Oil was imposing itself as the common denominator of development: Soon it emerged everywhere, through a global and somehow necessary dynamic of fractal flows leading from the wells to the most intense centers of commerce and industry, and then into every household.

The profusion of black gold at the edge of the Caspian by far exceeded anything seen before. The compact forest of derricks that blanketed the horizon and the thick smoke that emerged from two hundred small refinery workshops earned Baku the nickname "Black City." Development accelerated rapidly. In 1880, Baku had more than three hundred wells. In August 1883, the oil that sprang from a well named "Droujba" ("Friendship") broke its concrete form, shooting crude oil 7 meters high, pulverizing its derrick, and flowing for more than four months, up to 60 meters high, with an audible din that could be heard several kilometers away.

The workers, as Russian socialist writer Maxim Gorky observed, were the image of a "brilliantly executed picture of awful hell."[8] Tens of thousands of Azeri and Persian migrant workers labored almost naked in nightmarish conditions, then slept in unsanitary barracks. The paternalist philanthropy of the Nobel family was an exception to this pattern: It built permanent dwellings, redistributed part of its profits to the workers, and created free schools for their children.

Other local petroleum barons built villas inspired by postcards of Venetian palaces. To settle their differences, they engaged kotchis, mercenaries who, in their name, shot at each other during ritual duels. The most respected of these barons, Zeynalabdin Tagiyev, son of an illiterate bootmaker, became the "Azerbaijani Eunuch-Maker" in the Moscow newspapers after threatening to emasculate a Georgian prince who showed too much interest in his wife, the beautiful Sona, fifty-six years his junior. Tagiyev also created the first secular school in the Middle East for girls from wealthy families.[9]

The Rothschild Brothers and the Nobel Brothers, the World's Largest Oil Producers

In the early 1880s, Branobel succeeded in chasing Standard Oil from the territory of the Russian Empire, while John D. Rockefeller's company completed the conquest of the rest of Europe, an infinitely more significant strategic market. But Baku's production grew so that old rural Russia quickly became far too small of a market for the quantity of oil that flowed from the Black City. Two companies competing with Branobel sought partners to build a railway from Baku to the Black Sea, thus opening up a commercial alternative to the Volga route, which the Nobel brothers controlled with their tankers.

These two companies, Bunge and Palashkovsky, found their perfect partners in Paris: the barons Alphonse and Edmond de Rothschild. In 1882, the French branch of the celebrated banker family got around Russia's anti-Semitic May Laws. Signed that same year by Alexander III, these laws were behind the largest wave of Jewish emigration prior to the creation of Israel, and they forbade, among other things, the Tsar's Jewish subjects from investing capital in the lands of Holy Russia. The Rothschild brothers had already financed the construction of many railway lines in Europe, and they were no strangers to the black gold trade, since they owned a refinery in Marseilles, in the south of France, and another in Fiume, on the Adriatic.

When they went to Baku in 1883, the Rothschilds tasted the wild charms of the city. First, they were sold a fake oil well skillfully coated with crude oil. They refused to hire kotchis, who in return robbed them many times.[10] But the trains that began to run that year between Baku and the Black Sea almost instantaneously transformed the small port of Batumi, attached to the Russian Empire only five years before, into a mushrooming city. Batumi became as bustling as San Francisco, and no less dangerous. Almost everyone carried a revolver. In 1888, a wealthy nineteen-year-old Armenian from a powerful merchant family landed in Batumi; Calouste Sarkis Gulbenkian became one of the petroleum industry's legends. An engineer freshly graduated from King's College in London, the author of a thesis on oil exploitation, and the son of a man who earned a fortune by importing Russian kerosene into the Ottoman Empire, the elegant young Gulbenkian discovered a "filthy" city, where "the water stank of oil and drains and the inhabitants seemed to exist almost entirely by smuggling."[11] The United States consul, J. C. Chambers, quickly arrived on the scene; he was hired by Standard Oil to collect information.[12] Russia soon built a naval base in what for a time became the world's number one oil port.

The western route was now open. The Rothschilds founded the Caspian and Black Sea Oil Company, known by its Russian acronym: Bnito. To make their investments profitable, they accepted that most of the trains going back and forth between Baku and Batumi comprised the Nobel brothers' wagons. But the trains were slow because they had to cross the difficult Suram mountain pass, located 1,000 meters above sea level. In 1886, in spite of the initial reluctance of the Imperial Crown of Russia, which feared losing the large subsidies it collected on rail freight, the Rothschilds, the Nobels, and Zeynalabdin Tagiyev (the "eunuch maker") constructed a 68-kilometer pipeline across the Caucasus: 400 tons of Alfred Nobel's dynamite opened the mountain pass of Suram.[13]

On September 27, 1886, in the village of Bibi-Heybat, close to Baku, the Tagiyev workers saw a new fountain of black gold spring up, making all prior sources of crude look like mere trickles. The *Baku News*, the local newspaper, described "a colossal pillar of smoke, from the crest of which clouds of oil sand detached themselves and floated away a great distance without touching the ground." About 2 kilometers away, the Baku city hall square was flooded with oil. Before it was contained, the Bibi-Heybat well spat more oil every day than any other oil field in the world combined, more than even the twenty-five thousand American wells! There was an attempt to divert oil to other wells. It took fifteen days to contain the beast. During these two weeks, 1,400,000 barrels of crude oil flowed into the Caspian Sea; the oil "was lost forever to mankind," reported the *Baku News*.[14]

From the beginning of the 1880s, Branobel and Bnito established themselves as the two biggest oil producers worldwide. In the United States, Standard Oil, which dominated the refining market, had only four oil-producing properties and a handful of wells.[15] In 1891, Russia's share of world kerosene exports was 29 percent, while that of the United States was reduced to 71 percent.[16] Russia's share continued to grow in subsequent years. The war for control of the world's black gold market had begun.

In the United States, Opposing Standard Oil Becomes Futile

Standard Oil had, in the meantime, never stopped consolidating its position in the United States. John Rockefeller continued to mercilessly subdue allies and enemies. He took advantage of the railways' fear of wells drying up and offered to invest in railway infrastructure. By 1874, he controlled several essential train routes. From the late 1870s onward, Standard Oil dominated oil freight between Lake Erie and the Atlantic thanks to closer ties with two of the three largest railways, the Erie Railroad and the New York Central—a control that Rockefeller exploited without hesitation, to the detriment of his biggest competitors.

Tom Scott's Pennsylvania Railroad remained the most powerful railway company. In the spring of 1877, he joined forces with a Standard Oil competitor, the Empire Transportation Company, to try to counter Rockefeller's ever-increasing hegemony. But Rockefeller responded by simply diverting his freight away from the Pennsylvania Railroad lines. To this end, he shut down his own Pittsburgh refineries. Very quickly, the Pennsylvania Railroad was in dire straits, decreasing wages overnight and firing railway workers by the hundreds. This was followed by a general railroad strike that left dozens dead, one of the worst labor conflicts in US history. Hundreds of locomotives were burned. Tom Scott asked for mercy and, insult to injury, had to yield many assets to the Standard Oil Company to avoid bankruptcy.[17]

As for the oil regions, the local producers' crude oil was worthless without the network of short-distance pipelines owned by Standard Oil that collected the production of thousands of wells. On November 21, 1877, many producers met in Titusville to establish a "Petroleum Parliament." They attempted to pass a law in favor of the free use of pipelines. But Standard Oil killed the effort with large bribes to Pennsylvania's congressmen.[18]

Two independent long-distance pipeline projects were then launched. These were innovative projects, even experimental ones: Until then, no pipeline had been more than 50 kilometers long. Standard Oil, which until that point appeared to be a progressive force, did everything in its power to undermine the construction of these pipelines. The most ambitious and longest of the two projects, the 180-kilometer Tidewater Pipeline, was also the most threatening for Rockefeller. By connecting to an independent railway line, it would allow oil to be transported to the Atlantic coast without involving Standard Oil. It would require the oil to be pumped over an 800-meter-high pass, eight years before the Nobels built the pipeline that crossed the Suram mountain pass. The project was widely considered unfeasible.[19] Among Standard Oil's management, it aroused only sarcasm.[20] Construction was launched on November 22, 1878—one year and one day after the first meeting of the Titusville Petroleum Parliament. Rockefeller, referencing its backers, wrote testily in a letter: "I would have no mercy on them that don't deserve nor appreciate it."[21] His letter was addressed to the Standard Oil henchman Daniel O'Day, a solid Irishman who had distinguished himself during the Cleveland Massacre and displayed a scar on his forehead, a testimony to the strength of his determination to serve his boss.

To undermine the construction and break the spirit of its sponsors, Rockefeller intimidated the pipeline suppliers and bought out the last independent refiners who might become its customers. He acquired land along the route. When Standard Oil took possession of an entire valley, the pipeline's path was diverted into the surrounding hills. Ultimately, when the project was nearing completion, Standard Oil embarked on a new campaign of large-scale bribery of local elected officials. Admittedly, corruption was common and rampant in American democracy during that period.[22] The pipeline was completed in 1879. But in 1882, Standard Oil managed to take control of 88.5 percent of the pipeline traffic. At the same time, it completed the construction of four other pipelines leading to Cleveland, Buffalo, New York, and Philadelphia.

On January 2, 1882, all Rockefeller empire holdings were assembled within a new opaque and overpowering structure that received a great deal of attention: The Standard Oil "trust" was born.

Faced with the First Peak Oil in History, Standard Oil Becomes a Crude Oil Producer

Cleveland was too small and too far away from the heart of commerce and finance. On May 1, 1885, in New York, Standard Oil opened its new headquarters in an austere and massive building at 26 Broadway, on the tip of Manhattan, directly across from where the installation of the Statue of Liberty was nearing completion. From his office, John D. Rockefeller no doubt took time to contemplate what his empire had become: an immense intertwining of refineries, pipelines, railway lines, ocean freight, and investment banks—the most powerful company of the newly emerging American dominance.

Strangely, this empire was now vulnerable for the first time. This vulnerability was not due to competition from Russian oil. In Europe, Standard Oil's agents managed to contain the advance of Nobel and Rothschild kerosene via a fierce price war, while engaging in corruption and even sabotage.[23] Year after year, Standard Oil maintained a quasi-monopoly in Great Britain and France.[24] From 1885 on, first three and then ten of the leading French importing companies (soon known as the "Cartel of the Ten") agreed to buy oil only from Standard Oil. They all sold their petroleum for nearly identical prices, which guaranteed them generous profit margins.[25]

The danger was underground, invisible. "We have opened nothing of importance in the way of production this summer, and next winter must see a great reduction in the old Bedford and Allegheny fields," wrote John Archbold to Rockefeller in September 1885.[26] Rockefeller's former enemy, now second in command of Standard Oil, liquidated some of his company shares. While Standard Oil still supplied more than three-quarters of world production and controlled almost all American production, extraction of oil from the Pennsylvania oil regions fields unexpectedly began to decline sharply in 1881. The catastrophe was confirmed the following year, when the flow of wells from neighboring New York State collapsed in turn. The richest field, discovered in Bradford, Pennsylvania, in 1871, was drilled clean by avid prospectors. Many fanatics used nitroglycerin to stimulate their wells, a drastic remedy that generally did more harm than good. But the trend was unrelenting; the fields were on their way to being exhausted. In 1885, due to the lack of sufficient exploitable reserves, their decline was as rapid as their rise had been in the previous decade. The prospectors redoubled their efforts to delay the inevitable and relaunch the extractions one last time in 1887, with the help of newly discovered fields of secondary importance. But Pennsylvania crossed its last production peak in 1891, and faced an irremediable decline until the recent development of fracking.[27]

A quarter of a century after Edwin Drake began drilling in Pennsylvania, no other significant oil field had been developed in the United States. Rockefeller abruptly realized the vulnerability of his creation and beginning in 1884 declared an urgent need to build up large stocks. "Better that this stock be somewhat excessive than run the risk of Russian competition shutting us out," he told a dubious partner.[28] In the face of those who predicted the imminent exhaustion of the American oil boom, the president of Standard Oil pointed a confident finger in the direction of heaven: "The Lord will provide."[29]

And in May 1885, a few days after the inauguration of the Standard Oil Building on Broadway, the Lord provided. Near the small town of Lima, lost in the Ohio plains somewhere between Chicago and Cleveland, a prospector looking for natural gas for lighting and heating the village found something much better: large quantities of black gold. It was a new boom: 250 derricks were erected before the end of the year. Rockefeller immediately seized the chance to fill the cracks in his empire's foundation. Until the unexpected collapse of the oil regions, he had deliberately kept himself out of the riskiest part of the oil business, but now he invested millions of dollars in production. Standard Oil acquired and drilled many wells in Ohio's new intact crude fields, while winding its pipelines and railways around isolated Lima.

But Ohio's oil was of mediocre quality: Once processed, it was much less refined than Pennsylvanian crude oil, and because of a high proportion of sulfur, it "smells like a stack of polecats," wrote one journalist.[30] These crippling defects obliged Standard Oil to search for opportunities other than unsellable lamp oil. They tried to convince the railway companies and factories to use more and more oil, rather than coal, to run their boilers and stoves. In July 1886, Rockefeller hired chemist Herman Frasch to find a way to filter out the nauseating sulfur. Nicknamed the "wild Dutchman" because of his eccentric and vain character, Frasch was the first scientist hired full time by Standard

Oil. In October 1888, after two years of unfruitful attempts, he succeeded in ridding Lima's petroleum of its foul odor.

The registered patents yielded sumptuous profits to Standard Oil. This innovation allowed Rockefeller to invest even more in crude oil extraction. Rushing to acquire thousands of square kilometers of land not only in Ohio but also wherever new wells were being drilled in Pennsylvania, West Virginia, and Kansas, the Standard Oil octopus was now poised to absorb or stifle competition among producers, as it had already done with refiners. In 1891, it possessed the majority of the wells in the Lima region and no less than a quarter of American extractions. In 1898, Standard Oil controlled one-third of American oil production.[31] From then on, his fingers in all sectors of the industry—from wells to the sale of kerosene and oil, Rockefeller, pushed by necessity, created the "vertical integration" model that was imitated in many mining and industrial sectors around the world.

Scientific progress had, in the meantime, become a priority for Rockefeller. In 1892, the fervent Baptist inaugurated the centerpiece of his philanthropic work: the University of Chicago. Thanks to Standard Oil capital, it quickly became one of the most prestigious research centers in the world, mainly in the fields of physics, economics, and medicine.

For Standard Oil, a Competitor Appears on the Old Continent: Shell

In the early 1890s, as the commercial war between US and Russian oil began, Standard Oil, now one hundred thousand employees strong, had almost complete control of the US market, as well as a superior share of the European market. To sell their enormous stocks of black gold, Baku's producers had to look much further away, to the only remaining market left to conquer: Asia. Sent to the Far East in 1882 for a lengthy market study, William Herbert Libby, a sort of plenipotentiary ambassador of Standard Oil throughout the world, noted with surprise that oil had already somehow "forced its way into more nooks and corners of civilized and uncivilized countries than any other product in business history emanating from a single source."[32] From Siam to the most remote Chinese provinces, Standard Oil petroleum was bought and sold everywhere. However, Asia, located far from the docks of New York, was a secondary market where Rockefeller's company had not established control.

The opportunity offered there did not escape the Rothschilds. Anxious to find outlets outside the American and European continents, which were more or less under Standard Oil's exclusive dominion, Parisian bankers met Marcus Samuel, a thirty-seven-year-old London-based Jewish merchant. The father of this young acrobat of international trade had escaped from London's sordid impoverished districts by selling trinkets adorned with shells, and Shell was the name that Marcus Samuel gave his new company in 1897. Opposite Standard Oil, Shell positioned itself as an emerging force in the first struggle to control the world oil market at the turn of the twentieth century.

After their father's death, Marcus Samuel and his brother, Samuel Samuel, dramatically grew the family business: They expanded it to Yokohama. Like the Nobel brothers in Russia, the Samuels' export house played a major role in the industrialization of the Empire of the Rising Sun. At the Rothschilds' invitation, Marcus Samuel traveled to Baku in 1890. On his return to London, the meek but sharp-eyed businessman built a fearless stratagem of planetary scope. With the utmost discretion, two of the Samuel brothers' nephews, Joseph and Mark Abrahams, were dispatched to Asia, one in India and the other in the Far East. Their mission was to supervise oil reservoir construction and foster good relations with the Scottish merchants, who reigned over British trade from Calcutta to Hong Kong. Their task was to negotiate the conditions for the distribution of kerosene from Baku to Asia. Meanwhile, the Samuels and the Rothschilds, associated with other Parisian Jewish financiers from the Banque Worms, ordered tankers, as well as many wagons and specialized vans.

The most delicate and decisive aspect of the plan was to obtain authorization for tankers (vessels reputed to be dangerous, and rightly so) to cross the Suez Canal. In London, a secret struggle for influence ensued with the Foreign Office, which exercised control over navigation on the canal. On one side were Marcus Samuel, recently elected alderman in London, and the Rothschild family, whose timely loan enabled Benjamin Disraeli's British government to take control of the canal in 1875.[33] On the other side were solicitors manifestly under the control of Standard Oil.[34] An article in the *Economist* made an obscure reference to the opponents of a plan of "Hebrew inspiration."[35]

Marcus Samuel's tankers were authorized to navigate the canal on January 5, 1892: With this authorization, the Rothschilds gained weeks of passage to the Orient, and an advantage of thousands of miles compared to the route Standard Oil used to reach Singapore from New York. Leaving England, the *Murex*, Marcus Samuel's first tanker, loaded the Rothschilds' kerosene in Batumi on July 22, 1892. It crossed the Suez Canal a month later, en route to Bangkok. The *Conch* soon followed, then the *Clam*, and many other vessels, all baptized with the names of shells.

Rockefeller, Samuel, Rothschild: The First Ménage à Trois Ends Abruptly

While supervising their secret offensive in Asia against Standard Oil, the Rothschilds tried no less secretively to establish friendly relations with the company. In July 1892, while the *Murex* navigated the Black Sea, John Archbold made Rockefeller aware of a discreet visit Baron Alphonse de Rothschild had made to 26 Broadway. The "tentative agreement" that Archbold mentioned to Rockefeller did not succeed in bringing them together, despite the Nobels' wish to join them.[36] This was the first of many attempts to share the world market, made by the immense capitalist organizations that were in the process of fabricating or capturing the global web of black gold.

In 1894, Standard Oil embarked on a new price war against black gold originating from Russia, while at the same time proposing to absorb Marcus Samuel's shipping company. The latter rejected the offer. Rockefeller tried again with the Rothschilds and the Nobels, and on March 14, 1895, a major alliance was signed "on behalf of the petroleum industry of the US" by Standard Oil and "on behalf of the petroleum industry of Russia" by Bnito and Branobel. The agreement's protocol stated that 75 percent of world exports would be reserved for the Americans, while 25 percent would return to Russian companies.[37] But this, too, failed. Negotiations were leaked to the European press, where many cartoonists had fun drawing the awkward kiss between the octopus and the bear. Tsar Nicholas II's minister of finance, Count Sergei Witte, delivered a veto from St. Petersburg. Count Witte had the upper hand in the development of Russian industry. He relied on oil to realize this development: Baku crude was needed to avoid expensive imports of British coal. Notably, the locomotives of the Trans-Siberian Railway, the main section of which was opened in October of 1891, ran on petroleum.[38] *Mazout*, synonymous with heavy fuel oil in several European countries, is a Russian word. As in the United States, it was often used in addition to coal in blast furnaces. Russia's late but rapid industrialization, like America's before it, had its roots in rock oil as much as, if not more than, in coal.

The intricate struggle for global market control grew more vigorous. While Baku's extractions continued to expand, Standard Oil was confronted with a decline of the oil regions in the United States and began looking around the Pacific for opportunities that could open the door to the Asian market. It considered three promising areas where prospectors had worked in the 1880s: Southern California, where Standard Oil was outmatched by tough local competitors such as Union Oil of California (Unocal); Sakhalin, a Russian island in the Far East, located in the frigid waters of northern Japan; and finally, Sumatra, the largest of the Dutch East Indies. For reasons both geographical and technical, access to the crude oil of the Southern Californian desert, isolated and far away, was at that time almost as restricted as that of Sakhalin. This was not the case for the black gold whose excavation had just begun in Sumatra, opposite the port of Singapore—the Asian trade hub located on the other side of the very strategically situated Strait of Malacca.

The Brass Ring: Royal Dutch's Sumatran Oil

Sumatra's first oil well had been drilled in 1885 at the initiative of a Dutch tobacco grower, Aeilko Jans Zijlker, on the lands of the small sultanate of Langkat, near a navigable river that flows into the Strait of Malacca. With political support and capital from the king of the Netherlands himself, William III, the Royal Dutch Petroleum Company was launched on June 16, 1890. Zijlker died suddenly in Singapore six months later, at age fifty. In spite of the difficult conditions imposed by the tropical forest, the Dutch succeeded in rapidly developing the extraction of high quality black gold.

The oil of Sumatra was the decisive stake through which the balance of power between the world's leading oil companies was stabilized. It played a role comparable to that of Persian Gulf oil half a century later (though on a different scale). Marcus Samuel tirelessly sought to diversify his supply sources, to relieve his company from its total dependence on the Russian oil of the Rothschilds and Nobels. In the autumn of 1896, Standard Oil's spies observed a trip to the Dutch East Indies made by Mark Abrahams, Samuel's "clever" nephew.[39] Abrahams, who discreetly had taken the necessary steps to gain navigation rights for the *Murex* to reach Asia via the Suez Canal, was now on a mission to attempt to exploit the oil outcrops identified in Borneo, the large island neighboring Sumatra. But the jungle of Borneo proved much more hostile and unproductive than that of Sumatra. The offensive turned into a financial fiasco. Shell's Chinese laborers and their white foremen—constantly plagued by tropical fevers and entangled in a muddy forest where everything rotted—were unable to obtain substantial crude oil production in Borneo.

Early in 1897, Marcus Samuel approached the Royal Dutch leadership and offered to buy their company. In his April letter, he insisted on their common interest in avoiding the "ruinous competition," the exact same expression used by Rockefeller a quarter of a century earlier, just before the Cleveland Massacre.[40] Instead, the Dutch favored a simple commercial alliance with Shell. Over the summer, Royal Dutch also rejected an offer from Standard Oil.

The Dutch wells of Sumatra seemed inexhaustible, and the Royal Dutch became seemingly invincible. However, on December 31, 1897, while the Dutch company was celebrating the launch of a tanker named *Sumatra* with the sultan of Langkat, troublesome news was spreading. Many of the sultanate's wells were producing less and less oil, and more and more salt water. Their pressure had dropped dramatically. In an attempt to avert disaster, no fewer than 110 wells were drilled the following year on the sultan's lands, but these were all dry. Providence arrived for the Royal Dutch: Exploitable oil was discovered a hundred kilometers further north, in the small principality of Perlak. The relief was immense when, in 1900, Royal Dutch extractions began to grow again.

Shell Goes Big in Texas, but Stumbles on "Swindletop"

Marcus Samuel continued his anxious quest for new Shell supply sources. In January 1901, the perfect opportunity finally came. On Spindletop Hill, near Port Arthur and Houston—two small, isolated cotton markets along the coast of the Gulf of Mexico—a former Austrian Navy officer named Anthony Lucas discovered a formidable new source of crude oil.[41] Immediately, boreholes multiplied at a furious pace, thanks to the new rotary drill bits that enabled prospectors to drill faster and deeper. Moonshine, prostitutes, and crooks: The madness that had seized chilly Pennsylvania forty years earlier began anew in the sleepy southern United States. Texas made an explosive entrance on the world oil map. Shell's emissaries quickly undertook a long journey that led them from the banks of the Thames to the stifling, sometimes damp, sometimes dusty, and largely uninhabited Spindletop in Texas. There, they signed a tremendous contract with the biggest producer among the hundreds of derricks that had cropped up within a few weeks, nearly touching one another. A prospector from the oil regions, Colonel James Guffey, raised the necessary funds for Anthony Lucas to drill. Until then, Guffey had been unaware that Shell, which bought his oil, existed.[42]

Marcus Samuel finally possessed his great virgin source of crude, located, as a bonus, on American soil. Spindletop's oil was far from high quality. It provided little kerosene, but that did not matter; it was a good fuel, and Shell's boss was convinced that, increasingly, oil would take coal's place in running the engines of ships, locomotives, and the first automobiles. Samuel had little choice but to make a very generous offer, and he committed to buying 100,000 tons of oil every year for twenty-one years at a fixed rate. Focused primarily on supply and market access, Shell's management was not very concerned with geology, a science that in any case only provided uncertain answers. To invest (and, in this case, to invest significantly) in the production of an oil field was essentially a gamble.

Standard Oil was absent from the scene: Following a complaint filed in 1894, the state of Texas, one of the first to engage in the fight against cartels, had expelled the largest US company after finding that one of the principal local kerosene wholesalers had concealed its membership in the Rockefeller empire.[43]

In October 1901, Marcus Samuel went to New York to meet with John Archbold and the gentlemen of Standard Oil, hoping to achieve a new balance of power. Two months later, he flatly rejected their buyout offer. Samuel believed that the power of his company had just been regenerated, thanks to Texas oil. At age forty-nine, the Jewish merchant left London's East End with the intention of remaining the sole master of one of the most flourishing British companies. Samuel was recognized as a patriot: In 1898, he became the first British oil baron to receive the title of knight, after one of Shell's tankers towed one of the Royal Navy's best battleships and saved it from sinking.

Rather than being absorbed by the Yankee octopus, Marcus Samuel preferred to pursue his strategy of combining Shell and its distribution network in Asia with the Royal Dutch and its excellent new wells in Sumatra. A rapprochement in which he clearly believed he could retain leadership yielded a new structure—the "British Dutch," whose name expressed the order of precedence in the origin of its capital. Together, Shell and Royal Dutch founded the Asiatic Petroleum Company in June 1902, with the inevitable involvement of the Rothschilds and their Baku oil. This powerful new alliance occupied the strongest position in the oil trade east of the Suez Canal. Sir Samuel had become Standard Oil's sole equal worldwide: He now led the only competitor capable of seriously contesting the Rockefeller empire's dominance in North American and Europe.

Alas, the prospector rushed to drill too fast and too deep at Spindletop. To everyone's surprise, the Texas extractions started to collapse in the winter of 1902, just as they had twenty years earlier in Pennsylvania for Bradford, and four years earlier in Sumatra for Langkat. The chaotic drilling of Spindletop (renamed "Swindletop") was only the first foray into the many oil fields that remained to be discovered in Texas.

But Marcus Samuel had too hastily pinned the fate of his company on Spindletop. Bestowed with the honors and pomp of high Edwardian society (of which he delighted), Samuel was made lord mayor of London in 1902. Only now, this corpulent, newly rich man who rode his horse each morning in Hyde Park was caught between two fires on the business front. In Europe, Shell underwent a long, devastating price war in 1903, led by John Archbold, now in charge of the daily management of the Rockefeller empire. Standard Oil dominated the US market so well that it could maintain high enough prices to finance a dumping strategy on the Old Continent. Shell suffered the insult of being driven out of Germany thanks to the collusion of Deutsche Bank, the main partner of Standard Oil and the Reich.[44] On the other side of the globe, in the Dutch East Indies, the Royal Dutch saw its extractions increase again and again. The capital and mastery enjoyed by the Dutch even allowed them the luxury of producing in Borneo after 1901, succeeding where Shell had failed.[45]

Year after year, Royal Dutch dividends grew while those of Shell collapsed. In the end, Sir Samuel was forced to accept the conditions of a humiliating merger. He owned less than half, just 40 percent, of the new company created in April 1906 and ratified in February 1907: Royal Dutch Shell. This time the order of precedence was reversed: The name of the new company indicated the domination of Dutch capital. Due to the failure of his Texas poker game, Samuel was outclassed: The rules of the game were now dictated by a young, small, intractable and erratic Dutchman: Henri Deterding, appointed in December 1900 by the head of the Royal Dutch at only thirty-four years of age.

Deterding was from a high-society Amsterdam family that had fallen on hard times; his father was a naval captain who had died when Henri was six years old. Like Rockefeller, Deterding was above all a relentless accountant who had matured to become a methodical, merciless strategist. Who can say what caused his triumph at the end of a long decade of struggling for influence? Was it his managerial talents that triumphed over the bold intuitions of Marcus Samuel? Or was fate determined by the differences between the nature—almost indistinguishable with the tools and technical knowledge of the time—of the underground black gold reservoirs of Spindletop, Texas, and Sumatra?

The Birth of Two Major Texas Companies: Gulf Oil and Texaco, Standard Oil's Young Cousins

In Spindletop, the Mellons, the banker family who had financed James Guffey (prematurely declared in the press to be the biggest American oil magnate), managed to save their capital *in extremis*. They sidelined Guffey; he consoled himself through his role as a flamboyant figure in the Democratic Party. William L. and Andrew W. Mellon looked to John Archbold to hedge their bet. Heirs to a rich Pittsburgh dynasty, in 1894 the Mellons, then among Rockefeller's last notable competitors, had already agreed to sell their crude oil transport company to Standard Oil. But Standard Oil, now under attack by an all-out offensive against cartels and trusts, feared investing in Texas more than ever.[46] It thus allowed its only two competitors on American soil to flourish.

Fortunately for the Mellons, Spindletop was far from being the only source of crude oil in this part of the United States. Beginning in 1905, enormous oil fields were discovered on the plains of the Native American territories beyond the northern frontier of Texas, which two years later, after the Native Americans' ghost dances were unable to stave off white expansion, became the state of Oklahoma. There, overnight, important cities appeared in the middle of nowhere—notably Tulsa, which became known as the "oil capital of the world." By investing extensively in Oklahoma, Texas, and elsewhere in the southern United States, the Mellons managed to make their company, Gulf Oil, a small giant sitting at the foot of Standard Oil.

The Texas Company (or Texaco) was the second major company born in Spindletop to mature in the Texas crude oil fields. It was the last major American oil company that was able to subdue the world oil market through the first three-quarters of the twentieth century. The company was founded on March 28, 1901, by Joseph Cullinan, a former Standard Oil employee who prospered by selling petroleum products to cotton and cereal growers in the southern United States. Part of Texaco's initial capital came from an association of New York investors led by the company that dominated the US leather cartel, US Leather. The rest of the capital came from a group of Texans led by James "Big Jim" Hogg, who was the first Texas governor to be born in the state. Hogg was also the one who managed to expel Standard Oil shortly before the discovery of Spindletop. A reversal of fortune for the fierce Texan investors, protective of their independence, or a mere expression of the law of the fittest? Starting in 1909, Texaco's New York shareholders placed Elgood Lufkin, a young engineer from the East Coast, at the head of the company. He had made his mark as an employee of the Indiana Pipe Line Company, owned by Standard Oil.[47] His father was for a long time manager for all of the producing branches of the Standard Oil.[48] The dynasties of the major American oil companies' high executives formed an exclusive aristocracy, of which John D. Rockefeller seemed destined to remain the supreme center of gravity.

Revolts in Baku: The Black City Burns

It was not telluric mysteries like those in the oil regions and Spindletop that precipitated the fall of Baku but bloody political events. During the early years of the twentieth century, the Black City became the world's leading oil producer and exporter. In 1901, Russia produced 233,000 barrels a day, compared with 190,000 barrels a day in the United States. The Dutch East Indies were far behind, with 11,000 barrels a day.[49] Despite the shenanigans of Standard Oil's agents, Britain imported more Russian oil than American oil.[50] With more than 3,400 wells, Baku alone produced almost half of the world's extractions. It was the base for Branobel and the partnership that Shell, the Rothschilds, and Royal Dutch established in 1902. Its prodigious flow of black gold seemed inextinguishable, calling attention to Professor Mendeleev's hypothesis on the abiotic creation and thus inexhaustible nature of oil.[51]

In most of Russia's oil workshops and factories, the workday lasted from twelve to sixteen hours, even for children. The workers, often the children and grandchildren of freed serfs, were held quiet, disciplined with fines or sticks. In the port of Batumi, on the Black Sea, three large oil-refining factories were operated by some ten thousand workers. In March 1902, workers in the Rothschild factory went on strike. The tsarist police killed fifteen demonstrators. Among the leaders arrested was a twenty-three-year-old Georgian socialist activist who had spent the previous year in hiding. He called himself Koba.[52] Joseph Vissarionovitch Djougachvili had not yet earned his nickname, "Stalin."

After a year and a half in the jails of Batumi, Koba was exiled to Eastern Siberia. His official biographers claimed that he escaped on January 5, 1904 (in the middle of winter). A month later, Koba was back in Batumi. In June of the following year, he moved to Baku.[53] The population of the Black City had swelled with the level of its crude oil production. Its two hundred thousand inhabitants were mostly miserable, harassed Azeri Muslims. Revolts were frequent. They were often repressed by the Black Hundreds, terrifying pro-tsarist secret militias that carried out attacks against refineries whose workers were later accused of being responsible.[54] On December 13, four revolutionary adventurers, three brothers surnamed Sendrikov and their sister, launched a general strike and demanded an eight-hour day. They set fire to about twenty wells. After some concessions from the oil barons, the local Bolshevik committee, of which Koba was a member, called for a return to work. But with their overwhelming majority, the strikers refused. They were labeled as "the cold gray mass of workers" by the Bolshevik committee of the future master of the USSR.[55] The strike continued until December 31. On that day, Baku's oil workers obtained the first collective agreement in Russia's history, despite the conciliatory approach advocated by the Bolsheviks. They even wrested full payment for their two-week strike.[56]

The Baku strike's success was one of the major precursors to the 1905 revolution, which burned through the Russian nation after Bloody Sunday on January 22 of that year, when the Russian Imperial Guard fired at unarmed demonstrators in St. Petersburg. In the Black City, the revolution of 1905 led to ferocious interethnic massacres. The advent of brutal industrialization had undoubtedly exacerbated age-old resentment by Azeri and Tatar Muslims toward the Armenian Christians, the former constituting the bulk of the working class, while the latter thrived, comprised of traders rendered incredibly rich by black gold. Envy and greed increased the belligerents' ardor. In early February, in the days following St. Petersburg's Bloody Sunday, massacres began in Baku after it was reported that Armenians had killed a schoolboy and a Tatar shopkeeper.[57] Throughout the month, several violent outbreaks left hundreds, perhaps thousands of victims. Some Tatar leaders accused the Russian authorities of having encouraged them to continue to kill the Armenians (it is plausible that the tsarist regime, notoriously racist, considered the Armenian merchants with great suspicion).[58] After defeating a Tatar, one of the Adamoff brothers, founders of one of the richest Armenian oil companies in Baku, survived for three days during the siege of his residence, killing many Muslim assailants while shooting from his balcony. He died, along with his entire household, during the final assault, when his house was burned with kerosene taken from a store and street lamps in the neighborhood.[59]

Even worse massacres took place in early September. In the middle of the oil fields, the pro-tsarist Black Hundreds added to the situation's confusion by once again attacking the oil workers.[60] Pro-Russian militiamen fired on the Tatars, apparently hoping to incite them to take revenge by attacking the Armenians.[61] On September 3, a rumor of apocalypse crossed the barricades barring the streets of Baku. It was confirmed in the afternoon by thick smoke that masked the sun: Many oil wells were burning.[62] The next day, near the burning pits, Armenian refugees inside the Association of Petroleum Producers school shot at the crowd that had gathered on the platforms of the nearby station to welcome the Tatar chiefs. The Tatars attacked. A survivor recounted: "The flames from the burning derricks and oil wells leaped up into the awful pall of smoke which hung over the inferno. I realized for the first time in my life all that can possibly be meant by the words 'Hell let loose.' Men crawled or dashed out of the flames only to be shot down by the Tatars. . . . It was made worse than anything that could have taken place at Pompeii by the ping of rifle and revolver bullets, the terrific thunder of exploding oil tanks, the fierce yells of the murderers, and the dying screams of their victims."[63] The tsar's army did not interfere, but shot cannons in the distance. The fire spread rapidly, fueled by the crude oil of numerous cisterns and fanned by violent winds that swept onto the shores of the Caspian Sea.[64]

After these days of rage, the majority of Baku's oil wells were severely damaged. The local oil companies, starting with Branobel, never really recovered. Enormous investments were needed to revive production, but bankers were discouraged. Especially as new strikes broke out, in particular those of January and February 1908, after which Koba was once again arrested and sent into exile. Extractions from Baku stagnated; they did not exceed the record established in 1901 until the end of the 1920s, under Stalin's Soviet regime. The Nobels and the Rothschilds were accelerating their investments in two new Russian oil fields that had recently been exploited a few hundred kilometers from Baku, in the northern Caucasus, around the towns of Grozny and Maikop. As a result, the port of Batumi was abandoned. But the oil fields of Grozny and Maikop were not nearly as generous as those of Baku. Russia's share of the world market was collapsing.

American oil's advantage was turning out to be overwhelming: Standard Oil was practically unchallenged by competition in Europe. In November 1906, Deutsche Bank, the Nobels, and the Rothschilds set up the Europäische Petroleum Union (EPU) in Berlin, which was fueled by the rapid development of Romania's oil fields.[65] But Romania (where Standard Oil was also investing) remained a second-tier producer. The EPU and Standard Oil negotiated sharing the European market over the next two years. Free competition was still not the order of the day, and thanks to its now plethoric production, the Rockefeller empire assumed more than 80 percent of the Old Continent's market.[66] In 1910, Branobel dismissed a quarter of its employees.[67] As for Bnito, in 1911 it was absorbed by Royal Dutch Shell. The Rothschilds' oil affairs—which in the preceding years had risen to account for more than 80 percent of the Parisian bank's industrial investments, undoubtedly responsible for much cold sweat among its leaders—accounted for a majority of the Dutch-British company's holdings.[68] The Rothschilds remained among the leading shareholders of Royal Dutch Shell.

By the early 1880s, on the eve of the First World War, Standard Oil and Royal Dutch Shell had become the worldwide champion and the runner-up, respectively, of the global oil market. They remained secure in their leadership roles until the creation of the Organization of the Petroleum

Exporting Countries (OPEC) half a century later. As in the game of poker, the first spoils were rich in unpredictable hazards: the unexpected decline of the "oil regions"; bad then good surprises in Sumatra; Spindletop's collapse; Baku in flames; enormous discoveries in the Mississippi plains; much more modest discoveries in Eastern Europe and the Caucasus. But with petroleum prospecting preparing to rely on genuine scientific data, the rest of the game would be much less troublesome for Standard Oil and Royal Dutch Shell: Even more than in poker, the capital amassed by the strongest players seemed to make their positions pretty much unstoppable. But the biggest act of the world's oilmen was just beginning.

FOUR

The Automobile: American Oil Regenerates Capitalism

The petroleum lamp made it possible to lengthen the workday, before the light bulb slowly started to take its place at the end of the nineteenth century. Oil, grease, fats, and solvents derived from petroleum greatly helped to multiply the number of steam engines and increase their power. But at the dawn of the twentieth century, it was as a source of mechanical power, in other words as fuel, that oil began to release its full energy. Oil is more efficient than coal, just as coal is more efficient than wood.* It is a naturally denser, more malleable energy source: more powerful. For half a century, "king coal" remained the principal fuel of technical humanity, an enduring servant of "heavy" industry and the biggest contributor to the fields of metallurgy, electrical production, heating, and the merchant marine. However, even before 1914, oil became the number one source of energy, not in quantity but par excellence: It supplied motor vehicles, trains that went the furthest, the best warships, and the first planes. Coal continued to maintain a role at the base of industrial society, but it was oil alone that roused its expanding spiral. Men who enjoyed the power of black gold discovered a superhuman power. The capitalist organizations that wrested this power also controlled the destiny of human technology.

Gasoline Ignites the Automobile Era

Having demonstrated its superiority on heavy, steam-operated cars, the gasoline automobile was, in its early form, rivaled by electric traction. After

* Burning wood provides approximately 14 megajoules per kilogram (MJ/kg), compared to 27 MJ/kg for coke from coal, and 36 MJ/kg for regular fuel oil.

the success of electric trams in many cities in Europe and the United States in the 1890s, the first electric taxis appeared in New York and Philadelphia in 1897, and electric cabs appeared in Paris in 1899. On May 1, 1899, in the Paris suburbs, the first motor vehicle to cross the 100-kilometer mark was an electric car shaped like a torpedo: the Jamais Contente (or "Never Satisfied," in English), made in the workshops of the Jenatzy family, Belgian engineers who were also pioneers in the manufacture of rubber tires. But, again, the qualities and quantities of oil outperformed other technical options. The word *pétrolette* appeared in France in 1895, originally referring to a small automobile as well as a small motorcycle. The internal combustion car was much more autonomous, more robust, simpler, lighter, and much cheaper to produce in large numbers than the electric automobile. It remained to be decided, though, whether petroleum gasoline or fuels of agricultural origin would prevail. In 1903, a race called Fuel Circuit was organized in France. The winning car, a De Dietrich, ran on vegetable alcohol, but it was disqualified to the benefit of a gasoline-powered car, at the request of a jury member who was also secretary of the "Association for Petroleum," on the pretext that the De Dietrich's fuel was not homogeneous.[1] The agricultural fuels sector was in any case unable to compete with that of petroleum, either by price or by production volume.

In France, the United States, Great Britain, Germany, Belgium, and Italy, there were dozens of workshops making automobiles. France alone had several dozen manufacturers. In 1903, Renault, Peugeot, and Panhard briefly made France the world's leading carmaker.[2] But it was of course in the United States, the country of abundant oil, where the production of gasoline-powered vehicles grew fastest. In 1901, a thirty-eight-year-old engineer named Henry Ford, a former employee of Thomas Edison, helped create what eventually became the mythical Cadillac brand. The following year, the first assembly line to build mass-produced cars was set up in Michigan by Ransom Eli Olds, the founder of Oldsmobile. The Ford Motor Company was created in June 1903. By mid-December, the 12-horsepower gasoline engine installed by the Wright brothers on their airplane, the Flyer, was used to inaugurate the aviation era. And in that same year, in the rural state of Iowa, Charles Hart and Charles Parr made one of the first motorized agricultural vehicles to be baptized a "tractor."[3]

US production of gasoline-powered cars accelerated much faster than European production in the second half of the decade. General Motors was founded on September 16, 1908, in Flint, Michigan. The birth of the world's largest automaker was greatly facilitated by a $6 million loan from John D. Rockefeller, the latter having had the fruitful idea of being reimbursed in shares.[4] Two weeks later, Henry Ford presented his Model T in Detroit. This robust car was to become the first automobile manufactured by the millions. The immediate success of the Model T led Henry Ford to perfect his assembly lines in 1913, drawing on the work of American engineer Frederick Winslow Taylor, inventor of the "scientific organization of labor" (Taylorism). The work was standardized, every task timed, the free initiative of the workers proscribed. More than two hundred thousand Model T cars were sold as early as 1913. Total vehicle sales were close to half a million in the United States that year. The European automobile industry was now far behind that of the United States, with forty-five thousand vehicles sold in France, thirty-four thousand in Great Britain, and twenty thousand in Germany (electric cars were a small percentage of sales).[5] The automobile culture flourished most in the country richest in black gold and leading in world oil production.

Ford factories were soon able to churn out a car every fifteen minutes. The Model T was designed for the middle class. Thanks to its success and in order to combat turnover (few workers at the time were faced with monotonous standardization of their tasks), Ford doubled the pay of its skilled workers in January 1914, raising the pay to $5 an hour and a percentage of the profits. This decision spun the wheels of redistribution of wealth toward the emerging American middle class and consumer society.

The advent of the age of the gasoline engine was the vector of this wealth. It was greatly facilitated by a patent filed in1913 by Dr. William M. Burton of Standard Oil of Indiana. The process of refining petroleum by the "thermal cracking" of the longest hydrocarbon molecules allowed the useful portion of each barrel to expand by nearly half, making use of the crude petroleum that refiners had to dispose of before the arrival of the automobile. In 1914, in the United States, gasoline sales almost equaled those of kerosene, which was surpassed by heating oil. From then on, oil was mainly used—or, rather, burned—as fuel.[6] The thermal cracking process dispelled the specter of a gasoline shortage in the face of an explosive growth in the number of cars. It brought very comfortable profits to Standard Oil of Indiana, which remained the patent's exclusive owner until its validity was contested in 1921. Thermal cracking had already been patented as early as 1891, in Russia, by the brilliant engineer who designed the first Branobel pipelines, Vladimir Shukhov.[7]

Despite its rapid development, the automobile remained, before the war, an irresistible gadget of high society. The first trucks and tractors provided surprising utility, but they were still rare. Even the United States had only seventeen motor vehicles per one thousand inhabitants in 1914.[8] French journalist Octave Uzanne wrote in 1911: "Just as it is absolutely necessary for Madame to have her living room up to date, a man could not pass on the latest vehicle without feeling inferior to his neighbor."[9] From the outset, the automobile fascinated people and became an unstoppable sign of social distinction, an irresistible attraction. Thanks to it, humankind never again depended on the health of his animals or the whims of the wind. The slow loops between humans and nature, to drive the deployment of technical progress, were suddenly shortened: There was nothing left to do but put the pedal to the metal. Human desire seemed to have become the only true necessity, the measure of all things.

L'Auto, the ancestor of the daily French sports magazine *L'Equipe*, sold three hundred thousand copies a day in 1913, three times more than the number of gasoline-powered cars that circulated in France at the time. There were many fatal accidents: over four thousand in the United States alone in 1914.[10] In France, the writer Léon Bloy denounced a "kind of homicidal and demonic delirium," declaring, "This morning the coachman of our carriage showed me one of these machines that recently killed an old woman and seems ready to do it again. No punishment. The crusher gave a little money and that was that."[11] In Mexico, the engraver José Guadalupe Posada depicted bourgeois drivers running over "peóns" in the vengeful legend "La Autocratia Automovilista" ("The Autocratic Motorist").[12]

Energy Slaves

A 50-liter tank of gasoline contains the equivalent physical strength that one thousand men in top physical shape can produce during a very long day of labor. The energy released during the combustion of a single liter of gasoline in an engine is equal to the muscular effort needed to hoist a car twenty times to the top of the Eiffel Tower using a manually powered heavy-gauge pulley. The energy content of 100 grams of oil supplies about 1 kilowatt-hour, the mechanical energy equivalent of gravity releasing 10 tons of water from a height of 40 meters. By themselves, fossil energies in general, and oil more than any other, provide the industrial era with legions of "energy slaves": The combustion-release force is equal to those of many men.[13]

Before the industrial era, more than three-quarters of humanity lived, in one way or another, in a state of physical bondage to work.[14] The notion of *humanism*, falsely thought to date back to the Renaissance, is a nineteenth century invention, and more specifically, from the end of the nineteenth century, insists French philosopher Michel Foucault, who emphasized that the word *humanism* itself is absent from *Littré*, the great French language dictionary of the second half of the nineteenth century.[15] Until then, slavery, serfdom, sharecropping, and other forms of subjugation, more or less complete, were the banal realities found almost everywhere. Their omnipresence can be interpreted equally as the result of an ethical blind spot and as a consequence of a practical, physical constraint. This constraint arose from the limitation of technical possibilities to access renewable energy flowing from the sun (which includes the water cycle, wind, photosynthesis, and animal force).

Access to titanic solar-energy stocks of fossil fuels, coal, and hydrocarbons upset the equation. In the United States, the industrial and abolitionist North's victory over the agrarian and pro-slavery South was probably not made possible by morality and politics alone. There may have been another key influence, albeit an indirect one. "The arrival of the steam engine was probably a necessary condition for the abolition of slavery," says French historian Jean-François Mouhot.[16] From the United States to Russia (where serfdom was abolished in 1861), did the arrival of powerful machines diminish the unspeakable necessity of slavery? Did it present the opportunity for us to open our eyes to an age-old scandal? The evolution of large functional systems cleared pathways and opened or closed doors for the evolution of society, even where onlookers were content to believe that the human spirit was the only gateway to change.

With an increase in the number of oil-fueled internal combustion engines, energy slavery changed its potency level. Black gold's liquidity, energetic density, and maneuverability made it possible to disperse the work of innumerable new energy slaves everywhere: to more quickly move heavy loads on land, sea, and air; to introduce electricity to the steps of technology's empire thanks to generators; to pump more water from deeper underground; to irrigate fields more extensively; to sow and harvest more and longer; to heat greenhouses and trigger and maintain all kinds of large-scale chemical reactions at high pressures and temperature levels. Petroleum fuels and lubricants facilitated the routing, laying, and welding of the plumbing for the first networks of central heating, piped water, and sewerage at a time when taking a weekly hot bath remained a luxury, even for the bourgeoisie.

If coal and the advent of the steam engine favored the abolition of slavery and the progressive prohibition of child labor in workshops and mines, it may well be that the radical increase in energy slaves offered by oil was subsequently a sine qua non condition for the slow emancipation of various groups of proletarians, who hitherto had hardly been able to conceive of escaping from their alienating and more or less constrained work: the children of the countryside, the colonial semi-slaves, the wives, and the domestics (the latter disappeared at the beginning of the twentieth century, as electricity arrived in homes via coal and eventually oil). Left in place were the workers who served and controlled the machines. Except that thanks to these machines, the workers were "mechanically" replaced, particularly in coal mines and oil wells.

Oil greatly propelled the development of industrial machinery. Through it, sooner or later, all people found themselves committed to technology, in one way or another. In 1905, in the very last pages of *Protestant Ethics and the Spirit of Capitalism*, German sociologist Max Weber advanced this uncertain prophecy: "The Puritan wanted to work in a calling; we are forced to do so. For when asceticism was carried out of monastic cells into everyday life, and began to dominate worldly morality, it did its part in building the tremendous cosmos of the modern economic order. This order is now bound to the technical and economic conditions of machine production which today determine the lives of all the individuals who are born into this mechanism, not only those directly concerned with economic acquisition, with irresistible force. Perhaps it will so determine them until the last ton of fossilized coal is burned."[17]

The emancipation offered by the armies of energy slaves released by petroleum was not without a counter effect: the submission of humankind to the effects of a wide range of obscure necessities imposed by the machine's expanding domain.

Standard Oil Fortifies Its Stronghold, but Faces the First Antitrust Fever

The masters of energy flow were the ultimate masters of the new industrial world. The May 1911 US Supreme Court verdict requiring the dismantling of Standard Oil became a symbol of the inviolability of American public institutions confronted with the powers of money. Except that this verdict, the culmination of years of bitter judicial struggle, was little more than an illusion. In reality, American democracy was unable to impose its law on Standard Oil.

John D. Rockefeller had begun his fugitive career thirty-two years earlier. On April 29, 1879, a grand jury in Clarion County, Pennsylvania, charged Rockefeller, Archbold, O'Day, and six other Standard Oil managers with monopolizing the petroleum industry.[18] Rockefeller obtained a guarantee from the governor of New York, his home state, that he would never be extradited to Pennsylvania. But in July of the same year, the New York State Assembly created a committee to investigate clandestine agreements with the railways. Rockefeller spent the summer outside this committee's jurisdiction.[19] As investigations began to multiply in various US states over the next two decades,

Standard Oil's boss did everything in his power, within the limits of the law, to evade justice. Little by little the pitiless tactics Rockefeller employed to overcome his competitors were revealed. Tactics that were odious, admirable in their efficiency, that became a strong (albeit shameful) model for the entrepreneurial culture, and that still surface today in many sectors—in large-scale retail, for example. From the Great Plains to New England, the American territory under the Standard Oil empire was divided into fiefdoms. The monopoly of Standard Oil sales was allocated to wholesalers of various products whose membership in the parent company was often kept secret. It was with great pleasure that grocers accepted a 5 percent discount on the purchase of their stock in exchange for a commitment to sell only oil stamped by Standard Oil. When a Mississippi grocer persisted in his refusal to abide by this arrangement, he received the following warning from one of Rockefeller's main wholesalers: "We will start a grocery store and sell goods at cost and put you all out of business."[20] Rockefeller denied having knowledge of these practices, although his records show that, on the contrary, he encouraged them.

Standard Oil maintained a close network of informants. As soon as any rival petroleum railcar entered a territory controlled by the company, railway employees, paid for this purpose, revealed the intrepid competitor's identity. The interloper could be countered easily: Wherever a breach was attempted, oil was sold at cost for as long as it took to crush the rival. John Archbold pressed the company's executives to control about 85 percent of the market in the territories for which they were responsible.[21] But certainly not 100 percent: "We realized that public opinion would be against us if we actually refined all the oil," Rockefeller confided.[22]

During the 1880s, natural gas lighting was expanding rapidly in US cities, but Thomas Edison's electric lighting, requiring more infrastructure, was slow to spread. Seeking to sell more natural gas, a by-product of many oil fields, Rockefeller personally supervised the distribution of the bribes necessary to storm the public lighting markets, which were controlled by deeply corrupt officials. In Philadelphia, Daniel O'Day, crackerjack Standard Oil dealer, noted his astonishment that "all local Republican politicians" were shareholders of the local rival company. "They . . . feel now that unless they make some alliance with us that they will in all probability lose all their money," he reported with satisfaction.[23] In Toledo, Ohio, residents finally discovered that the two seemingly competing gas companies in the city were in fact each subsidiaries of Standard Oil.[24]

The first sign of citizen outrage over the many cartels that plagued entire sections of the US economy became apparent during the general elections of

1888. In February, John D. Rockefeller, then forty-eight, presented himself in top coat and top form before a New York State Senate commission of inquiry, "kissing the Bible with vehemence."[25] He artfully dodged all the traps set by the commissioners and, thanks to the hype of Joseph Pulitzer's papers, reached celebrity status, embodying a disturbing national archetype, the cold monster of big capital. The commission's report presented the Rockefeller empire as a "system that has spread like a disease through the commercial organization of this country" and that operates "in absolute darkness."

The first laws that targeted the trusts (the "cartels") were adopted in a number of American states after the 1888 elections. On July 2, 1890, President Benjamin Harrison signed the antitrust law filed by Ohio Senator John Sherman. The Sherman Act became federal power's weapon to fight cartels and monopolies. It was specifically targeted at Standard Oil. John Rockefeller publicly criticized the text, a very unusual move on his part. A Republican Party member, John Sherman was the younger brother of General William Sherman, the ruthless strategist of the Union army during the US Civil War. In his previous election to the Senate, John Sherman had received a $600 check from the Standard Oil boss in person. Unembittered, Rockefeller continued to contribute to Sherman's reelection in 1891. For many years, the Sherman law was hardly applied. It was nicknamed the "Swiss Cheese Act."[26]

Rockefeller Saves Wall Street for the First Time

Standard Oil's barrels were inexpensive for their tens of millions of customers, considering the immeasurable value of the varied services they rendered. But the highest favors were reserved for the industrialists, for whom crude oil cost much less. Based on the profits it generated, the oil industry was unequalled. The expression *black gold* was quite accurate. In and of itself, oil offered the capitalist unrivaled profits. To begin with, if opening an oil field required massive initial investments, these were not disproportionate to those required to open a mine or construct a blast furnace. Once the wells were drilled and the pipeline laid, fortune itself was almost guaranteed. After that, the year-to-year costs of extracting oil (what economists call "marginal" costs) had the delicious and unique property of being next to nothing. They were mainly limited to the simple maintenance of equipment. For a long time in the history of the petroleum industry, crude oil basically flowed on its own, whereas a large coal mine demanded thousands of workers. Often, without even resorting to pumps, black gold flowed out of a well as surely as golden water flew down the bed of the Pactole River. Moreover, the small amount of capital an industrialist had to extract from his own pocket in order to maintain the flow of crude oil was only equaled by the modesty of the labor required: an inestimable advantage, as oil does not require the mobilization of a large workforce. No other industry could spontaneously supply so many assets to those who exploited it. Rockefeller, convinced that oil was a gift from God, believed the benefits to Standard Oil would recur like the cows of Narayana.

1894 saw the first evidence that the American state was among the most voracious recipients of the power provided by the oil empire. Rockefeller's accumulated wealth was by then diversified across all the foundations of the American economy. He had significant holdings, notably in sixteen railway companies, nine real estate companies, six steel companies, six shipping companies, and nine financial institutions. "The Standard Oil Trust was really a bank of the most gigantic character," a journalist observed a few years later, "a bank within an industry, financing this industry against all competition and continually lending vast sums of money."[27]

At the center of this self-nourishing fungus, one Wall Street bank flourished particularly well. The National City Bank earned the nickname "oil bank" shortly after recruiting William Rockefeller, the younger brother who John D. had set up in New York, to be one of its directors. In 1893, this bank was unscathed by one of the worst financial panics in capitalism's history, which unleashed an endless wave of misery and unemployment, the brutality of which was surpassed only by the crisis of 1929. Depression began when, like Panurge's sheep, clients big and small rushed to sell their shares and get paid in gold, emptying bank coffers as word spread. In 1894, as bankruptcies multiplied and the United States approached bankruptcy, the US treasury secretary implored John Pierpont Morgan to attempt a rescue operation. The legendary banker replied that he no longer possessed the means to do that. J. P. Morgan then turned in haste to the National City Bank president, James Stillman. Stillman recounted that Morgan "was greatly upset and overcharged," that he "nearly wept, put his head in his hands and cried: 'They expect the impossible!'" With the haughtiness and irony of one who had triumphed, Stillman concluded his narrative: "So I calmed him down and told him to give me an hour, and by that time I cabled for ten million from Europe from Standard Oil and ten more from other resources."[28] Once invested and loaned, that $20 million (a colossal sum for that time) steered the situation into calm: the drain on the bank's gold ceased; black gold had stopped the hemorrhaging of the yellow metal.

J. P. Morgan, master of the world's most famous bank, assumed, *a posteriori*, the pose of savior for his country. Rockefeller, on the other hand, preferred to remain in the shadows, as usual. The most profitable of industries was already well on the way to supplanting the biggest traditional banks with its own leadership. Wells of black gold exuded the most fabulous source of opulence that man had ever brought to light.

Struggling with Muckrakers and American Justice, Rockefeller Saves Wall Street a Second Time

In 1900, Ida Minerva Tarbell, the great pioneer of investigative journalism, embarked on the Standard Oil trail. The robust, serious, and determined forty-three year old had spent her childhood in the oil regions. Her father was one of those independent oilmen who struggled with the rising Rockefeller empire. In the first installment of her long account, published in *McClure's* magazine in November 1902, Ida Tarbell recounted with emotion that "life ran swift and ruddy and joyous in these men" of her Pennsylvania childhood.[29] "There was nothing too good for them, nothing they did not hope and dare. But suddenly, at the very heyday of this confidence, a big hand reached out from nobody knew where, to steal their conquest and throttle their future."[30]

Every month for two years, Ida Tarbell's investigative pieces detailed much of the vileness that accompanied Standard Oil's unrivaled extension of power. In November 1904, the book that compiled the entire story was more than a media sensation: It was a major political event in US history. Until Watergate, in 1974, no other journalistic investigation made such an impact. For countless Americans in love with entrepreneurial freedom, the *History of Standard Oil* awakened the resentment felt for decades against the "robber barons" of industry—Andrew Carnegie, Cornelius Vanderbilt, Jay Gould, and Rockefeller, who was now the biggest tycoon. Their nickname referenced the Raubritters, those German nobles of the Middle Ages who levied undue taxes on those who wished to sail on the Rhine.

On November 8, 1904, the same month Tarbell's book was released, Theodore Roosevelt was reelected to the White House. The charismatic and thundering "progressive" leader of the Republican Party had made the fight against cartels his main battleground. There was more than a bit of bad faith in his campaign, with the Republican Party accepting many donations from industrialists to finance it, including more than $100,000 from Standard Oil. But Roosevelt ordered that the $100,000 be repaid and promised the American people a "Square Deal," which became the slogan of his mandate.

Roosevelt encouraged the journalists and writers who followed Ida Tarbell's footsteps to expose the fortresses of capitalism and politics. The US president was one of the first to refer to journalists' usually pejorative nickname, "muckrackers," in a respectful way. Standard Oil, long in the habit of presenting itself in the press with advertisements disguised as genuine news articles, now established what is often considered the first example of a modern, full-fledged communications strategy—a strategy gradually built by pioneering publicists like Joseph I. C. Clarke and Ivy Lee, skilled in the art of "press relations."

But the efforts of these early spin doctors proved to be in vain. In the aftermath of the 1904 elections, President Roosevelt's administration launched a far-reaching investigation of Standard Oil's practices. Legal proceedings for conspiring to restrain trade began two years later, based on John Sherman's antitrust law, adopted sixteen years earlier and now used for the first time. John D. Rockefeller, sixty-seven years old, saw his company targeted by seven federal and six local lawsuits.

The first battle was waged in the summer of 1907. On July 6, hundreds of curious people crowded into the stifling heat to wait for the wealthiest man of all time to appear in front of the court house in Chicago, where he had arrived the day before on his private train. People jostled to get a look him, pulling buttons off his jacket. Once in the courtroom, facing Judge Kenesaw Landis, Rockefeller kept his poker face. He even took the liberty of mocking several questions while remaining impassive, concealed behind an expressionless facade. But a month later, on August 2, Judge Landis took his revenge.

He created a sensation by condemning Standard Oil to pay an unprecedented fine: more than $29 million, almost one-third of the company's capital. Reporters noted that it was equivalent to half of the currency put into circulation each year by the US government, enough money for the US Navy to build five new battleships.[31] Rockefeller did not interrupt his golf game when he learned the amount of the fine, and stated that "Judge Landis will be dead a long time before this fine is paid." He was right.

In the days following the announcement of the judgment, Standard Oil stocks fell heavily. The company's weight on Wall Street was so great that it dragged a good part of the stock exchange down with it. The unexpected downturn in Wall Street's central pillar was the trigger for the new financial crisis that occurred in the autumn of 1907. The recent financial euphoria had created a bubble of loans secured by speculative financial bonds. The purge was violent. While banks were running out of money, Rockefeller deposited funds in several New York establishments that could be pledged as collateral for government bonds (for this new Wall Street safekeeping operation, he enjoyed a comfortable 2 percent margin).[32] On October 21, the crisis turned into panic when one of Wall Street's biggest banks, the Knickerbocker Trust, ruined itself by vainly attempting to corner the copper market. On the night of October 22, the US Treasury again turned, as in 1894, to J. P. Morgan. The latter was responsible for saving the situation and to this end was entrusted with $25 million in government funds. Once again Morgan assumed the posture of the providential man, but once again it was Rockefeller who advanced more funds than any other private source.[33] The next morning, the Standard Oil boss called the director of the Associated Press to let him know that the US economy was strong and that, if necessary, he would give half of his fortune to replenish it. The worst of the storm had already passed when, on October 24, a horde of journalists invaded the golf course on Rockefeller's property in Pocantico, on the banks of the Hudson River, asking him if he would really be willing to give up half of his fortune. Never before had a man promised to save an economy such as that of the United States single handedly, nor been taken seriously when making such a promise. "They always come to Uncle John when there is trouble," said Rockefeller.[34]

The Arrogance of the Rockefeller Empire Precipitates Its Condemnation

Following this new divine intervention by the emperor of oil, Theodore Roosevelt observed a brief truce in his campaign against Standard Oil. The American president declared that the disputes between "Uncle John" and "Uncle Sam" were, in fact, settled out of court. But at the end of October, Rockefeller's right-hand man, John Archbold, had the awkward task of letting Roosevelt know that Standard Oil's financial support was assured as soon as a compromise was found.[35] This misstep ended the truce. In January 1908, Roosevelt denounced Standard Oil's merciless and "unscrupulous craft": the US federal justice system was in pursuit again. It accumulated a file worthy of the gigantism of Standard Oil: 1,374 exhibits, 444 testimonies. Meanwhile, Rockefeller's prediction about Chicago's judge was confirmed: In July, the record $29 million fine from eleven months earlier was annulled on appeal. Disappointed, President Roosevelt quipped with his customary refrain that the court of appeal's judgment "hurt the cause of civilization."[36]

The denouement was near when press tycoon William Randolph Hearst, Joseph Pulitzer's great rival, succeeded in stirring up mass outrage, thanks to his consummate penchant for scandal. In Ohio, at the political meeting of one of his protégés (an axle-grease manufacturer who one day refused to be bought by Standard Oil), Hearst began to read signed letters from John Archbold, addressed to two members of the Congress. The letters show that these elected officials had been bribed by Standard Oil for years. In front of the crowd, Hearst claimed that a stranger had handed him the letters a few hours earlier in his hotel room. In fact, it had been four years since an office boy named Willie Wilkins, employed at Standard Oil's New York headquarters at 26 Broadway, had looted Archbold's correspondence and sold it to Hearst. With the $20,500 they received, Willie Wilkins and his accomplice opened a saloon in Harlem.[37]

Mistrust of the Rockefeller empire reached its height in 1909. The US Navy suspended their oil purchases from Standard Oil. Newly elected William Taft, formerly Roosevelt's war secretary, decided to ban the exploitation of several newly discovered oil fields. In order to preserve these reserves of crude oil for the future, the corpulent President Taft placed these fields under the responsibility of the US Navy. The preservation of crude oil reserves became, for the first time, a national security issue.

Finally, on November 20, 1909, the verdict of the St. Louis federal court concluded that Standard Oil had violated antitrust laws. The Rockefeller holding company, based in New Jersey, was given thirty days to divest its thirty-seven subsidiaries. Teddy Roosevelt, then on safari in Africa, on the White Nile, said it was "one of the most decisive triumphs of decency ever achieved in our country."[38] The lawyers of Standard Oil immediately appealed.

From "Duck Hunting" on Jekyll Island to Creating the Central Bank of the United States

Sixteen months elapsed before the US Supreme Court delivered its judgment. During this long interval, one evening in November 1910, several of Wall Street's most prominent bankers discreetly boarded the private railcar of Nelson W. Aldrich, an influential Republican senator from Rhode Island. Throughout the night, the train carried them far south of New York, to a private island in Georgia, for a so-called duck hunt. The Jekyll Island hunting club, which was restricted to one hundred of the richest figures in the United States, included William Rockefeller as a founding member. Except for the years when yellow fever raged in the area, members of the Rockefeller, Morgan, and Vanderbilt families took refuge in their sumptuous cottages during the winter months. Senator Aldrich was at the heart of the Wall Street establishment. Elected to the smallest American state, he was regarded by his opponents as the mouthpiece of the "money trust" in Washington. In October 1901, his daughter Abby married John D. Rockefeller Jr., the Standard Oil founder's only son and his principal heir.

The financiers who had boarded Senator Aldrich's private car secluded themselves for ten days in utmost secrecy on the island. Together, they outlined the Aldrich Plan, the original scheme of the Federal Reserve system that still governs the operation of the central bank of the United States.[39] Of all the financial institutions that these men represented, only J. P. Morgan's were not closely linked to the Rockefeller financial empire. Morgan controlled tens of thousands of kilometers of railway in the United States, though, so he benefited greatly from the expansion of the oil industry. The Aldrich Plan's main designer, Paul Warburg, was one of the partners of the powerful investment bank Kuhn, Loeb & Company (and the son-in-law of its founder, Solomon Loeb). In 1911, Kuhn, Loeb & Co. took control, in association with John D. Rockefeller, of the Equitable Trust Company.[40] With John D. Rockefeller Jr. as its first shareholder, the Equitable Trust merged with the Chase National Bank in the aftermath of the 1929 crash to create the first bank on Wall Street and, hence, the first worldwide bank.[41] Benjamin Strong, another Jekyll Island "duck hunt" guest, was vice president of the Banker's Trust Company, of which John D. Rockefeller had been one of the principal shareholders since its inception in 1903.[42] Also present was Frank Vanderlip, who since 1909 had chaired the National City Bank, the "oil bank" of which William Rockefeller was the key shareholder. William's sons married the daughters of James Stillman, Vanderlip's predecessor at the head of the National City Bank, who in 1894 had come to Wall Street's rescue with Standard Oil money. Before joining the "oil bank," Frank Vanderlip occupied the number two position at the US Treasury from 1897 to 1901. In this capacity he negotiated the $200 million fundraising campaign to finance the Spanish-American War in 1898, which resulted in the United States taking possession of Cuba, Puerto Rico, Guam, and the Philippine archipelago: the outposts of the American Empire off the American continent. Vanderlip then appointed the National City Bank to take over paying the $20 million granted to the Spanish crown in compensation, two years before being recruited by James Stillman as vice president of that same bank.[43] In the United States, politics and war are often the continuation of business by other means. Like Chase, National City Bank (now Citigroup) remains one of the world's leading banks to this day.

A single Taft administration representative was present on Jekyll Island: A. Piatt Andrew, number two to the secretary of the treasury, one of Vanderlip's successors.[44] In his memoirs, Vanderlip noted that the "essential points" of the Aldrich Plan "were all contained in the plan that was finally adopted" by the American Congress.[45] The law passed on December 22 and 23, 1913, creating the central bank (Federal Reserve system), the most powerful nongovernmental face of political power overseeing American financial institutions, designed by the leaders of banks. Banks whose power and capital germinated from the fabulous profits of Standard Oil's black gold.

The Dissolution of Standard Oil: A Smoke Screen

Finally, on May 15, 1911, at four o'clock in the afternoon, Judge Edward White rendered a verdict of the Supreme Court of the United States. Many congressmen crowded in to hear the judgment, which Justice White read for three-quarters of an hour in such a weak voice that, several times, it was necessary to ask him to speak louder. The dissolution of the Standard Oil Company was confirmed. The company that had marketed more than 85 percent of the kerosene produced in the world had six months to separate from its subsidiaries. After hearing the Supreme Court ruling, the leader of the populist Democratic Party, William Jennings Bryan, stated that Justice White "waited fifteen years to throw his protecting arms around the trusts and tell them how to escape."[46]

John D. Rockefeller once again learned of the decision while playing golf on his private Pocantico course. His first reaction was to advise his partner that day (a Catholic priest) to buy stock in Standard Oil. Once again, it was an insightful recommendation. Was it because of the legal proceedings that the capitalization of Standard Oil proved to have been hitherto extremely conservative? Subsidiaries that were now independent due to the judicial decision proved to be bursting with hidden capital. By January 1912, the stock values of the Rockefeller empire's offspring skyrocketed. The former parent company, Standard Oil of New Jersey, saw its stock increase by 65 percent between January and October. That of Standard Oil of New York, the ex-financial arm of the dismantled giant, climbed 123 percent. The value of Standard Oil of Indiana, number one producer of gasoline and holder of the very lucrative patent on the thermal cracking of crude oil, took flight, rising more than 171 percent. During the next ten years, the total value of companies bought by the parent company quintupled.[47] A former J. P. Morgan associate, George Perkins, told a friend that all of Wall Street was "laughing in its sleeve at what has been going on."[48]

The dismantling of Standard Oil appears to have been only a formality. The subsidiaries sold their products under the same brand and divided the sales territories; during the next two or three decades, there was virtually no perceptible competition between them. The presidencies of the two main offshoots, Standard Oil of New Jersey and Standard Oil of New York, were entrusted to John Archbold and Henry Folger, two historic pillars of the former company.

Above all, the main shareholders remained the same as before, beginning with John D. Rockefeller, who retained about one-quarter of the shares of each of the thirty-three companies created after the Supreme Court ruling. His fortune, estimated at $300 million on the eve of dismantling, was radically increased: In 1916 Rockefeller became the first person in history to become a billionaire in US dollars. The value of the capital he controlled continued to swell so much that he remained the richest man of all time.

British philosopher Bertrand Russell wrote in 1934: "Two men have been supreme in creating the modern world: Rockefeller and Bismark. One in economics, the other in politics, refuted the liberal dream of universal happiness through individual competition, substituting monopoly and the corporate state, or at least movements toward them. Rockefeller is important, not through his ideas, which were those of his contemporaries, but through his purely practical grasp of the type of organization that would enable him to grow rich. Technique, working through him, produced a social revolution; but it cannot be said that he intended the social consequences of his actions."[49]

John D. Rockefeller, at seventy-two years of age, wearing a wig and with cheeks as hollow as ever, left the presidency of the company to which he had devoted his life—the company dissolved, in appearance, by the highest judicial body in the United States—in December 1911. He nevertheless continued to invite the leaders of the new companies spawned from Standard Oil to meet at 10:30 every morning at 26 Broadway, in order to maintain friendly relations and exchange information. Written exchanges were not encouraged.[50] His dragon had not been slayed, he just changed skins; he was bigger than ever.

FIVE

The Tank: American Oil Feeds the Victorious Fighting Machines of the Great War

Just when black gold began to transform the United States into a budding imperialist economic power, the dawn of the Oil Age revealed an improbable weakness in the British Empire: The vast domain that the Royal Navy (thanks to British coal) had imposed its rule upon—"the empire on which the sun never sets"—appeared to be almost devoid of hydrocarbons.

The Royal Navy Does Not Want to Be at the Mercy of Royal Dutch Shell

The island of Great Britain, so generously endowed with coal mines, had only a tiny bituminous shale extraction industry in Scotland. It was also enterprising Scots who drilled crude wells in the jungle of Burma, almost the only oil wells in the empire at the turn of the twentieth century. But the relatively small reserves of Burmah Oil Company (founded in Glasgow in 1886) supplied only half as much crude oil as those under Dutch control in Sumatra and Borneo.

It quickly became obvious that if the British Empire wanted to retain its supremacy—particularly that of its fleet—finding its own oil was imperative. In London, at the turn of the century, many Crown strategists were keenly aware of this vulnerability. None of them wanted America's domination of crude oil to continue. Nor did they want to be surpassed by their French rivals.

The American navy was, since the Civil War, the first to use oil as fuel.[1] In 1887, the Russian admiralty launched the construction of a battleship, the boilers of which burned liquid fuel, in order to end their dependence on English coal in case of conflict.[2] But it was during the naval-buildup race that preceded the First World War that the change took place (and this was one of the factors that led to that war). Oil had many decisive advantages for war fleets. First, it produced less smoke, making ships more difficult to spot. Second, its fluidity made it possible to fill tanks and feed boilers and motor cylinders by gravity flow, compared to the numerous sailors—called "stokers"—who were occupied with transporting and shoveling coal day and night. Above all, petroleum's higher calorific value made it possible to design more compact engines and consequently to manufacture lighter, more powerful ships, capable of a superior operating range, an increased acceleration rate, and faster cruising speeds for tactical advantages.

As early as 1899, Sir Marcus Samuel, Shell's founder, set about convincing the Royal Navy to replace coal with oil in the boilers of Her Majesty's warships. He quickly became a strong ally to the admiralty. Admiral John Arbuthnot Fisher, veteran of the Crimean War, the Second Opium War, and the Anglo-Egyptian War of 1882, won the nickname "oil maniac" from his detractors.[3] With a smooth face and a round, sagging mouth like a grouper fish, John Fisher was fully convinced of the need to modernize the fleet as quickly as possible. He encouraged Marcus Samuel to diversify his supply sources, and Samuel offered, as a pledge of confidence, to make the admiral a trustee of Shell. But the other members of the admiralty were suspicious: Even if ennobled by Queen Victoria, Sir Samuel was still a Jewish merchant associated with Parisian bankers and running a company that bought Russian and Dutch oil and counted the Empire of Japan among its biggest customers.

In both France and Italy, the construction of small torpedo boats, fueled with a combination of coal and oil, began. But fuel oil was expensive, and in Paris one was especially afraid of being left at the mercy of the "trusts" in case of conflict.[4] That fear that proved to be well founded, justifying their perilous wait-and-see stance.

In London, naval buildup accelerated with the 1906 launch of a formidable class of British battleships, the dreadnoughts, highlighting more than ever the major strategic challenge posed by the switch to oil. Fuel was tested in the burners of an earlier battleship, the HMS *Hannibal*. During tests carried out in Portsmouth, England, Marcus Samuel took the opportunity to demonstrate to the admiralty the superiority of his fuel over coal. But the *Hannibal*'s boilers were obsolete: A thick black smoke appeared; the bridges were covered with soot. This first experience was a humiliating failure.[5] Although the *Hannibal* actually converted to oil in 1908, the Royal Navy remained wary of Marcus Samuel.[6] The Indian office expressly kept Shell away from Burmese concessions, fearing that Burmese wells would end up in foreign hands. Once it became Royal Dutch Shell, the company Marcus Samuel had founded looked more suspicious than ever. When, in 1910, the Royal Navy decided to accelerate its conversion to oil, it entrusted its main contracts to the Scots of Burmah Oil Company.[7] However, many of the oldest Burmese wells, some exploited for more than twenty years, were beginning to dry up. Although Burma was constantly drilling new wells, the Burmah Oil Company proved incapable of extracting more oil fast enough or in great enough quantities to meet the admiralty's demands. The oil fields operated in Burma were quite far from the Thames, and their production remained too small to meet the growing needs of the English navy.

British Empire Extends Its Hand to Persian Oil

London had already begun to bet on a promising new source of black gold: Persia. In May 1901, an ambitious and very wealthy English investor named William Knox D'Arcy acquired prospecting rights for an immense territory, twice the size of Texas, from the grand vizier of Tehran and the king of Persia, Mozaffar ad-Din Shah Qajar. D'Arcy had made a fortune by bringing an old gold mine in Queensland, Australia, back to life with mechanical power, making it one of the world's richest mines (like the revived mines of Rio Tinto). The naphtha outcrops of Persian arid lands, located far from any industrial center, had always been known to exist, but D'Arcy intended to inject the capital necessary for their development, as he had done in the Australian mine.

For the British Crown, the sixty-year concession D'Arcy acquired constituted a victory in the "great game" against Russia for the domination of Central Asia. London could not have immediately appreciated the full extent of this victory: D'Arcy had ravished one of the main future sources of crude on the planet right under Russia's nose, just a thousand kilometers from Baku. The Persian government initially made an offer to Royal Dutch Shell, through Calouste Gulbenkian—the brilliant thirty-two-year-old Armenian who came to London to trade in Baku oil and while there, through the Rothschilds, became an intimate friend of Henri Deterding. At the beginning of his long career, Gulbenkian, the man who was poised to become one of the most skilled and perhaps the most influential negotiators in the history of capitalism, made a rare blunder—his only blunder perhaps, and certainly his biggest. Following the advice of Deterding, three years his senior, the young Armenian merchant refused to acquire the concession that the Persian government presented to him on a silver platter, because he considered it too risky for its price: £20,000. From this experience, Gulbenkian drew the leitmotiv that never ceased to guide him: Never give up an oil concession![8]

It was true that for the prospectors D'Arcy engaged, the isolation, the complexity of the subsoil of Persia and the harshness of its climate made discovering sites that were favorable to the first boreholes a long, laborious process. London, however, was keenly interested. A detachment of cavalry guards was sent from India in 1907 to protect the prospectors. The Scotsmen of Burmah Oil Company, who were already collaborating with the Royal Navy, brought the capital D'Arcy lacked, his funds exhausted as much by the unsuccessful drilling in Persia as by the luxury that the obese millionaire liked to surround himself with in his London residence. Finally, on May 26, 1908, when the Glasgow sponsors became impatient and said they were ready to give up, George Bernard Reynolds, a self-educated geologist employed by D'Arcy, came across barren land that had been cracked by exposure to the sun, near an ancient temple of fire, Sar-Masjed, that seemed to date back to the time of the Achaemenid Persian Empire. The Masjed-i-Soleiman oil field was located in the Maidan-i-Naftan, an "oil plain" not far from the bountiful Karun River that flows into the Persian Gulf some 300 kilometers to the south. In 1909, William Knox D'Arcy and Burmah Oil Company founded the Anglo-Persian Oil Company. British oil extracted in Persia was marketed by a company baptized British Petroleum (BP) beginning in 1916. The Middle East had arrived on the world oil chessboard.

The ultimate objective toward which empires were expanding their power was changing. The aim of territorial conflicts (for humans as for other species) is ultimately to control the resources transformed by energy supplied by the sun.[9] In civilized societies before the Industrial Revolution, these resources were partly crops and timber and partly manual labor, or slaves; dominating them required conquering extended territories. Western industrial powers were able to emerge thanks to the coal available in their own subsoil, through which they gained possession of immense colonial domains that provided new arable land and new workers, more or less servile.

Not only did oil change the game, it transformed its very goals. This form of concentrated solar energy, even more powerful than coal, was all but absent from British territory, as well as from other European industrial powers. But the very concentration of the energy contained in petroleum, the small amount of manual labor it took to extract it, and the unequaled number of "energy slaves" they procured from those extractions, made it possible for those Western industrial powers to set their sights on much smaller, but strategic, territories. Moreover, the incomparable profit margins offered by black gold made ruling an empire easier: It proved superfluous to enslave whole peoples, it was now at least as efficient and much cheaper to bribe some rulers, who are often corrupt, in order to control the black gold and other precious resources of their countries. Extensive and, above all, military, the modes of imperial control were heading toward becoming intensively and essentially capitalist.

The British Navy, Root of the First National Oil Company

In July 1911, the blockade of the Moroccan port of Agadir by the German gunboat *Panther* convinced London and Paris that German emperor Wilhelm II's expansionist aims would lead, sooner or later, to war. On the banks of the Thames a young, brilliant aristocrat, elected ten years earlier to the parliament of Westminster, realized how urgent it was to speed up the conversion of His Majesty's fleet to oil. Three months after the "coup d'Agadir," Winston Churchill, at thirty-six years old, agreed to assume the responsibilities of first lord of the admiralty. The Royal Navy already had several dozen oil-powered destroyers and submarines, but its key assets, the battleships, still relied on coal. Churchill adopted the cause of Admiral Fisher, the oil maniac. The latter never ceased to point out that the American navy had ordered its first oil-powered battleships in 1910, and that the vessels being made at the German navy shipyards were also increasingly oil-powered. In 1912, the Royal Navy decided to launch the construction of super-battleships, equipped with twenty-four oil-fired steam boilers, dubbed the Queen Elizabeth class. It was a radical strategic bet, a "fateful plunge," as Churchill later called it.[10] The power of the British navy was no longer based on Britain's coal mines but on crude wells located more than 10,000 kilometers, by sea, from the shores of Albion. A coincidence unnoticed at the time: 1913 was the year the United Kingdom reached the peak of its coal mining, with a production record (287 million tons) that was never beaten. Stanley Jevons's 1865 prediction in *The Coal Question* was now realized: Lacking sufficient reserves of economically exploitable minerals, the United Kingdom became the first industrial country to experience the long, inevitable decline of its mines. But in 1913 the English were still not concerned about the question of coal. They were preoccupied with oil. The surge in the number of automobiles, particularly in the United States, and the explosion of the needs of His Majesty's fleet were raising the prices of gas, fuel oil, and lamp oil (oil lamps remained much more common than electric light). In London, taxis went on strike and demonstrators denounced a "petrol ring" through which industrialists arranged to fix prices at the highest level.

For the admiralty, it became clear that the Royal Navy must have its own supply sources. The Anglo-Persian Oil Company did not have the necessary capital to exploit Persian oil quickly enough. Built at the mouth of the Tigris, the Euphrates, and the Persian Karun River, in the long delta of Shatt-al-Arab ("Arvandrud" in Persian), a refinery on the island of Abadan opened in 1912 with serious technical problems. At Westminster, on June 17, 1914, Winston Churchill rose to parliament to defend a bill authorizing the government to acquire 51 percent of the Anglo-Persian Oil Company's shares. Noting that Standard Oil and Royal Dutch Shell had shared the world market, Churchill criticized the "long and steady squeeze by the oil trusts all over the world." He stressed that at all costs, the Persian oil fields must not be "swallowed up by the Shell or by any foreign or cosmopolitan companies," a calculated allusion to the Jewishness of Sir Marcus Samuel, founder of Shell. The bill was passed by an overwhelming majority. Eleven days later, on June 28, Archduke Franz Ferdinand of Austria was assassinated in Sarajevo. Germany declared war on Russia on August 1, then on France and Belgium, who were allied with England, on August 3. On the same day, at Luneville, in Lorraine, a German airplane accomplished the first aerial bombardment of the Great War.

Prewar Maneuvers for Black Gold

Oil played an important role in Germany's imperialist tendencies leading up to the First World War. A role that was demonstrated clearly by the Western powers' enormous appetite for the riches of the Ottoman Empire. And a role that may explain the tragic death of the creator of the most robust motorization concept the world had seen. German engineer Rudolf Diesel, who invented the engine of the same name in 1897, disappeared at sea on September 13, 1913, at age fifty-five. That evening, on his way to England on business, Diesel boarded a German steamer in Anvers, Belgium. The famous industrialist retired to his cabin at 10:00 in the evening and asked to be awakened at 6:15 the next morning. He was never seen again. Murder, suicide, or mere accident? No one will ever really know. The multifuel diesel engine, which initially operated on coal dust, could eventually be fueled either with oil alone or with vegetable oil. It was less prone to fire and was particularly needed for submarines.

In 1901, a German exploration team in Mesopotamia reported that the area north of Baghdad, in the arid steppes around the ancient city of Mosul (ancient

Nineveh), was situated on a "veritable 'lake of petroleum,' of almost inexhaustible supply."[11] Just like Persia, where William Knox D'Arcy bought the rights to prospect on immense territories that same year, Mesopotamia had virtually no modern infrastructure. In 1903, German industrialists joined an already existing project: the construction of the Baghdadbahn, the Berlin-Istanbul-Baghdad railway across the vast Ottoman Empire, then an ally of Germany. Having recently imposed themselves as protectors of the small emirate of Kuwait, the British were in position to thwart the Germans' dream to construct this line leading to the Persian Gulf, a land route to the East Indies—an alternative to the English maritime route. Once the Royal Navy blocked this route, the railway promoters suggested an outlet through the Turkish Empire. Deutsche Bank leaders realized that the line would also make it possible to transport oil from Mesopotamia to Europe without the need for the British-controlled Suez Canal.

But the project progressed slowly, encountering many difficulties, behind some of which, many believed, was the hand of the British empire. The Germans were facing off with the English oilmen and bankers, who themselves were at odds, all seeking exclusive access to Mesopotamian oil. Heavily indebted, the Ottoman Empire became a hunting ground open to the voracious appetites of the most powerful banks in Europe. In 1908, Uncle Sam invited himself to the party. A former American vice admiral, Colby Chester, used bribes to obtain from Constantinople the exclusive right to exploit its oil. The following year, the revolution of the "Young Turks" deposed Sultan Abdülhamid, and invalidated everything.[12] Only the German shareholders of Baghdadbahn still had valid mining rights: a strip of 20 kilometers on either side of the railway they were building. After years of negotiations, a compromise was finally signed on October 23, 1912. That day, the Turkish Petroleum Company (TPC) was born. A quarter of its capital was held by Deutsche Bank, a quarter by Royal Dutch Shell, and most of the remainder was owned by the National Bank of Turkey, controlled by an English financier, Sir Ernest Cassel. This agreement was the work and the first coup of the ultimate shareholder in TPC: Armenian merchant Calouste Gulbenkian, for whom the oil of Mesopotamia became the immense little deal of a lifetime.

Small, with a round face and fine mustache, sagacious, patient, and extremely suspicious, Gulbenkian learned the subtleties of Eastern bargaining during his childhood in the shadowy courtyards of the wealthiest houses in Constantinople's bazaar. He became indispensable as an intermediary of black gold dealings in the Middle East. His father and uncle were bankers to the sultans of the "Sublime Porte": Every transaction between Baghdad and Constantinople passed through them.[13] During the massacres of the Armenian population in 1896, Calouste Gulbenkian had to flee Turkey. He settled in Europe to develop his business, the forefront of which was the petroleum trade, staying with his family sometimes in the Savoy Hotel in London or the Ritz in Paris.*

Not only did Gulbenkian help Henri Deterding and Royal Dutch Shell gain control of the Mesopotamian "lake of petroleum," he also succeeded in convincing the Germans and the English to unite against Admiral Chester and the Rockefeller empire. However, the first capital sharing of the TCP did not last long: The Royal Navy accelerated its conversion to oil and, in London, Winston Churchill demanded that the British position be strengthened. Sir Ernest Cassel (whose granddaughter married Lord Mountbatten, future governor of India) complied without delay. On March 24, 1914, the Anglo-Persian Oil Company combined forces with the capital of the TPC, acquiring almost half its shares. To Deterding's great fury, Shell remained content with its quarter of the shares, the same amount held by the Deutsche Bank. As for Gulbenkian: He was the official financial adviser of the Ottoman government and so was in a good position to defend his interests. He was awarded 5 percent of the capital. The skilled Armenian, who throughout his life sought to escape publicity, was soon known as Mr. Five Percent. He always reckoned that it was better to have a small slice of a big cake than a big slice of a small cake; he continued to cling to his 5 percent but hardly ever went near an oil well.[14]

Finally, on June 28, 1914, Grand Vizier Said Halim Pasha assigned all of Mesopotamia's oil concessions to the TPC.[15] The same day in Sarajevo (for a long time a Turkish province), Archduke Franz Ferdinand of Austria was assassinated. Not a single drill site existed when on August 4, the day the United Kingdom entered the war against Germany, London placed Deutsche Bank's shares under sequestration. There were only a hundred kilometers left to go before the final leg of the Berlin-Baghdad line would be completed and reach the old city of Mosul, where, after the war, the first oil derricks were built in what would later become Iraq.

The Engines of the Great War

If machines multiplied the size of the war in 1914–1918 as well as the number of its victims, it is primarily because they mobilized and supplied many more soldiers. The dominance of the German train, with its ample coal resources,

* Opened in 1898, the famous Grand Parisian Hotel was one of Calouste Gulbenkian's first investments. During their Christmas dinner in 1900, he and his only son, Nubar, were served one of the four feet of an elephant ordered by an eccentric American, and acquired by the maître d'—in haste and at God knows what price—from the zoo at Paris's Jardin des Plantes. (See Nubar Gulbenkian, *Pantaraxia*, Hutchinson, London, 1965, p. 13.)

was eventually surpassed by the mobility and power of motor vehicles (trucks, tanks, automobiles) sent to the front lines by the Allies in ever-increasing numbers over the years of conflict. Almost insignificant in 1914 in the eyes of the officers, who tended to prefer traditional movements and cavalry charges, the problem of access to oil was increasingly important from 1918 on.

Starting in the first days of September 1914, the German Army was at the gates of Paris. Horse-mounted divisions of the terrible Uhlans were spotted only a few dozen kilometers from the capital's suburbs. The French general staff attempted to speed up troop replenishment by any means. General Joseph Gallieni, military governor of Paris, requisitioned more than a thousand taxis from the city, as well as buses, on September 6 and 7. Gathered at Les Invalides, a large complex of military buildings in Paris, taxis and buses transported about five thousand soldiers to the front, where the Battle of the Marne began. The French army duly paid them for each of their runs. Contrary to popular belief, the "taxis of the Marne" did not decide the outcome of this battle, which led to the Germans' withdrawal and made it possible to bar them definitively from the road to Paris. Dozens of divisions were engaged, and the vast majority of the hundreds of thousands of French soldiers who participated were transported by train. Nonetheless, the taxis-of-the-Marne episode remained the first indication of upheaval in the war's tactical and strategic foundations, which the internal combustion engine soon upended.

Within the French army, artillery officer Jean-Baptiste Étienne was considered the father of the armored tank. Just before the war, he led one of the earliest groups of reconnaissance planes for the French army. On August 2, three weeks after the fighting began, Colonel Étienne declared to the officers of his regiment, who were riding beside him: "Gentlemen, victory in this war will belong to the first of the two combatants who mounts a 75 mm field gun on a car capable of moving on all terrain."[16] English colonel Ernest Swinton had the same idea in October. An engineer by training, Swinton closely followed the British army's experiences transforming agricultural tractors into off-road vehicles capable of firing artillery or breaking through barbed-wire defenses. In turn, Colonel Étienne learned of experiments conducted across the Channel.[17] Swinton presented his ideas to his staff, and Winston Churchill promptly seized the matter and formed, in February 1915, a "Landships Committee." Many new prototypes were tested. In the secret exchanges of the British War Office, a generic code name was assigned to them: tank. For his part, in December 1915 Jean-Baptiste Étienne made contact with Joseph Joffre, general of the French army. On January 31, 1916, Joffre ordered the first four hundred tanks from Schneider.[18]

Tanks played a decisive role in ending trench warfare. British Mark I tanks were deployed for the first time on September 15, 1916, during the Battle of

Somme. They were slow and capricious. Many breakdowns occurred during the offensive, and they were easily destroyed by German artillery. But some of them succeeded in spreading death and terror across enemy lines. On April 16, 1917, French Schneider tanks participated in the Dantesque offensive at Chemin des Dames. "Assault artillery," as tanks were called in the French army, was rapidly improving. By the end of 1917, Renault's factories had manufactured more than three thousand FT light tanks, the first assault tanks with modern configurations: a rotating turret and a motor placed in the back. British tanks achieved a historic breakthrough during the Battle of Amiens, on August 8, 1918—described as the "black day" of the German Army by the Reich troops' general-in-chief, Erich Ludendorff, the day after the battle. The United States entered the war in 1917 and manufactured more than nine hundred FT tanks under a license agreement from Renault. In the middle of September 1918, the young American major George Patton victoriously launched two battalions of French tanks in the assault of the Salient of Saint-Mihiel, near Verdun. The German Army was too late to realize the fatal progress of the Allied tanks. It wasn't until March 1918 that they brought their slow and massive model A7V, which weighed more than 30 tons, to the battlefield. Besides the tanks, in 1918 the Allies had at their disposal hundreds of thousands of land vehicles—trucks, cars, and motorcycles—where, at the beginning of the war, there had been only a few hundred, for the most part requisitioned. In the trenches, paraffin was vital for protecting food and ammunition from moisture. The purring of mobile, oil-fueled electric generators made it possible to operate field telephones even in the most exposed areas of the front lines. Finally, search planes, bombers, and reconnaissance were counted by the tens of thousands during the armistice, on both of the Allies' fronts. They completed the war's intensification, and confirmed the starring role of oil in the outcome of armed conflicts.

In War, Fuel Becomes as Necessary as Blood

As early as 1915, Germany saw its industry hampered by a shortage of lubricants. Beginning in 1916, it was no longer able to secure access to oil. In August, Romania's entry into the war, on the side of the Allies, severed the Reich from its main supply source. In November, the Allies succeeded in blowing up most of the Romanian oil installations just before the German Army arrived. After peace was established with young Bolshevik Russia on March 3, 1918, by the Treaty of Brest-Litovsk, Berlin redeemed Lenin's access to the Baku oil that Germany would never obtain. The Turkish allies besieged the city on the banks of the Caspian Sea at the end of July and laid hands on certain wells. But in the middle of August, the United Kingdom sent an expeditionary force through Persia, and expelled the Turks from the Black City.

To protect the Abadan refinery, the British army assembled in November 1914 from the Mesopotamian port of Basra on the estuary of the Tigris and the Euphrates. It then took Baghdad in March 1917, thanks to Indian troops, and advanced as far as Mosul in November 1918. Meanwhile, in Persia, the Anglo-Persian Oil Company engineers briskly developed production. However, it reached only twenty-three thousand barrels per day in 1918, barely equaling Burmese production.[19] In spite of the precautions Winston Churchill had taken on the eve of the war, to endow his country with the first national oil company in history, the still low volumes of crude oil drawn by the Anglo-Persian Oil Company only met a meager part of the British army's needs. As for France, in 1914 it was cut off from its supply sources in Russia and Romania, which until then had supplied almost half of its imports.[20]

In sum, about 80 percent of the oil fueling the Allied war machine was American. In essence, it was supplied by Standard Oil companies and, to a lesser extent, by Royal Dutch Shell (which remained faithful to the British Crown despite Churchill's 1914 insinuations to the contrary). The total maritime war triggered by German submarines in January 1917 led the United States to go to war on April 6. A month later, Woodrow Wilson's treasury secretary, William Gibbs McAdoo, announced triumphantly that Kuhn, Loeb & Co. and Standard Oil of New York ranked first among the major underwriters of the loans issued by Washington.[21]

In the weeks and months that followed, shipwrecks of tankers carrying American oil and an explosion of fuel consumption in the United States led the Allied armies to the brink of a shortage. The Wilson administration's appeal to the American people to stop driving on Sundays had no effect. An unverified suspicion, which endured in the upper French administration, arose: that Standard Oil had decided to deliberately divert its tankers from Europe and play the Reich against the Allies.[22] The rumor was born after the revelation of secret, exclusive dealings between Paris and London for the partition of the Turkish Empire and (especially) the black gold of Mesopotamia.* In fact, oil tankers and American sailors dreaded entering the English Channel. In addition, French ports were ill equipped to accommodate large tankers.[23]

Companies importing American oil sent a distress signal to the French government in December 1917: Gasoline stocks had fallen to 34,000 tons, barely three weeks of consumption, the previous month.[24] A little too late, France

* See *infra*, chapter 7.

understood that it had made the mistake of practically ignoring what was not yet called energy independence. Failing to accelerate supplies, the reserves threatened to dry up completely as early as March 1918. On December 15, Georges Clemenceau, named prime minister a month before, gave the American ambassador a note in which he fretted about the risk of surrender: "Any disruption in the gasoline supply would cause an abrupt paralysis of our armies and could rush us into an unacceptable peace for the Allies." Clemenceau urged US president Woodrow Wilson to send him one hundred thousand tons of oil as soon as possible.[25] The plea signed by the "Tiger" was transmitted immediately by cablegram across the Atlantic. His last sentence became famous: "If the Allies don't want to lose the war, France must fight in the hour of the supreme Germanic confrontation, and must possess the gasoline that will be as necessary as blood in tomorrow's battles." Woodrow Wilson sent no reply but followed up on the request: 117,450 tons of petroleum products were delivered to France in early January 1918, by nineteen tankers from the United States.[26]

In London, ten days after the armistice of November 11, 1918, Lord George Curzon, former viceroy of India and future British secretary for foreign affairs, declared: "We might almost say that the allied cause had floated to victory upon a wave of oil."[27] In reality, the First World War made that wave swell even more, in the same way that the US Civil War had spurred its first undulations. The wave then crashed and spread out across a planet that was in between two world wars.

Assassinated in Paris on July 31, 1914, three days before the start of the Great War, the pacifist Jean Jaures, leader of the socialist movement in France, made the following statement to the chamber of deputies nearly twenty years earlier: "Because this tormented society, to defend against incessant worries that arise from its own depths, is perpetually obliged to thicken armor against armor; in this century of unlimited competition and overproduction, there is also competition between armies and military overproduction. Industry itself being a fight, the war becomes the first, the most excited, the most feverish of industries. . . . Always your violent and chaotic society, even when it wants peace, even when it is in a state of apparent rest, carries within it the war, as the sleeping cloud carries the storm."[28]

As dawn rose on April 20, 1914, in Ludlow, an outpost on the high plains of Colorado, in the center of an American nation in "apparent rest," National Guard soldiers sprayed machine-gun fire across the tents of coal miners who had been striking against the Rockefeller empire for months. Drunken soldiers (many of whom, according to the unions, were employees of the mines enlisted by the army) set fire to several tents using torches soaked with gasoline. The strikers were demanding decent working and living conditions, as well as the recognition of their union. The Ludlow Massacre killed about twenty people, including two women and eleven children. Unyielding since the strike began, before the beginning of winter, John D. Rockefeller Jr., heir to his father's empire and a generous philanthropist, was surveying his immense property on the banks of the Hudson River when he heard the sad news. He was reviewing the layout of the gardens with his wife Abby Rockefeller (born Aldrich). His press secretary, Ivy Lee, initially tried to attribute the cause of the fire to a poorly extinguished stove.

SIX

The Roaring 1920s: Consolidating the New Empires

The successors of Standard Oil naturally paved the path the whole industry followed. Their power shaped the most important banks on Wall Street—just a few steps from the imposing, obsidional building erected by John D. Rockefeller at the tip of Manhattan on 26 Broadway. Unequalled, the fertile riches that had accumulated around the petroleum empires pushed the bounds of the thermo-industrial society further and further. At the end of the Great War, the industry extracted 1.5 million barrels of petroleum per day—barely 2 percent of the consumption that was reached at the end of the century of oil. The spiral of technical progress led to a rapid parallel expansion of energy demands. In concert with this expansion, the quest for new sources inevitably accelerated. In this pursuit of new oil fields, the imperialist nations were sometimes the partners and more often the zealous auxiliaries of their major oil companies.

Man and Machine: The Consummated Union

Soldiers of the First World War felt the power of the mechanical energy unleashed by petroleum. Throughout 1917, mutinies and desertions in the trenches of Lorraine, worker demonstrations in the great cities of Germany, and October's Bolshevik revolution placed the bourgeois social order of the Industrial Revolution at the edge of the precipice. During the final days of combat, between November 1 and November 4, 1918, the German sailors of Kiel mutinied, refusing to follow their officers in a pitiful and suicidal last stand against the Royal Navy. The servants of the most formidable death machines ever built by man were at the forefront of a new attempt at revolution, just as the Russian sailors of Kronstadt had been a year earlier. German emperor Wilhelm II knew that he could no longer count on the troops or the factory workers. On November 5, he called members of the Social Democratic Party of Germany (SPD) to meet with government authorities, to negotiate an armistice. On November 7, in Munich, the revolutionary council of soldiers and workers took the SPD reformists off guard and the next day proclaimed the People's State of Bavaria. Other labor councils took control of Berlin and several other large German cities when Willhem II abdicated on November 9, the day after the armistice.

The German revolution nevertheless failed. On the same day, the SPD's leader, Friedrich Ebert, who never stopped supporting the war, leaned on the military, police, and judges to put the Communist revolution in check and safeguard the established order. Anarchist Ret Marut cried out in a Munich theater on December 28, 1918: "Think! You do not need anything else. Become aware of the serene passivity inside of you, in which your invincible power is rooted. With a calm and carefree heart let economic life crumble down; it never brought me happiness and neither will it bring you any. Consciously, let industry rot, otherwise it will rot you."[1]

As Marut had seen, a great divvying up of viewpoints on technical progress had taken place. Social democrats, socialist totalitarians, and fascists shared one fundamental vantage point: the belief that the oil industry and its perpetual expansion, like an eternal drill's bore, was the only thing capable of extracting humanity from its material torments. The ideology of the SPD, following the path of least resistance, gave rise to the most prosperous political order of the twentieth century. The leaders and partisans of the SPD, which originally assumed a Marxist program, were convinced from the end of the nineteenth century (in the midst of the thermo-industrial boom) that it was entirely possible and desirable to collaborate with the bourgeois and their prodigal machines. By negotiating their place in the established industrial order, the Social Democratic proletarians planned to form a two-headed technical empire alongside their bosses; their goal was to control and benefit from machine labor almost as much as their bosses did.

On both sides of social democracy, the totalitarian utopias of socialism and fascism thrived on the disgust felt by the European intelligentsia for its bourgeoisie, its own class.[2] In Moscow, in March 1919, Lenin famously quipped: "Communism is Soviet power plus the electrification of the whole country." That electrification was primarily generated by fossil fuels, coal, oil, and natural gas (but particularly oil in Russia). Lenin marveled at the "beginning of a very happy period" in which politics and politicians took a backseat and engineers and agronomists had the floor. "Marxism is philistine. . . . The father of Marxism is steam," proclaimed German anarchist Gustav Landauer, who was assassinated in the final hours of the aborted revolution in Bavaria.[3] In Bolshevik Russia, there was no longer any question, as there was in the eyes of Karl Marx, about the risk of workers becoming "a mere appendage of the machine": Perfecting the industrial order was, on the contrary, the very means of liberating the proletarians.[4]

Fascism, finally, did not cease to idolize the industrial machine. The very idea of fascists evokes the concentration of energy. In the film *Metropolis* (1927), Austrian filmmaker Fritz Lang brought to the screen the disturbing and ambiguous figure of the "robot" (a word coined in Prague in 1920 and rooted in the Slavic word for *work*). The film's scriptwriter, Thea von Harbou, Lang's wife, became a prolific author, one appreciated by Adolf Hitler. She eventually joined the Nazi Party. *Metropolis* concludes with a rekindled alliance between the "hands" (the workers) and the "head" (the boss) thanks to the "heart" (the son of the boss), to better subdue the perilous power of, but ultimately benefit, the machine. Lang's radically innovative graphic vision was inspired by his discovery of the Manhattan skyscrapers in 1924. "I spent an entire day walking the streets," he later recounted. "The buildings seemed to be a vertical sail, scintillating and very light, a luxurious backdrop, suspended in the dark sky to dazzle, distract and hypnotize. At night, the city did not simply give the impression of living: it lived as illusions live."[5]

Fascism, socialism, and social democracy: Perhaps they were tributaries flowing from a singular original experience—the one in which, in the aftermath of the First World War nightmare, everyone perceived the promise of an era of limitless energy powerful enough to change the world, without pain. The effort required for the rapid rise of technology intruded into every ideology as a self-evident necessity. This technology boom became the key element in our universal understanding of "progress." Everywhere, as Nietzsche had foretold, the pursuit of industry monopolized human conscience.[6] The acquisition of energy outweighed its lavish expenditure. On the eve of the Second World War, despondent German Jewish Marxist philosopher Walter Benjamin observed: "Nothing corrupted the German working class so much as the notion that it was moving with the current. It regarded technological developments as the fall of the stream with which it thought it was moving." A systematic misunderstanding in the Marxist vulgate, this idea of work being "the Messiah of modern times," explains Walter Benjamin, "does not linger long on the question of how its products benefit the workers themselves, as long as . . . [its products] are not at their disposal. It only wishes to see the progress in the control of nature, and not the regressions of society. We already recognise here the technocratic traits that will be later be revealed in fascism. . . . Work, as conceived henceforth, is equivalent to the exploitation of nature, which we opposed with a naive satisfaction to the exploitation of the proletariat."[7]

Those few renegades who denounced the necessity of technical expansion—pointing to its false neutrality, its unlikely harmlessness—did so in the name of far-flung, uncommon ethical choices, often purely personal. Their views may have challenged the mass productivist ideologies, but they never tore those ideologies apart or changed their course.

Anarchist Ret Marut was one such renegade. In January 1919, the German Social Democratic governments violently crushed the Spartacist revolution in Berlin, then crushed the ephemeral Bavarian Council Republic in spring of the same year. To do this, they relied on the Friekorps, German volunteer units that foreshadowed the Nazi militias. Fleeing repression, Marut boarded, perhaps as a coal stoker, a steamer departing for the American continent. From this experience, he wrote *The Death Ship*, a novel he published in 1926 under the soon-to-be-known pseudonym of B. Traven. On a ramshackle vessel, miserable sailors "smeared with oil, coal dust, and petroleum" shoveled for fourteen hours a day to supply the machine with fuel and maintain the pressure inside its boilers—all relics of the old industrial regime of coal, which was about to be replaced by that of oil.[8] The stateless writer then landed in the port of Tampico, Mexico's black gold capital, where, as in Maracaibo, Venezuela, US oil companies had arrived at the beginning of the century. Traven sometimes worked on oil wells, and in 1929 he published *Rosa Blanca*, his longest novel.[9] In it, he told the supposedly true history of how land was stolen from Native Americans through violence and corruption, all to benefit an American petroleum company that was already drilling all around it. Oil was then on the verge of establishing itself as the principal form of energy for maritime and land transport: An indispensable good transported to the far corners of the world, black gold had become the very vector of the industrial vascular network, the blood of human technology.

Shell and US Oil Companies Assist the USSR's Birth

Bolshevik leaders and capitalist oilmen collaborated to revive Baku, the Black City, from its ashes. First, in the aftermath of the 1917 revolution, the oil majors bet on a rapid collapse of Communist Russia during the civil war between the Red Army and the White Army that was supported (with tortuous ulterior motives) by the Western expeditionary forces. In 1918, Royal Dutch Shell president Henri Deterding tried to regain control of Baku. Deterding did not owe his nickname, "the Napoleon of petroleum," only to his small size: Shell was negotiating an agreement with the Democratic Republic of Azerbaijan, which the British army was supposed to be protecting. Shell's dreaded boss, however, was not the only one to covet the Black City. An American giant, more than six feet tall, initially cordial but stern, stood in his way: Walter C. Teagle, just thirty-nine years old, had been named president of Standard Oil of New Jersey a year earlier and was acquiring new concessions on the shores of the Caspian Sea. But the British governor general of Baku vetoed the Standard Oil contract. Meanwhile, London was trying to negotiate with Moscow so that the Anglo-Persian Oil Company would export oil from Grozny in Chechnya, controlled by the Communist troops. But, again, negotiations failed.[10] As the Red Army approached victory, Wall Street also tried to get closer to the Bolsheviks. The linchpin of this rapprochement was the American International Corporation, an alliance of merchant banks charged with developing American interests abroad and very well connected to American diplomacy. Founded in 1915 by JP Morgan and the National City Bank (the bank of William Rockefeller), the American International Corporation, headquartered at 120 Broadway, was a stone's throw from Standard Oil and counted one of William Rockefeller's sons, Percy, among its directors.[11]

The capture of Baku by the Red Army in April 1920 was a turning point in the Russian civil war. It became clear that the Communists would be in power for a long time. The civil war ended in 1921, but Russia was worn out. On March 21, Lenin resolved to launch his New Economic Policy (NEP), which consisted of opening the door to free enterprise and foreign capital in time to put the economy back on its feet. Moscow's new masters had a pressing need for the oil resources of Baku, Grozny, and Maikop: Black gold was indispensable to industrialization and constituted the main (and one of the few) export resources.

The international conference of Genoa in April 1922 confirmed the normalization of relations with Bolshevik Russia. Lenin had dispatched an old and courteous industrialist, Leonid Krassin, who spoke the language of capitalists. The great Western companies and their diplomat protectors arrived determined to form a "united front" against him, and Henri Deterding imposed himself as the front's natural leader: It was mostly Shell's capital that the Bolsheviks had confiscated in Baku. Uncompromising toward Moscow, British prime minister David Lloyd George began by demanding that the wells of Baku return for ninety-nine years to their rightful claimants. It was then that Henri Deterding manipulated the proposal, restricting it to agreements reached before the revolution—essentially sidelining Standard Oil and the French and Belgian industrialists who had advanced their pawns in the

Caucasus after 1917. The negotiations in Genoa failed to partition Russian oil, not because of the Soviets, who wanted to revive Baku as quickly as possible, but rather because Shell's boss wanted to overtake its competitors.

The united front would not last long; it soon became everyone for himself. Moscow easily exploited the antagonism between Royal Dutch Shell and the American oilmen. An offshoot of Standard Oil, the Vacuum Oil Company, hastened to buy half a million tons of very low-priced Soviet kerosene on the Indian market with too little discretion. Henri Deterding hastened to follow suit.[12] Standard Oil of New Jersey sent a representative to the Kremlin. Soviet Russia succeeded in attracting a share of the capital needed to rescue the Black City's oil industry, which had suffered from the consequences of the war and never really recovered from the bloody upheavals of 1905. Calouste Gulbenkian was among the lenders: In exchange for a 50,000-pound loan, the omnipresent Armenian merchant obtained a one-year monopoly on Caspian caviar. Also on the list of creditors who were in a hurry to gain access to Baku oil was William Averell Harriman, who owned one of the most powerful investment banks in New York, created with the money of his father, the railroad "robber baron" Edward Henry Harriman.[13] The bank was presided over by a brilliant internationalist financier, George Herbert Walker, maternal grandfather of the future US president George Herbert Walker Bush.

Among the Western oilmen, a maverick emerged as the Bolsheviks' best ally in the rebirth of Russia's black gold industry. Concession hunter Henry Mason Day, a multilingual American colossus, arrived in Baku shortly after the conference in Genoa. Under the noses of Deterding and Teagle, who were busy undermining their mutual interests, he succeeded in negotiating a fifteen-year agreement to rehabilitate the Black City's wells and drill new ones. In exchange, Day received a comfortable percentage on crude oil sales. His first team of oilmen arrived in Baku in June 1923. The team included six American engineers who introduced the modern rotary drills and pumps that revived Baku from its ashes. Other US companies and European leaders also rushed into the breach opened by Henry Mason Day. They allowed the Soviets to equip themselves with efficient industrial infrastructure, replacing in particular the old Baku-Batumi pipeline built by the Nobel brothers.[14]

Thanks to his spectacular success, Henry Mason Day was hired as an intermediary by the second richest US oilman after John D. Rockefeller: Harry Ford Sinclair, president of Sinclair Oil. With his fleshy jowls, round eyes, and drooping lips, Sinclair invariably evoked the image of a toad among his contemporaries. Sinclair Oil's prospectors had traveled the whole world in search of opportunities, even venturing to Angola, the remote Portuguese colony in southern

Africa.[15] But it was Baku, above all, that sharpened Harry Sinclair's appetite. In 1922, the businessman embarked for Russia with one of Teddy Roosevelt's sons, Archibald, at his side. In exchange for $365 million in investments and loans, the Kremlin offered Sinclair no less than a forty-nine-year partnership in Baku and Grozny. In addition, there was a monopoly on the development of the remote Sakhalin island off eastern Siberia. (When Sinclair sent a team to Sakhalin, it was stopped by the Japanese, who were trying to attach the entire island to the emerging Nippon empire, devoid of black gold).[16] The Bolsheviks finally pressed Sinclair for a small favor: to obtain from the United States the full diplomatic recognition of Communist Russia. Harry Sinclair, very close to President Warren Harding and several members of his government, was able to boast about being "Russia's salesman to the world."[17] But, as we shall see, the fall of Harry Sinclair followed quickly on the heels of President Harding's death in 1923, during one of the biggest political scandals in US history.

Once Lenin disappeared in 1924, Stalin quickly closed the door to Western investments in Soviet oil, and one by one, annulled the agreements that had been reached. But he never closed the door to engineers from Europe and the United States. On the contrary, he appealed to them in large numbers and imported Western materials destined for several strategic sectors—first and foremost those linked to oil—when he launched his five-year forced industrialization plan in 1928.[18] Thus, for example, American industrialist Fred C. Koch sold about fifteen state-of-the-art refining facilities to the USSR between 1930 and 1932, which the Soviets were quick to copy.[19] (Koch was the father of Charles and David Koch, the two US petrochemical billionaires who became famous in the late 2000s as the main sponsors of the ultraconservative and climate-skeptic Tea Party and other right-wing movements.) But few in the West boasted of this decisive intervention for the sustainability of the Soviet Union. And it was understood that Soviet propaganda would not say a word during the mass trials against the "saboteurs" in the early 1930s. Starting in 1930, Baku once again became one of the most productive oil fields in the world. After helping to resuscitate it, the capitalist industry was deprived of control over its crude oil for the next sixty years.

The Roaring Twenties

Unable to settle down and smoothly adapt its production to the new conditions of peace, the oil industry experienced an intense overproduction crisis after 1918. In the United States, after a decade of rapid development, the car market became saturated and, in 1921, General Motors' sales collapsed. But the crisis was short lived. In particular, with the profits from the war, Wall Street became the world's foremost financial center; its capital was irrigating US industry in a staggeringly vigorous process of urbanization, standardization, and electrification that drove the advent of the American model of mass production and consumerism. The obstacles along the way were overcome by decisions made in Washington, where two Republican presidents (Warren Harding from 1921 to 1923, then Calvin Coolidge from 1923 to 1929) lowered the taxes of the rich and strived to defend the private sector against government interference. The Federal Reserve (the US central bank conceived in 1910 on Jekyll Island) opened the floodgates of cheap credit by setting very low interest rates. It thus allowed money to be created rapidly, and banks to have outstanding loans that far outweighed their own money reserves. Millions of Americans went into debt, consuming without fear, and when possible, without limits. They bought cars that they parked proudly in front of new suburban homes, and began to acquire all sorts of amazing electrical appliances. Confidence in the future meant pumping more and more crude oil.

In France, the 1920s were called "Les Années Folles," or the "Crazy Years," years of transgression in both manners and the arts. In the United States, which in the course of this decade became the world's leading economic power, the 1920s were the Roaring Twenties: They roared like gas motors or the lion of Metro Goldwyn Mayer studios, founded in 1924 in a suburb of Los Angeles, the epicenter of the most prolific oil boom of the decade. The American population had tripled since the Civil War, exceeding one hundred million in 1920. The great American cities displayed the pride of human technology. New wonders of the world, skyscrapers erected in Chicago, Detroit, and, of course, New York showcased humanity's new capability of living high above the ground, on the verge of triumphing over the forces of nature—previously just a fantasy. The art deco style seen in the Chrysler tower or the Empire State Building magnified the industrialists' hubris. Destined to "dazzle, attract attention and hypnotize," the design of modernist buildings avoided any reverence, or even any reference, to previous cultures.[20] If these tours de force were now possible on the gigantic scale of the new American metropolises, it was because of new possibilities offered by concrete, steel, and glass—three materials whose mass manufacture, distribution, and assembly set immense resources of energy, in the form of heat and motor power, in motion with an ease that would have amazed the planners of the old, less energy abundant Europe.

In 1924, inspired by the "steel rhythms" of a train (probably fueled by oil) while traveling between New York and Boston, American composer George

Gershwin captured the promise of a people ad-libbing their way to a destiny of power: With "Rhapsody in Blue," he said he had produced "a musical kaleidoscope of America, of our vast melting pot, of our unduplicated national pep, of our metropolitan madness."[21] Where did this unparalleled vitality and urban madness that had seized the United States come from? What was its vital, fertile source? A little later, in his *Journey to the End of the Night*, published in 1932, French author Louis-Ferdinand Céline brought his character Bardamu to explore the dark side of this landscape. Céline described the large, squat, glassed-in buildings of Detroit's Ford factories as "endless fly traps, in which men could be discerned to stir, but hardly move, as if they were struggling only feebly against some impossible thing."[22]

The Teapot Dome Scandal, and the Furious Quest to Prevent the Decline of US Oil Production

President Warren Harding's administration counted among its ranks many who were intimately linked to the oil industry. Walter Teagle, head of Standard Oil of New Jersey, was one of the Republican president's poker partners.[23] In the days following his inauguration in 1921, Harding traveled to the very rich oil fields of Oklahoma, in what was then the heart of the American petroleum industry. In a speech he said: "Next to agriculture and transportation the petroleum industry has become, perhaps, the most important adjunct to our civilization and well-being."[24] Harding failed to mention that transportation and agriculture, like construction, were all becoming highly dependent on oil.

There was just one danger believed to be capable of jeopardizing the growth of the American economy (far outpacing Europe's): the exhaustion of the oil fields. The United States was by far the world's largest crude oil producer. However, its economy swallowed up almost everything that could be extracted from American wells. After exporting more than two-thirds of their petroleum products in the early decades of the industry, and a quarter more in 1914, the United States was for the first time transformed into a net importer at the end of the First World War.[25] Apart from two tragic interludes, the crisis of 1929 and the Second World War, the largest oil nation remained an oil importer, increasingly, throughout the remainder of its history.[26] By growing upward, a body weighs down its own vital supports. The American economy's appetite for energy grew in step with its power. More than any other, the economy of the American nation, whose power became the focus of the whole world, including the USSR, had become "addicted" to its own oil. The United States was the country of black gold; without it, it would not have been able to profit so easily from the magnificent, inert riches present on and beneath its land.*

To satisfy its oil addiction, the United States needed to find new sources of crude. And in 1920, there was an emergency. The old oil fields of Pennsylvania, New York, West Virginia, and Ohio, which supplied almost all American production at the turn of the century, continued to decline. Fields discovered in Indiana, Kansas, and Nebraska turned out to be less than generous. And many oil engineers were concerned that the much larger extractions of Oklahoma and Texas were not enough to compensate for the decline of the oldest fields. Some industrialists planned to heavily exploit oil from bituminous shale, abundantly present in the Rocky Mountains, but such an undertaking "would require an industrial organization greater than our entire coal-mining organization," one American expert told the *New York Times*.[27] Another expert, from the US Geological Survey, told the *New York Times* he was concerned about the "probability that within 5 years—perhaps 3 years only—our domestic production will begin to fall off with increasing rapidity, due to the exhaustion of our reserves."[28] Also in 1920, the director of the US Geological Survey concluded: "The position of the United States in regard to oil can best be characterized as precarious."[29] This prediction could have come true if the United States had stopped at the Rocky Mountains. But the oil fields of the Pacific Ocean, whose rapid exploitation began the same year in the remote harbor of Los Angeles, soon proved to be of prodigious and unexpected abundance. Thanks to California's oil, the American economy continued on its path, and accelerated again.

Replacing the reserves of mature or declining oil fields by constantly acquiring new intact fields was vitally important in the crude oil industry. While all the major companies were rushing to Los Angeles to take part in the upcoming frenzy, behind the scenes in Washington, two of America's most powerful independent oilmen maneuvered to get their hands on the prewar fields that had been under US Navy control since the height of the offensive against Standard Oil.

One of these two oilmen was the audacious Harry Sinclair, who at the same time was struggling to become the master of Bolshevik Baku. The other was Edward L. Doheny. Fortune had long remained aloof from this son of an Irish immigrant from Tipperary, who for a time had become a

* Consider, in comparison, the humble fates of Australia and South Africa, former British colonies almost devoid of oil, certainly much smaller than the United States, but infinitely rich in minerals and conducive to agriculture.

whaler after fleeing his island during the great famine of the 1850s.[30] Perhaps more than any other famous oil pioneer, Doheny was a tough guy, full of resources, and inventive. He liked to tell how at the beginning of his career as a prospector he had taken advantage of the fact that he had broken both legs at the bottom of a mine to get his law degree.[31] The story became part of his legend, except that Doheny never was licensed to any bar.[32] In 1893 he was the first to successfully drill in the asphalt outcrops near a crossroad of farms called Los Angeles. He used a rudimentary tool, indicative of the few resources that drilling required at the time: the cut and hardened point of a eucalyptus tree trunk.[33] However, California crude oil was heavy and difficult to refine with the techniques then available. Almost fifty years old, this son of a rugged, obstinate Irish immigrant was one of the first gringos to build a colossal fortune in Tampico in the early twentieth century, during the Mexican oil boom.

Both Harry Sinclair and Edward Doheny showed great interest when, in 1921, President Warren Harding instructed his secretary of the interior, Albert Fall, to authorize the concession of three intact crude fields previously reserved for the US Navy. Two of these fields were located in California, north of Los Angeles. The third, located in Wyoming, bore a name that evoked its geological form: the Teapot Dome. To snatch the concessions they coveted, Sinclair and Doheny each resorted to paying ample bribes. Albert Fall received a total of more than $400,000, later revealed in court. As the evidence accumulated in the 1920s, the Teapot Dome scandal became a symbol of collusion between industry and political power in the years preceding the 1929 crisis. It remained the most resounding political scandal in the history of the United States until the Watergate affair.[34]

Warren Harding died on August 2, 1923, shortly after news of the scandal broke, on his return from a trip to Alaska. A few days earlier, on July 27 in Seattle, during his final speech (written by Herbert Hoover, the eminent geologist who became president in 1929), President Harding marveled at the "measureless oil resources" recently discovered above the Arctic Circle in the very north of Alaska, the largest and most northerly of the American states purchased from Russia in 1867.[35] One of Harding's last decisions, made to calm the fury over the scandal,[36] placed the Alaskan oil fields under US Navy control, as President Taft had done for the Teapot Dome and the California reserves. But it was a paltry gesture, since it was technically unfeasible to exploit America's crude oil in the North Pole at the time; that would not occur for half a century. The immediate future of American oil was found a little closer to home.

Los Angeles: The Archetype of the Twentieth-Century Metropolis Emerges from Oil Fields

It was inside an oil crucible that the American dream was born, in the hot, arid climate of Southern California. Developed at a fierce pace throughout the 1920s, the extraordinarily productive oil fields along the beaches of Los Angeles, more than 4,000 kilometers from New York and the oil regions, gave rise to the twentieth century's first megalopolis that was designed for the automobile. In one decade, California crude oil extractions tripled, accounting for almost a third of US production.

As early as August 1919, nearly two hundred US Navy warships were anchored in San Pedro Bay, facing Long Beach. On the eve of the oil boom, Los Angeles became the base for the American navy in the Pacific. Some of the newest American battleships traveled to California via the Panama Canal, which opened in 1914.[37] They defended the United States' commercial interests in the Pacific and China, in the face of Japanese imperialism, of which the Allies were just endorsing the first step.* San Pedro remained the Pacific fleet's principal port until it was moved to Hawaii's Pearl Harbor in 1940. The arrival of battleships encouraged the oilmen to press forward: As a client, the US government was as important as it was faithful. In the ten years of an oil boom, Los Angeles jumped from the tenth largest city in America to the fifth, rising from 500,000 in 1920 to 1.2 million in 1930. The City of Angels was populated by "retired farmers, grocers, Ford agents, petty hardware and shoe merchants from the Middle West."[38] Hundreds of thousands arrived by steamships, by oil-fueled locomotives from the Southern Pacific or Santa Fe Railroad, and even, more and more, by car. These working class arrivals "live in a tiny bungalow with a palm in front," ready to enjoy an eternal spring.[39] As early as 1925, Southern California had a car for every 1.6 inhabitants, a ratio that was not reached in the rest of the United States until the end of the 1950s, and even later in Europe.[40] Faster than anywhere else, the first petrol stations around Los Angeles were networking and nurturing the first distant suburbs of automobile civilization, with its shopping centers surrounded by parking lots. The city quickly welcomed major automobile assembly plants and tire factories, which lured workers deprived of trade-union rights.

---

* With article 156 of the Treaty of Versailles, signed on June 28, 1919, the United States, the United Kingdom, and France allowed Japan to take ownership of "the railways, the mines and underwater cables" that Germany had established in the strategic Chinese region of Shandong, initially promised in Beijing by President Wilson.

Images promoting Los Angeles showed the first American surfers but kept the oil wells that stood behind them out of the frame. Yet every weekend, the people of Los Angeles arrived by the thousands to stroll along Long Beach, Seal Beach, and Huntington Beach, where hundreds of wooden derricks were spaced a few meters apart. Between the Pacific and the city center, Signal Hill—nicknamed Porcupine Hill—was ground zero for California oil drilling. The wells there, at the very heart of this growing megalopolis, reached their absolute peak of production very rapidly, in 1923.[41] As the Los Angeles fields became exhausted one after the other over the years and decades that followed, drilling was moved further and further inland, and eventually, beginning in the 1960s, out to sea, in order to continually increase California's production capacities until their decline in 1986. One of the biggest port terminals in the world, San Pedro, was dug at the site of the nearly dry wells of Long Beach.

Los Angeles had abundant oil but a shortage of water. Large investments were made to lay pipes that transported water from farther and farther away. The companies that pumped it to supply the city (and irrigate immense citrus plantations) were often the same ones that drilled for oil. It was because they found water where they were looking for black gold that developers built the posh city of Beverly Hills.[42] Oil money partly founded Hollywood, the American "dream factory." It financed the creation of one of the best research centers in the United States: the California Institute of Technology (Caltech) founded in 1921, in Pasadena, thanks to California oil from the Mudd and Kerckhoff families.[43] This private research center quickly became home to world leaders in physics, including Albert Einstein and Robert Oppenheimer. In addition to playing a decisive role in the emergence of the American military and civil aeronautics industry, Caltech also trained geologists and engineers, on whom the major oil companies increasingly relied as they evolved. Caltech's first president, Nobel Prize–winner Robert Millikan, remained in charge until 1945. As with many American establishment figures at the time, the physicist believed in social Darwinism and the natural supremacy of the white race.[44]

William Gibbs McAdoo, former treasury secretary and President Woodrow Wilson's son-in-law, was the political champion of the oil industry and the developers who built their fortunes around the Los Angeles oil boom. McAdoo presided over the establishment of the Federal Reserve in 1913, and then sat on its first board of directors before requesting that Standard Oil be among the first underwriters of the First World War. In 1922, this ambitious lawyer left to continue his political career in California, still one of the most remote and underdeveloped territories of the United States. There he accepted the support of the Ku Klux Klan, an inescapable force in a state where blacks and

Mexican immigrants had settled in large numbers, and where Communist militants from the First World War were harassed by the police. McAdoo did not reject support from the racist Protestant group during the Democratic Party's race for the 1924 presidential election. Nor did he refuse the generous financial support of Edward Doheny, the Irish Catholic oil magnate who was entangled in the Teapot Dome case.[45] During the July 1924 Democratic Convention, McAdoo, the longtime favorite, lost badly; his opponents chanted "Oil! Oil! Oil!" during each of his press appearances.[46]

The city of Los Angeles continuously consumed and absorbed oil money, even when the money was only a mirage. The fantasy of profits without limits allowed a flamboyant scammer named Courtney Chauncey Julian to raise tens of millions of dollars, beginning in 1923, to fund oil wells that would never be drilled. Among the victims were a few local celebrities, including the Hollywood king himself, Louis B. Mayer, but Julian particularly preyed on the thousands of small savers eager to take advantage of the black gold boom.[47] After this scandal broke in 1926, a local columnist commented: "The crooks have taken the money from the fools. What difference does it make? The money remains in Los Angeles. It is helping to build the city."[48] This scam, along with the Teapot Dome scandal, remains one of the major symbols of the "decade of debauchery and collective egoism" that President Franklin Delano Roosevelt denounced during the Great Depression of the 1930s.[49]

Half of California's oil production remained under the control of companies whose capital came from the East Coast.[50] Principal among these was Standard Oil of California (SoCal), created in 1900, when Rockefeller's tankers still frequently rounded Cape Horn to reach San Francisco. SoCal acquired a substantial share of California's well sites beginning in 1920. Later known as Chevron, it remained one of the biggest global players. In his 1927 novel, *Oil!*, Upton Sinclair was one of the first writers to describe the parasitic growth of the City of Angels, born from oil and built, despite its lack of water, by men who came from afar.[51] To others, the city evoked the image of Laputa, the city in the clouds of *Gulliver's Travels*.[52] LA was not a pipe dream, however; it had ten times the production level Pennsylvania ever attained. California more than offset the oldest US oil regions' production decline. As spectacular and unexpected as it was, though, the thundering Californian oil boom was not the only factor stimulating the growth of the US machine. In Texas and Oklahoma, the oil industry continued to enjoy robust progress thanks to far greater resources than anyone had dreamed of or planned. The spiral of energy and complexity (of opulence and ingenuity, of nature's generosity and human excess) was accelerating once again. Drilling accuracy was radically enhanced thanks to the first seismographic tests, which revealed, through great dynamite blasts, the location and size of crude oil reserves. Extraction rates increased with the emergence of the first modern oil pumps, called nodding donkeys for their resemblance to bobbing donkey heads. In 1927, Cyrus Avery, a businessman who made his fortune from Tulsa black gold, began constructing Route 66: The "mother road" (nicknamed by John Steinbeck in *The Grapes of Wrath*) soon connected Chicago to Los Angeles via Tulsa, the new oil capitol of the world.[53] Dotted with a network of gasoline pumps along its 4,000-kilometer journey and covered end to end by an extraordinary amount of bitumen extracted and refined close by, it wound from Illinois to California, through Kansas and Oklahoma, a catalyst for the development of rural America. It initiated the process by which automobiles and trucks supplanted the railway system in the United States.

Petroleum on the American market was like honey on an anthill. Total US crude oil extractions doubled in the 1920s. This considerable amount was still not enough to supply the explosive growth of the US economy, which extracted but also consumed nearly three-quarters of the world's crude oil production. US drilling alone was barely enough to quench the energy thirst of the rapidly expanding US population, whose average income was rising and quickly surpassed the British citizen's.[54] Oil consumption had begun to exceed the consumption of coal, whose market share had slid from 72 percent to 58 percent in a decade. From its beginning, the roots of the American middle class were steeped in oil; four-fifths of the world's vehicles were located in the United States. Would this have happened without the opulence flowing up from under Southern California's desert floor? Were the American people responsible for their status as the globe's leading nation, or was it the American land, with its mountains and its minerals, its plains and—first and indispensably—its hydrocarbons, that made it the richest country in the world? In 1926, President Calvin Coolidge, who occupied the White House during most of the Roaring Twenties, said, "It is even probable that the supremacy of nations may be determined by the possession of available petroleum and its products."[55] When he wrote those lines, the United States was just beginning to reveal its irrepressible need for foreign oil.

—◂ SEVEN ▸—

Birth of a Petrol-Nation: Iraq

"No nation should seek to extend its polity over any other nation or people," declared President Woodrow Wilson, who believed that the imperialist ambitions of each warring party were equally responsible for their conflict. It was, however, the American economy's ever-increasing thirst for oil that pushed Washington to interfere in France and the United Kingdom's postwar wrangle over splitting up the crude oil of Mesopotamia.

The Ottoman Empire's Spoils of War

The haggling centered around an agreement signed in London on December 21, 1915, between François Georges Picot, a skinny French diplomat with a reedy voice who was convinced of the "civilizing mission" of his country, and an English diplomat from the highest aristocracy, Sir Mark Sykes. Together, Sykes and Picot plotted out the division of the Middle East, drawing a line on a map that ultimately determined the equilibrium and disequilibrium of this region of the world. The line divided up the best lands of the Ottoman Empire. It began at the port of Haifa, Palestine, and ended at Persia's border, between the cities of Mosul and Kirkuk, the location of the "lake of petroleum" that the Germans desired before the war. The "Red" zone, south of the line and stretching to the Arabian desert, was promised to the British Crown. North of the line, a "Blue" area, encompassing Syria, a section south of what is now Turkey, Lebanon, and the region of Mosul (a city that Sir Sykes hated), went to France.[1] In order to avoid sharing control of Palestine, the English decided to support the young Zionist cause to create a buffer zone that blocked access to the Suez Canal, where oil was the main freight. According to its author, English diplomat Herbert Samuel, this strategy also served to win the support of the two million Jews living in the United States and thus put pressure on Washington, which was still refusing to enter the war.[2*] The agreement between Britain's Whitehall and the French Quai d'Orsay was finalized on May 16, 1916.

The following month, Hussein, the wily sharif of Mecca, head of the Hashemite dynasty, attempted to incite the Holy City's Arab population against its Turkish occupiers. On October 16, an English intelligence officer, barely twenty-eight years old, arrived in Jeddah and set off into the Arabian desert five days later, full of uncertainties. He began talks with the Hashemites to organize the uprising of the Bedouins from the Hijaz region of Mecca. This eccentric young officer, Thomas Edward Lawrence, was quick to aspire to the role of liberator of the Arab people and was treated like a romantic hero by the Anglo-Saxon press, which dubbed him Lawrence of Arabia. For the time being, T. E. Lawrence assured Sharif Hussein, as well as Emir Faisal, Hussein's most capable son (or at least the one who had been least compromised with the Turks), that the English only wanted to help the Arabs become independent. A British delegation used similar language with a fiery young Arabian prince, sworn enemy of the Hashemites: Abdulaziz Al Saud. The existence of the Sykes-Picot agreement was kept secret. It was only later disclosed to chancelleries of the Russian and American allies. But after the capture of Winter Palace in Petrograd in October 1917, the text was published by the Bolsheviks in the columns of the *Pravda*. To soften Emir Faisal, and continue to use the Bedouin tribes in the war against the Turks, Britain's Arab bureau in Cairo opened its purse strings more widely to the Hashemites, and promised them a throne in Damascus.

On October 3, 1918, Turkish troops abandoned Damascus, and Faisal (encouraged by T. E. Lawrence) entered the old Arab capital ahead of the regular British troops. The Arab leaders who had fought for the British expected the latter to keep their promise. Three days later, on a visit to Paris, British prime minister Lloyd George declared his intention to retract the Sykes-Picot agreement, not because doing so would fulfill their promise of Arab independence but to renege on Britain's promise to give France the city of Mosul, which was under British control. This sudden interest in the northern Mesopotamian city was sparked by a Royal Navy memorandum sent to the prime minister, declaring that the steppes that surrounded Mosul contained "the largest undeveloped resources [of oil] at present known to the world."[3]

* This strategy was confirmed on November 2, 1917, by the Balfour Declaration, an open letter addressed by the British minister of foreign affairs and Lord Arthur Balfour to Baron Lionel de Rothschild, by which London declared itself favorable to the establishment of a Jewish national home in Palestine.

The admiral announced: "The Power that controls the oil lands of Persia and Mesopotamia will control the source of supply of the majority of the liquid fuel of the future."[4] Lloyd George's cabinet secretary insisted: Securing access to such a source of crude oil before "the next war" constituted a "first class British War Aim."[5] As the allies prepared to carve up the Ottoman Empire, the eternal flames that burned around Mosul and Kirkuk fueled their greed.

On November 9, 1918, a week after the Ottoman Empire's capitulation, Paris and London signed the Anglo-French Declaration, which stipulated that the two capitals lend their support to future governments promoting free "choice of the indigenous populations." But London indicated that its support concerning Germany's restitution of Alsace-Lorraine (not to mention France's claim on Ruhr), an ultrasensitive question for Paris, was in no way guaranteed if ever France refused to make concessions about Mosul. The two richest treasures of the First World War were directly related to energy, and Downing Street did not hesitate to use the coal of Lorraine and Ruhr as leverage to gain the black gold of the Ottoman Empire. During a visit to London on December 1, when Georges Clemenceau asked Lloyd George what he wanted, the latter responded, "I want Mosul."

"You shall have it. Anything else?" replied the French government's leader.

"Yes, I want Jerusalem, too."

"You shall have it," promised Clemenceau, but indicated that his minister of foreign affairs, Stéphen Pichon, might "make difficulties about Mosul."[6]

With its military resources exhausted, France could not occupy distant northern Mesopotamia. The following month, Quai d'Orsay transmitted a memorandum to the British confirming the cessation of French claims on the city. But the French called for equal access to oil in Mesopotamia, where not a drop had yet been extracted. The lessons of war were clear, and in their aftermath, access to black gold became a major strategic priority for France. In March 1919, Lloyd George resumed negotiations in Paris. He was accompanied by Lord Arthur Balfour, the secretary for foreign affairs. T. E. Lawrence was also there. But Lawrence of Arabia, who was now becoming world famous, was promptly sidelined by the English diplomats. Spotting Lloyd George and Balfour sitting under a balcony of his hotel, Lawrence of Arabia retaliated by unraveling rolls of toilet paper onto them.[7]

Reasons for discord between France and the United Kingdom remained numerous, and negotiations were threatened several times. The mistrust led former allies Clemenceau and Lloyd George to heated altercations, a common occurrence throughout the two rival colonial powers' tumultuous history. Claiming he could break the deadlock, American president Woodrow Wilson created a commission to investigate the desires of the Turkish empire's people. When the English discovered that one of the commissioners Washington sent had been a Standard Oil representative before the war, they discouraged the rest of the commission from getting involved in Mesopotamia. London then transmitted testimonies that naturally emphasized the affection that inhabitants from Basra to Baghdad felt toward their British occupiers.[8] In Syria, as in Mesopotamia during this time, pockets of independence movements organized and grew impatient. In May 1919, the Arabs brandished the Anglo-French Declaration and publicly demanded that it be respected. Six months later, Arnold Wilson, the British Empire's commissioner to Mesopotamia, forcefully declared the declaration "moonshine."[9]

Mesopotamia's oil was England's loot. In September 1919, a British petroleum banker, Edward MacKay Edgar, noted: "All the known oil fields, all the likely and probable fields outside of the United States itself, are in British hands or under British management or control, or financed by British capital."[10] Sir Mackay Edgar then lobbied in the press for British Parliament to approve an increase in appropriations for Mesopotamia's occupation. He emphasized the prevailing pessimism among geologists from Washington. Then he rejoiced, a little quickly: "The time . . . is, indeed, well in sight, when the United States . . . will be nearing the end of some of its available stocks of raw materials on which her industrial supremacy has been largely built. America is running through her stores of domestic oil and is forced to look abroad for future reserves."

London Takes the Matter in Hand: The Creation of the State of Iraq

The April 1920 conference in San Remo, on the Italian Riviera, gave France a mandate that placed Syria and Lebanon under its authority. The British mandate over Palestine and Mesopotamia was also confirmed. A specific "agreement on petroleum" was signed on April 24 by Lloyd George and Alexandre Millerand, minister of foreign affairs and president of the French council.[11] In accordance with the prewar arrangement, nearly half of the shares of the Turkish Petroleum Company (TPC)—which had nothing to do with Turkey—returned to the British national company, the Anglo-Persian Oil Company. Royal Dutch Shell retained a quarter of the capital. Lloyd George consented to yield Germany's shares in the TPC, sequestered by London at the beginning of the war, to France—its first direct access to a potential crude oil source. Once again, Armenian trader Calouste Gulbenkian played a decisive role. He convinced his friend Henri Deterding, in need of an ally and protector, to introduce Royal Dutch Shell to France. To foster the alliance between France and Shell, Gulbenkian had the support of French senator Henry Bérenger, former editor and journalist with an elegant Gallic moustache, appointed "commissioner general of gas and combustibles" by Clemenceau shortly before the war's end. Since the war, Calouste Gulbenkian had been in charge of representing Shell's Paris negotiations for the supply of French oil; his son Nubar, attached to the Ministry of Supply, wrote the checks.[12] The agreements signed in San Remo also earmarked 20 percent of the profits for the Mesopotamians, but this provision was annulled before the first drilling project.[13] Standard Oil of New Jersey counterattacked as soon as it caught wind of this horse-trading.

In May 1921, several senior industry representatives met in Washington with the secretary of commerce and future president Herbert Hoover, who assured them that although the White House could not represent any single American company, the government pledged to intervene on behalf of a group of companies.[14] So Walter Teagle, head of Standard Oil of New Jersey, was charged with organizing a consortium of major American petroleum companies. In 1911, the American Petroleum Institute, unofficially organized under the banner of Standard Oil of New Jersey, requested that the government claim its share of Mesopotamia's black gold.[15] American diplomats faced lively protests in London, where they were accused of attempting to dominate the global crude oil market.[16]

Several Arab insurgent movements broke out in the weeks following the San Remo conference. On July 14, 1920, in Damascus, General Henri Gouraud, the one-armed high commissioner of France appointed by Clemenceau, chased Emir Faisal from the throne of Syria on which the French had consented to let him sit just four months earlier. Gouraud, a hero of the Great War and the "colonial party" who considered himself a new crusader, criticized Faisal for his failure to maintain order and accused him of being paid by the English.

For the British, things were progressing less smoothly in Mesopotamia. In the midst of an uprising, the British Crown's representative in Baghdad, Arnold Wilson, who before the war was sent from India to Persia to protect William Knox D'Arcy's oil prospectors, was relieved of his duties for refusing to recognize the seriousness of the situation. In order to preserve his reputation he was then knighted and made responsible for overseeing the Anglo-Persian Oil Company's activities in the Middle East. Soon, one hundred thousand combatants, members of Shiite tribes and Arab Sunni nationalists, allies of Emir Faisal, harassed the seventy thousand British soldiers (primarily Indians sent in as emergency reinforcement).

In London, Winston Churchill, now aviation secretary, authorized the Royal Air Force to assess the possibilities of using mustard gas against the "recalcitrant natives."[17] Churchill stated, "I do not understand this squeamishness about the use of gas," asserting, "I am strongly in favour of using poisoned gas against uncivilised tribes. The moral effect should be so good that the loss of life should be reduced to a minimum. It is not necessary to use only the most deadly gasses: gasses can be used which cause great inconvenience and would spread a lively terror and yet would leave no serious permanent effects on most of those affected."[18] Without the proper equipment in place, the British army finally could not use chemical weapons. Instead, it was content to use its usual tactics: aerial bombardment and conventional punitive torching of villages.[19] T. E. Lawrence was critical of these methods. Equally romantic and pragmatic, the latter wrote in a letter: "Bombing the houses is a patchy 'way' of getting the women and children. By gas attack the whole population of offending districts would be wiped out neatly; and as a method of government it would be no more immoral than the present system."[20]

The insurgency continued until it was curbed in March 1921. It left 8,400 dead on the Arab side and killed 2,300 Indians and English.[21] Until then, the merits of this war for Mesopotamia's black gold were hardly questioned in London. Only its profitability was debated, and the government had justified the war by prioritizing future oil profits.[22] But the *Times* seriously questioned the logic, "to force the British public to provide £40,000,000 a year for Mesopotamian semi-nomads will overtax the patience of the country."[23]

A less wasteful solution was quickly adopted: Grant Mesopotamia's inhabitants the status of a country and give them a king. The borders and institutions of Iraq were drawn up during a symposium held in Cairo, March 13–20, 1921, under the aegis of Winston Churchill and in the presence of T. E. Lawrence. Forty British Middle East experts gathered in the Semiramis Hotel. They were immediately nicknamed the "forty thieves," much to the amusement of Churchill, who visited the pyramids during the tedious discussions, for which the principle results already had been determined.[24] To brighten up the meeting, two lion cubs were brought in from Somaliland. A single Iraqian was present at Iraq's founding: Jafar Pasha, one of Emir Faisal's generals. The English were considering entrusting Faisal, who the French had chased from the throne of Damascus, with another throne, this time in Baghdad. Churchill sent Lloyd George this telegraph: "Feisal offers hope of best and cheapest solution." The country's pacification was entrusted to the Royal Air Force, which was less costly than using ground troops. Winston Churchill, who was trying to build political capital in order to seek high political office, believed using the

Royal Air Force could save Britain £20 million per year. T. E. Lawrence pressed Faisal to return from Mecca, where he was a refugee, and "on no account put anything in the press."[25]

On August 23, 1921, Faisal the First of Iraq sat on the throne the English had carved out for him. In 1922 and 1924, the Royal Air Force was deployed to squelch several Iraqi Kurd rebellions. A majority of the Kurds, who lived in northern Iraq, were denied the right to self-determination, despite commitments made by the August 1920 Treaty of Sèvres, which divided up the Ottoman Empire. The modern state, established on the site of one of civilization's birthplaces, had been created under dubious motives. By design, Iraq was a divided nation, almost an illusion of a nation, invented in order to allow the English, unchallenged, to reign over what lay under the ground within its borders, which had been drawn up by foreigners.

Deprived of Russian oil, Royal Dutch Shell relied completely on the protection of the British government in Iraq, where Shell claimed a quarter of the oil. The Dutch-British company's leaders argued in favor of a merger with the Anglo-Persian Oil Company, which claimed almost half of Turkish oil, but lacked Shell's global trading network. The project garnered many supporters in London. It had the advantage of shifting the control of Shell from the Dutch to the British. Who did Shell call on to defend its interests? None other than Winston Churchill, who was temporarily out of political office. He, who had been Shell's defiant opponent at the inception of the Anglo-Persian Oil Company, charged £10,000 for his lobbyist services. However, the first Labor government in British history blocked what would have been the creation of a mammoth English company capable of prevailing on the totality of black gold in the Middle East.[26]

The question of energy was increasingly important to public opinion because everyone understood that coal extractions, the foundation of British power, were in irreversible decline. Great Britain had 100,000 more miners now than at its historical production peak eleven years earlier, in 1913; in 1924, coal industry workers numbered 1,200,000, and an increasing number of these were now out of work.[27]

Paris and Washington Contest the Loot: Creation of the Compagnie Française des Pétroles, Total's Ancestor

The United Kingdom firmly controlled the future of Iraqi oil. This was an unsatisfactory situation for France, which had to be content with a quarter of the Turkish Petroleum Company, and still less satisfactory for the United States, who were "never backward in coming forward," according to Nubar Gulbenkian, but from the outset, the United States was sidelined from the dealings between Paris and London.[28] The oil industry and the US government were saddled with the fear of an imminent decline in US oil production. In the wake of the treaty of San Remo, the tone of the proceedings between the head of American diplomacy and his British counterpart, Lord Curzon, soured.[29] Again, oil's history took a cryptic path. Even after the creation of Iraq, London suspected France and the United States of encouraging the Turks to seize Mosul, in the hope of obtaining better access to black gold. Turkey supported the Kurdish separatists of northern Mesopotamia against the British troops.

On the basis of spy reports, in February 1922 Churchill spitefully concluded, "So long as the Americans are excluded from participation in Iraq oil, we shall never see the end of our difficulties in the Middle East."[30] In August, the British bequeathed 12 percent of the Turkish Petroleum Company to the Americans, who eventually got double that. The companies who got their foot in the door of Middle East oil were the four most powerful in the United States: Gulf Oil, Texaco, Standard Oil Company of New York (Socony, which became Mobil), and, of course, Standard Oil of New Jersey (the future Exxon, called Jersey Standard or simply Jersey). Allen Dulles, head of American diplomacy in the Middle East and future director of the CIA, followed events very closely. He was in constant contact with Walter Teagle, the young, tireless head of Jersey Standard.[31] Quickly, Mosul became less chaotic. Anxious to banish the specter of a shortage and to consolidate US industrial power, the government urged American companies to take the lead. However, one of them, Gulf Oil, was reluctant to invest in Iraq, a remote and dangerous country. The representative of the Texan company testified: "Representatives of the industry were called to Washington, and told to go out and get it," referring to the Iraqi oil.[32] Gulf Oil and Texaco threw in the towel: During the 1930s, their shares were absorbed by Jersey Standard and Socony.

In order to exploit its 25 percent of the Turkish Petroleum Company, France sought to create its first sizeable oil company, the Compagnie Française des Pétroles (CFP). France rejected the offer, which was communicated via Calouste Gulbenkian, to merge with Royal Dutch Shell. The nationalistic French leader Raymond Poincaré wanted a wholly owned French company. The following year, he entrusted his creation to a finicky polytechnician with an intense gaze, Ernest Mercier. Injured by shrapnel during the war while defending oil fields in Romania, after the armistice Colonel Mercier distinguished himself as an industrialist by taking charge of German factories, and especially by occupying a central role in the organization and development of the French electrical dams and power plants. From its inception, CFP, ancestor of the present-day Total company, was entirely private, although the French government had the power to approve its directors. The Paris branch of Rothschild bank (still a major Shell shareholder) figured among its key shareholders.

Finally, in 1925, an all-inclusive team of twenty-eight geologists (including Frenchman Pierre Viennot), authorized by English, American, Dutch, and French shareholders of the Turkish Petroleum Company, began to search for the best drilling sites.[33] The same year, the government of King Faisal the First of Iraq agreed to grant a concession under conditions imposed by TPC, shortly thereafter renamed Iraq Petroleum Company. The concession was set up to continue until the year 2000. The cancellation of the commitment made at the 1920 conference in San Remo, to reserve 20 percent of the oil for the Iraqi government, caused the dismissal of two of Faisal's ministers.[34] For Western oilmen, Iraqi petroleum served as a model in the Middle East: The shared boards of directors offered nothing voluntarily and made only minor concessions to their host countries. For four decades, the latter were content merely to receive royalties.

The Red Line Agreement: For Iraqi Oil, the Birth of the Colonial Partnership between the Victors of the Great War

On October 14, 1927, about fifteen kilometers from Kirkuk, oil sprang up forcefully, killing two prospectors. The location of the drilling, Baba Gurgur, was near one of the eternal flames known in the region since biblical times. The Iraq Petroleum Company's major shareholders still had, however, a small thorn stuck in their side: Calouste Gulbenkian. Not only did the Turkish Petroleum Company's prewar architect claim his famous "5 percent," but the Armenian businessman, as secretive as he was opinionated, bald, and boasting huge bushy eyebrows, insisted on enforcing a clause in the company original statutes adopted in 1914. This clause stipulated that throughout the whole territory of the Ottoman Empire, none of the partners could acquire new concessions without giving benefits to all the others. Although the Ottoman Empire had ended and Calouste Gulbenkian was at odds with his old accomplice, the autocratic Henri Deterding, he was not deterred. In Paris, from his suite at the Ritz, or from his imposing mansion at 51 Avenue d'Iéna, behind the curtains where he collected youthful mistresses and works of art (including two self-portraits of Rembrandt and an exquisite Diane sculpted by Houdon and acquired by Catherine of Russia, siphoned from the collections of the Ermitage Palace by the Bolsheviks in payment for his services), Mr. Five Percent was doing just fine.[35*] At the end of three years of relentless, subtle, and torturous negotiations, of promises and threats, with his army of lawyers going up against the richest industrial companies in the world, supported by the most powerful governments of the world, he prevailed. The agreement eventually was concluded on July 31, 1928, in Ostend, Belgium, at the Royal Palace Hotel. The elegant Belgian seaside resort was chosen for fiscal reasons, and because it was in neutral territory. The Anglo-Persian Oil Company, Shell, the Compagnie Française des Pétroles, and the consortium of American companies (dominated by Jersey Standard and Socony) each took 23.75 percent of the Iraq Petroleum Company. Gulbenkian retained his 5 percent.

As if that was not enough, the fifty-nine year old, already the richest negotiator in existence, achieved the greatest coup of his prodigious career. He played a major role in imposing this Byzantine equilibrium of powers on London, pushing the Americans and sometimes his French friends, who offered him the Grand Cross of the Legion of Honor in appreciation for his inestimable services.[†] In Ostend, in the course of one of the ultimate arbitration meetings, Gulbenkian is said to have grabbed a red pencil and drawn a line around the former Ottoman Empire, going from the Bosphorus down to the extreme south of the Arabian peninsula, bypassing the small independent emirate of Kuwait. All shareholders were already in agreement: None of them alone, without inviting the others to participate, could invest anywhere in the vast territory that included the biggest known or yet to be discovered oil fields on the planet. By this treaty and to Gulbenkian's great benefit, this Red Line Agreement linked the vital interests and competition of the British, French, and American oil companies around the Persian Gulf. The United States, which, since the end of the war, consistently had called for an "open door" for their companies, "hermetically sealed" the door behind themselves in Iraq, noted Gulbenkian, who furthermore observed with intimate knowledge: "Oilmen are like cats; you can never tell from the sound of them whether they are fighting or making love."[36]

---

* His only son wrote: "It was upon medical advice that he himself had one mistress, of no more than seventeen or eighteen, whom he changed every year until he was eighty." (See Nubar Gulbenkian, *Pantaraxia*, op. cit., p. 38–39.)

† Stubbornly disdaining honor and attention, Calouste Gulbenkian refused, as he again refused after the Second World War, in 1953, to be knighted by the British Crown. It was his son Nubar, far too young to receive the Grand Cross, who was made Legion of Honor commander by Poincaré. For a long time, he remained the youngest recipient of this very high French Republic distinction.

The pipeline route from northern Iraq up to the Mediterranean still remained undecided. France was determined to have its own oil. The CFP argued for the immediate construction of a pipeline along the shortest possible path: that of Syria and Lebanon (under French mandate). The English and American shareholders did not agree, and dragged their feet. The arrival of Iraqi oil threatened to plunge the market barrel price. Iraq was assigned, for the first time, the role of market adjuster for global black gold. The country was called upon to perform this role many times, at its own expense.

In March 1928, the Anglo-Persian Oil Company's director, Sir John Cadman, stated that the pipeline construction was "premature."[37] London demanded that the crude oil passing through its territories be under British army control and be directed to the Palestinian port of Haifa. In Paris, on the other hand, Ministry of War strategists came to the opposite conclusion: With a pipeline through Syria, "If . . . England declared war on us, we could, for a while at least, block their supply of Iraqi Petroleum."[38] It was Walter Teagle, American director of Standard Oil of New Jersey, who finally forced the French diplomats' hand. In August 1930, the Quai d'Orsay tried to intervene with the help of French banker Horace Finaly, director general of Paribas bank, which included the Jersey Standard's French subsidiary among its biggest customers.[39] But Teagle was in no mood to compromise because, in order to facilitate the development of the French petroleum industry, Paris had imposed import quotas over the last two years, and favored the construction of refineries in Le Havre and Marseille, which foreign companies would be forced to use. Abruptly, Teagle responded to Finaly that France should not wait for American companies "to furnish a friendly support in London [while] she harasses them at home."[40]

After offering Faisal a throne in Syria, Paris ended up accepting a compromise.[41] In March 1931, the Iraq Petroleum Company's shareholders came to an agreement: The pipeline construction would start in Kirkuk, splitting off to both Haifa, Palestine, on the English side, and Tripoli, Lebanon, on the French side. The construction of the two longest pipelines ever created at that point in time began by crossing through 1,000 kilometers of desert. American Mack trucks carried thousands of steel tubes coated with a bituminous anticorrosion layer by a special machine, which were then buried by teams of Bedouin workers supervised by foreign foremen. It was the largest construction site in the world at the time.[42] After thirteen years of effort, the victors of the Great War were finally going to be able to touch their most prized loot. On July 25, 1931, in Paris, the French Parliament ratified a convention raising CFP's government capital 35 percent and assigning 40 percent of the voting rights to the administration council. Across the small, dusty old city of Kirkuk, the British governor general built an ostentatious boulevard, a bombastic symbol of the construction cost in Iraq, the subject of great controversy around Westminster parliament. At the beginning of the Second World War, *Time* magazine called the section of the pipeline from the Euphrates to Haifa the "carotid artery of the British Empire."[43]

—◀ EIGHT ▶—

The Majors Band Together: A Secretly Planned Industry Weathers the Great Depression Unscathed

At the same time the Great War victors' oil companies were finally reaching an agreement on the division of Iraqi black gold, the major Anglo-Saxon companies were negotiating a secret agreement of much greater scope: the organization of a worldwide petroleum cartel.

The second half of the 1920s saw a reversal of the state of the oil market: Fear of postwar shortage disappeared. In the United States, while the pace of automobile sales hit their stride by 1925, the massive resources of crude oil in production on American soil (California, Texas, Oklahoma) and in the rest of the world (Iran, Venezuela, Mexico) created unexpected overproduction. Sporadic incursions of large quantities of Soviet oil further aggravated the situation. After the large companies' failed attempt to maintain a united front to resume control of Russian oil at the 1922 conference in Genoa, each of them tried to negotiate privately for export contracts with the new masters of Baku. A price war arose between Shell and Standard Oil of New York, both attempting to sell oil purchased from the USSR on the Indian market ("the soft under-belly of the industry," according to journalist and historian Anthony Sampson).[1] All other large companies were sucked into the battle, which stretched quickly to all continents, as overproduction grew. This overproduction placed companies that extracted the bulk of their crude oil in the United States—namely, Standard Oil—in a difficult position: US oil was more expensive to produce than Mexican or Iranian, because the wells in the United States were on average less productive (and the wages of workers, who were white for the most part, were better than elsewhere).

The As-Is Agreement: A Secret Pact between Major Oil Companies

At the beginning of August 1928, after long months of negotiations and just a few days after the signing of the Red Line Agreement, the heads of the three largest oil companies in the world met in Scotland at an austere castle in the Highlands. The imposing Achnacarry Castle had been rented by an insatiable amateur hunter and knight of the British Crown: Henri Deterding. Walter Teagle of Jersey Standard and Sir John Cadman, president of the Anglo-Persian Oil Company, gathered with him for a pheasant hunt. Hunting was really just a pretext though, like Jekyll Island in 1911: What they were about to decide was integral to modern economic history. Despite precautions, reporters caught wind of the meeting. One of them marveled at the "impenetrable fortress which harbors one of the most interesting groups of silent personalities in the world."[2] Walter Teagle later complained that the pheasant hunt turned out to be "lousy."[3] But the trophy that the directors of Shell and future companies Exxon and BP truly sought was more important. Deterding, Teagle, and Cadman realized the dream that John D. Rockefeller and Marcus Samuel failed to achieve more than thirty years earlier.[4] They were soon joined by the heads of other large American companies, including William L. Mellon, president of Gulf Oil. By a tentative agreement, soberly titled the "pool association," the heads of Shell, Jersey Standard, and the Anglo-Persian Oil Company agreed to exploit the worldwide black gold market. Endorsed on September 17, 1928, without a single signature, the Achnacarry agreement expanded to include the fifteen largest American companies and to encompass the quasi-totality of world crude production outside of the USSR. Three subsequent agreements, adopted between 1930 and 1934, clarified the provisions in detail. The existence of the agreement of Achnacarry was known to the British government (main shareholder of the Anglo-Persian Oil Company) and was certainly not unknown to the American administrations that followed (according to a public inquiry in the United States in 1952).*

The Achnacarry agreement was also known under the name "As-Is" because it froze the major Anglo-Saxon market shares and production levels all around the globe. The purpose: to guarantee profits by maintaining steady prices, avoiding overproduction, and resisting external competition. The conspirators justified their conspiracy in the name of a "responsibility" (the monstrous burden, in their eyes, of keeping the economy afloat for the hundreds of millions of humans who consume oil): The original text of the agreement of Achnacarry

---

* The report of this investigation is titled *The International Petroleum Cartel* (op. cit.).

notes that "the petroleum industry has not of late years earned a return on its investment sufficient to enable it to continue to carry in the future the burden and responsibilities placed upon it in the public's interest."[5]

The draft established at Achnacarry allowed members of the cartel to agree on almost everything: production, prices, construction of new infrastructure, redemptions, and even marketing. Its first measure froze the balance of power between the large companies in a status quo that continued during the following four decades. It stipulated "the acceptance by the units of their present volume of business and their proportion of any future increase in consumption."[6] The members of the oligopoly then agreed on the construction of new refineries, the distribution of production and the division of markets in each geographical area. The precise instructions, adopted between 1930 and 1934, allowed the determination of production quotas and set the price of crude. A mechanism for rigging prices, called "phantom freight," built a protective dike around the US industry: Thanks to this system, as clever as it was indistinguishable, the courses of crude coming from Venezuela or Iran were adjusted to the higher production costs of the barrels from Texas.[7] On the selling side, local representatives of the cartel companies organized monthly meetings between them; independent auditors ensured that everyone respected their share of the commitments.[8] Consultations were required prior to the purchase of external competitors, to arrange the way of integrating them into the As-Is Agreement.[9] The strategy of the hidden alliance, the same way that John D. Rockefeller left his competitors in the dust half a century earlier, this time was extended to include the entire capitalist world.

Cartels were common at that time in many industries on both sides of the Atlantic, and none were without flaws or clashes or betrayals. But none had the scope of the cartel of Achnacarry. None benefited from its degree of refinement. A full-time secretariat was established and divided into two committees, one in London, the other in New York.[10] The last enactment of the cartel was activated on January 1, 1934, designed to eliminate any real competition on the retail selling price and to strictly limit the use of advertising. Among the forms of advertising to be "eliminated" or "reduced" or "kept within reasonable limits" included display panels and signs, inserts in newspapers, gas station signs, sponsorships of race car drivers, and even gadgets such as cigarette lighters offered in service stations.[11] No one could attempt to destabilize the delicate balance of the cartel through advertising: Bias that exacerbated competition was not permitted.

An eternal source of big capital, able to regenerate and multiply constantly, oil often costs less to extract than water. With the Achnacarry agreement, the major Anglo-Saxon companies took full advantage of the divine generosity of black gold. Faced with a natural abundance of oil, the proposition of competition was not attractive, it could only lead to a fight to the death: a luxury that none of the majors were willing to risk. The cartel, following a human norm, chose the path of least resistance.

Henri Deterding often liked to quote the old Dutch proverb: "De eendracht maakt macht" ("The union is strength").[12] The philosophy of the bosses of the largest oil companies had remained rigorously consistent since the time when John D. Rockefeller and Marcus Samuel scoffed at the "ruinous competition." In the text of its preamble, the agreement created in the middle of August 1928 behind the walls of Achnacarry denounced "excessive" and "destructive" competition.[13] This competition had "resulted in the tremendous overproduction of today," which undermined the positions of the oligarchs. The masters of black gold anticipated the crisis of 1929 brilliantly.

The Crisis of 1929: A Trap of Abundance and Greed

The terrifying economic and financial crisis that broke out in October 1929 was, essentially, a crisis of overproduction. In the United States, rapid productivity gains thanks to the abundance of oil and electrification, as well as the organization of mass production, caused a 33 percent increase in industrial production between January 1920 and January 1929. At the same time, household income rose by at least 20 percent, making the American population the richest in the world.[14] During the same period, Wall Street, the Dow Jones index of industrial values, rose 200 percent. The avid confidence in a bright future—which owed without a doubt much of its candor to the scoop of abundant energy—encouraged the United States to achieve a global level of debt that climbed to 300 percent of GDP, a level that wouldn't be attained again until the 2000s.[15] The banks loaned ten times more than what they had in their coffers. These were the first instances of consumer credit, and people were borrowing to buy shares. Values on the stock exchange increased faster than the profits of the companies, profits were rising faster than the production levels of plants, production increased faster than the wages of workers.[16] The staggering progression of the working class in the United States was aroused and overtaken by an industrial machine that was physically capable of speeding up even faster, in a thrust of great energy surpassed only by the frenzy of desire that it generated in appetites and investment portfolios. When the Wall Street house of cards collapsed on

October 24, 1929, President Herbert Hoover was taken completely by surprise. Arriving in the White House seven months earlier without ever having been elected previously, the former secretary of commerce for presidents Harding and Coolidge was in perfect continuity with the policies of his predecessors: absolute priority to business. Orphaned at the age of ten, this self-made man, hoisted up to the summit of elite American industrialists, Hoover was perhaps the only head of state who was a trained geologist and mining engineer (one of the first graduates of the University of California Stanford in 1895). Before his entry into politics and up to the First World War, Hoover traveled the globe for a variety of large mining and oil companies. A high-level prospector, he sold an oil field located at the foot of the Peruvian Andes to Walter Teagle, the future president of the Standard Oil.[17]

President Hoover experienced some misgivings in the face of the natural consequences the Wall Street crash—layoffs, expropriations, and mass misery. Leading the hardline approach within his government was Andrew W. Mellon, irremovable secretary of the treasury, founder of Gulf Oil. The more moderate Hoover described Mellon's approach as a single-minded quest to "liquidate labor, liquidate stocks, liquidate the farmers, liquidate real estate."[18] A powerful banker and industrialist (involved in the birth of the energy-intensive American aluminum company Alcoa), Mellon believed that the economy should be bled-out, without assistance of any kind.* The "invisible hand" of the market must fight hard in order to accomplish the wholesome work that will allow the strong to survive: This is what Andrew Mellon, the main shareholder of Gulf Oil, preached; his nephew, William L. Mellon, president of Gulf, was one of the conspirators at the Achnacarry Castle.

During the Worst of the Great Depression, Texas Sees the Most Savage of the Black Gold Rushes

No industry suffered more from overproduction as phenomenal as that of Big Oil during the crash. To prevent the aggravation of this overproduction, President Hoover, on April 15, 1930, enacted a prohibition on the exploitation of shale.[19] The causes of oversupplying the market with black gold stemmed not only from an abundance of crude extracted in the United States and elsewhere since the 1920s; incomparable profits were also a big driver. From the fields of Texas to California, at the periphery of the most fertile areas, which the large companies almost always ended up owning, the small, independent,

* Alcoa is still one of the world leaders in aluminum, along with Rio Tinto.

isolated "wildcat" wells, fared better than they would if they were pumping the leftovers of the majors.

Many prospectors did not drill vertically under the concession. The practice of directional, or slant, drilling, developed at the beginning of the century, exploded in the United States in the 1920s. The first horizontal drilling was accomplished in Texas in 1929.[20] The purpose was to reach the more productive pockets of crude in nearby ground, like putting a straw in a neighbor's glass. It is difficult to know what is happening under the ground and the direction of a drill once horizontal drilling is underway. Of course, these oblique drillings were often at the origin of violent altercations between oil prospectors. But their illegality was difficult to demonstrate, since the plots were narrow and tight against each other in Los Angeles and Texas. Under US law, the "rule of capture" prevailed most of the time. This legal principle was established in the oil regions in 1889 by a Pennsylvania supreme court judge. The ruling came from old English customary laws that governed the rights of hunting: As with game for the lords of the Middle Ages, black gold belonged, in the United States, to those who were able to extract it from their land.[21] In Texas, surveys assessed the tens perhaps even hundreds of millions of dollars' worth of oil stolen in this way over the years.[22]

The Great Depression that swept across the capitalist world from 1930 only aggravated this spurious crude drilling: With the barrel price collapsing at the same time as demand fell, desperate wildcatters attempted to scrape together an income any way they could. Overproduction took on even more magnitude when in October 1930 a new "elephant"—a giant oil field—appeared in a Texas ravaged by poverty and drought. Its discovery caused the worst of the chaotic rushes toward black gold ever seen in the United States.

In the middle of the pines, on October 3, 1930, near the Texas-Louisiana border, a seventy-year-old man, hunched back and miserable, named Columbus Joiner, discovered a monstrous gusher in a place where all the geologists had concluded that there could not have been a drop of black gold. The old wildcatter followed the advice of a wacky charlatan doctor (yet another in the history of oil) who was roaming the routes of the South selling his potions of rock oil. By charming gullible widows, Columbus Joiner raised the funds to pay for third-rate drilling equipment and a group of vagabonds to do the work. Daisy Bradford 3 (the prodigious well was named after the widowed farmer living there) was the first foray into one of the most enormous petroleum fields ever discovered: the 90 kilometer long East Texas field called the "Black Giant." Just as Edwin Drake before him, Columbus Joiner, nicknamed "Dad" Joiner, ended up penniless, his oil field bought at a discount with gambling money by the man who became the richest millionaire in Texas, Haroldson Lafayette Hunt.*

The boom of East Texas launched by "Dad" Joiner turned quickly into anarchy. In its epicenter, the small dusty town of Kilgore, lands were split up into tiny plots only a few meters apart. Drilling rigs were literally touching each other, each one hastening to pump crude as quickly as possible, at the risk of crashing the flow of all wells in the area. Inevitably, many wells dried up at the end of a few months, due to lack of pressure. Despite this, many more were instantly erected a little further away. The "rule of capture" unleashed an unbridled, even insane exploitation of crude by July 1931; nine months after Daisy Bradford 3, 1,200 wells produced more than 900,000 barrels of crude oil per day, one-sixth of the American production total.[23]

The abundance of black gold in the United States was undermining the industrial infrastructure erected on the oil fields. The crisis was quick to amplify, with the price of the barrel collapsing completely, from $1 in 1930 to sometimes less than 30 cents. The crude of Kilgore, skimmed off under suffocating and murky conditions, provided a poor-quality fuel called Eastex that sold well even during those tough times. Oil is no less vital in a time of crisis: Hundreds of thousands of Americans fled the economic crisis and drought, following the California mirage, via Route 66.

With the Blessing of Standard Oil, US Oil Becomes an Arranged Economy

The political authorities of Texas, the ultimate frontier of free enterprise, decided to intervene. A commission that supervised rail traffic attempted to impose limits on the production of crude oil. The Railroad Commission of Texas was created in 1891 by the governor, "Big Jim" Hogg, the scourge of Standard Oil of Texas, in order to regain control of rail traffic. This ordinarily lethargic commission, normally limited to organizing the segregation between whites and people of color in railcars, found itself propelled into the eye of the hurricane.[24] The obscure council soon became the authority overseeing, during the following four decades, quota compliance of oil production across the United States.

But, at the beginning of the summer of 1931, the Railroad Commission of Texas was still unable to enforce the limits favored by Walter Teagle, the head

* At the end of the 1970s , H. L. Hunt was the inspiration for the character of J. R. Ewing in the famous series *Dallas*. (See Alex Hannaford, "Dallas: The Feuding Family that Inspired the TV Series," *The Telegraph*, September 5, 2012.)

of Jersey Standard, and William Stamps Farish II, Humble's president. Humble Oil, an affiliate of Standard Oil of Texas, threatened to shut down wells in East Texas until the barrel price came back up. In August, production exceeded one million barrels per day. Passions flared and there was talk of blasting noncompliant wells with dynamite. Finally, the governor of Texas intervened, at the request of producers in favor of quotas. The governor, a wealthy member of the Democratic Party, Ross S. Sterling, was none other than the founder and ex-president of Humble Oil. He declared East Texas to be in a "state of insurrection" and "open rebellion."[25] On August 17 he sent in the National Guard and the Texas Rangers to restore order and close down the 1,162 wells that were flagged within the spine of the "Black Giant."[26] Some of the wells were allowed to reopen in February 1932. Ten months later, Ross Sterling signed a law adopted by the Congress of Texas, which formally authorized the Railroad Commission of Texas to impose quotas. By the end of that year, the barrel price was already back to $1. While the rest of the economy floundered, the large American oil companies began to reconnect with their usual comfortable profits.[27] So much for the "rule of capture" and the dreams of the more humble wildcatters. The official prerogatives of the Railroad Commission of Texas fill appeased the bosses of Big Oil: They completed, in the United States, the provisions of the secret agreement of Achnacarry, which spread across the rest of the capitalist world.*

On March 4, 1933, Democrat Franklin Delano Roosevelt moved into the White House. Not only did the administration of the New Deal approve the quotas of the Railroad Commission of Texas, but it also ensured that they were fully implemented and pushed their extension to the rest American production. Unlike his uncle Teddy Roosevelt, who despised Standard Oil, Franklin D. Roosevelt did not hesitate to support the powerful dynasties of US oil that were carrying on the work of John D. Rockefeller (free of any active role, the latter died peacefully in 1937 at the age of ninety-seven). Roosevelt appointed the president of Standard Oil of New Jersey, Walter Teagle, to the advisory committee to implement the keystone of the New Deal, the National Industrial Recovery Act, the anti overproduction act adopted on June 16, 1933. Section 9 of the National Industrial Recovery Act authorized Washington to organize the transport of oil and to take control of any pipeline in violation of the law. Roosevelt entrusted the important position of secretary of the interior to Harold Ickes, a tenacious lawyer from Chicago, who was not an insider. But, upon taking office, Ickes appealed to James Moffett, the vice

* "Big Oil" is the nickname given to the more or less unified entity, formed by the largest oil companies in the United States.

president of Standard Oil of New Jersey, who opened the doors of industry for him. Moffett helped Ickes in the particularly delicate task of convincing the ferocious independent oilmen to allow Washington to take charge.[28] Thanks to the active support of Jersey Standard, Ickes's men were able to investigate contraband oil sold by producers who refused to respect the quotas imposed by Washington. This fight against so called "hot oil" was similar to the fight against alcohol smuggling in the prohibition era. Federal officers inspected wells, tanks, gauges, and even dug up pipelines. In 1935, these efforts succeeded, and things began to return to order: The price of a barrel was firmly established back at the $1 level.[29] Expanded across the United States by the New Deal, the authority of the Railroad Commission of Texas survived the Roosevelt administration and governed production up to the eve of the first oil shock in 1973. A quarter of a century after the New Deal, the principles on which the Railroad Commission of Texas was founded directly inspired the fiercest competitors that US industry ever faced: the founders of OPEC, the Organization of the Petroleum Exporting Countries.

With the New Deal, the Petroleum Networks Establish Roots in Washington

Franklin Roosevelt did more than just rely on his de facto allies, the heads of the major oil companies, in an effort to gain control of the oil industry and stop its independent rogue producers. Through men such as Walter Teagle and James Moffett, Roosevelt also aligned with the heirs of the aristocracy in the opaque industrial and financial establishment network of Wall Street.

James Moffett was the son of a vice president of Standard Oil of New Jersey and became the youngest vice president of the company in 1924.[30] He accompanied Walter Teagle at the negotiating table behind the walls of Achnacarry Castle.[31] Moffett was rare in his class; he was a rebel who didn't hesitate to act on his own accord. In July 1933, shortly after having been recruited by Harold Ickes to join the Roosevelt administration, Moffett was asked to resign from Jersey Standard's board of directors by Walter Teagle and William Stamps Farish II. He was probably too wedded to the cause of the New Deal.[32] Moffett remained an important asset for Roosevelt, who named him head of his administration's housing program. Despite being removed from the board of Jersey Standard, Moffett remained a shareholder in the main offshoot of the Rockefeller empire and benefited from its significant outreach. Around Christmastime in 1933, for example, at the height of the Great Depression, he threw lavish balls where gifts were distributed to a few needy children, aboard his yacht and at his Palm Beach, Florida, home. Moffett invited more than a hundred guests, including the wife of newspaper magnate William Randolph Hearst, members of the Gould clan, and even the influential financier, Joseph Kennedy, father of the future president.[33]

Walter Teagle, the boss of Jersey Standard at age fifty-five, was the grandson of Maurice Clark, the first associate of John D. Rockefeller, who prompted him to engage in the petroleum industry. Renowned for his autocratic demeanor, Teagle learned the value of the family secret in the black gold business at a very young age. His father was a partner in one of the main competitors of Standard Oil, but in reality secretly associated with it. The competitor, Scofield, Schurmer & Teagle, included among its associates William Scofield, father-in-law of John D. Rockefeller's youngest brother, Frank: two men that the founder of Standard Oil hated profusely. Frank, the youngest son of "Devil" Bill Rockefeller, was regarded as a lame duck by his elder brother, who criticized him in front of the press at the time of the Cleveland Massacre.[34] As for William Scofield, he persuaded Scofield, Schurmer & Teagle to cheat on the secret quotas, forcing Standard Oil to attack legally. This attack created backlash in the Cleveland court: In 1880, the company of the father of Walter Teagle caused Standard Oil to face its first legal hindrance when the judge ruled that by assigning production limits to a competitor, Standard Oil had executed a contract in restraint of trade. Initiated to the intricacies of the Rockefeller empire, the brilliant Walter Teagle knew the path to the best seat in the building at 26 Broadway. The president of the Standard Oil of New Jersey was the most powerful of the American oil bosses and, without a doubt, the most important person Roosevelt had to come to terms with. Smiling, inflexible, and obese, Walter Teagle's career escalated in nine years, until 1942, when he was the head of Standard Oil and director of the Federal Reserve Bank of New York, the main branch of the central bank of the United States.

The successive generations of the Teagle, Moffett, Farish, Mellon, and Rockefeller families occupied a choice place at the summit of WASP (white Anglo-Saxon Protestant) society. In the 1930s this small society, where they married among themselves and where intruders were kept at bay, was the breeding ground of a new emerging dynasty of oil: the Bush family. More on that later. "The trouble with this country is that you can't win an election without the oil bloc, and you can't govern with it," complained President Franklin Roosevelt.[35]

In the year 2000, the governor of Texas, George W. Bush, snickered with a crowd of wealthy presidential campaign supporters by telling them, "Some call you the elite, I call you my base."[36] The need to cozy up to the heirs of the industrial and financial powers of oil never ceased to forcefully exercise its constraints on American politics.

Crisis and Resurrection of Abundance of the Industrial Sector in the Land of Black Gold; Defiance

Because black gold was so abundant, the largest oil companies, paragons of the successes of free enterprise, decided to regulate its flow with the Achnacarry agreement and the Railroad Commission of Texas. A day would surely come when oil would become rare, and where consequently the largest companies would have less interest in avoiding competition.

During the 1930s, Harold Ickes, secretary to the interior under Franklin Roosevelt, was one of those who dared to confront the possibility of a future depletion of black gold, just as the problem of overproduction appeared to be under control. In leading the fight against the smuggling of "hot oil," the "oil czar," as he was called, had something else in mind than the restoration of Big Oil profits. In a magazine article published in 1935, the short lawyer with round glasses and a notoriously rank demeanor warned: "Without oil, American civilization as we know it could not exist."

Ickes was obsessed with the need to preserve the windfall from US oil. "Our total dependence on oil is undeniable. . . . It is true that there is a reasonable expectation of discoveries of new oil pools. That is taken for granted. If this were not true, we would indeed be in a sad state. But there will sometime be an end to new discoveries of oil, just as the finish of our present known oil reserves is in sight. Our low-cost oil resources, known and unknown, may last for ten years or twenty years or even for thirty years, but they will not last indefinitely. Inevitably there will come a day of scarcity of cheap oil. We must, with prudent thrift, conserve our rapidly dwindling low-cost reserves."[37]

When Ickes wrote that article, the American economy was emerging from the Great Depression and the collective mood was no longer filled with pessimism. His warnings were ignored. Ickes, champion of the public, was widely despised within the American black gold industry. No doubt this detestation was hypocritical, or in any case unfair, because the quotas set by the Railroad Commission of Texas, to which Washington granted legal status and whose federal agents imposed compliance, were doing the bidding of the majors. The quotas represented the best guarantee of the majors' continued domination and sustainability within the territory of the United States. They facilitated more than just the resumption of US demand for crude, which was back to its 1930 level.[38] They also prevented the return of overproduction, while sealing the door to any new unwanted competitor. Once the nightmarish bleed of the Great Depression was complete, the fuel of growth was still there, available under the earth. The valves were reopened, and the imperturbable up and down of nodding donkeys resumed as if nothing had happened.

The crisis left deep traces in the collective memory—perhaps nowhere more than in Los Angeles, at the end of Route 66. The brutal stop in the miracle of the City of Angels gave birth to a dark, new genre around Hollywood.* Its most popular writer, Raymond Chandler, began writing after being laid off in 1932 from a major Signal Hill oil company, where he was promoted from accounting officer to vice president.

Oil permeated daily life, allowing movement of people and goods with unmatched power and flexibility. Little by little the rich earth was covered with petroleum. Roads were continuously added and coated with asphalt, right up to the outskirts of the Forbidden City in Beijing, where the luxury cars of the lords of war chased the rickshaws. Much better than the railways, the roads increased the possibilities for exploitation of the land exponentially. They deployed a master network of the modern economy, where long strips of concrete covered with asphalt determined the future of businesses, neighborhoods, cities, entire regions; oil established the conditions of their "worth." The United States possessed the abundance of energy necessary for cooking great quantities of cement, and for transporting billions of cubic meters of concrete and tens of thousands of workers who completed major projects with the New Deal, like the erection of large hydroelectric dams. On September 30, 1935, the fantastic Boulder Dam, later renamed Hoover Dam, was inaugurated by Franklin Roosevelt and Harold Ickes at the border of Nevada and Arizona, not far from the small village of Las Vegas; at least 112 workers died at this enormous worksite, led by a family company from San Francisco that was called upon to erect many other mega-energy infrastructures around the world and to take a leading role in the implementation of the American geostrategy: Bechtel. In retaining the Colorado River and in capturing its energy, the Hoover Dam made "the desert bloom" and brought a powerful new breath into the development of the American Southwest, hitherto largely desolate.†

* For example, Horace McCoy's novel *They Shoot Horses, Don't They?* (1935) tells the story of a man and woman who, at the height of the Great Depression, participate to the point of utter exhaustion in a dance marathon on the beach in Santa Monica, in hopes of winning $1,500.

† Inscribed on a monument honoring the workers who died building the Hoover Dam: "They died to make the desert bloom."

Benito Hoover, Bernard Marx, Polly Trotsky, and Morgana Rothschild, the robotic characters of *Brave New World*, the science fiction novel by Aldous Huxley published in 1932, were linked by a universal faith in the machine that was instilled at birth. Henry Ford was its prophet, and it was this faith, surmised the English author, that would halt each of the great political currents of the twentieth century (capitalist democracy, communism, fascism). The oilman, the man of abundant energy, had begun to falter.

Oilmen Resume Work Around the Shores of the Persian Gulf

The As-Is Agreement of Achnacarry Castle considered almost all possible sources of discord that could arise between the major Anglo-Saxon companies within their web of empire. On the border of this empire, however, skirmishes over carving out new fiefdoms were frequent. Accused by Congress of favoritism toward several companies within his empire, Andrew Mellon hastily left his position as secretary of the treasury at the peak of the crisis, in February 1932, in order avoid impeachment. To save face, Herbert Hoover then appointed him ambassador to London. There, Mellon set out (at the age of seventy-seven) to expand Gulf Oil into a whole new hunting ground for oilmen: the Persian Gulf. In 1927, Gulf Oil repurchased various exploration rights in Kuwait and on the island of Bahrain through an officer in the British army who was born in New Zealand, Frank Holmes, an adventurer who the Arabs remember as the father of oil, "Abu Naft." When geologists deemed oil discovery on the Arab side of the Persian Gulf hopeless, Holmes considered redeeming his concessions with the Anglo-Persian Oil Company, but they refused because their geologists decided that there was no hope of discovering oil on the Arab side of the Persian Gulf. But in 1928, Gulf Oil took part in a consortium preparing to exploit oil in northern Iraq, Iraq Petroleum Company, and found itself bound by the Red Line Agreement: The British strongly objected to the Texas company drilling alone anywhere on the territory of the former Ottoman Empire. Kuwait was outside of the Red Line. But the island of Bahrain was not: Gulf Oil sold its rights to a new entrant in the Persian Gulf, another American company eager to develop new resources: Standard Oil of California. When the American engineers found oil on Bahrain, located off the Saudi coast, on May 31, 1932, the boss of the Anglo-Persian Oil Company, John Cadman, attempted to repair the fatal blunder committed by his geologists. Bahrain's black gold was of poor quality. Better prospects seemed to exist in the emirate of Kuwait, where Gulf Oil eagerly cozied up to the ruling party, and perhaps also in Saudi Arabia, in which British intelligence learned that several American oil majors were interested.

Andrew Mellon, one of the richest men in the United States, did not hesitate to use his position of ambassador to pressure Washington to keep the English out of Kuwait.[39] But Sheikh Ahmad relied on the Royal Navy for the security of his small country. The heir of the Al Sabah family, who had reigned over Kuwait since the eighteenth century, was particularly anxious to find a compromise that would preserve their fortune from pearl trading, which had been declining since a Japanese company developed the production process for the perfect spherical cultured pearl. Therefore, in December 1933, the Americans of Gulf Oil and the British of the Anglo-Persian Oil Company agreed to a joint venture to exploit Kuwaiti oil in equal shares. The world economy was at the bottom of the Great Depression, and the global market for black gold was overstocked. The black gold of Kuwait and Saudi Arabia was waiting. But the oil business required anticipation. The onset of the Second World War revealed the importance of the path opened by these American and British oil companies at the edge of the Persian Gulf. In Persia, John Cadman and the Anglo-Persian Oil Company faced an authoritarian and very determined King Reza Shah, former colonel in the personal guards of the royal family of Iran, who became leader in 1925, founding the doomed Pahlavi dynasty.

Very concerned by the effects of the recession on his subsidies, Reza Shah had the audacity, after lengthy negotiations were unsuccessful, to cancel the concession of the Anglo-Persian Oil Company on November 16, 1932. In five months, at the cost of a few round-trips to Tehran, John Cadman, the inflexible Englishman, coaxed the shah into an amenable agreement. The latter obtained a modest part of the profits (20 percent of dividends, conditional) and granted in exchange a new concession for forty years.

Venezuela, Bolivia, Mexico, and a Defeat for the Majors

Since the First World War and even more so since the end of Bolshevik Russia, Latin America became a decisive playing field for the majors. On its rugged roads, trucks were still rare and unable to go faster and farther than mules; in the absence of an electrical grid, oil remained especially useful for lamps. To the south of the United States, the oil countries of Latin America played a strategic role despite their lack of technology and modernity. Consumption of petroleum products there remained very limited. The capacity of Latin American exports to supply the northern hemisphere with fuel was very strong compared to the United States.

Venezuela turned out to be the crown jewel of Latin America, a jewel that fell into the hands of Walter Teagle. In 1928, Standard Oil of New Jersey acquired control of Creole Petroleum, the first company to have dared to drill wells in the water, developing the very rich Lake Maracaibo (before we called them "black tides," they were neither rare nor minimal, or regarded as tragedies).[40] Jersey Standard then acquired Pan-American Petroleum in 1932, the largest US producer outside the borders of the United States. The security of the As-Is Agreement and a $100,000,000 bargain price for the Pan-American Company eased Jersey Standard out of the 1929 crisis in much better shape than many of its competitors. Pan-American belonged to one of its younger sister companies, Standard Oil of Indiana. Teagle ignored warnings about the risk of another antitrust suit in the United States. Thanks to the purchases of Creole and Pan-American, Jersey Standard became the biggest oil producer in the world during the 1930s, surpassing Shell and preventing the British from dominating the world's oil supply. In 1935, the administrative council of Creole entrusted the reins of the company to a young man of twenty-seven, full of ambition for the future: Nelson Aldrich Rockefeller, the grandson of John D. Rockefeller, named in honor of Nelson Aldrich, his great-maternal father, the Republican senator and founder of the Federal Reserve. His uncle, Winthrop W. Aldrich, became president of Chase National Bank, created in the aftermath of the 1929 crash and instantly the richest bank in the world; Chase held important interests in the oil of Latin America and Rockefeller was the most powerful shareholder.

There was oil in the vicinity of the little known war that happened between 1932 and 1935 in two of the poorest countries in the world, Bolivia and Paraguay: a war which can rightfully be called the most inept of the twentieth century and the most deadly ever in South America. Conducted with the best armaments from Europe and the United States, the Chaco War (also called the "War of Thirst") was fought over control of Chaco Boreal, a semiarid, almost uninhabited region, inhospitable and devoid of all wealth except perhaps, some hoped, black gold. On the Bolivian side, near the area of the fighting, Standard Oil of New Jersey had a handful of drilling sites. Located at the foot of the Andes, far from the sea, those wells produced little results since a lean first discovery in 1924. The area of the wells was not part of the territories officially claimed by the enemy, Paraguay. It was only when the complete collapse of the Bolivian army happened during the summer of 1934 that the Paraguayan army attempted to take control of Standard Oil's Bolivian wells.

However, the fact that Royal Dutch Shell had acquired some modest concessions from Paraguay goes a long way to fuel speculation that the Chaco War was about oil, with Standard Oil financing Bolivia on one side and, on the other side, Shell supporting Paraguay.[41] This rumor was born of reciprocal accusations launched by the belligerents themselves (none of whom assumed responsibility for this senseless war, which killed one hundred thousand), and nourished by the Communist propaganda of the Comintern.[42] In 1950, Communist Chilean poet Pablo Neruda alluded to the Chaco War in his poem "Chant General," dedicated to "Standard Oil Co." In it, he attacked the "obese emperors" and "smiling assassins." Finally, there was scarcely any oil available to fight the Chaco war.

After the war, the Bolivians relinquished control of Chaco in Paraguay and, in 1937, perhaps playing the role of scapegoat in a country ruined by its generals, Standard Oil was accused of fiscal fraud: Its Bolivian wells were confiscated by the junta in Sucre, the nation's capital. This nationalization, the first in oil's history, regardless of its lack of economic import, raised anger and rancor on Wall Street, and had a great impact across Latin America.

The impact was particularly powerful in Mexico, along the "Golden Lane" of fertile petroleum fields around Tampico, where since January 1936 more than ten thousand workers enacted multiple strikes protesting the sad conditions of their lives. The Mexican workers were paid miserable wages and lived in slums apart from the Americans, according to segregationist rules that were rampant in Mexico, Iran, Texas, and California.[43] In addition, a long-standing bitter dispute already pitted Mexico against the American and British oil companies. In order to understand this dispute, we must go back nearly thirty years. At the request of General Porfirio Diaz, Weetman Pearson, a strong English entrepreneur, came to Mexico in 1908 to construct a railway to compete with the Panama Canal. The oil resources of the country soon proved to be very important. In making this appeal to a British businessman, Diaz, the crafty old dictator, succeeded in pulling the rug out from under the Yankee pioneers who sought Mexican oil, much to the Americans' chagrin, starting with Edward Doheny. From that point on, Weetman Pearson, discreet and cunning, dominated the black gold of Mexico (despite the brigands that the American oilmen were accused of employing to blow up its pipelines and wells).[44] This domination persisted during the Mexican Revolution that began in 1910, and then in the course of the civil war that followed. American businessmen pressured President Woodrow Wilson to secretly finance general Victoriano Huerta (called "the usurper"), the dictator who reached the height of his power in 1913.[45] The civil war was ended by the adoption of the Constitution of the United States of Mexico in 1917. According to article 27 of this constitution, the underground resources of Mexico belonged not to the companies who exploited them but to the state. This provision dismayed Wall Street almost as much as the confiscation of the wells in Baku by the Bolsheviks that same year.

In 1919, the business climate was no longer favorable. Weetman Pearson (who had become Lord Cowdray) relinquished the essential shares of his oil company, the Mexican Eagle, to Royal Dutch Shell. Fabulously rich, Pearson acquired 45 percent of the shares of the Lazard Bank's London branch that same year, then laid the foundations of a media empire, the Pearson Group, which long owned the two linchpins of the worldwide economic press, the *Financial Times* and the *Economist*.[46]

Shell put an end to some of the most backward working conditions hitherto imposed by Mexican Eagle, which preferred to use men to carry sacks of dirt on their backs rather than to use bulldozers, judged too expensive.[47] Alas for Shell, even before the sale of Pearson's shares took place, in London there was a new worry, but one that was difficult to interpret: Salt water rose into the wells of the "Golden Lane." During that period, people did not understand that the extraction of a well must be regulated, that pumping crude oil at full speed would damage the oil field and shorten its life span. In vain, a young American geophysicist, Everette Lee DeGolyer, put Mexican Eagle in custody (his fortune was assured when people came to understand that he was right).[48] Like all the other oilmen, Henri Deterding and Calouste Gulbenkian lost a lot of money before they figured out what was happening.[49] Shell (which controlled 60 percent of the extractions), as well as Jersey Standard and SoCal (who controlled the bulk of the rest), stubbornly continued to accelerate the pumping of crude oil during the years that followed, making Mexico the second largest producer in the world, that is, until the country's oil production declined in 1922, as dramatically as it had risen.[50] As a result of the premature depletion of a number of its wells, Mexico, which had acquired a role of strategic exporter during the First World War, lost its ability to export crude oil for a half century. In Mexico, many suspected the gringos had deliberately pumped all the crude oil that they could in retaliation for the establishment of article 27 of the Mexican constitution. The 1936 worker strikes rekindled the dispute between Mexico and the major oil producers. Those of 1937 kindled still more worker strikes. The Mexicans workers were often paid half as much as the gringos, for the same work.[51] But, in the English and American press, the oil companies called these claims "absurd" and collectively refused to negotiate anything. Lazaro Cardenas, the popular Reformer president elected three years earlier, attempted to introduce a mediator. No non-Communist petroleum-producing country had ever appeared so irreverent and combative. If the Anglo-Saxon oil companies yielded to the Mexican government, they feared they would create dangerous precedent. They called for the arbitration of the Supreme Court of Mexico, which ruled against them. In spite of last-minute attempts at compromise, the impasse was now complete. On March 18, 1938, President Cardenas crossed the Rubicon: He nationalized the seventeen foreign companies that exploited Mexican oil. This decision was heralded as a new day of independence. A monument was erected in Mexico, to which foreign diplomats were instructed to pay their respects. It was the first resounding victory of a southern country rebelling against the major industrial powers of the north. A fundamental revenue source for the state, and object of innumerable bribes for its political leaders, Pemex, the national Mexican oil company, soon became the financial pillar of the Institutional Revolutionary Party that governed Mexico until the present day, almost without interruption.

The major oil companies strongly protested the nationalization of their Mexican companies. The British government provided Shell with unfailing support. But the Roosevelt administration, which had some years earlier adopted a "good neighbor" policy in respect to Mexico, was much more measured in its approach, angering American petroleum companies. The British and American companies boycotted Mexican oil, refusing to commercialize. After many years, they did receive significant compensation from the Mexican government: $160 million in total. However, due to this boycott, Mexico turned to Nazi Germany, which became its number one client until the Second World War began and the boycott ended. Standard Oil fueled a press campaign in the United States in order to counter a stream of articles published in American newspapers, which criticized its stinginess and its contempt for Mexican workers. Caricatures depicted Mexico as a thief in rags, wearing a sinister smile beneath a large sombrero, a la Emiliano Zapata. Nevertheless, few articles in favor of Standard Oil repeated the company's insistence that Mexico sold oil to the Third Reich.[52] Without a doubt this was because, on the eve of the Second World War, Germany bought eight times more oil from the United States and the major oil companies operating in Venezuela, than from the new Mexican national company.[53]

NINE

The Persistent Alliance of Big Oil with Nazi Germany

Four months after the appointment of Adolf Hitler as chancellor of Germany, in New York, the Mexican painter Diego Rivera was forced to interrupt the creation of what he considered his masterpiece. On the night of May 29, 1933, he was given the order to remove a scaffolding erected in the middle of Rockefeller Center's grand hall, where he had created a 20-meter mural high on the wall. Like many artists the fascists considered to be "degenerates," Diego Rivera never concealed his Communist sympathies. Along with Picasso and Matisse, Rivera was a celebrated artist of his generation, infinitely more illustrious than his artist wife, Frida Kahlo; his benefactor was Nelson Aldrich Rockefeller's mother. Rivera played a practical joke on the richest family in the world, the principal shareholder of the offspring of Standard Oil and a myriad of other companies.

The Rockefeller family had commissioned a mural by Rivera for Rockefeller Center, an art deco complex inaugurated in 1930 in Manhattan, the heart of US finance. The theme they requested was "Man at the Crossroads, Looking with Hope and High Vision to the Choosing of a New and Better Future." In the center, Rivera painted a worried worker manipulating levers of an omnipotent machine. To his right advancing soldiers wore gas masks, the bourgeoisie were feasting, and policemen on horseback were clubbing demonstrators. To his left, people were marching. In the middle were the faces of Marx, Trotsky, and Lenin. Between Lenin's hands were men and women of all races, under the furious gaze of a British officer. Rivera was fired and his fresco was covered with a sheet. At the end of several months, workers at Rockefeller Center received the order to destroy the mural with hammer blows.

Fascism, according to Mussolini, is the fusion of power between business enterprise and the state. Big Oil, like many other industries, expressed a manifest preference between the "brown plague" and "red cholera" during the 1930s. In particular, the leaders of Jersey Standard and Shell showed no scruples about doing business with the Nazi regime, until the war obliged them to cut ties with Berlin.

Big Oil and the Fascists

The idea of a natural hierarchy of races was common in the 1930s. The social theory of eugenics, in vogue throughout the West, advocated genetic selection to improve the human species. Eugenics enjoyed particular success within the Anglo-Saxon establishment, from London to New York. It provided immeasurable racist pride with an adulterated scientific alibi. Eugenics enjoyed the political protection of Lord Barfour, and the intellectual backing of biologist Julian Huxley (the brother of Aldous Huxley) and young and brilliant economist John Maynard Keynes. On Wall Street, the financial capital of the planet, eugenics enjoyed the support of patrons of the highest capitalist dynasties, starting with the Rockefellers and the Harrimans.[1] Even after the Nazis took power, the Kaiser Wilhelm Institute for Anthropology, Human Heredity, and Eugenics continued to receive subsidies from the Rockefeller Foundation. Located in Berlin, this institute included among its researchers a certain Josef Mengele, the future "angel of death" of Auschwitz.[2]

In the 1930s, this euphoria for eugenics had many willing victims at the top of the Western social elite. Signatory to many almost epileptic anti-Semitic pamphlets widely disseminated in the United States and Germany, Henry Ford was considered by Adolf Hitler to be a source of full-blown "inspiration." Two years before his rise to power, Hitler confirmed to a Detroit journalist that on his desk he had a photo of the largest automotive manufacturer in the world. Admirer of the Third Reich, Ford later received the highest Nazi distinction for a foreigner: Four months after the Anschluss, in July 1938, he was made a Grand Cross of the German Eagle.[3]

The Napoleon of petroleum, Henri Deterding, head of the Royal Dutch Shell for more than thirty years and lord of the British Crown, also proved to be an ardent supporter of the National Socialist Party. After marrying a white Russian and developing a visceral hatred toward the Bolsheviks following its setbacks with the USSR, the absolute master of the major Anglo-Dutch company publicly displayed his admiration for Hitler beginning in 1933. The following year, the stocky Dutchman was for four days the guest of the Nazi chancellor in his Berchtesgaden residence, to try to negotiate a monopoly on the distribution of oil in Germany.[4] Deterding was not removed from the directorship of Royal Dutch Shell, although many leaders of the company were embarrassed by his stance. In December 1936, just prior to his retirement at age seventy, he made the Nazi regime a huge gift of livestock and various other agricultural products, the value of which was estimated at £1 million sterling: This was intended to support Germany in its fight against Bolshevism.[5] Deterding died on February 4, 1939. He was buried in Germany. Hitler and Hermann Göring, the second-most-powerful leader of the Reich, dispatched emissaries to his funeral, during which he was hailed as an enemy "of the pan-Jewry and Bolshevism."[6]

The boss of another major oil company also exploited his position to advance the cause of national socialism: flamboyant and authoritarian president of Texaco, Torkild Rieber. Of Norwegian origin, named captain of a tanker at only twenty-one years of age, Rieber was among the American oilmen who opened Saudi Arabia to Big Oil in the beginning of the 1930s.* In 1937, during the Spanish Civil War, Rieber was threatened with prosecution for conspiracy by the American justice system: Via Belgium and Italy, Texaco secretly refueled General Franco's army with gasoline, diesel, and oil.[7] In January 1940, in the midst of the Battle of Britain and after a lengthy meeting with Göring, Torkild Rieber visited Roosevelt, likely in order to convince Roosevelt to support Germany's peace plan for the surrender of Great Britain.[8] Seven months later, Rieber provided for the needs of an emissary from Berlin, Dr. Gerhardt Westrick, who came to Washington to convince the United States to stop supplying the British army with fuel (because clearly, the triumph of the Third Reich was assured). Westrick left the United States a few days later, embarking in Los Angeles with his wife and two children on a Japanese ship. As for Torkild Rieber, he was pushed to resign from the presidency of Texaco on August 23, 1941, less than four months before the attack on Pearl Harbor and the United States' entry into the war.

Standard Oil Ahead of Benito Mussolini in Ethiopia

At the end of the summer of 1935, just after the conclusion of the Chaco War, some directors of Standard Oil fed their unholy greed regarding another isolated corner of the planet: Ethiopia. Better known then under the name of Abyssinia, the empire of the Negus Haile Selassie, and coveted by Benito Mussolini, Ethiopia was invaded by the Royal Italian Army in October 1935, in what remains today as the first step in fascism's expansion.

* See *infra*, chapter 11.

During the period between the two wars, oil prospectors omitted practically no region of the globe in their search for signs of black gold. If oil spurted out often where no one had expected it, heavy disappointments also followed. Against this backdrop, on August 31, 1935, when an Italian invasion seemed all too imminent, a sensational dispatch arrived in the Ethiopian capital of Addis Ababa: The Negus Haile Selassie, emperor of Ethiopia, had just granted Standard Oil a petroleum concession of seventy-five years, covering more than half of its immense country, "seeking to stop an expected Italian advance into Ethiopia," the Associated Press reported.[9]

Yet, in Washington as in Rome, diplomats were taken aback. The territory granted, located in the city of Harar, lay exactly in the middle of the region's different forces: between Addis Ababa, the imperial capital of the negus; Eritrea, controlled by Italy; the French territory of Djibouti; and the British protectorate of Somaliland. In Addis Ababa, the beneficiary of the concession (the first that the negus had granted to a foreigner) was revealed to the press. His name was Francis Rickett. This British promoter merely explained that he represented American interests, without clearly confirming that he worked for Standard Oil. The agreement, which bore the seal of the imperial "Conquering Lion of Judah," had been signed on a desk and with a pen that were both manufactured in the United States, the Associated Press reported. When asked if he expected that the rights of the concession would be respected in the probable case of an Italian invasion, Francis Rickett responded: "Absolutely." While still in the capital of the negus, Rickett didn't hesitate to add that he had personally helped Mussolini, "an old friend," negotiate the purchase of some Iraqi oil. With respect to its concession in Ethiopian, Rickett swore that Ethiopia was "as rich as the Kirkuk region."[10]

Although it had never been named by Rickett, the Standard-Vacuum Oil Company—a subsidiary shared by Standard Oil of New Jersey and Standard Oil of New York, the oldest branches of the Rockefeller empire—immediately denied that it was meant to be the beneficiary of the Ethiopian concession. But on September 4, a new drama emerged: During a surprise press conference in Washington, DC, Roosevelt's chief diplomat, Cordell Hull, flatly stated that he had summoned the president and the director of Standard-Vacuum and had forced them to admit that their company was the holder of the enormous Ethiopian concession, and ordered them to relinquish the concession.[11] Cordell Hull explained to journalists that exploit of this concession would be a "cause of great embarrassment" not only for the United States, but also for all countries then trying to intercede to prevent the war. Rickett's allusion to his "old friend" Mussolini embarrassed the US secretary of state.

On September 8, at Suez, a crowd of journalists awaited Francis Rickett's return from Addis Ababa at the Egyptian port's largest hotel, whose business was better than ever under the British administration.[12] The United Kingdom, Haile Selassie's protector against Mussolini, was taken aback by Selassie's agreement with Standard Oil. Rickett was placed under close police surveillance. But the British businessman, uncowed, reaffirmed that his concession would remain valid no matter what might happen. Did Standard Oil attempt to short-circuit the Roosevelt administration by prematurely putting the agreement forth as accomplished fact? The next day, in New York, on his return from a trip to Europe, Walter Teagle distanced himself from Standard-Vacuum in front of journalists and appeared offended when asked if he had been aware of the negotiations with Ethiopia.[13]

Three weeks later, on October 2, 1935, Mussolini's troops penetrated Abyssinia. The Royal Italian Army aligned 250 aircraft and 5,000 land vehicles against Ethiopian soldiers with outdated equipment. Events unfolded so rapidly that "the Duce" quickly ordered the use of chemical weapons. The League of Nations envisioned sanctioning Italy by an embargo on several raw materials, and in particular, petroleum. Mussolini knew that his army was dependent on imported oil. He took precautions: In Rome, buses started running on coal instead of gas.[14] But the negotiations at the heart of the League of Nations were impeded. Pierre Laval, leader of the French government (and future collaborator of the German occupiers), contributed to dragging out the negotiation process. Despite their official position, it was clear the British were sparing Mussolini: The Anglo-Iranian Oil Company (Persia was now called Iran) consistently supplied the Italian Navy, and had refueled their ships that December, according to the *New York Times*, while Mussolini rejected the part of the Franco-British peace plan that pertained to him.[15] In Paris as in London, the threatening rants of "the Duce" raised fears that the small colonial war in Abyssinia could degenerate into open conflict in Europe.

In the United States, Roosevelt did not want to kill negotiations by embracing the petroleum embargo. He simply asked the American oil companies to observe a strict neutrality in respect to Italy—to maintain their deliveries at the usual level. After establishing the semblance of an embargo against Mussolini, the oil czar of the New Deal, Harold Ickes, had to retract it.[16] In December, the press noted that a mysterious report circulating in London and Washington affirmed that Standard Oil was engaged in replenishing Italy's fuel supply, in opposition to the embargo, in exchange for a thirty-year monopoly on petroleum sales in that country. Once again, Walter Teagle was offended: According to him, this was a "preposterous piece of propaganda" coming from who knows where. A battle of short news briefs between Standard Oil and the Roosevelt administration followed.[17] Twenty-six Broadway, Standard Oil's headquarters, announced that the deliveries of Standard Oil to Italy were normal, "with the single exception" of an unspecified volume of oil sold in September, just before the invasion.[18] In response, Washington provided the press with figures that showed US exports to Italy and its colony of Eritrea had in fact skyrocketed since the beginning of the invasion.[19]

The scandal would never fully explode. Francis Rickett traveled to Rome in January, where he proclaimed to the press that the concession of Standard-Vacuum remained valid.[20] His presence was announced on March 21, in Addis Ababa.[21] Two months later, the capital of the negus was overtaken by the Italian troops, completing the conquest of Abyssinia. In Italy, Mussolini held onto his triumph. "The Duce" was at the pinnacle of his glory. Later, he explained: "If the League of Nations . . . had extended the economic sanctions to oil, I would have had to withdraw from Abyssinia in a week. That would have been an incalculable disaster for me."[22] Alas for the hopes of Mussolini and Standard Oil, Ethiopia, even more than Chaco, lacked an abundance of black gold: Few drops of oil were ever extracted.

The Irresistible Marriage between Jersey Standard and IG Farben

Many alliances were forged between the Anglo-Saxon oil companies and the German industry starting in 1924, when American banker Charles Dawes's plan permitted a nearly bankrupt Germany to regain its hold on the coal from Ruhr, and he organized its bailout by Wall Street banks. Desperately lacking in capital up to that point, Germany's chemical industry had a lot to offer the prosperous American petroleum industry.

Since the beginning of the century, German researchers had achieved multiple major advances in the field of organic chemistry. This central branch of industrial chemistry, which used high temperatures and pressurized the atoms most common to the Earth's surface (hydrogen, carbon, nitrogen, and oxygen), had high-energy needs.* Prior to oil, the industry used coal, of which Germany had an abundance. But after 1930 the organic chemistry processes developed in Germany were sold or copied. They migrated most significantly to the United States, as a result of agreements with US industrial

* Industrial chemistry was born with the creation of the company CIBA in Basel, Switzerland, in 1859, the same year "Colonel" Drake first drilled.

giants, primarily with Standard Oil of New Jersey. In the new factories built in the United States starting in the 1930s, hydrogen atoms and carbon from American petroleum and natural gas took the place of coal. They gave birth to a new branch of chemistry, and created a major new industry: petrochemicals. Petrochemicals were more efficient, more profitable, and almost identical, from a scientific viewpoint, to carbochemistry, the chemistry of coal that had been invented mainly in Germany.[23] Thanks to the abundance of their black gold and their other assets, the United States remained for more than half a century the uncontested master of the petrochemical industry, and petroleum became the Swiss army knife of industrial chemistry.

Upon its creation, IG Farben, the conglomerate formed in 1925 by the merger of the largest German chemistry companies (in particular BASF and Bayer), undertook to develop on a grand scale what may have been the most decisive technical innovation in modern history, without which the demographic explosion of the second half of the twentieth century certainly would not have occurred. In 1909, the German chemist Fritz Haber had invented a process that made it possible to affix, with the help of gasified coal, the most abundant element in the atmosphere: nitrogen.

Implemented in 1913 by BASF thanks to the remarkable engineering of German Carl Bosch, this Haber-Bosch process made it possible to synthesize ammonia. The industrial ammonia Germany possessed during the First World War gave it the means to factory-produce large quantities of nitric acid, necessary for the preparation of gun powder and nitroglycerin. After 1918, the Haber-Bosch process patent was confiscated by the Allies, and the secrets of BASF were fully exploited. But despite the discovery of various techniques, Carl Bosch's system was not supplanted. The synthesis of ammonia provided emerging modern agriculture with huge quantities of chemical nitrogen fertilizers, a substitute for natural nitrogen fertilizers, such as extracts of saltpeter or guanoa, which had come from the Latin America's Pacific coast.* The Haber-Bosch process also supplied the pharmaceutical industry with the entire variety of ammonia derivatives, the chemical building blocks in a great number of drugs. At the end of the 1920s, thanks to the coal of Saxony and Ruhr, IG Farben was by far the number one worldwide producer of ammonia. But beginning in 1930, in San Francisco Bay, California, Henri Deterding's Royal Dutch Shell was the first to create an ammonia production factory no longer reliant upon gasified coal but instead having access to the lighter forms of hydrocarbons, naphtha—also a basic

* From 1879 to 1884, the "war of the saltpetre," which involved Chile, Peru, and Bolivia, made the latter lose its only access to the sea. With the Second Boer War (1899–1902), it is one of the first cases of modern warfare motivated primarily by the access to raw materials.

component of gasoline—and natural gas.[24] Thanks to the possibilities offered by hydrocarbons, any obstacle to production of ammonia (and, therefore, to production of fertilizer, pesticides, explosives, or drugs) was lifted.

Synthesized ammonia was not IG Farben's only treasure. In 1926, Walter Teagle was invited to visit the factories and laboratories of the German firm. The president of Standard Oil of New Jersey and senior executives who accompanied him observed with their own eyes the tour de force that they feared could revolutionize the business of black gold: German chemists had developed an inexpensive way to mass produce liquid fuel from high-quality coal brought to high temperatures and pressures.[25] IG Farben had access to a technology patented in 1913 by German chemist Friedrich Bergius: hydrogenation of coal. This technology enabled the Kaiser's army to possess small quantities of synthesized gasoline, compensating in small part for Germany's petrol shortage during the First World War. The first important hydrogenation plant was launched in 1924 near Mannheim, at the edge of the Rhine, thanks in particular to capital from BASF and Royal Dutch Shell.[26] When Teagle visited the IG Farben conglomerate, it had just completed the construction of a much more significant hydrogenation plant in Leuna, in eastern Germany, close to the Reich's most important lignite quarry. By 1931, IG Farben could produce 2.5 million barrels of synthesized fuel. The same year, Friedrich Bergius and Carl Bosch received the Nobel Prize in Chemistry.

The confidential agreements established between Standard Oil of New Jersey and IG Farben, following Teagle's voyage to Germany, continued under the Nazi regime. In part brought to light by the American courts in the midst of the Second World War, this alliance became a stain that the army of lawyers and publicists serving the world's number one petroleum producer had a hard time erasing from public memory.

IG Farben, a "State within the Nazi State"

In November 1932, the National Socialist Party swept the legislative elections, which potentially opened the door to power in Germany. While the context of the Great Depression encouraged IG Farben to consider abandoning its synthetic fuel-production program, two of its leaders visited Adolf Hitler in Munich to learn how the Nazi leader could help their firm if he were elected chancellor. More precisely, according to records from the Allies' prosecution of IG Farben's leaders during the Nuremberg trials, two emissaries of the German chemical giant approached Hitler to ask whether he intended to support the development of their hydrogenation plants.[27] The company's leaders left satisfied: Their factories played a key role in Adolf Hitler's projects.

After his ascension to the chancellery in January 1933, thanks to the generous support of IG Farben and the largest industrial and German bankers, Hitler guaranteed the price and sale levels of synthetic fuel for years to come. The Nuremberg trial records indicate: "Farben concentrated its vast resources on the creation of the German military machine of war, invented new production processes and produced huge quantities of war materials, including synthetic rubber, synthetic gasoline, explosives, methanol, nitrates and other critical material. Without them Germany could not have initiated and waged aggressive war."[28]

The synthetic rubber referred to here was the last great wonder of the German chemical industry. The latex yields had become insufficient for the growing needs of the tire market and, in addition, the rubber trees grew quite far from Europe and the United States. In 1935, IG Farben launched the industrial production of synthetic elastomers from a technique patented in 1909 by a team of Bayer chemists, which enabled the manufacture of artificial rubber with lime, water, and gasified coal. This new miracle material was called *buna*, a contraction of *butadiene*, a complex synthetic hydrocarbon, and *natrium*, the German name for *sodium*.

In 1936, in Stuttgart, German engineer Ferdinand Porsche built the prototype for the future "Beetle": Development of the "car of the people" (Volkswagen) was spurred on by the Führer, who wanted Germany to become a "nation on wheels." The same year, the "four-year plan" of economic development initiated by Hermann Göring launched the construction of the first motorways (autobahnen). It also accelerated the development of the chemical industry, far beyond the needs of a peacetime country. Nazi leaders were fully aware that the power of a nation was equal to the energy it was capable of possessing.

IG Farben was by this time totally Nazified. The Jews in its management were sidelined. Carl Bosch was also judged to be too critical. IG Farben found itself "promoted to governmental status" of the Third Reich, according to Hitler's Albert Speer: "the State within the State."[29] IG Farben's high officials were promoted to SS officer status to run the conglomerate, notably including one of the two emissaries Hitler sent in 1932: Heinrich Buetefisch, who was part of Himmler's circle of friends, directed the giant factory in Leuna. Eleven other plants of various sizes were under construction at this time.[30] Those plants provided almost all of the high-quality fuel necessary for German Air Force planes. Berlin also acquired a dozen factories enabling the manufacture of fuel from coal using another technique invented in Germany: the Fischer-Tropsch process, which provides heavy petroleum substitutes (grease and diesel fuel for tanks, trucks, and ships).[31] Ingenious as they were, these processes, which transformed coal into liquid fuel, rendered mediocre energy output, due to the amount of energy necessary to heat and pressurize the coal. It was necessary to invest one joule of energy in order to produce two joules of fuel, while the energy investment required to produce the same fuel from crude oil wells at that time was negligible.

Jersey Standard and IG Farben: The "Sinister Figures of the Cartel"

At no time during the course of the 1930s did Standard Oil of New Jersey and IG Farben halt their business. In April 1929, IG Farben launched the American IG Chemical Corporation in the United States. This US branch of the German firm was headed by Hermann Schmitz: Elected to the Reichstag in 1933 under the Nazi party banner, two years later Schmitz became the president of IG Farben and remained so until 1945. He was later sentenced to four years in prison for war crimes and crimes against humanity by the Nuremberg tribunal. Financed by the largest banks of Wall Street, the American IG Chemical account included on its board of directors banker Paul Warburg, the designer of the Federal Reserve; Edsel Ford, Henry Ford's only son; and the boss of Jersey Standard, Walter Teagle.[32] During the 1930s, all of these bigwigs liked to meet at the Cloud Club, located on the top three floors of Manhattan's Chrysler Building, with a Tudor lounge, a barber, a Renaissance marble staircase, and a restaurant with walls that were covered in idealized representations of American industry. Amid the wafting smoke of Havana cigars, Teagle maintained regular contact with Hermann Schmitz, the Nazi dignitary at the head of IG Farben.[33]

On November 9, 1929, Jersey Standard and IG Farben reached a secret agreement by which the American company obtained a number of patents, in return for which IG Farben received $35 million from Standard Oil. By an agreement similar to the As-Is Agreement of Achnacarry, the world's number one petroleum and chemical producers colluded to avoid competition in their respective industries.[34]

After the outbreak of the Second World War in September 1939, the Roosevelt administration began to pay close attention to the relations between IG Farben and Jersey Standard. Little by little, the administration discovered the companies' collusion, perpetuated as "business as usual" after the Nazis seized power. In 1941, the US Department of Justice launched two antitrust procedures against Jersey Standard. The first involved a conspiracy to control the flow of US pipelines. The second targeted the restrictive agreements with IG Farben. Dismayed, Teagle wrote to Roosevelt to try to justify his company's actions. But the American president refused to intervene. The rest of the Jersey Standard board of directors preferred to reach a compromise with the Department of Justice. They agreed to modify their practices, especially their commitment to exclusively and secretly share patents with IG Farben, in exchange for a symbolic fine of $50,000.[35] The case sparked a media outcry when, on March 26, 1942, the top Justice official in charge of the antitrust procedure, Thurman Arnold, testified in front of the Senate committee investigating national defense, chaired by Democratic senator and future American president Harry Truman. Immediately the scandal blew up and made an oil blot in the US news, in the furious year of 1942. As the United States entered the war, the leaders of Jersey Standard found themselves accused of having refused the US Navy access to the synthetic rubber patent because of their alliance with IG Farben![36] A year earlier, Jersey Standard had announced the future launch of the first American synthetic rubber plant, which used a process developed with the help of IG Farben, replacing coal with oil.[37] At the end of Thurman Arnold's testimony, a reporter asked Harry Truman: Could this be a case of treason? "Why, yes, what else is it?" Truman replied.[38]

Five days later, journalists rushed to hear the defense that the new president of Jersey Standard, William Stamps Farish II, provided the Senate. The Texan oilman, who took the reins of the company in place of his friend and hunting partner Walter Teagle, explained at length. His line of defense was effective: "Whether the several contracts made with I.G. [Farben] did or did not fall within the borders set by the patent statutes or the Sherman Act, they did inure greatly to the advance of American industry and more than any other thing have made possible our present war activities in aviation gasoline, tulol and explosives and in synthetic rubber itself."[39] The next day, the *New York Times* headline proclaimed: "Farish Says German Deals Speeded Our War Industry."

The Jersey Standard president conceded very little. But his response sometimes appeared weak when the questions were precise. When senators asked whether his company had helped Germany to undertake the construction of hydrogenation factories in occupied France, William Stamps Farish II replied that he preferred to put his interests in France in the hands of IG Farben rather than in the hands of a German commissioner. He continued: Standard Oil had abandoned those fuel factory projects, clearly destined for the German Army, after the British Shell company, which from the beginning had also engaged in that enterprise, indicated that they no longer wanted to participate. William Stamps Farish II specified that his last employee had left France in January 1941.[40] For six months, the collaborationist Vichy regime had been in place.

Business is business. In November 1939, US Congress adopted an embargo against Germany on products manufactured in the United States. But in Romania, Jersey Standard nonetheless continued to extract oil, of which the larger share was returned via the Danube River to Germany. This transport of Romanian oil continued until the end of January 1940 at least—and likely up to December 4, 1940, the date of the nationalization of the pro-Nazi Bucarest government.[41]

After William Stamps Farish II's deposition to the Senate, the US Department of Justice continued the investigation. Its chair, Thurman Arnold, was brought before the Senate's Truman Commission once again in June 1942. The Commission accused Farish of hiding and distorting facts. The top Justice official specified that Standard Oil was not allied with the Germans for "unpatriotic motives" but in pursuit of their old transgressions: with the "sole motive" to "get a protected market" and "eliminate independent competition."[42] William Stamps Farish II, a pious man with a volcanic temperament, viscerally attached to his company, was affected by these accusations. He died six months later of a heart attack at the age of sixty-one, while resting at Walter Teagle's mansion. Teagle had become nervous, distracted, and irritable, and resigned from his various positions at the end of 1942.[43] After having had, for almost three decades, the directorship of the richest private empire in the world, Teagle decided to leave before his terms were up.

In 1943, under pressure from Roosevelt's secretary of the interior, Harold Ickes (the oil czar during the New Deal), the prosecution against Standard Oil of New Jersey was suspended. Most members of the Petroleum Administration for War (PAW) that Harold Ickes directed were, in effect, connected to the black gold industry and vital to the war effort. Yet on September 13 Roosevelt's vice president, Henry Wallace, made a public declaration, a symbolic gesture in front of the Department of Justice. Wallace began by reading a Jersey Standard internal memo sent in April 1938, in which it appeared that the company was engaged in blocking the development of synthetic rubber in the United States because "our partners," IG Farben and Berlin, were opposed. In another letter dated in November 1939—two months after the Führer's troops invaded Poland—a Jersey Standard company manager said he had been pleased to ensure that a US Navy representative who had visited the synthetic rubber laboratory had departed without the chance to photograph sensitive installations. The symbolic condemnation delivered by the American vice president during the most uncertain time of the Second World War was a heavy badge of dishonor. Henry Wallace stated that "behind all this subterfuge, concealment and double-dealing was the sinister figure of the cartel of Standard Oil and I.G. Farbenindustrie. Standard was forced to choose between the interests of the United States and the cartel with I.G."[44]

IG Farben Justifies Its Cooperation with Jersey Standard to the Gestapo

Shortly after the war's end, a new story about the relations between Jersey Standard and IG Farben surfaced in the rubble of the Third Reich. It consisted of a report transmitted by IG Farben to the Gestapo. During the war, in Berlin, people of course read of the scandal published in the American press. The Gestapo's directors called for explanations and were certainly dissatisfied to learn that, to defend himself, the head of Jersey Standard bragged that the industry and the US Army had learned a lot thanks to IG Farben. In the report transmitted to the Nazi secret police, IG Farben's directors presented a perfect counterexample: the technology of tetraethyl lead, supplied before the war by the US industry to the German chemical firm. Tetraethyl lead increases the octane rating of gasoline: It is essential to guarantee that the highly compressed gasoline in aircraft engines doesn't ignite prematurely. "Without tetraethyl lead the present method of warfare would have been impossible," argued August von Knieriem, IG Farben's board lawyer. The report, dated June 6, 1944 (the day of the Normandy landing), continued the argument: "The fact that since the beginning of the war we could produce tetraethyl lead is entirely due to the circumstances that shortly before, the Americans presented us with the production plans, complete with their know-how."[45] This report was later placed under the magnifying glass by those Department of Justice officials responsible for conducting the Nuremberg trials for war crimes and crimes against humanity directed toward IG Farben officials, during which August von Knieriem was charged but released for lack of evidence. In the Nuremberg trial indictment, the question of Germany's access to tetraethyl lead just before the war fact surfaced briefly.

At the beginning of the summer of 1938, the minister of the German Air Force (the Luftwaffe) contacted IG Farben about an urgent problem: The Luftwaffe had realized that it would not have enough tetraethyl lead if the Führer failed to annex Sudetenland without fighting, and that the German

Army would be overburdened with the general conflict in Europe. Therefore, per order of the minister of the air force, three of the highest IG Farben directors—von Knieriem, Hermann Schmitz, and chemist Carl Krauch—traveled to London to urgently negotiate an exceptional command of no less than 500 tons of tetraethyl lead with a Standard Oil affiliate, which was duly delivered just before the invasion of the Sudetenland in September 1938.[46] It would have been clear to Standard Oil executives that the only possible recipient of such an order was the German Army, in the middle of the disastrous diplomatic crisis concluded with the agreements in Munich on September 29 and 30. The oil companies were, however, complex institutions and errors could have occurred.

Operational at the end of 1939, IG Farben's tetraethyl lead factories had been developed through a 1935 agreement between a Jersey Standard subsidiary and General Motors, the Ethyl Gasoline Corporation, thanks to the discharge granted after a long arbitration by the Roosevelt administration's War Department. On December 15, 1934, a high official of the DuPont Company sent a letter to the chairman of the Ethyl Gasoline Corporation, to try to prevent the alliance with the German firm. Unlike Jersey Standard, DuPont, the American chemistry giant and a main shareholder of General Motors, was IG Farben's direct competitor. However, Dupont's precautionary attempt appeared to transcend the strict purposes of business: "It has been claimed that Germany is secretly arming. Ethyl lead would doubtless be a valuable aid to military aeroplanes. I am writing you this to say that in my opinion, under no conditions should you or the Board of Directors of the Ethyl Corporation disclose any secrets or 'know-how' in connection with the manufacture of tetraethyl lead to Germany."[47]

The Nuremberg trial indictment provides an overview of the agreements maintained after 1933 between IG Farben and many industrial companies within the Third Reich's future national enemies. However, some of the leaders of these companies demonstrated more scruples about doing business with Nazi Germany than those in charge of Standard Oil of New Jersey.

TEN

The Enablers of the Second World War

The First World War had killed eighteen million military personnel and civilians. If the human losses were four times higher during the Second World War, this was not solely due to the cruelty of the belligerents, nor because the conflict lasted longer. The energy power available to the military forces of the war of 1939–1945 multiplied their deathly power and expanded their arena of action. The bombing of the Spanish village of Guernica by German Condor Legion aircraft in April 1937 and the Japanese air raids on the civilians of the Chinese city of Chongqing in May 1939 foreshadowed the unleashing of the fury of internal combustion engines.

During the months preceding the outbreak of the Second World War, Germans, Italians, and Japanese searched everywhere for petroleum sources for their countries. The companies representing the interests of Berlin, Rome, and Tokyo rushed to appeal to Saudi Arabia, where oil had been discovered for the first time in February 1938 by the Americans.[1*] The same year, the Japanese attempted to acquire concessions from the Mexican government, including the possibility of building a pipeline to reach the Pacific coast.[2]

These three aggressor countries of the Second World War shared one major fault: They scarcely possessed any crude oil within their own territories. Despite the horrendous consequences of their atrocious efforts, this vulnerability was not offset by German production of synthetic petroleum, and still less by the much more limited program later developed by Japan. Italy, which had to import nearly all of its fuel, had been unable to build a synthetic fuel factory because it lacked coal; quickly short of fuel, it saw half of its fleet definitively stranded in port as early as February 1941.[3] Access to oil played a key role in the strategic choices of the Third Reich and the Nippon Empire,

* See *infra*, chapter 11.

since without it, they could not pursue the main objectives of their most important offensive moves, which turned out to be fatal.

The Allies, on the other hand, were able to count on the United States, the dominant producer of black gold in 1939, controlling 60 percent of global extractions, with 3.5 million barrels per day for a total world production of 5.7 million barrels per day—a sixteenth of actual production. The oil remained above all an American affair and secondly a British affair. Outside of the USSR (the second largest producer in the world, far behind the United States), Jersey Standard, Shell, and the other Anglo-Saxon companies ruled approximately 80 percent of the worldwide market: After eighty years of industrial development, they hardly left crumbs to the competition. The United States and Venezuela, the empires controlled by the Anglo-Saxon oil companies, achieved 70 percent of worldwide oil production from wells on the American continents alone.[4] Throughout the Second World War, the production of crude oil, refined oil products, and petrochemical products proliferated west of the Atlantic and in the United States, and were the most decisive material factor in the Allied victory.

Blitzkrieg: Necessary Tactic for an Army with Limited Fuel

The tactics of the Blitzkrieg, through which the German Army triumphed everywhere it advanced at the beginning of the conflict, consisted of a maximum concentration of the offensive flow of mechanical energy. It meant achieving the highest possible impact from tactical gains, energy investment, and the materials required for the attack. The improved performance of assault tanks and aviation machines rendered traditional German tactics particularly fierce. As it was defined in 1935 in a German military periodical, *Deutsche Wehr*, the Blitzkrieg was an explicit response to the constraints imposed on states with few raw materials: They must "finish as quickly as possible with the war, attempting at the outset to make a decisive attack with all of their offensive power."[5] When Hitler launched the Third Reich troops to the Poland assault on September 1, 1939, Germany only had six months of reserves of gasoline, diesel fuel, and fuel oil: The Nazi army had to win quickly, or run out of gas.

It took much more fuel to continue the Blitzkrieg on the Western Front. Petroleum rationing was mandated among civilian populations; this became more drastic over the course of the war. Car trunks and bus roofs carried gasifier machinery—heavy, rustic, ineffective devices invented in the nineteenth century to collect carbon monoxide, a very toxic gas derived from an incomplete combustion of wood or coal, emitted by the vehicles. Berlin maneuvered to increase its resources of crude oil. As soon as September ended, the USSR claimed the eastern part of Poland, following the protocols of a secret German-Soviet pact signed the month before. Stalin took hold of the best oil fields of Galicia but pledged to deliver to the Reich the exact equivalent of their production: 7,500 barrels per day.[6] This was a low quantity for the USSR, for which wells, principally those of Baku, each day produced more than half a million barrels. Finally and most importantly, in December 1939, a special group of the Abwehr, the secret services of the German Army, managed to secure the petroleum installations in Romania. The Wehrmacht then took control of the main source of German crude oil (just before a French petroleum engineer from the Belgian company Petrofina, Léon Wenger, who had already participated in the 1916 sabotage of Romanian wells, failed in his attempt to execute a new plan of destruction).[7] On May 10, 1940, the German Army thus possessed an ample supply of fuel to commence Blitzkrieg on the Western Front, making its entry in Paris as early as June 14, scarcely more than a month later. On June 18, in London, before launching his historical call to resistance, General de Gaulle summarized the issue on the BBC: "Vanquished today by mechanical forces, we will be able to overcome them in the future by a superior mechanical force. The destiny of the world is here."

The Battle of England, Aided by the Superior Quality of American Gasoline

The Battle of Britain was the first failure that the German Nazis experienced in the Second World War. In July 1940, Göring sent the Luftwaffe across the English Channel in order to destroy the aeronautical infrastructures and ports of the United Kingdom, and to prepare for the invasion of Great Britain. Berlin committed to battle some 2,500 aircraft; the Royal Air Force (RAF) had less than 2,000. German aviation had already trained a number of pilots, including several experienced pilots. Its losses, however, were substantially higher than those of the RAF. Among the factors determining young British fighter pilots' success, in their heroic defense of the English sky, was the superior quality of the gasoline that flew their planes. During the 1930s, Royal Dutch Shell was the first company in the United States to launch the production of 100 octane fuel, high performance and resistant to catching on fire. American aviation forces converted to this fuel shortly before the war, and in July 1940, an emergency import program allowed the RAF to be ready in time for the largest air battle in history. Until then, the old

Hurricane and the new Spitfire of the RAF ran on 87 octane fuel, exactly like the German Messerschmitts. By converting to 100 octane, via a simple adjustment of the carburetor, their Rolls-Royce Merlin engines went from 1,030 to 1,310 horsepower![8] Thanks to their new fuel, British fighter pilots could fly faster and longer. Their Spitfires became more maneuverable and gained altitude more easily than the Messerschmitts, which were significantly lighter.[9]

The German hydrogenation plants, despite their giant size, never succeeded in producing high-octane, high-performance gas in quantities equal to the refineries of the United States, especially since they lacked the abundance of American black gold. Far from it. Large quantities of oil were absorbed during the refinement of these fuels: The higher the desired octane, the more fuel was required to make it. Massive reliance on tetraethyl lead, which to some extent increased the octane level, could not solve the problem on its own. At the beginning of the war, the Luftwaffe absorbed more than 300 tons each month, and was constantly threatened by shortages until the creation of several new plants in Germany and the activation of a plant in Paimboeuf, near Saint-Nazaire, once France had been conquered.[10] In 1941, IG Farben succeeded at producing limited quantities of 95 to 97 octane gasoline, but the German fighter plane engines were not converted to make the best use of this new fuel: Performance gains proved disappointing, and the window of opportunity for invading the British Isles had already closed.[11] When Hermann Göring asked one of the ace Luftwaffe pilots what he needed to better protect the bombardiers, he responded, "Better engines and 100 octane." Göring took note of this: "This will be done," he assured the pilot.[12] But the scope of his promise exceeded the extent of natural and artificial German petroleum resources.

The Japanese Empire Seeks to End Its Reliance on Californian Oil

The ground beneath Japan is essentially volcanic, poor in fossil fuels. During the archipelago's industrial growth, the Japanese were quickly forced to go to elsewhere to seek the energy sources necessary to accomplish the Nippon Empire's ambitions. The knot of war between Tokyo and Washington, DC, leading to the Pearl Harbor attack on December 7, 1941, was tightened much earlier in response to Japan's problems accessing oil.

In 1930, the government of a Chinese republic, at the edge of chaos and facing the economic and military pretensions of Japan, decided to impose a 400 percent increase in its tariffs on Manchurian coal.[13] It was the *casus belli* that the Japanese imperial government expected: Not only did the Japanese industry have a vital need for coal from this vast northern Chinese territory, but they also hoped to extract oil shale in Manchuria, and if possible, to find black gold. The Chinese government was fragile, and Japan used the tariff as an excuse to invade Manchuria at the end 1931, and the following year to install the puppet government of the Manchukuo (led by the last Chinese emperor, Puyi). From then on, all Manchurian coal was reserved for Japan. But the search for oil was unsuccessful and, despite high ambitions, the Japanese industry was unable to develop large-scale production of artificial fuel or oil shale. The imperialist projects were watered down: Japan needed control of its own oil supply if it hoped to effect long-lasting power using its military fleet. Between 1931 and 1939, the archipelago's oil consumption doubled, to reach one hundred thousand barrels per day. But 80 percent of this oil was imported from California.

The war between China and Japan, which began in August 1937, only increased the Empire of the Rising Sun's crude oil consumption (in practice, the Japanese fleet, the third in the world, alone absorbed more than half of the archipelago's resources) and at the same time increased its dependence on US oil. This fact worried American oilmen less than Japanese admirals.[14] In September, while preparing to board Roosevelt's yacht for lunch on the Hudson, James Moffett—at the time head of a joint venture between SoCal and Texaco, whose interests extended from the new fields of Saudi Arabia to the ports of the sea of China—told the press that he was not ready to abandon to anyone his massive commercial interests.[15] In 1938, as public opinion and the American media screamed of bombardments aimed at the civilian population of Canton, a "moral" embargo was imposed on the sale of aircraft to Japan. But no embargo was placed on fuel. In Washington, American strategists knew very well that halting the sales of petroleum could also suffocate the imperialist ambitions of Japan.[16] Beyond California, the Japanese imported crude oil from Sakhalin, north of the archipelago, off the coast of Siberia; controlled by the USSR, it was a meager, strategically unreliable source. Further south, but still within reach from the Japanese fleet, were the Dutch East Indies. Controlled by Royal Dutch Shell and the American oil companies, the black gold production of Sumatra, Borneo, and Java was sufficient to meet the entire needs of Japan. Thanks to it, the Empire of the Rising Sun could survive without American petroleum.

From the beginning of the war in Europe, Berlin had pressed Tokyo to take possession of the oil wells in the Dutch East Indies. However, Japan also needed American and British products. The Japanese government did not want to sever the cord: For the time being, the moderates in government prevailed. Tokyo declared itself neutral but benefitted just the same from the strategic weakness of the Dutch, to reclaim a substantial share of the oil of the Insulinde. In an irony of fate, some weeks before the war exploded, the corpulent, austere prime minister of the Netherlands, Hendrikus Colijn, formerly second-in-command at Royal Dutch Shell, was responsible for directing the company when pro-Nazi Royal Dutch Shell president Henri Deterding took long hunting and skiing vacations.[17] During the months that followed, the United States transferred the bulk of their Pacific fleet from Los Angeles to Pearl Harbor, in Hawaii. During this time, the Japanese tripled their purchases of American gasoline intended for aviation, which they had delivered primarily to the ports of South China! Not fooled, several members of the American government (with London's firm support) tried to convince Roosevelt to implement a total embargo. But Roosevelt still hoped to loosen the knot of war. On July 22, 1940, he decreed a limit on aviation fuel sales, without, however, imposing an embargo on crude oil for Japan. Japanese fuel imports increased.

The Intractable Knot of War between Japan and the United States

On September 26, 1940, when Japan had just invaded the French colony of Tonkin, north of Indochina, and was preparing to sign the Tripartite Pact with Germany and Italy, Washington hardened its restrictions on exports of iron and steel but still did not impose an oil embargo. On November 13, in a letter to his wife, Eleanor, Roosevelt summarized the dilemma which tormented him, and the inevitable outcome the White House feared: "If we forbid oil shipments to Japan, Japan will increase her purchases of Mexican oil and, furthermore, may be driven by actual necessity to a descent on the Dutch East Indies. At this writing, we all regard such action on our part as an encouragement to the spread of war in the Far East."[18] During the fall and winter, while accelerating its effort to arm itself, Japan continued to exert pressure on the Dutch, demanding total control of the economy of the Dutch East Indies.

At the beginning of 1941, a report of the US War Department described the only two options that remained. These options were like the two strands of the knot of the war, tightening week after week despite the efforts of the rational leadership on both sides of the Pacific, looking to avoid confrontation. The first option: The Japanese could quickly declare war on the United States and the United Kingdom, but they would face acute fuel shortages after three years. The second option: Japan could continue its hegemonic policy in Asia while avoiding war with the Anglo-Saxon powers, but the latter would strengthen their embargoes in response. In such a case, American strategists estimated, "Japan's natural resources would be markedly diminished. The shortage of liquid fuels, in particular, would deal a fatal blow to the nation."[19] For the Empire of Japan, the best option would be to go on the offensive as early as possible, the report's authors concluded. Admiral Isoroku Yamamoto, commander in chief of the Japanese fleets, had at this time already begun to plan the attack on Pearl Harbor.

Fully aware of the limits imposed by the lack of oil, Yamamoto knew that he had only a year remaining to take on the United States before the US fleet grew too big and Japan ran out of fuel. Yet, throughout 1941, Admiral Yamamoto himself fueled serious doubts as to the outcome of a conflict with the United States. He had lived in the United States for several years and asserted that "anyone who has seen the auto factories in Detroit and the oil fields of Texas knows that Japan lacks the national power for a naval race with America."[20] On June 24, after confirmation that Japan was about to invade French Indochina, Washington decided to freeze Japan's US assets. There was still no official embargo on fuel; however, shipments of crude oil could only be authorized on a case-by-case basis. Roosevelt left this ultimate diplomatic opening ambiguous, but this was only symbolic. In actuality, the secretary of the interior, Harold Ickes, did effectively establish an embargo: Japanese tankers were no longer permitted to fill their tanks in California.

The knot of the war appeared impossible to undo. Prince Fumimaro Konoe, the Nippon prime minister, who until the last minute had attempted to avoid war with the United States, was replaced on October 18 by General Hideki Tojo, called "The Razor." Tojo immediately authorized the attack on Pearl Harbor; he refused to allow his country to become a "third-class nation."[21] Launched at dawn on December 7, 1941, the surprise attack on the American naval base in Hawaii, located 3,500 nautical miles from Tokyo—one-sixth of the circumference of the Earth—required the development of specialized techniques for refueling at sea.[22] The plan's bold purpose was to send an advanced Japanese fleet deep into the Pacific, to protect Japan's left flank for a decisive offensive: the invasion of the Dutch East Indies, their oil fields and rubber plantations. There were more than two thousand American deaths, four of the eight American warships anchored in Hawaii were put out of service, and nearly two hundred aircraft were destroyed: The audacity of the Pearl Harbor attack allowed the Nippon Empire to gain the time necessary to corner the crucial energy source they needed.

By January 24, 1942, Japan had seized the wells of Borneo. On February 14, approximately fifty Lockheed Hudson patroller airplanes bearing British and Australian army insignia descended on Palembang, the main oil port of Sumatra. In their flanks, they transported four hundred Japanese navy paratroopers. The Dutch hesitated to launch their anti-aircraft defenses and, despite very heavy losses in the first wave of assault, the Japanese troops seized refineries and wells before Dutch teams had time to sabotage them so as to cause irreparable damage. Any damages were rapidly repaired by teams of Japanese engineers and technicians who landed in Sumatra in the days that followed. Thanks to the seventy thousand barrels of crude oil that the Dutch Indies wells supplied each day, Japan enjoyed an uninterrupted succession of victories during the months that followed.

Stalingrad: The Decisive Battle for Caucasus Oil

The German-Soviet pact signed in August 1939 gave the USSR free rein to expand its empire. It was during the course of the painful invasion of Finland during the winter of 1939–1940 that the Finns, little intimidated by the infinitely superior forces deployed by the Red Army, baptized "Molotov cocktails," named after Stalin's minister of foreign affairs: the bottles of gasoline ignited when hurled at Soviet tanks.

Moscow could not escape a confrontation with Nazi Germany. In July 1940, Adolf Hitler decided it was necessary to capture the Soviet Union's petroleum fields. On the eve of the largest military offensive in history, in June 1941, the Führer declared: "The course of war shows that we have gone too far in our efforts to achieve autarky. It is impossible to produce all that we lack by synthetic processes."[23] In particular, the failure of the Battle of Britain had proved the inadequacy of German factories to produce synthetic fuels made from coal. Hitler provided Göring with three strategic reasons that the invasion of the USSR would be essential: (1) Stalin would attack sooner or later; (2) It was necessary to take control of the crops of the Ukraine; and (3) "We must break through to the Caucasus in order to get possession of the Caucasian oil fields, since without them large-scale aerial warfare against England and America is impossible."[24] What followed showed that Germany lacked the fuel to take possession of the oil it coveted.

When Operation Barbarossa was triggered on June 22, 1941, Berlin believed (after having sharply raised their initial estimates) that the 144 divisions engaged on the Eastern Front required more than 150,000 barrels of black gold per day.[25] That was twice as many barrels as the number required by the victorious offensive on the Western Front a year earlier.[26] One division of Panzer tanks consumes 2,400 liters of fuel per kilometer, but consumption could reach 5,000 liters in off-road travel.[27] To the extent that the German Army forged ahead thousands of kilometers across the Russian steppes, the supply lines had to stretch further, along muddy paths often impassable for the trucks. In order to conserve gasoline, the Wehrmacht increasingly used horses to assure its logistics. Winter approached, and the German general Heinz Guderian had lost most of his Panzer tanks when, near Moscow, he wrote to his wife at the end of November: "The ice cold, the lack of shelter, the shortage of clothing, the heavy losses of men and equipment, the wretched state of our fuel supplies, all this makes the duties of a commander a misery."[28] Guderian reported that the fuel transformed into black ice inside the gas tanks. The engines themselves froze, or had to run day and night to avoid freezing, more quickly exhausting the meager supply. They used the antiseptic alcohol intended to treat wounded troops to preheat their airplane engines. The oils were transformed into tar.[29]

Unable to take Moscow, in 1942, the Nazi army focused its efforts on the Caucasus. Considering it the strategic and symbolic key of the offensive, on June 23, Hitler ordered the capture of Stalingrad, more than a thousand kilometers north of Baku; the city extended to the banks of the Volga, the river by which Russia routed Baku petroleum. But barely a month later, two army groups (A and B) sent toward the south began to run out of fuel. An emergency aerial bridge was established. Field Marshal Ewald von Kleist, the head of Group A, which had its sights on the Maïkop and Grozny oil fields, later said: "A certain amount of oil was delivered by air, but the total which came was insufficient to maintain the momentum of the advance, which came to a halt just when our chances looked best."[30] However, the Soviet Army, itself faced with gasoline shortages, prepared to regroup. Taking advantage of the Soviets' weak resistance, on August 15 the Germans seized a portion of the petroleum fields located around Maïkop, near the Black Sea. But Maïkop was only a secondary trophy, and the Soviets had taken care to fill its oil wells with concrete as they withdrew. A Red Army five times more numerous than the German soldiers on this section of the front defended the black gold of Grozny and Baku: It remained irretrievably out of reach. Insofar as the Wehrmacht allowed the rest of its forces to be sucked into the battle of Stalingrad, by the end of November the debacle had been transformed into a nightmare due to the significant fuel shortages faced by armored vehicles, and the even greater shortages faced by aircraft.

When the order to evacuate Maïkop was given in January 1943, the "technical brigade of oil" long ago constituted by Göring (formed by eight thousand workers and German engineers trained to fight fire and sabotage, as well as some seven thousand prisoners and specialized Russian workers) reached seventy barrels per day.[31] On June 1, 1941, three weeks before the beginning of Operation Barbarossa, Hitler had predicted with some foresight: "If I do not get the oil of Maïkop and Grozny, then I must end the war."[32] The Nazi's unrealistic plans were no less astonishing: Even if the German Army had been able to take Grozny and retain Maïkop, it would still have been necessary to route the oil by the Black Sea, which was wholly under the control of the Soviet fleet.[33] The German Army proved unable to see the war as anything more than a linear succession of battles fought over time. The Soviet effort to recapture a front spread across nearly 3,000 kilometers invoked a whole new military approach: the operative art, also called operational warfare. Defined in 1926 by a professor of Moscow's military academy, Alexandre Svetchine, operational warfare envisioned war as the sum of a complex combination of simultaneous activities (maneuvers, combat, but also logistics, industrial production, intelligence, psychological warfare, and so on) in which all must work together to achieve victory.[34] Soviet operational warfare appeared in response to the new magnitude of war, hugely increased by the energy that could now be invested in it. In all the major countries, it spurred strategic reflection after the war.

Rommel and the Inaccessible Middle East Oil

The Nazis dreamed of capturing the oil fields of the Middle East. Hitler told an ambassador of the Reich in Paris: "We will link up with the Japanese in Basra!"[35] In April 1941, Erwin Rommel and the Afrikakorps marched across the deserts of Cyrenaica with the ultimate goal of taking the Suez Canal. At the same time, on the other side of the Mediterranean, Axis forces advancing in Greece reached Rhodes: From the airfields of the island that served as their rear base, the oil fields of Iraq and Iran were in range. Berlin quickly moved in a group of engineers and oil technicians. On April 1, in Baghdad, the Iraqi generals took power by a coup. They blocked the flow of the Kirkouk-Haifa pipeline, "the carotid artery of the British empire," and soon requested Germany's assistance. In Iran, the sovereign Reza Shah—for a long time in a delicate relationship with the English, for whom he had torn away only a meager share of the profit from his country's black gold—opened the doors of Tehran wide to German advisors. The Vichy government offered the Nazis free access to its Syrian bases, still under French control. By mid-May, the

Pétain regime began to supply arms to the Iraqis who faced the British, while the Luftwaffe dispatched dozens of fighter aircraft and bombers in Mosul.[36]

But the Middle East was a little too far away for the Axis. The Royal Navy imposed a blockade along the coast of Lebanon. In early June, the British occupied Mosul. While the scant forces of a free France faced troops loyal to Pétain in Lebanon and Syria, Australian soldiers took Damascus on June 21. Deprived of the supply by the Syrian bases, the Iraqi Golden Square junta collapsed and its generals escaped. On August 25, Churchill and Stalin launched a joint invasion of Iran, rapidly taking possession of the country with minimal losses. As in 1914, troops from India occupied Abadan's vital refinery. Less than a month later, on September 16, Reza Shah was forced to abdicate, officially for health reasons, which left the throne to his highly suggestible son, Mohammad Reza Pahlavi, who was only twenty-one years old. Taking advantage of logistical capabilities that surpassed those available to the United Kingdom, the United States made a resounding entry in Iran. Beginning in 1942, the American army had enlarged ports, built roads, and built a bomber assembly factory in Abadan. Using this Persian corridor, Roosevelt delivered hundreds of thousands of tons of material to the Red Army: electric generators; aluminum and steel bars; toluene (a petroleum derivative used to make TNT); airplanes and Studebaker trucks that proved invaluable to Soviet soldiers on the central Russian front, particularly when serving as platforms for the famous Stalin rocket launchers; and even the makings of a synthetic rubber plant, as well as six ready-to-assemble refineries.[37]

After exhausting his forces between April and November of 1941, the following year Rommel found himself confronted with increasingly acute refueling problems in Cyrenaica, around the port of Tobruk. The fuel supply and spare parts for his tanks depended on the perilous transit of freighters between Naples and Tripoli, which were relentlessly harassed midway by the British, who would not relinquish control of Malta. In June 1942, in spite of all this, Rommel succeeded in taking Tobruk: He finally had a deep-water port in close proximity to Egypt. But the following month the Royal Navy sank three-quarters of the supply vessels intended for Afrikakorps. In August, Rommel considered his fuel needs to be 255,000 barrels, but he had only 68,000.[38] Confronted by the "Desert Fox," the British armored divisions never went for long without fuel, thanks to the tankers that arrived from Abadan, as well as the modest but valuable quantities of crude oil supplied by Egypt's wells. Rommel's armored divisions continued to advance; they transformed into scavengers: Their mobility depended more and more on oil deposits and British vehicles that they managed to capture as they passed by.[39] At the end of August, in the Battle of Alam el Halfa, Rommel tried once more to penetrate the British lines in Egypt.

On August 27, the field marshal declared that "the outcome of the battle will depend upon the delivery of this fuel at the proper time."[40] On September 5, the battle was lost. "We were permanently short of fuel," Rommel wrote.[41]

Two months later, the Third Reich's ultimate hope was to take control of the crude oil that was needed to continue the advancement of its armed forces into the sands of El Alamein, 250 kilometers from Cairo. Won by the British in early November, the Battle of El Alamein, along with the Battle of Stalingrad, turned the tide for the Axis forces at the beginning of winter, 1942. At the battle's beginning, on October 26, a fuel convoy coming from the tanker *Prosperina* and two freighters, carrying a total of thirty-seven thousand barrels of oil and defended by four Italian destroyers, came in view of Tobruk after having already endured two Royal Air Force attacks. At dusk, a group of bombers from Cairo arose out of nowhere. The *Prosperina* exploded. Its debris landed all over the port. When the smoke dissipated, nothing remained of the two freighters. "Rommel's last hope of victory" disappeared in that moment, observed a German officer who witnessed the disaster.[42] After the defeat of El Alamein, Rommel said: "We could attempt no operation with our remaining armor and motorized forces because of the fuel shortage; every drop that reached us had to be used for getting our troops out."[43]

The American Oil Machine Is Activated

In the United States, cooperation between the government and the oil industry, born during the duress of the Great Depression, became even more valuable during the war: It became the key to meeting the material needs for an Allied victory. This cooperation was initiated a little before the Pearl Harbor attack. It was cemented when Roosevelt created the Petroleum Administration for War on December 2, 1942, to which he naturally appointed Harold Ickes as director. Even more than he had during the New Deal, Ickes called on the major petroleum industry leaders from SoCal, Jersey Standard, and Socony to implement the PAW's fundamental tasks. The US Army used nearly five hundred different petroleum products: fuels; aerodrome asphalt; toluene for TNT bombs; incendiary weapons; smoke screens; synthetic rubber; petroleum coke for the manufacture of aluminum; wax to pack the ammunition, weapons, and food; and of course greases and lubricants for machinery and canons. By guaranteeing enormous purchase volumes and advantageous prices, the PAW facilitated the radical acceleration of extractions and industry exploration. Stable around 3.7 million barrels per day from 1939 to 1942, the American production reached

4.7 million barrels per day in 1945: An increase of more than a quarter in only three years! Louisiana became an important producer, while the moribund fields of Pennsylvania and New York were given new life by the introduction of expensive new recuperation techniques such as injecting pressurized water into the wells. Starting from zero, the production of synthetic rubber reached 750,000 tons in 1945. Construction began on no fewer than thirty-two new refineries entirely financed by the government, while the petroleum industry assumed three-quarters of the construction costs for two hundred high-octane airplane fuel refineries.[44] By the end of 1941, the United States had the ability to refuel the Red Army's aircraft in the Pacific Russian ports.

To accelerate the flow of oil and avoid the risks posed by transporting fuel to the major Atlantic ports, between which, in 1942, the coast was still patrolled by German submarines, the United States equipped itself with two new pipelines: "Big Inch" for crude oil and "Little Big Inch" for refined products; in 1942 and 1943, these were assembled across American territory at a rate of more than 10 kilometers a day, linking the fields of south Texas to the terminals of New Jersey.[45]

The US government encouraged industry to return to using coal in order to reserve the oil for Allied armies. US citizens were not fully responsive to the government's entreaties to economize, despite the inventiveness of propaganda posters such as one that encouraged carpooling with the slogan, "When you ride ALONE you ride with Hitler!" Emerging during the 1920s, the American dream was bathed in petrol, and everyone in the United States had come to consider access to oil, like access to water, an inalienable right: In Washington, many elected officials doubted the necessity of the restrictions Harold Ickes imposed. In May 1942, at the same time that ration coupons were issued, the black market appeared.

These difficulties were not, however, sufficient to seriously curb the rise in power of the American war economy. Starting in 1942, the shipyards manufactured more tankers that were able to sink Admiral Dönitz's U-boats. The Liberty class cargo ships, produced in less than a month at the end of the war, were powered by diesel-fired boilers that were as simple as they were effective.

The Destiny of the Third Reich, Precipitated by the Decline of Romanian Oil and the Bombing of Fuel Factories

The major disaster for the Third Reich's crude oil supply was neither in the Middle East nor on the Russian front; it was the decline of its first source of black gold: the wells of Romania. Until the final shock of 1944, this decline owed nothing to the Allied attacks. On August 1, 1943, the American Operation Tidal Wave, which aimed to bombard the Romanian oil installations with aircraft launched from Libya, caused one of the heaviest rates of loss ever sustained during an air raid: 53 of the 175 B-24 aircraft that left Benghazi never returned to the base. The anti-aircraft defenses that the Germans installed around the Ploiesti reservoirs and wells were more than solid. The cause of the decline of Romanian oil production was located elsewhere, under the drilling rigs taken in vain as targets by American pilots. The Romanian extractions had peaked shortly before the war, in 1936, and had decreased since then, year after year.[46] Despite all of the German engineers' attempts, nothing could reverse this natural evolution (Romanian production was relaunched in the 1950s thanks to the new techniques of pressurized water injection, before reaching its final production decline in the 1970s). Beginning in April 1944, the Royal Air Force managed to dump thousands of floating mines in the Danube, drastically reducing the Romanian oil influx that could travel up the river toward Germany, before Ploiesti fell to the Allies in September.

The ultimate blow began on May 12, 1944, a month before the invasion of Normandy, when, after years of unsuccessful attempts, the Allied air forces were finally capable of efficiently bombing the German synthetic fuel factories. Albert Speer, Hitler's architect and then minister of the Third Reich's armament, wrote in his memoirs: "That day decided the outcome of the technical war. The attack that day, by 935 bombers of the 8th American Fleet . . . meant the end of German armament."[47] On May 23, Albert Speer traveled to the Berghof, Adolf Hitler's residence in the Bavarian Alps, close to Berchtesgaden. "Dazed and absent," according to his architect, when he was told about the situation, the Führer appeared "concentrated" and "realistic" when he observed: "In my opinion, the fuel factories of Buna [synthetic rubber] and oxygen [produced by hydrogenation] constitute a particularly vulnerable point for the direction of the war, the raw materials needed for armaments being produced in a limited number of plants."[48]

Having finally hit the Achilles heel of the Nazi war machine, the Allies attacked it incessantly in the weeks that followed. After reaching thirty thousand barrels per day in May of 1944, its highest level ever since the war's beginning, German synthetic fuel production dropped to two thousand barrels per day in October, under American and British bombardment.[49] In 1944, the most important of the Allies' air operations alone, which daily mobilized more than a thousand American bombers, consumed on average more fuel each day than the Luftwaffe's entire daily gasoline supply![50] Up to the end of the war, 1,300,000 tons of bombs—a tenth of the total bombs dropped on the Reich in 1944 and 1945—were aimed at fuel facilities.[51]

Berlin prepared to reverse the situation. It was a monstrous task. The Nazi "Night and Fog" decree of December 1941 filled concentration camps with prisoner labor to produce equipment and, in particular, fuel for its army. It took at least ten times more steel to produce a barrel of gasoline in the synthetic fuel factories than it did to draw and refine a barrel of crude oil, and the work was infinitely harder.[52] In these plants, explosions, fires, and deadly gas leaks were common; deportees and prisoners formed at least one-quarter of the Third Reich's workforce starting in 1943.[53]

Established in October 1942, in the vicinity of the Auschwitz extermination camp, the Monowitz fuel and synthetic rubber plant was one of IG Farben's most important operations. Its location was chosen for its proximity to Polish coal mines. One of the most famous Auschwitz survivors, Primo Levi, described the industrial complex of Monowitz this way: "This huge entanglement of iron, concrete, mud and smoke is the negation of beauty. . . . Within its bounds not a blade of grass grows, and the soil is impregnated with the poisonous saps of coal and petroleum, and the only thing alive are machines and slaves—and the former are more alive than the latter."[54] At the end of 1944, in a desperate effort to repair their fuel plants and create new ones, the Nazis attempted to mobilize 120,000 workers in mines, quarries, and caves scattered throughout Germany and Austria. Half of them were slaves.[55] The forced workers of Nazi Germany, however, were not sufficient to offset its deficit of "energy slaves"—in other words, the energy that crude oil would have provided.

On the Western Front, the American Army Unleashes Its Unequaled Energy Power

During the June 6, 1944, Normandy landings, the superiority of Allied aviation was crushing to a German army that was frequently short of oil. On June 10, the Seventeenth SS Panzergrenadier Division was bogged down by a fuel shortage near Saint-Lô, where it sustained heavy losses. Shortly after, Field Marshal Gerd von Rundstedt issued the following order: "Move your equipment with men and horses; don't use gasoline except in battle."[56]

By 1942, the Allies had begun to accumulate fuel in preparation for the landings. On D-Day, stocks reached two million barrels. By August, the British Petroleum Warfare Department accomplished an unprecedented feat:

In Operation Pluto, they installed the first underwater long-distance pipeline across the English channel, between the Isle of Wight and the French port of Cherbourg. But the operation was continually faced with grave technical difficulties. Tankers provided the essential fuel supply, anchoring a few hundred meters from the Normandy beaches; in the course of a few hours, they discharged their gasoline through pipes to fill tanks scattered behind the shore. As the Allies advanced, teams of American petroleum technicians successfully placed up to 80 kilometers per day of preassembled pipeline, 10 centimeters in diameter, along the French roads. Day and night, the trucks of the Red Ball Express also transported fuel, as well as moving the rest of the Allied equipment across Normandy, between Saint-Lô and Chartres, on a circuit of one way roads originating in Alençon and returning via Mayenne. After the August liberation of Paris, the Red Ball Express was extended to Soissons, and remained the critical supply path until the Allies took control of the port of Antwerp in mid-November. American troops revealed themselves to be more mechanized than the much-feared German Army. No less than half of the tonnage of equipment sent from the United States across the Atlantic and Pacific (vehicles, weapons, ammunition, food, etc.) was petroleum.[57] When the Allied soldiers, who could progress in trucks and jeeps, took the Florence region of Italy in the summer of 1944, they found hundreds of horse cadavers floating in the Arno River. "The discovery really opened our eyes to the limitations on German mobility," an American infantry officer testified.[58] In the fall, the German Army reduced the use of trucks for troop transport to a strict minimum. The German retreat occurred mostly on foot or by bicycle.

In France, however, the advancement of Allied troops began in September. On the Red Ball Express, provisioning of fuel remained limited; the longer the route, the greater the proportion of fuel intended for the front was used before it reached its destination. These difficulties enraged General Patton, who, at the head of the Third Army, predicted they would cross the Rhine by October. His soldiers diverted convoys to steal gasoline, and sent reconnaissance aircraft hundreds of kilometers, from the front lines back to the rear, to locate fuel depots.

The logistical problems the Allies encountered and Germany's terrifying efforts to revive its synthetic fuel production allowed Field Marshal von Rundstedt to engage in what was an important effort of the Third Reich war machine: the Ardennes counteroffensive. The German Army concentrated the best of the remaining mechanized forces, throwing into the offensive many more tanks and artillery pieces than the Allies had at their disposal on this part of the front. Their purpose: Retake the port of Antwerp. On December 21, the Germans encircled the Belgian city of Bastogne. But the counteroffensive fizzled out quickly, again due to lack of fuel. Hitler had promised two thousand aircraft. Hardly more than three hundred were able to fly on the Western Front. Around Christmas, three of the five divisions of the Forty-Seventh Panzer Corps were wholly immobilized, as were many artillery and refueling units, which became easy prey for the ubiquitous aviation of Allies.[59] Patton's tanks moved in on January 1 and took Bastogne; four thousand American soldiers died during the battle, but the fate of the German Army on the Western Front was irrevocably sealed. Two weeks earlier, at the very beginning of the counteroffensive, the Wehrmacht had come closer to a unique opportunity to reverse the situation, at least for a time. On the morning of December 18, at the entrance of the village of Stavelot, somewhere between Bastogne and Liege, 142 tanks of the First SS Panzer Division had miraculously halted their advance less than 500 meters from a huge Ally fuel depot. Thanks to a simple wall of gasoline-fueled fire set in haste by the American soldiers, the Allies succeeded in dissuading Lieutenant Colonel Joachim Peiper, former warrant officer of Himmler and veteran of the Russian front, from advancing his tanks. The wall of fire consumed 500,000 liters of gasoline but saved 9 million liters, which, had they been taken, could have changed the battle's outcome.[60]

The Final Collapse of Japan and Germany, Drowned by a Tide of Oil

In the Pacific, the day after Pearl Harbor, Japanese tankers became a priority target for the US Navy. As the Americans progressed across the islands, their submarines and aircraft caused increasing damage. The sea was vast, and the opportunities for attack were numerous during the ten-day journey across the 2,600 nautical miles separating Sumatra from Japan. With only 2 tankers sunk in 1942, the American military score rose to 23 the following year, then to 131 in 1944.[61] That year, Japanese fuel resources fell by half.[62] Japanese naval shipyards were unable to compensate for the losses. The Imperial Japanese Navy resorted to ingenious but nevertheless desperate tactics: Submarines transported the crucial fuel, and some ships had to tow large rubber pouches filled with several hundred barrels of black gold, making them an easy mark for American fighter planes.[63]

As the American War Department had predicted, the consequences of the gasoline shortage were severe. Missing, in particular, fuel for its aviation, the Imperial Japanese Army had no other choice than to reduce the training hours of young pilots, who were being massacred in combat by veteran American pilots. In June 1944, military engagements around the Mariana Islands archipelago became known in the US Air Force as the Great Marianas Turkey Shoot. It was also during the summer of 1944 that squadrons of "kamikazes" appeared: young students, novice pilots, most of whom voluntarily hurled their planes into American warships. Their rudimentary training did not necessarily include landing; in any case, their airplanes did not have enough fuel to return to base. Also in mid-1944, the petroleum freighter from Insulinde effectively became nonexistent. American submarines had dropped so many floating mines around the oil port of Balikpapan, Borneo, that the Japanese abandoned it in December. In 1945, Japanese planes on average flew no more than two hours per month.[64] In a vain effort, the Imperial Japanese Navy launched a program to create gasoline from pine roots; this method only produced three thousand barrels by the war's end.[65]

Japanese fighter pilots were effectively grounded when, on March 9, 1945, having realized that Japanese cities were essentially built of wood, the Americans began to bomb them with incendiary weapons. More than one hundred thousand citizens of Tokyo burned and died in the fire, which reached 1,000°C (1,832°F). "The heart of Tokyo is gone," reported the *New York Times*.[66] This was probably the most lethal air raid in history. The new B-29 long-distance bombers dropped phosphorus and napalm bombs, weapons based on the gasoline gel also used by flamethrowers, developed in 1943 by Louis Fieser, an organic chemist from Harvard University, better known for his work on synthetic cortisone. The production of napalm was monopolized by Standard Oil of New Jersey.[67] The bombing of Tokyo was such a success, and American losses were so minimal, the US Air Force decided to launch similar attacks on more than sixty Japanese cities (Yokohama, Nagoya, Osaka, Kobe, Kawasaki, etc.) in the following weeks. The devastating chemical energy of oil continued to be adapted and perfected in every conceivable form. The night of May 25 to 26, Tokyo was again the target of a monstrous raid. Five hundred B-29 bombers dropped 3,251 tons of incendiary bombs, which razed an area of the Japanese capital equivalent to half of Paris. The head of American aviation, Curtis LeMay, a man of few words, had access to 348 tons of enormous M76 bombs manufactured in Henry Kaiser's Los Angeles steel mills.[68] Kaiser was the industrial magnate at the head of the prodigious shipyards that built Liberty ships. These bombs were filled with an extremely flammable mixture of magnesium and bitumen known as "goop."[69] The long bombing campaign preceded the dropping of atomic bombs on Hiroshima and Nagasaki in August 1945 and was much deadlier.

In early 1945, Hitler decided to transfer his dwindling troops on the Western Front, not toward Berlin to defend the capital against advancing Soviet tanks but toward Hungary, in order to attempt to protect the meager oil production of the Lake Balaton region, the Nazis' last remaining substantial source of crude oil. In order to make their V-2 missiles work—the first ballistic missiles that Germany rained down on Antwerp and London (designed by Wernher von Braun, the future father of the American space program)—the Nazis resorted to potato alcohol.[70] The Reich had even less means to develop large-scale assembly lines and produce fuel for its ultimate new miracle weapon: the first jet aircraft. It was much too late. On April 30, 1945, as the Red Army won control of the ruined city of Berlin, street by street, enough oil remained to immolate the bodies of Adolf Hitler and his companion Eva Braun in a bomb crate, in accordance with instructions left by the Führer just before his suicide, when he learned of the fate reserved for the mortal remains of Mussolini, who had been shot two days earlier and hanged by his feet at a gas station in Milan.

After Japan's capitulation on September 2, 1945, Nippon's major daily newspaper, *Asahi Shimbun*, wrote: "It was a war begun as a fight for oil and ended by the lack of oil."[71] During the course of an interview, a professor at Tokyo Imperial University declared, "God was on the side of the nation that had oil.[72]

Delusions of power, born of the oil machine, dissipated for lack of energy, like the dream of an alchemist's apprentice who squanders all his gold to make gold. Part of the delusion of the Berlin and Tokyo masters had been to lay claim to a material force whose limits they were incapable of admitting, or even seeing.

From this inability, a blindness caused by a kind of immeasurable juvenile pride, the best brains of the American army were not exempt. In April 1944, for example, the first mission of the formidable new B-29 bombers, Operation Matterhorn, began an attempt to bomb the Japanese steel mills from China. To do this, the Americans planes first flew from Kansas to India. Then, loaded with as much fuel as they could carry, the B-29s made roundtrips over the Himalayas and landed on runways specially constructed by thousands of Chinese workers, in order to stock the fuel that was needed to bomb Japan. American officials realized too late that to undertake the trip back to India and reload fuel, the B-29 had to empty the very same deposits they had traveled to China to fill. "To make a long story short, it wasn't worth a damn," later explained Robert Strange McNamara, one of the authorizing officers in the operation, and future president of Ford and secretary of defense for Presidents Kennedy and Johnson.[73]

The United States, however, possessed energy resources sufficient to recover from such an intercontinental mess, and learned their lesson. The US delusion of power gave birth, just before the war, to icons such as Superman, the "man of steel" who *Look* magazine portrayed as early as February 1940 flying to Berlin to catch Hitler by the scruff of his neck.

Yet it was not merely delusion. It was a logical outcome of the contradictions embodied in Operation Matterhorn, Operation Barbarossa, and Japan's decision to take control of Sumatra's petroleum. It was what caused the Soviets to redefine war strategy as operative art: Human activity and human desire had been subjugated to the technological complexities of machines whose very needs had come to govern us. Even the sharpest minds had failed to understand this power of the machines, precisely when they fell under their rule.

The energy-complexity spiral swelled abruptly at the onset of the Second World War. Periodically or perhaps absolutely, the spiral already seemed to exceed the human intelligence that claimed to shape it. From 1939 to 1945, the war expressed itself in every aspect of peoples' lives. Humanity could only change the course of history within physical limits that eluded us more and more. The vortex of energy and complexity bestowed an unbreakable grip on corporations and governments, themselves the greediest and most complex energy consumers. Technology was deployed thanks to the energy it required—in war, and soon also in peace. The war had revealed the manner in which the new world would function—a world where it was the masters of technology who would govern. The ultimate masters were beyond the trivial realm of men: they were the laws and limits of physics.

The "triumph of the will" that the Nazis had foretold for themselves could simply not come to pass. An unprecedented energy avalanche was unleashed after Pearl Harbor, and it never ceased to gain strength afterward. Its hold on the major societal bodies precipitated chaotic and unpredictable consequences, irresistible and irreparable, that were radically difficult to understand at a time when it was believed that humans would triumph over nature and barbarism through technical innovation. And so, structuralist thinking was about to vainly pronounce the "death of man" as an exclusive and driving force of history.

PART TWO

SPRING

1945–1970

ELEVEN

After Yalta: The United States and Saudi Arabia Seal Their Alliance

In the spring of 1941, while in Washington the Roosevelt administration investigated the relationship between Standard Oil of New Jersey and IG Farben; a smattering of American pioneers, geologists, engineers, and technicians found themselves trapped in the Saudi Arabian desert, away from the tumult of war. The United States had not yet entered the war and the Axis forces had almost reached the pinnacle of their power. The Axis occupied almost the whole of Europe, and Field Marshal Erwin Rommel was about to set foot in North Africa: The war made it impossible to route crude extracted from the east of the Suez Canal, where steel was in short supply. Soon, these American technicians decided to close the first oil wells of the young kingdom of Saudi Arabia. The few women and children who had accompanied them were evacuated, awaiting better days.

Roosevelt Declares the Defense of Saudi Arabia "Vital"

Everything had started well. In May 1939, four months before the outbreak of the Second World War, the first American cargo of Saudi Arabian crude extract had been loaded aboard the American cargo ship *D. G. Schofield*. The future looked bright for the California-Arabian Standard Oil Company (CASOC), a partnership since 1936 between Saudi Standard Oil of California (SoCal) and Texaco, the two second-most powerful American oil companies.

Located in the midst of the British Empire's "zone of influence," far from Texas and California, the enclave of Saudi Arabian and American petroleum was in peril. The greatly feared King Abdulaziz Al Saud, who had secured control of a domain with Enfield rifles and scimitar strikes, was short of gold. His two main sources of revenue had dried up: The royalties paid by CASOC were even less than the taxes levied on the rare pilgrims who managed to forge a path to Mecca. The king was left to rely soley on aid sent by British India. In early 1941, American CASOC directors feared that their promising beachhead would escape them and be expropriated to the benefit of British oilmen, who had dominated most of the rest of the Persian Gulf for almost a half century.

On April 9, 1941, SoCal and Texaco sent an emissary to plead their cause with Franklin Delano Roosevelt. They appealed to an old New Deal ally of the president, James Moffett, who had taken charge of the business of the two companies in the Persian Gulf, on the island of Bahrain. Moffett had helped Roosevelt raise millions of dollars during the election campaign of 1940.[1] But their plan did not work. Moffett's plea failed in the face of categorical opposition from Roosevelt's secretary of commerce, Jesse H. Jones. The latter, then in charge of the strategic political financing of the Allies, later justified: "The national interest was not going to be served by extending financial assistance to a backward, corrupt and non-democratic society like Saudi Arabia."[2] Roosevelt quickly defended his secretary of commerce. On July 18, 1941, he wrote: "Jesse, can you tell the British I hope they can take care of the king of Saudi Arabia. This is a little far afield for us. FDR."[3]

A year and a half later, however, Franklin D. Roosevelt enacted the Lend-Lease Act with the monarch of Saudi Arabia. Washington's credit assistance program was supposed to be reserved for Allied democracies. However, on February 18, 1943, the American president solemnly declared "I hereby find that the defense of Saudi Arabia is vital to the defense of the United States." In the meantime, the United States had entered the war, and in December 1942 Harold Ickes, the all-powerful US secretary of the interior, championed a program vital for the war effort: the Petroleum Administration for War. Again, as he had been during the New Deal, Ickes became the "oil czar," writing that: "An honest and scrupulous man in the oil business is so rare as to rank as a museum piece."[4] The "old curmudgeon" (as Ickes liked to be called) certainly had not changed his opinion on the men of the oil industry. But he knew them very well, and he knew how to differentiate between a solid argument and one designed to pull the wool over his eyes. As soon he took the reins of the Petroleum Administration for War, Ickes received the leaders of SoCal and Texaco, who came to request the US Navy's assistance to protect their concessions in the far-off desert of Saudi Arabia, more than 6,000 miles from Washington. This occurred during the strongest thrust of the German advance toward Egypt; at the head of Afrikakorps, Rommel was threatening to reach the Suez Canal.

Ickes must have been convinced by the figures Texaco and SoCal placed before him. To verify these numbers, he quickly dispatched the best American geologist of the era, Everette DeGolyer, to the fields of Saudi Arabia.* The assessment of the oil reserves revealed fabulous, dizzying resources, unequalled on Earth, including in Texas: Saudi Arabia had twenty-five billion, maybe one hundred billion barrels of crude oil.[5] At the end of the day, it turned out that the Saudi Arabian oil field contained four times more oil than the total proven reserves American companies had controlled up to that time![6] On February 3, 1943, two weeks before Roosevelt's decision to finance Saudi Arabia, a State Department analyst and member of the DeGolyer mission sent Washington a cable in which he wrote: "The oil in this region is the greatest single prize in all history."[7] When the war began, barely 5 percent of the crude oil produced in the world came from the Middle East—63 percent was extracted in the United States. But after the DeGolyer mission and increasingly starting in 1943, many diplomats and senior leaders became aware of the fact that everything was about to change. In his final written account of 1944, Everette Lee DeGolyer announced: "The center of gravity of the world oil production is shifting from the Gulf-Caribbean areas to the Middle East, in the Persian Gulf area."[8]

In the Name of Anti-Imperialism, Big Oil Gains the Upper Hand

The American army remained engaged in the largest military deployment in history, which spread across three continents and two oceans. And the United States provided up to 90 percent of the oil the Allied forces required. Except that since the 1930s, discoveries of new, important oil fields under American soil had become more and more rare. One of Harold Ickes's men, the director of reserves of the Petroleum Administration for War, diagnosed profound consequences: In the United States, "[a]s for what's left of the discovered oil, the jackpot is just about finished."[9] Needless to say, Mexico had nationalized its petroleum just before the war, causing the gringos to lose control of one of

* DeGolyer had definitively established his reputation as a prospector before the First World War by making the first discoveries that led to the development of Mexico's "Golden Path," then by cautioning against the consequences of their excessively rapid extraction rate. World-renowned, Everette DeGolyer's geophysics company became Dallas's uncontested industrial giant and in 1951 was the originator of Texas Instruments, one of the pioneer American electronics and armament companies.

the biggest world production sources, with the advantageous position of being a friendly neighbor to the United States.

In August of 1943, Harold Ickes wrote to Roosevelt: "Next to winning the war, the most important matter before us as a nation [is] the world oil situation."[10] In January 1944, in an article published in a widely circulated magazine, Ickes warned the American people: "We're running out of oil!" The "old curmudgeon" unambiguously added: "If there should be a World War III it would have to be fought with someone else's petroleum, because the United States wouldn't have it. . . . America's crown, symbolizing supremacy as the oil empire of the world, is sliding down over one eye." He concluded: "We should have available oil in different parts of the world. . . . The time to get going is now."[11]

The Roosevelt administration's reaction was dramatic. In a memo sent to the president in June 1943, the undersecretary for the navy, William Bullitt, said: "to acquire petroleum reserves outside our boundaries has become . . . a vital interest of the United States."[12] According to Bullitt, the United States urgently needed to follow the British Crown's example of exerting absolute control over Persia's oil, initiated on the eve of the First World War by lord of the admiralty, Winston Churchill. "We are forty years late in starting—but we are not yet too late," wrote the head of the US Navy.[13]

From the Roosevelt administration's point of view, Saudi petroleum, hardly yet exploited, became too important a commodity to be left in the hands of the oil industry. The idea of nationalization contradicted, of course, American traditions and principles. A piece in the *New York Times*, for example, condemned an "approach leading to imperialism," not comprehending why American taxpayers should participate in the financing of oil drilling at the other end of the world: "Why does it [the US government] not use the $135,000,000 to $165,000,000 estimated for Saudi Arabia in financing the development of oil deposits on the north and south coasts of Alaska?"[14]

Discovered as early as the 1920s, although less abundant and above all infinitely more difficult to develop than those of Saudi Arabia, the fields of Alaskan crude oil were put into production much later, when the industry was forced to do so, at the end of the first petroleum crisis. The American oil industry and the Roosevelt administration embraced the obvious: It appeared to them simply much easier and more profitable, and therefore necessary, to seek petroleum on the other side of the world (even if this meant aligning with a strange regime), rather than extracting a more difficult source of power, even though it was within US territory. Always and everywhere, material power expresses itself by flowing, like water, down the slope of least resistance: Men mobilize this power, but are bound by the laws of nature, and ecological limits, which surpass by far their own reasoning or will.

For the Pentagon, necessity was the law of the land. In July, during a meeting at the White House, Franklin Roosevelt, Harold Ickes, Secretary of War Henry Stimson, and Secretary of the Navy Frank Knox decided that Uncle Sam must acquire 100 percent of SoCal and Texaco's Saudi Arabian concession. The two company's leaders concluded later that they had managed to attract Washington's attention only too well. "As for the companies, they had gone fishing for a cod and had caught a whale," commented a State Department advisor.[15]

But the grandiose plan designed to give the US government direct control of the stunning Saudi reserves never saw the light of day. At least not in the way Harold Ickes and the US Navy had imagined. At the end of 1943, Rommel lost the battle of North Africa. The shadow that menaced the Middle East was disappearing. SoCal and Texaco put forth an intense campaign, lobbying the White House to prevent the US nationalization of Saudi oil. Harold Ickes repeated that other powerful countries all possessed nationalized oil companies; Roosevelt's secretary of the interior needed to face the fact that he depended on Big Oil to win the war. He rapidly threw in the towel. Or almost. He attempted a compromise, proposing that the government build a 2,000-kilometer pipeline at its own expense, in order to route the Saudi Arabian crude oil to the Mediterranean without passing through Britain's Suez Canal. In exchange, the shareholders of CASOC, SoCal, and Texaco would be required to reserve a fifth of their future production for the US Navy, at a friendly price.

First obstacle: It was impossible to construct this kind of pipeline, through Jordan and Palestine to Lebanon, without the United Kingdom's full consent. So, counseled by Harold Ickes, Roosevelt proposed to Churchill that the United States and Britain share the Persian Gulf oil. "Are dollar diplomacy and economic imperialism to prevail?" the *New York Times* worried. London was very suspicious. Churchill responded to Roosevelt, "There is apprehension in some quarters here that the United States has a desire to deprive us of our oil assets in the Middle East on which, among other things, the whole supply of our Navy depends."[16] The defiance was reciprocal. Roosevelt replied that he himself had received reports indicating that British Empire agents were in the process of maneuvering to "horn in" on the Saudi concessions.[17] Despite all this, on August 8, 1944, America and Britain came together in a very ambitious joint project. This Anglo-American oil agreement was based on the same logic as the secret agreements of oil market "regulation"—the As-Is Agreement of

Achnacarry and the Red Line Agreement negotiated by the major oil companies at the beginning of the 1930s. But there was an important nuance. It amounted to negotiating, this time very openly, the rates of Middle East oil production, between the large Anglo-Saxon companies on both sides of the Atlantic. A monumental stumbling block for the latter was that the preexisting Anglo-American oil agreement provided that the regulation of worldwide production would be entrusted to the American and British governments, through a hypothetical International Petroleum Commission, instead of to the oil companies.

The Anglo-American oil agreement, just like the "state pipeline" proposed by Harold Ickes, was met with hostility in the United States. In October 1944, the independent petroleum companies of Texas and Oklahoma violently objected to the agreement when it went before the Senate. They proclaimed their fear that such a treaty would generate a "vicious cartel" around the owners of CASOC—which had changed its name at the beginning of the year to become the Arabian American Oil Company, or Aramco.[18] An Oklahoma senator allied with the oil industry went so far as to call the pipeline that Washington wanted an "imperialist venture." And a report funded by the American oil industry bluntly stated that the whole project represented a "fascist approach."[19] Faced with hostility from the Senate, in January 1945 the White House withdrew its proposal to formally share the black gold market with Britain. On his return from Yalta the following month, the American president was pleased to secure an open door for American oil companies in Saudi Arabia.

As Soon as the Kingdom of Saudi Arabia Is Founded, the American Oil Industry Arrives on the "Pirate Coast"

On the evening of February 12, 1945, the sovereign who embarked on the American destroyer *Murphy* was not a tranquil man. It was perhaps only the third time in his life that His Majesty Abdulaziz Al Saud had left the Kingdom of Saudi, which he founded and which bore his clan's name. The US Navy had to lead the monarch from Jeddah, the port where the pilgrims from Mecca landed, up to the Great Bitter Lake at the mouth of the Suez Canal 1,000 kilometers north, across the Red Sea. The Arab king was on his way to join President Roosevelt, who the evening before had participated in the closing ceremonies of the Yalta Conference in the USSR, in the company of Winston Churchill and Joseph Stalin.

About forty people accompanied King Abdulaziz Al Saud, including his astrologer, his chef, his food taster (and his assistant), his chamberlain, as well as ten guards armed with swords, daggers, and English rifles. The king also wished to take along a herd of eighty-six sheep, intended to be butchered in honor of the American president, but US Navy officials insisted that only eight animals be brought on board.[20] The king also brought hostages from two unruly tribes, the Mutaïr and the Bani Khalid, to ensure their tranquility during his absence—because the risk of insurrection was real and Abdulaziz had to stay on his toes.[21] The reign of suzerainty that he had established, and through which he had been able to inaugurate his kingdom thirteen years earlier, in 1932, remained fragile, in particular when gold was lacking. When the USS *Murphy* left the port of Jeddah, all aircraft sirens were silenced, so the departure went unnoticed. During navigation in the Red Sea, radio contact with Mecca was maintained. Every half hour, a radio operator called the Holy City, asking simply: "OK?" Mecca responded: "OK," then the communication was cut.[22] During the secret voyage of the House of Saud's leader, rumors circulated claiming that he had abdicated or been abducted.

Early in 1945, as the Allies neared their final victory, the foundations of the young Saudi kingdom remained precarious. With his giant stature (he was about 2 meters tall and weighed 130 kilos [187 pounds]), King Abdulaziz had the temperament of a cunning war chief. Sometimes called the "Leopard of the Desert," and in the West often called simply Ibn Saud, the king had known many reversals of fortune before reaching his throne. His carved cane and unsteady approach reminded everyone that his body had been bruised by the legacy of a dozen serious war injuries. To become king, he had waged war, riding on camelback through the desert for the last twenty years, from the shores of the Persian Gulf to the Red Sea. The roots of his state ran to the deepest depths of the Nedj Desert, in the center of the Arabian peninsula: There, in 1744, Abdulaziz Al Saud's ancestor Mohammed Ibn Saud, leader of a small village of mud huts near the present capital, Riyadh, sealed the unwavering alliance of his clan with Mohammed Ibn Abd al-Wahhab, the founding Imam of "Wahhabism," the branch of radical Islam that the present day Al Qaeda terrorist movement claims as their lineage.[23]

As soon as the first decisive battles of conquest by the House of Saud began, over the course of the decade beginning in 1910, warriors with eyes encircled in black, wearing white robes and long, pointed beards, lined up beside Ibn Saud: the fearsome combatants of the Ikwan. This elite body, following the example of the Prophet Mohammed's Islamic legions, was required to form the heart of this new kingdom's army. According to Ibn Saud, the Ikwan should also constitute the basis of agrarian farming communities capable of unifying the fickle Bedouin tribes around a few large dusty plains where the king dreamed of seeing an unceasing profusion of fertile fields.

Raids between tribes had always been part of the vagaries of nomadic life in the great Saudi desert. During these attacks, relatively few died. But the members of the Ikwan, many from the tribe of the Mutaïr, behaved differently: They overrode Ibn Saud's orders, razed villages, and killed women and children they judged to be insufficiently pious. The troubled pact between the Saud family and the religiously fanatic Ikwan remained both pillar and poison of the kingdom that became the first world oil power. On March 29, 1929, during the decisive battle of Sabilla, Ibn Saud ordered a devastating charge against the Ikwan, which soon quashed them permanently. The founder of the royal Saudi line had subdued the revolt of his henchmen in cold blood. Hostile to technological progress, many were mowed down on their camels by Western machine guns. Throughout its history, the House of Saud did not cease, depending on the circumstances, to punish, woo, and, as often as necessary, encourage the zealots of barbaric Wahhabism.

The first American prospectors landed in Saudi Arabia in 1933, only one year after the kingdom was founded. And it was a former English diplomat, Jack Philby, who opened the door to this new country, subsidized by the British Empire and located between its two strongholds, Egypt and India. After the First World War, in profound disagreement, like his rival T. E. Lawrence, with Great Britain's policies in the Levant, Harry Saint-John Bridger ("Jack") Philby left the king of England's service to become Ibn Saud's personal advisor. By Ibn Saud's side, Philby witnessed in 1925 the conquest of the Hijaz and the flight of the Hashemites, Lawrence of Arabia's protégées. Philby—who had become Sheikh Abdullah by this time—was quickly convinced that the Arabian peninsula contained immense mineral wealth. As early as 1930, scarcely after the chiefs of the Ikwan revolt had been massacred or imprisoned, Philby pushed Ibn Saud to bring foreign prospectors and to sell them concessions. The monarch hoped that the infidels would discover water rather than oil. "Oh, Philby, the king replied, if anyone would offer me a million pounds, I would give him all the concessions he wanted."[24]

However, the king had to wait before hitting the jackpot. After the crisis of 1929, the oil market had been saturated. Ibn Saud, as a war chief, had never hesitated to rely on the English, ever since the beginning of good relations with the Iraq Petroleum Company (IPC), which was devoted to British interests.[25] But the Bedouin king wanted gold, while London told him condescendingly that the sterling and the rupee were worth just as much as precious metal.

The Compagnie Française des Pétroles (CFP) had the opportunity to try its luck, but Colonel Mercier, a man of compromise, refused to risk breaking the Red Line Agreement.[26] Taking advantage of this lack of willingness from both the English and the French, the Californians of SoCal—who were not shareholders of the Iraq Petroleum Commission and not bound by the famous Red Line Agreement—acquired the first Saudi concession in August 1933, with a modest loan equal to £50,000 and an annual rent of £5,000, all paid in gold. Jack Philby was gratified by the salary SoCal paid him: 1,000 pounds annually. Ibn Saud's adviser (father of Kim Philby, the future celebrated British double agent in the USSR's employ) was quite satisfied by the privileged opportunities Americans afforded his imperialist compatriots.[27] According to Philby himself, when he announced to London's representative in Jeddah that SoCal had won the concession, the latter appeared to be "thunderstruck, and his face darkened with anger and disappointment."[28]

The SoCal engineers, with limited means for prospecting, were joined in 1936 by those of Texaco, which had acquired 50 percent of the shares of the California-Arabian Standard Oil Company, the future Aramco. The first successful drilling produced black gold on March 4, 1938. Like many others that followed, this well—Damman 7, shut down in the early 1980s—was located a few kilometers from the Persian Gulf, in the Hassa region, along a shoreline long ago nicknamed the "Pirate Coast" for its role as a sheltered point where pirates could disembark and attack rich sailing vessels, stealing their shipments of Arab pearls and Persian fabrics. In May 1939, for two days, banquets were held to celebrate the first loading of crude oil onto the tanker *D. G. Scofield*, which bore the moniker of SoCal's first president. King Ibn Saud himself opened the valves. He was so satisfied that, shortly afterward, he increased the exclusive American concession to more than half the total of his young kingdom. Not a single American diplomat was present to celebrate the event and thank the sovereign: All the dealings had taken place between a country and a company.

On the Return from Yalta, Roosevelt Seals the Alliance with Abdulaziz Al Saud

On the Great Bitter Lake on Valentine's Day of 1945, the US Navy Destroyer *Murphy*, carrying His Majesty Abdulaziz Al Saud, pulled up alongside the USS *Quincy*, where they awaited President Roosevelt's return from Yalta. On February 14, 2005, during the sixtieth anniversary of the meeting, a joint communiqué of President George W. Bush and the Crown Prince of Saudi Arabia Abdallah did justice to the event, pronouncing: "In six hours President Bush's predecessor and the Crown Prince's father established a strong personal bond that set the tone for decades of close relations between our two nations."[29] Witnesses of the meeting remarked that there seemed to be a reciprocal understanding between Franklin Roosevelt and Ibn Saud. The first, stricken by poliomyelitis, moved with the help of a wheelchair, while the injuries of the second caused him to walk with difficulty, relying heavily on his cane. In the middle of the meeting, the Arab king told the American president that they were "twins" because of their similar ages, their interest in agriculture, and, finally, because of the infirmities that had struck them.[30] "You are luckier than I because you can still walk on your legs and I have to be wheeled wherever I go," said Roosevelt. "No, my friend, you are more fortunate. Your chair will take you wherever you want to go and you know you will get there. My legs are less reliable and are getting weaker every day."[31] Roosevelt had an extra wheelchair aboard the *Quincy*. He decided to give it to Ibn Saud. Although he was too stout to use it, the king of Saudi Arabia liked to say that this was one of his most valuable assets.[32] Yet it was difficult to imagine a greater dissimilarity than that between these two friends—Roosevelt, leader of the most advanced democratic nation on the planet, and Ibn Saud, autocratic sovereign of superstitious camel riders, "backward Arabian nomads," almost ignorant of modern science and of the world around them, as the *New York Times* reported.[33]

Roosevelt wanted to talk about oil.[34] Strangely, official American reports seem to have kept a prudish veil over the topic. Perhaps the two men skirted the dossier, too obvious or too trivial, and already finalized, to concentrate on forging an alliance between the two cultures and political regimes, separated by far more than geographical distance. The conversation centered on the British Empire's politics with respect to Palestine. Before their meeting, Roosevelt had on several occasions written to the king to ask his counsel regarding the increasingly frequent clashes between Jews and Palestinian Arabs. Each time, Ibn Saud responded that the only solution was to put an end to Jewish immigration. According to Ibn Saud, if the Jews returned to Palestine, "the Heavens will split, the earth will be rent asunder, and the mountains will tremble at what the Jews claim in Palestine, both materially and spiritually."[35] Aboard the *Quincy*, the Arab king, protector of the most holy Islamic sites, reiterated his warning: A conflict in Palestine would be inevitable, unless the American president envisioned resettling the Jews in Europe, where they had been wronged. At the end of the interview, Roosevelt made two promises:

The American government would never change its policy in Palestine without consulting both the Arabs and the Jews, and they would not undertake any action hostile to the Arabs. Though these promises were not binding, they were enough to satisfy Ibn Saud. The American spy responsible for translating the conversation (and who continued to facilitate communication between Washington and Aramco) said that, in conversation with Roosevelt, the Saudi king did not cease to complain about Churchill's inflexibility on the subject of returning Jews to Palestine.[36] This was an additional sign that the king had decided to make the United States his new main ally, to the British Empire's detriment. Roosevelt had achieved his essential mission. Even if one ignores what the president and the king specifically said on the subject of oil, the Cairo *New York Times* correspondent remarked immediately after the meeting: "The immense oil deposits in Saudi Arabia alone make that country more important to American diplomacy than almost any other smaller nation." He cited an "observer" who believed that "except for the Philippine Islands, Saudi Arabia may well prove to be the most interesting foreign area to the United States in this century."[37]

Learning of the meeting, Winston Churchill "burned up the wires to all his diplomats in the area, breathing out threatenings and slaughter" if they did not arrange an interview with Ibn Saud.[38] The British prime minister rushed to Egypt and, three days after Roosevelt and Saud's meeting aboard the *Quincy*, Churchill met with the Arab sovereign in the south of Cairo. Aware that he might already be too late, Churchill behaved condescendingly toward the king. The latter indicated that he could not tolerate anyone smoking or drinking alcohol in his presence. At the banquet that followed the meeting, the "British Bulldog" ignored his request. Churchill explained later without remorse, and with what appeared to be retrospective frustration: "I was the host and I said if it was his religion that made him say such things, my religion prescribed as an absolute sacred rite smoking cigars and drinking alcohol before, after, and if need be during, all meals and the intervals between them."[39] During his ascent to the throne, Ibn Saud had always feared the English and while seeking their support had endured their harassment, having seen the fallout of Ikwan punished by British machine guns and planes. Without a doubt, he enjoyed Churchill's discomfiture; this faux pas in February 1945 foreshadowed the erosion of the United Kingdom's power in Saudi Arabia.

Exhausted by his voyage, Roosevelt died two months later in the United States, on April 12. During the years preceding Abdulaziz Al Saud's death in 1953, Harry Truman's personal physician made several trips to Riyadh to try to treat the old, lame king, who had become senile and blind.

Big Oil and Uncle Sam Share New Roles

With the war over, and a strong alliance sealed aboard the *Quincy*, SoCal and Texaco had nothing more to do than to exploit the Earth's largest petroleum source, which was also the least expensive to extract. All around the Dhahran complex, the sprawling Aramco headquarters on the shores of the Persian Gulf, the excellent "Arabian Light" crude oil sprang forth almost by itself, under the force of fantastic pressure that had built up over the course of geological time. It fed the trans-Saudi pipeline and filled the tankers moored at the Ras Tanura terminal, opened in 1945 and located less than 100 kilometers from the main wells.

In the war's aftermath, Europe had a thirst for oil. The Marshall Plan was under development and oil was its centerpiece. But SoCal and Texaco were almost absent in Europe. The two companies lacked easy access to European markets and sufficient capital for rapid development of Saudi Arabian oil production. That is why, one year after the war, they decided to invite the other principle American companies, richer and already ideally situated in Europe, to the party.

Initially, the idea to ally with the most powerful American oil companies was unappealing to SoCal's leaders. During the summer of 1946, several senior leaders of the San Francisco-based company pled their case for taking the offensive: SoCal should take advantage of the Saudi oil treasure to defeat the competition. But SoCal refrained from this perilous strategy. Its directors in effect ordered a study that showed that the company could make more money opting for a more modest percentage of Saudi petroleum as part of a broader alliance. Aramco would derive more benefit from selling their Arabian crude oil through Standard Oil of New Jersey's network than by attempting to confront that giant of giants, Exxon's ancestor and the earliest offshoot of John D. Rockefeller's Standard Oil company. And so in September 1946, negotiations began with Jersey Standard and Socony-Vacuum (formerly Standard Oil of New York and Vacuum Oil, and soon to become Mobil). Anticipating immense postwar energy needs, Jersey Standard and Socony-Vacuum were eager to acquire new crude oil sources. Jersey Standard feared competition with Aramco, which had the potential to saturate Europe with unlimited quantities of Arab oil sold at a low price. In light of this, an alliance seemed to be the wisest path to take. And, in December 1946, SoCal, Texaco, Jersey Standard, and Socony-Vacuum decided to negotiate their association within Aramco.

Middle East oil production therefore remained concentrated in the hands of the major American companies that could harvest the fruits of Washington's initial Second World War efforts to expand the base of American oil power beyond US borders. The Trans-Arabian Pipeline for which Harold Ickes had lobbied was built by Aramco rather than through the auspices of the US government. Aramco's Trans-Arabian Pipeline Company (Tapline) was established in July 1945. Aramco broke ground on the initial site in 1947. Construction was entrusted to the Bechtel, the Californian industrial firm that had built the Hoover Dam on the Colorado River in the 1930s and that had grown considerably thanks to army contracts during the war. The pipeline was a monumental structure, hundreds of thousands of tons of steel extending more than 1,200 kilometers across the desert, bordered by an asphalt road and regularly surveilled by aircraft. Forty water wells were drilled along its route.[40] It was the largest pipeline in the world. The enormous amount of energy absorbed by its construction amounted to a small fraction of what it could carry (up to one-third of Saudi Arabian crude). The route ended at the port of Sidon, Lebanon, on the shores of the Mediterranean, and traveled through Syria, bypassing the plateau of the British-controlled Golan territories of Palestine, which would soon become the state of Israel. Thanks to Aramco's petro-dollars, Ibn Saud could generously support Housni al-Zaïm, a Syrian officer who had defended Damascus against Anglo-Gaullist troops in 1941, and who came to power in March 1949 in the first coup d'état of the modern Arab world.[41]

The construction of this pipeline was the first of many subordinate acts by a Washington aimed at growing Big Oil in the Middle East. An American engineer, a pioneer in Saudi petroleum, had predicted in 1948 that although the pipeline was controlled by private interests, "it committed our government to a fixed foreign policy for at least twenty-five years."[42] The American military received its share of benefits, too. The Dhahran Aerodrome, built by the American oil companies in the vicinity of their wells, was used during the last months of the war to transfer equipment from Europe to the Japanese front; in January 1946 it became an authentic military base.[43] In February 1949, on the first round-the-world nonstop flight, the B-50 bomber Lucky American Lady II refueled mid-flight above Dhahran.

Could Washington take direct control of Saudi oil? No, said George McGhee, a Texas petroleum industry insider who played a central role in the oil diplomacy of the United States beginning in 1949: "It's not the American way."[44] But in answer to this same question, Abe Fortas, a former collaborator of Harold Ickes who later sat on the Supreme Court, exclaimed that it would have been legitimate for the American people to take their share of the biggest treasure in history: "At least it would have given the government a seat at the poker-table."[45]

1948: Americans Bar France's Entry to Saudi Arabia, and the Red Line Is Erased

Though many feared that the association of four of the five largest American oil companies in Saudi Arabia would one day break America's antitrust laws, the Truman administration saw no reason to worry. Quite the contrary: The US Navy's secretary of state, James Forrestal, cared little about "which American company or companies developed the Arabian reserves," as long as these companies were American.[46] In 1936, Forrestal, one of Texaco's lawyers, had been the primary architect of the SoCal-Texaco alliance in Arabia. However, Socony's lead lawyer warned its president in 1946: "I cannot believe that comparatively few companies for any great length of time are going to be permitted to control world oil resources without some sort of regulation."[47] What followed proved that this lawyer's concern was unwarranted.

A more difficult obstacle stood in the Alliance's way: the Red Line Agreement. The thick line drawn around the map of the Middle East by Armenian businessman Calouste Gulbenkian forever prohibited any shareholder of the Iraq Petroleum Company (IPC) to privately prospect for oil in the former Ottoman Empire: Jersey Standard and Socony could not invest in Saudi Arabia, a former Turkish province, without breaking their agreement. They first had to obtain the consent of IPC's shareholders.

BP and Shell were not difficult to win over. To these two firms, holders of the British Crown's petroleum interests, Saudi Arabia already seemed like a lost cause. During his interview with Roosevelt, the Saudi king had made it clear that he intended for Saudi petroleum to remain under exclusive control of the Americans, viewed as untainted by imperialism, particularly in the Arab world. Uncle Sam fostered the myth of freedom by allowing American companies to pursue their own agenda, beyond government control.

Jersey Standard offered BP advantageous long-term contracts to sell Iranian and Kuwaiti petroleum. Socony placated Shell with a similar tactic. It was more difficult to obtain the final agreement of the other two shareholders. From his suite at the Paris Ritz, Calouste Gulbenkian would have none of it. For Mr. Five Percent, the IPC represented an annual income of $20 million. The Compagnie Française des Pétroles (CFP), for its part, decided to join Gulbenkian in what was to be perhaps the bitterest negotiation in the history of oil, and certainly the one whose consequences ultimately proved the most significant.

The Americans believed they had a foolproof reason to break the Red Line Agreement. During the war, London had seized the assets of both the CFP and Calouste Gulbenkian, since they were in enemy territory. Moreover, Mr. Five

Percent had followed the Pétain government at Vichy, before prudently retiring to the neutral country of Portugal in April 1942, where he could meet with American and English oilmen en route to the Middle East. CFP and Gulbenkian holdings were returned when France was liberated. But Jersey Standard and Socony lawyers calculated that they could demonstrate the Red Line Agreement they had signed in 1928 was null and void as a result of this wartime property seizure, and would have to be renegotiated.[48] A Socony executive said he was confident that the French and Gulbenkian would "be careful to wash the linen within the family circle."[49] He was wrong. Immediately after the war, General de Gaulle, head of the provisional government, was angered by the weak level of oil extractions in Iraq and the slow pace of negotiations through which France was to receive compensation for oil production, which it had been unable to access during the war.[50] In early January 1947, CFP's lawyers launched an assault on IPC headquarters in London. And in Washington, France's ambassador very publicly complained about the American petroleum industry's claim.

One of the pillars of American diplomacy, Paul Nitze, proposed a skillful compromise: Socony could buy Jersey Standard's shares in the IPC, while remaining separate from Aramco. Jersey Standard therefore would not be bound by the Red Line Agreement, and would thus obtain free reign in Saudi Arabia. A sign of the degree to which the American government and the petroleum industry were intertwined: In 1936, Paul Nitze, then a business banker, had assisted James Forrestal with the Texaco-SoCal agreement. Nitze, who became one of the great strategists of the cold war, had worked for Nelson Rockefeller in the federal office sending propaganda to Latin America during World War II. His plan was nonetheless rejected by both Jersey Standard and Socony.

In Paris, senior French officials could clearly see that the stakes were enormous. They were ready to pay any price in order to obtain Saudi oil access for the CFP. But the directors of the American oil industry eventually convinced France to back down, by promising to provide a significant portion of the capital necessary for the rapid development of Iraqi extractions. This compromise addressed General de Gaulle's initial grievance. With the creation of the Marshall Plan underway, Paris could not afford the luxury of endless negotiations with Washington. Because of CFP's low propensity to find its own oil, the Americans mocked the French company, nicknaming them: "Can't Find Petroleum." Calouste Gulbenkian was the last to let go. Even after learning that his French friends, including Victor de Metz, the CFP boss he used to frequent in Vichy, had abdicated, the hardy Armenian, nearly eighty years old, refused to bend to the world's most powerful industrialists. For over a year, squadrons of negotiators made the round-trip between Manhattan and the Grand Hotel Aviz in Lisbon, where Gulbenkian (who still shunned the public eye) became a tourist attraction in spite of himself, seated at his own personal table, set on a platform in a corner of the room to elevate him above the other diners. Unshakeable to the last, Gulbenkian raised the stakes again. He feared that the Americans, despite all their discourse about freedom and competition, had monopolized Saudi Arabia's oil production and restricted the Iraqis.[51] Finally, at 2 a.m. on a November night in 1948, Mr. Five Percent erased the Red Line. The contracts signed that night were abominably (and doubtless to a certain extent deliberately) abstruse. "No one will ever be able to litigate about these documents because no one will be able to understand them," a Gulbenkian lawyer joked.[52]

A US Senate report would later state: "Although Exxon and Mobil eventually reached an IPC settlement, the French never forgave the Americans for keeping them out of Saudi Arabia."[53] In March 1957, nine years after CFP's failed attempt to get its foot in the door of the Saudi Arabian kingdom, General de Gaulle exclaimed from the sands of another desert, this one the Sahara, that the late discovery of oil in Algeria, then in the midst of its war for independence, could "change the destiny" of France.[54] What would have happened to this destiny if the ancestor of the French oil firm Total had captured a fraction of the Saudi oil? What would the consequences have been if Paris had succeeded in thwarting the exclusive alliance between American petroleum and the House of Saud, poised to become the spine of the contemporary energy world? These were idle questions: France probably lacked the means to achieve such ambitions.

The Power of Standard Oil Is Resurrected in the Arabian American Oil Company

Finally, in December 1948, SoCal and Texaco sold 40 percent of their shares in Aramco to Jersey Standard and Socony, for nearly half a billion dollars. A colossal amount during that era, this sale in fact proved to be the bargain of the century. It was the most important capital divestment in history: 40 percent of more than one-fifth of the total crude oil reserves ever discovered on Earth. Jersey Standard, SoCal, and Texaco each took 30 percent of Aramco's shares. The initial plan was for Jersey Standard and Socony each to acquire 20 percent of the shares. But Socony's directors waffled. They finally chose, prudently, just 10 percent. This blunder contributed significantly to the future Mobil company's relegation to a second-class role in the world petroleum market.

Jersey Standard, Socony, SoCal (the three most vigorous branches of the Rockefeller empire), and Texaco were separated in 1911 by the US Supreme

Court in response to antitrust laws and the ire of American public opinion. Jersey Standard and Socony's new possession of Aramco shares in some ways revitalized the core of Standard Oil's power in the Arabian desert. "Aramco was, in effect, the neurotic child of four parents, subject to the whims, qualms and jealousies of each," wrote Aramco's first public relations manager, Michael Sheldon Cheney (no relation to Dick Cheney). Despite their bickering, beginning in 1948 these four American companies once again shared a narrow common interest that keeps them, even today, among the most powerful industrial organizations. The American oil kings took the queen—Saudi Arabia—as it became the new hub of the global energy chessboard.

Half a century later, in December 1998, Jersey Standard and Socony, which had become Exxon and Mobil, respectively, formed the ExxonMobil group, commonly referred to simply as Exxon.* In the early 2010s, ExxonMobil remained one of the world's top—and sometimes the top—private company.† In France and elsewhere, Exxon was known as Esso, one of its brand names. The name Esso (pronounced *S. O.*, for Standard Oil) was an elegant marketing ploy that allowed the company to separate itself from its problematic history without denouncing it. For its part, Standard Oil of California (SoCal) was rebaptized Chevron after its 1984 merger with Gulf Oil. In 2001, Chevron absorbed Texaco, the other major Texas player.‡ Its profits kept the fused company in the top ranks of the largest international firms.

In all respects, 1948 was a great year for Aramco: It was then that its engineers brought forth the "greatest single prize in all history." Located two hours down a dusty road from the Ras Tanura terminal, in the Persian Gulf, Ghawar was the largest oil field in the world. It remains unequalled. Thirty kilometers wide and 300 kilometers long, it represents 60 percent of Saudi extractions to this day. During the Second World War, in spite of the United States' top-producer status, the American oil companies in reality owned only a tenth of the world crude oil reserves. If their control extended to half of the world oil stock in 1950, it was thanks to Ghawar and Saudi Arabian petroleum. A year later, Aramco supplanted Creole, Jersey Standard's affiliate in Venezuela, to gain the rank of number one oil producer on the planet.[55] Transposed, transfigured, and increased, the power of John D. Rockefeller's old empire perpetuated itself through the matrix of Saudi oil, like a dragon changing its skin, extending its reach to a new, fabulous treasure.

---

* See *infra*, chapter 27.

† According to the rankings of *Forbes* magazine.

‡ See *infra*, chapters 23 and 27.

TWELVE

Washington Gives Absolute Power to American Petroleum

The men charged with supporting the United States' power as it emerged from the Second World War had to recalibrate their geostrategic compass. In his memoir, President Roosevelt's chief diplomat described the new fundamental premise of American foreign policy: "Iran (once known to us as Persia), Iraq, Saudi Arabia, Lebanon and Syria began to appear more and more in the American print, not as lands of the ancients but as cogs in the machine of war," wrote Cordell Hull, who headed the US State Department from 1933 to 1944.[1*]

Machine of war, machine of peace: The reconstruction plan developed under the auspices of General George C. Marshall was the great tool of "pax Americana" in Western Europe. In 1953, the man who had been US Army chief of staff throughout the Second World War won the Nobel Peace Prize for it. The Marshall Plan's objective was to raise a wall of capitalist wealth capable of curbing the Soviet Empire's expansion. Oil was the cement of this dike. Throughout the implementation of the Marshall Plan, from 1948 to 1952, the governments of countries situated to the west of the Iron Curtain, who received loans from Washington, invested not less than a fifth of that money in petroleum products, half of which were purchased from American companies using American dollars.[2]

In Washington, the new Truman administration had not lost sight of Harold Ickes's warning at the war's apex. An internal White House memo dated 1945 described the simple but far-reaching plan implemented after the fall of the Reich: "The Navy wants Arabian oil developed to supply

* A significant share of the extractions in Iran, Iraq, and Saudi Arabia were transported toward the Mediterranean by pipelines that crossed Syria (a relatively minor producer of crude oil) and Lebanon.

European commercial demand, replacing western hemisphere oil which might otherwise go to Europe, thus conserving supplies which are subject to U.S. military control."[3]

The recent enormous discoveries of oil in the Middle East allowed the preservation of American crude oil fields so they could continue to irrigate the American economy and supply the American army. As for the United States' European allies, their economies were reconstructed with Arab crude oil piped to the Old Continent and extracted primarily by Anglo-Saxon companies. The tectonic movement was heavy: In 1946, 77 percent of the oil consumed in Europe still came from the American continent. But as early as 1950, even before the Marshall Plan ended, Western Europe became (and remained for a long time) almost exclusively dependent on the supply from the Persian Gulf, essentially controlled by American and British companies. In effect, the oil giants "castled," thus opening the period of American hegemony in the West.*

The Americans were spared the dependence on Middle East oil experienced by the Europeans throughout the tremendous growth of the Western economy that French economist and demographer Jean Fourastié later called the "Thirty Glorious Years." The United States remained firmly installed on the first rung of world crude oil producers. The American economy, which had become by far the most powerful in the world, could still count on perfect energy self-sufficiency. Or, more specifically, almost perfect: The same year that the Marshall Plan was implemented, a major economic event posed a threat that, alone, could justify Washington's strategy. Since the 1920s, the bloated production of United States crude oil was no longer sufficient to cover their full needs. Imports, however, remained modest. But in 1948, despite the frantic work to increase production during the Second World War, the country for the first time joined France and the United Kingdom in the vulnerable realm of net importation, becoming partly dependent on foreign oil. US oil imports continued to grow: Uncle Sam's power was challenged by its thirst for petroleum. In 1948, however, the problem seemed modest: Only 6 percent of the oil required to run US motors, electric turbines, and factories was extracted outside US territory; even this percentage was extracted by American companies. All but about 8 percent of this import came from the new American crude oil stronghold at the edge of the Persian Gulf: Saudi Arabia.[4]

This extraordinarily solid geostrategic position was defended without hesitation. A text drafted in 1948 by George Kennan, one of the architects of

---

* Castling: a chess move in which the king swaps positions with the rook, the rook is exposed, and the king is no longer vulnerable.

the Marshall Plan, outlined the strategic logic and ethics that animated the ambitions of the new superpower: "We have about 50% of the world's wealth but only 6.3% of its population. . . . In this situation, we cannot fail to be the object of envy and resentment. Our real task in the coming period is to devise a pattern of relationships which will permit us to maintain this position of disparity without positive detriment to our national security. To do so, we will have to dispense with all sentimentality and day-dreaming; and our attention will have to be concentrated everywhere on our immediate national objectives. We need not deceive ourselves that we can afford today the luxury of altruism and world-benefaction. . . . The day is not far off when we are going to have to deal in straight power concepts."[5]

Voiced by one of the great figures of the cold war, this pragmatic, cynical concept of America's new role was already in place in order to fortify and grow the "cogs of the machines of war"—the positions acquired by American oil companies outside the United States.*

A 50/50 Split of Oil Profits: The Venezuelan Precedent

The 1938 nationalization of Mexican oil was still a raw memory for American and British oil companies. That same year, when Venezuela petitioned for more equitable oil profit sharing, Jersey Standard, Shell, and Gulf Oil, the three dominant companies in the large South American oil producing country, quickly resolved to make concessions. Beginning in September 1939, Washington and London strongly encouraged those concessions because during the war Caracas played a strategic role as little known as it was decisive. Being at once sparsely populated and rich in crude oil, very distant from theaters of operation and under American control, Venezuela instantly imposed itself as the number one global crude oil exporter. The Creole Petroleum Corporation, a subsidiary of Jersey Standard, was for a time the number one oil producer of the world.† Jersey Standard sought American diplomatic support to address Caracas's growing demands. It was a futile attempt: The Venezuelan oil pumped by Jersey Standard and the other major companies was indispensable to the war effort. They were forced to make concessions. Moreover, Jersey Standard kept a low profile, since it was under the fire of embarrassing critique because of its close affiliation with IG Farben.

* George Kennan inspired the US doctrine of "containment," and was ambassador to Moscow in 1951 and to Belgrade in 1961.

† Extractions from Creole reached 450,000 barrels per day in 1946.

In order to negotiate a quick agreement, the Roosevelt administration dispatched renowned geologist Herbert Hoover Jr. to meet with the Venezuelan government. (Hoover was eldest son of the American president Herbert Hoover, a geologist who built his fortune prospecting for minerals and investing in petroleum and who was in power during the 1929 stock market crash.) Finally, the oil companies made a reasonable deal with Caracas: They agreed to pay more royalties in exchange for a forty-year extension of their concessions. New oil legislation was adopted in March 1943 by the junta in power. In theory, this act established a 50/50 profit share between Venezuela and the oil companies. But this was in theory only: A young militant democrat, economist, and lawyer named Juan Pablo Pérez Alfonso, who later founded OPEC, demonstrated that in reality the oil companies continued to capture the bulk of the profits and leave only crumbs to their host countries.

In 1945, Acción Democrática (Democratic Action), the Reform Party to which Pérez Alfonso belonged, rose to power through a military coup. Romulo Betancourt, the "father of Venezuelan democracy," was initially designated president, before handing over his power in December 1947 to the first democratic government elected in the history of Venezuela, led by the Acción Democrática. In 1945, Pérez Alfonso became Venezuela's minister of development; on November 12, 1948, he won the passage of a new oil law, just twelve days before Acción Democrática was ousted by the same military personnel who had put him in power three years earlier. Considered the mastermind of the democratic government, Pérez Alfonso was exiled to the United States, and then to Mexico. His law was enacted anyway, without tarnishing relations between the new junta and the American oil industry. The latter expanded production considerably and transformed Venezuela, according to Pérez Alfonso, into an "oil factory," until Betancourt and Acción Democrática's triumphant return to power in January 1958. Four months later, the young American vice president Richard Nixon was nearly killed by the crowd during an imprudent visit to Caracas.

Pérez Alfonso returned to become minister of oil. His strategy of sharing the oil windfall proved adroit. According to Romulo Betancourt, the augmentation of taxes collected from the oil companies was as much as, if not more than, Mexico had received by nationalizing their oil, while preserving Caracas from diplomatic embarrassment and troubles with industry. Betancourt later affirmed: "Tax income was increased from then to such a degree that nationalization was unnecessary to obtain maximum economic benefits for the people of the country."[6]

The Golden Gimmick, or How Washington Agreed to Fund Saudi Arabia *in Perpetuity*, Decisively Sowing the Seeds of Globalization

The 50/50 profit sharing that the first democratic government of Venezuela established with the oil companies in 1948 rapidly became an established practice in the other developing nations that produced oil; among them, only Mexico exercised direct control of its own oil. This sharing, however, did not prevent Big Oil from remaining, more than ever, an entity that reaped historically unequaled profit levels in the industry.

The king of Saudi Arabia was among the most eager to claim a share of profits that matched the share Venezuela obtained. Abdulaziz Al Saud could point to a precedent. In 1948, American independent oilman J. Paul Getty landed a deal in the neutral zone between Kuwait and Saudi Arabia, accepting conditions much less advantageous than those granted to Aramco's partners. Those conditions did not prevent Getty (a man little known for his generosity) from becoming the new richest man in the United States a few years later. Getty's deal proved that in spite of their objections, Aramco and other Western companies installed around the Persian Gulf could indeed pay substantial royalties without bankrupting themselves.

In 1950, the same year the US military machine arrived in Korea en masse, Washington was prepared to show that they would make big sacrifices in order to ensure their country's petroleum companies long-lasting Saudi oil access. During the torrid Arab summer, the State Department dispatched to Riyadh thirty-seven-year-old diplomat George McGhee. Born in Texas in 1912, McGhee had made his oil fortune in the American company Conoco before age thirty. He then married the daughter of Everette DeGolyer, the geologist who had confirmed the magnitude of the fantastic Saudi reserves, one of the pillars of the Yankee petroleum establishment.[7] For McGhee, it went without saying that the major oil companies entrenched around the Persian Gulf had no other choice than to accept the 50/50 profit share with their host countries. During a meeting in Washington, DC, on September 18, 1950, the young, loquacious diplomat with a hoarse voice recommended to the American companies' representatives that they should consider "rolling with the punch."[8] But ultimately, it was the US government, represented by McGhee, that rolled with the punch in place of the oil companies.

The compromise to which Washington and the Aramco shareholders agreed three months later remained known as the "golden gimmick." Resting on the back of the American taxpayer, this gimmick had massive consequences that extended well beyond the petroleum industry. From that moment on, the revenue Aramco paid to Saudi Arabia in tax monies was exempt from taxation in the United States. In 1949, Aramco paid $43 million to the US Treasury and $39 million to Saudi Arabia. Two years later, while Aramco's oil extractions and sales grew sharply, the share paid to the US Treasury fell to only $6 million, while Saudi Arabia collected $110 million.[9] In accordance with a timely (and very generous) reinterpretation of 1918 tax regulations on profits reaped by American companies on foreign soil, American democracy accepted "what in effect would amount to a subsidy of Aramco's position in Saudi Arabia by U.S. taxpayers."[10] Through the golden gimmick, Washington provided continuous financial support to a kingdom whose fidelity to America had already become indispensable, without having to ask every year for the support of the US Congress and its many pro-Israeli legislators. But this technique of disguising subsidies paid to a foreign regime had vaster implications. The golden gimmick was adopted by Aramco's American competitors, before extending to all US companies located abroad, and ultimately was imitated in Europe. This spurred Western firms to expand operations in developing countries, where taxation in general was more lenient, or even obliging, through the systematic corruption of local potentates.

The golden gimmick of 1950 cemented Aramco's position in Saudi Arabia: Those corporations—called multinationals today—do not necessarily pay taxes in their country of origin, where their headquarters are usually located, but only where they make their profits. Aramco's gimmick partly explains how for more than a half-century the foremost French company, Total group, as well as one-fourth of the largest French companies, pay virtually no tax to the French government.[11] Pure, hard economic logic: Without the golden gimmick, the headquarters of the multinationals would have all moved to tax havens a long time ago.

The tens of millions of tax revenue dollars that the US Treasury agreed to forego, year after year, greased the "cog" named Saudi Arabia. To its misfortune, the British Empire was not as successful in their own corner: Iran.

The British Empire Clings to Its Last Cornerstone, Iran

After India won independence in 1947, the old English military and bureaucratic empire was little more than a shadow of its former self. Its commercial empire endured. The sole extractor of Iranian crude oil, the Anglo-Iranian Oil Company, 51 percent of which belonged to His Majesty, was then the first

British overseas company and the Crown's primary income source outside Great Britain. The Royal Navy purchased its oil at a price much lower than market value. Forty years after Lord of the Admiralty Winston Churchill's decision that the Navy would run on petroleum, Iran remained the energetic source of Britain's power. Defending this capital was all the more vital since London had been unable to stop the advance of American petroleum everywhere else around the Persian Gulf. The Americans were implanted in Iraq, held half of Kuwait's oil and all of Bahrain's, and, especially, had taken from right under the noses of their English cousins the incredible treasure hidden under Saudi Arabia's desert sands. Only the fields of ancient Persia remained under exclusive British control. But not for long.

By the end of the 1940s, in spite American engineers' rapid success in Saudi Arabia, the Anglo-Iranian Oil Company was still the world's third-largest oil producer and the largest in the Middle East, supplying no less than 40 percent of the region's extractions. In southeastern Iran, near the shores of the Persian Gulf, the giant oil field situated close to the city of Gachsaran remained the world's most productive. And on the island of Abadan, near the Persian Gulf, the Anglo-Iranian Oil Company had what was at that time the largest oil refinery in existence.

But the resentment and fear that the Iranians had for the British Empire were immeasurable. When, in 1941, the shah of Iran, Reza Pahlavi, refused to expel the thousands of Germans to whom it had granted key positions, London and Moscow decided to invade Iran in order, on the one hand, to protect the Abadan refinery and, on the other hand, to ensure the Soviet Union's oil supply. With Churchill's and Roosevelt's full discharge, Stalin's troops requisitioned the crops grown in the north of Iran, causing famine. Forced into exile, Reza Shah died on July 26, 1944, in South Africa at sixty-six years of age. In his place, Churchill installed the young son of the flamboyant sovereign, Mohammad Reza Pahlavi, on the peacock throne. The latter for his entire life was haunted by the fear of British authority and maneuvering—as well as, no doubt, by the shame of having served as the Westerners' puppet. Surely there must have been a degree of restraint in the way Shah Mohammad Reza later relayed his perceptions of British petroleum companies to journalist and English historian Anthony Sampson: "Then we were hearing that the oil company was creating puppets—people just clicking their heels to the orders of the oil company—so it was becoming in our eyes a kind of monster—almost a kind of government within the Iranian government."[12] The tens of thousands of Iranian oil workers were barred from enjoying the comfortable living conditions of some two thousand British Anglo-Iranian Oil Company employees. The director of the National Iranian Oil Company later testified:

"Wages were 50 cents a day. There was no vacation pay, no sick leave, no disability compensation. The workers lived in a shanty town called Kaghazabad, or Paper City, without running water or electricity. . . . In winter the earth flooded and became a flat, perspiring lake. The mud in town was knee-deep, and . . . when the rains subsided, clouds of nipping, small-winged flies rose from the stagnant water to fill the nostrils. . . . Summer was worse. . . . The dwellings of Kaghazabad, cobbled from rusted oil drums hammered flat, turned into sweltering ovens. . . . In every crevice hung the foul, sulfurous stench of burning oil. . . . in Kaghazad there was nothing—not a tea shop, not a bath, not a single tree. The tiled reflecting pool and shaded central square that were part of every Iranian town, . . . were missing here. The unpaved alleyways were emporiums for rats."[13]

Iran Rebels against the English "Demons" and Nationalizes Their Petroleum

Churchill's promise to restore Iran's sovereignty after the war, solemnly pronounced in 1943 at the Allies conference in Tehran, was not enough to convince the increasingly numerous and audacious nationalists. On November 4, 1944, the *Times of London* published an editorial that proposed that the United Kingdom, Russia, and the United States share Iran after the war. A month later, a fiery Iranian nationalist, the rich aristocrat Mohammad Mossadegh, brandished this *Times* editorial to submit to his colleagues in the Majlis (Iranian parliament) a proposition that prohibited all oil-related negotiations with foreign powers.[14]

The nationalists' ardor redoubled after the announcement of the 50/50 petroleum profit-sharing agreements obtained by Venezuela and subsequently by Saudi Arabia. But the tough Scottish president of the Anglo-Iranian Oil Company, Sir William Fraser, baron of Strathalmond, had no intention of accommodating the nationalists' demands, despite the calls for conciliation launched by American diplomacy and delivered by the ubiquitous voice of George McGhee. Between 1945 and 1950, the Anglo-Iranian Oil Company had garnered £250 million in profit, while only paying Tehran royalties of £90 million: The Anglo-Iranian Oil Company paid much less to Iran than to the British Crown, its majority shareholder.[15]

In the spring of 1949, the Anglo-Iranian Oil Company proposed a "supplementary agreement" revising the terms of the concession granted by Tehran. But the nationalist parliamentarians' hostility was such that for many weeks, the shah's government dared not submit the agreement to the Majlis. In June, the parliamentary petroleum commission, chaired by Mohammad Mossadegh, rejected the British company's offer. The nationalist leader openly called for the nationalization, pure and simple, of the Anglo-Iranian Oil Company.

Mohammad Mossadegh, who the conservative British journalists depicted as a fanatic buffoon, did, it is true, appear eccentric at times. He sometimes received interviews while lying in bed, wearing pajamas. He knew how to shed a tear to emphasize his point. Mossadegh, trained in international law at the Sorbonne and holding a doctorate from the Swiss University of Neuchâtel, was above all a statesman. He was also a fervent democrat, one of the few opposed to Reza Shah's 1925 power grab.

When, in 1950, the Anglo-Iranian Oil Company was at last ready to concede to the 50/50 profit share, Mossadegh, who had become the undisputed leader of the opposition to the shah and his British patrons, pushed his advantage to the utmost in order to achieve nationalization. The Anglo-Iranian Oil Company, he said, was for Iran the sole source "of all the misfortunes of this tortured nation."[16] The majority of the Majlis members openly rebelled against the English "demons." The shah's prime minister, General Ali Razmara, several weeks earlier had reluctantly taken the risk of personally defending the Anglo-Iranian Oil Company's proposal. On March 3, 1951, he declared to parliament that Iran could not legally cancel the Anglo-Iranian concession. Five days later, this former army chief of staff, trained in the French military academy of Saint-Cyr, was murdered by a young carpenter prodded by religious leaders to rid their homeland of "British stooges." The next day, deputies voted to nationalize Iran's oil. On March 20, the high chamber in turn voted in favor of nationalization—the day remains a holiday for the Iranians, a symbol of their emancipation.

The shah at first refused to implement the decision voted in by the Majlis. On April 28, following weeks of strikes and martial law, Mohammad Mossadegh was appointed head of the government, the first to be chosen by parliament and not imposed by the sovereign. Four days later, the shah acceded, and signed the law, which abolished the Anglo-Iranian Oil Company. London immediately levied an embargo on Iran, going so far as to threaten prosecution of any cargo-ship owner willing to load oil "stolen" by the Iranians. The oil wells and the Abadan refinery closed. It was the beginning of a long power struggle; the British, no less than the Iranians, could not afford to forego Persian black gold. On September 25, Mossadegh gave the last British present in Abadan a week to vacate the country.

The exchanges between London and Washington during the last week of September were intense. British Prime Minister Clement Attlee of the Labor Party informed US President Harry Truman that a plan to invade Iran was at the ready. Truman responded that it was out of the question for the United States to support such an initiative. He strongly encouraged London to resume negotiations. The heart of Western power had crossed the Atlantic. Attlee resolved to inform his cabinet that, "in view of the attitude of the United States Government, [he did not] think it would be expedient to use force to maintain the British staff in Abadan."[17] On October 4, 1951, carrying with them tennis rackets, fishing rods, and golf clubs, the last British in Abadan embarked on the HMS *Mauritius*. In spite of their hopes, they never returned. When the Royal Navy cruiser began its slow, upriver journey in the direction of the port of Basra, Iraq, the crew's orchestra slowly played the opening notes of the British settlers' anthem: the Colonel Bogey March.[18] Four years after India gained independence, the British Empire underwent its second, most decisive defeat.

Unlike London, Washington was not hostile to Mossadegh; quite on the contrary. Four days after the HMS *Mauritius* departed from Abadan, the first Iranian prime minister landed in New York to defend Iran's position to the United Nations. He was received by President Truman. As a badge of honor, he was even invited to Philadelphia to place his hand on the Liberty Bell, a symbol of the American battle for independence against the British. In 1951, Mossadegh was *Time* magazine's man of the year. The conservative American journal paid an ambivalent tribute to the "Iranian George Washington," who "in his plaintive, singsong voice . . . gabbled a defiant challenge that sprang out of a hatred and envy almost incomprehensible to the West."[19]

During his trip, Mossadegh spoke with the young American oil prodigy and master of oil diplomacy, George McGhee. After eight hours of discussion, McGhee thought he glimpsed the possibility of a compromise: The Anglo-Iranian Oil Company would be permitted to acquire and commercialize Iranian petroleum, and Shell, less directly linked to the British Crown's interests, would be allowed to buy the Abadan refinery. Mossadegh insisted on a prohibitive subsidiary clause: No British would be authorized to work in Iran. With anger and contempt, London closed the exit door Washington had opened. Anthony Eden, minister of foreign affairs for Churchill's new conservative government, was personally outraged, and partially ruined. A Farsi speaker who was passionate about Persia, Eden had just watched a large portion of his fortune vanish, since in order to eliminate accusations of conflict of interest, he had resolved to sell his shares in the Anglo-Iranian Oil Company, whose value collapsed following the British embargo.

Throughout 1952, London and Tehran kept each other in check. The United Kingdom did not want to make any concessions but could do nothing without US support, apart from bribing candidates in Iran's spring legislative elections and fomenting unrest, including in Abadan.[20] The increased disorder was such that the Mossadegh government was forced to suspend the electoral process upon the return of their delegation to the International Court of Justice in The Hague, before which the Anglo-Iranian Oil Company attempted to assert its rights.[21] For its part, Iran was on the edge of bankruptcy, deprived of all possibilities to sell its oil other than by contraband routes in the mountains in the north of the country. In July, the Italian-run *Rose Mary* was one of the last cargo ships that dared try to force its way through the Royal Navy's blockade. It loaded its crude oil in Abadan but was compelled to berth at Aden, the British Army's base in Yemen. The fledgling Iranian democracy's fate hung on the fate of its oil.

Washington Launches Operation Ajax

The United States' attitude shifted radically with General Dwight D. Eisenhower's arrival in the White House in January 1953. Elected by the Republican Party, Eisenhower championed a significantly more aggressive foreign policy than his predecessor and entrusted it to two brothers, John Foster Dulles and Allen Dulles, New York lawyers and sons of an austere Presbyterian minister. As head of the State Department, the eldest, John Foster, took the reins of diplomacy, while his younger brother, Allen, was appointed director of the CIA. John Foster and Allen Dulles were associated in one of the leading New York law firms, Sullivan & Cromwell, which counted among its customers many companies related to the Rockefeller empire. In July 1953, the approach of the American secret service changed when Allen Dulles launched Operation Ajax. The plan, developed jointly by the CIA and MI6 and approved by Eisenhower and Churchill, aimed to "bring about the fall of Mossadegh" by any means.[22] A few weeks earlier, during a meeting of the National Security Council in Washington, John Foster Dulles had warned President Eisenhower: The Mossadegh regime would not hesitate to transform itself into a dictatorship that would fall like a ripe apple into the Russian hands (the Iranian Communist Party, the Tudeh, was indeed a powerful, albeit inconsistent, ally of Mossadegh).* The head of the State Department alerted Eisenhower that there would be catastrophic consequences if Iran entered the Soviet sphere: "Not only would the free world be deprived of the enormous assets represented by Iranian oil production and reserves, but the Russians would secure

* The National Security Council is the White House policy group that determines American foreign policy.

these assets and thus henceforth be free of any anxiety about their petroleum resources. Worse still . . . if Iran succumbed to the Communists there was little doubt that in short order the other areas of the Middle East, with some 60 percent of the world's oil reserves, would fall into Communist control."[23] Together, Eisenhower and Churchill resolved to save the Iranian democracy from Communist dictatorship by imposing a dictatorship of their own. Joseph Stalin had died in March: It was the ideal time to act.

In Tehran, Kermit "Kim" Roosevelt, architect of the conspiracy against Mossadegh, was already at work. Kim Roosevelt, thirty-seven years old, was none other than President Franklin Roosevelt's cousin and the grandson of President Theodore Roosevelt. The young spy was exceedingly excited when, in early July, he crossed into Iran via an isolated route on the Iraqi border to begin the adventure of his life: "I remembered what my father wrote of his arrival in Africa with his father, T.R., in 1909 on the African Game Trails trip. 'It was a great adventure, and all the world was young!' I felt as he must have then."[24]*

Operation Ajax enjoyed the support of the British MI6. But the CIA was in charge. Allen W. Dulles, the director of central intelligence, approved $1 million that provided the Majlis with $11,000 per week as a bribe.[25]

Upon his arrival in Tehran, Kim Roosevelt sent General Norman Schwarzkopf to meet the shah in his palace. This American general was a hero in the eyes of the faithful shah. After the Anglo-Soviet invasion of Iran, between 1942 and 1948, Schwarzkopf had been tasked with training and organizing the dreaded Iranian gendarmerie, the imperial police force. In particular, he had established a secret security brigade whose forceful, effective operations placed the young Mohammad Reza Shah eternally in his debt.

The Schwarzkopf family occupies a unique place among the American families who established the US energy empire. Norman Schwarzkopf was renowned in the United States for leading the investigation of the famous kidnapping and assassination of aviator Charles Lindbergh's baby in the 1930s. But his name is famous today because of his son, General Norman Schwarzkopf Jr. (called "the Bear"), who in 1990 planned Saudi Arabia's defense during the invasion of Kuwait by Saddam Hussein's troops (Operation Desert Shield), then later served as commander of allied forces during the Gulf War (Operation Desert Storm).

On August 1, 1951, Kim Roosevelt directed the elder General Norman Schwarzkopf to brief the shah. During the interview, the shah appeared restless and only spoke to the general in whispers, sitting on a table he had pulled into the middle of a ballroom. Without a doubt, he feared the microphones

* Theodore Roosevelt.

of the British, oblivious to the fact that the latter were working hand in hand with the Americans. Schwarzkopf had a mission to convince the shah to sign two firmans (imperial decrees): the first to dismiss Mossadegh; the second to name in his place General Fazlollah Zahedi, the ex-minister of the interior sacked two weeks earlier by Mossadegh after the bloody repression of an anti-American demonstration.[26] During Reza Shah's rein, Zahedi, Washington's favorite, had been the long-time governor of the Khuzestan province, the heart of the Iranian oil industry. After the 1941 Anglo-Soviet invasion, he was imprisoned by the British as pro-Nazi, which now worked in his favor: He could not easily be suspected of being in British employ.[27] But the shah, terrified by the potential consequences of the firmans crafted by the CIA, refused to sign.

Kim Roosevelt was not a man who gave up easily. The same evening he met with Schwarzkopf, at midnight, he met with the shah, concealed in the back seat of a car, and attempted to force his hand.[28] He assured him of Eisenhower's full support. But it was not enough: The shah feared that Churchill might make an underhanded move. And so, to prove that the British were cooperating with the Americans, Kim Roosevelt predicted that the next evening, instead of announcing as usual "it is now midnight," the BBC presenter would say "it is now exactly midnight." The code, actually broadcast by the BBC (at Churchill's express request), was still not enough to reassure the shah, nor to convince him to accept the coup d'état they offered him.[29]

On August 4, Mossadegh won a referendum that triggered chaos in the country. By this vote, Iranian electors accorded the government the right to dissolve the Majlis. But it was a parody of democracy, which proved disastrous for the prime minister: There were separate urns for the *yeses* and the *nos*; the *yeses* carried 99 percent of the votes. Mossadegh was accused of deceiving the people, and several newspapers and religious leaders called for his resignation. The CIA hastened to throw oil on the fire. It distributed five hundred bundles of rial banknotes to pay rioters in Tehran's poorer districts. CIA agent Donald Wilber later said that antigovernment propaganda "poured off the Agency's presses and was rushed by air to Tehran."[30]

On August 13, after several more meetings between Kim Roosevelt and the shah, the latter finally agreed to sign the fatal firmans. On August 15, at midnight, the commander of the imperial guard, Nematollah Nasiri, appeared at Mossadegh's home to tell him he would be removed from office. Doubtless warned in advance, Mossadegh was conveniently absent, and it was Nasiri who found himself arrested. The morning of August 16, Kim Roosevelt went to general Zahedi's apartment. Zahedi, leader of the government appointed by one of the CIA firmans, escaped by hiding under a blanket in Kim Roosevelt's car and was transported to a safe house. The coup failed. Washington halted the operation. When Kim Roosevelt returned to the CIA office in the American Embassy, a high State Department official handed him a message from Eisenhower saying that the United States would probably now need to "cozy up to Mossadegh."[31]

Mossadegh Is Ousted by CIA Shenanigans

But Kim Roosevelt refused to abandon his mission and follow the President's recommendation. Moreover, General Zahedi gave him an excellent idea: distribute copies of the firmans the shah had signed, in particular in the poorer quarters of Tehran, south of the capital, where the CIA recruited its rioters.[32] On August 17 and 18, Iranian men paid by American agents spread through the city to bribe police officers, elected representatives, and religious leaders. Kim Roosevelt's agitators rioted in the center of Tehran, claiming their allegiance to Mossadegh and communism, striking passers-by and firing shots at mosques. Among them, two faithful CIA operatives, Ali Jalili and Farouk Keyvani, were reluctant to deal with the scale and risk of their task. Kim Roosevelt gave them a choice between receiving $50,000 in cash for themselves and their cohorts, or being shot in the head. The two men took the money.[33]

Early on August 19, 1953, the streets of the center of Tehran were filled with protesters screaming "Long life to the shah!" and "Death to Mossadegh!" Many of these protesters were "bazaar thugs and bully-boys," according to the *New York Times*.[34] "'That mob that came into north Tehran and was decisive in the overthrow was a mercenary mob,' asserted Richard Cottam, who was on the Operation Ajax staff in Washington. 'It had no ideology, and that mob was paid with American dollars.'"[35] In the morning, the shah's army encircled Mossadegh's neighborhood with American Sherman tanks. By the end of the afternoon, intense combat began around the prime minister's home. At the same time, Kim Roosevelt decided it was time to bring General Fazlollah Zahedi back on the scene. Roosevelt found Zahedi in his underwear, in the cellar where he had hidden three days earlier. Zahedi barely had time to don his uniform before other high-ranking officers faithful to the shah, rounded up by Roosevelt, escorted him from the cellar and paraded him through the streets of Tehran on an American tank. Kim Roosevelt believed that in this historic moment it was better that he himself went unnoticed: When the Iranian officers entered Zahedi's cellar, the CIA agent hid behind a boiler.[36]

At the end of the day, the Mossadegh residence was taken by force. The prime minister managed to escape by climbing over a fence. The next day, he agreed to surrender. General Zahedi took power (it is said that, before he gave his radio speech, a technician who believed he was doing the right thing played the first measures of the American national anthem in error). On August 22, the shah returned from Rome, where he had been a refugee during the two weeks that decided the fate of democracy in Iran. Mossadegh was sentenced to death for high treason by a military tribunal. Zahedi and the shah chose to commute his sentence: After three years in solitary confinement in a military prison, Mossadegh, the father of the Iranian democracy aborted by the CIA, ended his days under house arrest, where he died on March 5, 1967.

On the evening they stormed Mossadegh's residence, Kim Roosevelt went to the Iranian military officers club. He didn't recognize all the officers who enthusiastically saluted him, and who, for their part, all seemed to know who he was. Here is what he later said he told the men that were about to regain control of Iran: "Friends, Persians, countrymen, lend me your ears! I thank you for your warmth, your exuberance, your kindness. One thing must be clearly understood by all of us. That is that you owe me, the United States, the British, nothing at all. We will not, cannot, should not ask anything from you except, if you would like to give them, brief thanks. Those I will accept on behalf of myself, my country and our ally most gratefully."[37]

Mohammad Mossadegh would perhaps have been chased out of power without the help of the Americans. His style of governing by arbitrary decree, his decision to dissolve Parliament, and his alliance with the Tudeh Communists were sufficient to make him many enemies in Iran, in the bazaar, and among religious leaders. Nonetheless, CIA involvement in the 1953 coup was more than decisive in the reestablishment of the absolute authority of the shah, who remained the leader of a bloody dictatorship until his downfall in 1979, after the victory of Ayatollah Khomeini's Islamic revolution.

In November 1979, the first declaration of Khomeini's supporters, who seized the American Embassy in Tehran and took its personnel, including many CIA agents, hostage, was to demand Washington's apologies for overthrowing Mossadegh. Expert young Iranian rug makers spent years reconstructing, fragment by fragment, the embassy archives that American diplomats hastily shredded. It wasn't until 2009, in a gesture of appeasement to the Islamic Republic of Iran, that US President Barack Obama formally recognized US involvement in the coup of 1953.[38]

Washington sent the son of American President Hoover, Herbert Hoover Jr., to Tehran, where the expert oilman who had negotiated the 50/50

Venezuelan accord arbitrated the lengthy negotiations that determined how the spoils of Iranian black gold would be divided. Of course, US oil companies grabbed the largest share.

On October 29, 1954, the shah signed an accord that instituted the new order of Iranian petroleum, which remained in place until the victory of the Islamic Revolution and Ayatollah Khomeini's rise to power. The national oil company that Mossadegh had created continued to own the largest oil fields and the Abadan refinery. But real control escaped it: A consortium of Western companies were the only ones empowered to buy and sell Iranian petroleum. The British had to be satisfied with only 40 percent of the consortium's shares. The Anglo-Iranian Oil Company name disappeared, rebranding itself under the name of the company that sold its oil in Great Britain: British Petroleum. American companies claimed 40 percent of the capital. Among these were the four companies already associated with Aramco (Jersey Standard, SoCal, Socony, and Texaco), joined by Gulf Oil (the fifth major American oil company) as well as an alliance of nine "small" independent American producers. Shell obtained 14 percent of the shares. Finally, Paris and the Compagnie Française des Pétroles managed to wrangle a meager 6 percent of the pie.

Suez 1956: Eisenhower Leaves London and Paris "to Boil in Their Own Oil"

Concerning the role Western powers played in Mossadegh's fall, even the most critical observers of the era only considered the inflexible blockade Britain had imposed for two years, and did not immediately recognize the hand of Washington. The American oil empire courted success after success, from the precious compromises forged with Venezuela and Saudi Arabia to the secret maneuvers during the Iranian coup. But this was not the end of the road. Among the Western powers, Washington was the only one to emerge victorious from the 1956 Suez Canal crisis.

Yet, at the outset, it was Washington's categorical refusal to allow the World Bank to lend Colonel Nasser the funds to build the Aswan Dam that pushed the Egyptian dictator to nationalize the Suez Canal on July 26, 1956. John Foster Dulles, head of the American State Department, had refused the loan because he believed Gamal Abdel Nasser's attitude about the Communist block was ambiguous: His big mistake was to approve arms sales to Czechoslovakia in 1955.

The decision of the United Kingdom and France to take control of the Suez Canal zone (with Israel's secret support) of course had to do with oil. Oil tankers comprised two-thirds of the vessels that traversed the canal. The bulk of the Persian Gulf oil, vital for Europe, used the canal to reach the ports of Southampton, Marseilles, and Rotterdam.

On October 29, 1956, Israeli troops easily invaded the Gaza Strip and the Sinai Desert, then turned in the direction of the canal. As previously agreed with Jerusalem, London and Paris gave an ultimatum demanding that the belligerent Israelis and Egyptians withdraw from the canal zone. Nasser refused. On October 31, the United Kingdom and France bombed the Egyptian airfields. On November 5, French paratroopers were dropped in the vicinity of the canal, followed the next morning by the Royal Navy commandos (who were using attack helicopters for the first time). The Egyptian troops fell back, and France and the United Kingdom seized the Suez Canal.

But the overwhelming military victory instantly turned into a diplomatic fiasco. The USSR threatened to send troops to Egypt. London and Paris, who believed they could count on Washington's support, discovered their error too late. A petroleum shortage threatened Europe. A Mediterranean pipeline that transported oil from Iraq was sabotaged by Syrians in solidarity with Nasser. Yet Eisenhower refused to allow American companies to supply oil to France and Great Britain, and encouraged Saudi Arabia to join this embargo, the second of the modern era, once again imposed by a Western power. The American president told a Pentagon official: "I'm inclined to think that those who began this operation should be left to work out their own oil problems—to boil in their own oil, so to speak. They will be needing oil from Venezuela and around the Cape and before long they will be short of dollars to finance these operations and will be calling for help. They may be planning to present us with a fait accompli, then expect us to foot the bill. I am extremely angry with them."[39] At Washington's injunction, the International Monetary Fund refused to grant an emergency loan London claimed it needed in order to finance the delivery of an oil shipment via the Cape of Good Hope. And on December 15, the French and British began to withdraw their troops.

As in Iran, the relevance of London's brutal strategy was once again refuted. Historically close to the interests of the British Crown, Shell's directors were dismayed by the Suez Canal invasion. Two years later, the company operated undisturbed in Egypt, even as diplomatic relations between Cairo and London remained suspended. The major oil companies once again found themselves no less capable of conducting more effective diplomacy than their powerful governments. On the occasion of the Suez crisis, the United States (and, by extension, the American oil companies) greatly increased their prestige with the governments of the Arab world, and in particular with the Gulf countries. Uncle Sam could be trusted. Yet even if the history of the CIA's interventions in Iraq in 1963 and Indonesia in 1965 remains sketchy, the reality of these interventions is now well established, thanks in particular to many documents subsequently declassified by US Congress.

How Kennedy and the CIA Helped Saddam Hussein Get in the Saddle

The tumultuous history of the alliance between Saddam Hussein and the United States did not begin with Ronald Reagan and the Iran-Iraq War in the 1980s. This story begins in 1963, when the leader of the Iraqi government, General Abdul Karim Kassem, was executed at the end of a coup marking the first incursion of the Baath Party's power in Iraq.*

General Kassem had seized power five years earlier with the support of his Communist allies (executing or turning a blind eye to the execution of young King Faisal II of Iraq, grandson of Faisal I, Lawrence of Arabia's protégé). In a sensational article published by the *New York Times* only four days before the US Army invaded Iraq in March 2003, American historian Roger Morris wrote: "Forty years ago, the Central Intelligence Agency, under President John F. Kennedy, conducted its own regime change in Baghdad, carried out in collaboration with Saddam Hussein."[40] In this article, Morris, one of Henry Kissinger's National Security Council collaborators, tells how the CIA "marshalled" opponents of General Kassem's regime in the early 1960s, thanks to the intelligence agency's operational base near Kuwait, which according to Morris was used for "radioing orders to rebels" in Iraq.[41] The "health alteration committee" of the CIA first attempted to assassinate Kassem by sending him a poisoned scarf. Then, on February 8, 1963, the Baath Party took power after a coup. After a speedy trial, Kassem was shot. "Almost certainly a gain for our side," a National Security Council analyst wrote to Kennedy that same day.[42]

According to the former Baathist leader Hani Fkaiki, among party members colluding with the CIA in 1962 and 1963 was young Saddam Hussein.[43] He was then hiding in Cairo (after having participated, four years earlier at the age of twenty-two, in a failed attempt to assassinate Kassem on October 7, 1959; an attempt that was organized by the CIA).[44] In the weeks following the Baath Party's power grab, the CIA supplied the Baathists with lists of

* This pan-Arab socialist party, founded in 1947, had two separate principal branches: one in Syria, the other in Iraq.

intellectuals suspected of being Communists: doctors, professors, engineers, and lawyers who would be assassinated by the hundreds—murders to which Saddam Hussein seems to have lent a hand.

The United States was not hostile to General Kassem when he gained power in 1958 after the murder of young King Faisal II and his family; on the contrary. Roger Morris noted: "From 1958 to 1960, despite Kassem's harsh repression, the Eisenhower administration abided him as a counter to Washington's Arab nemesis of the era, Gamal Abdel Nasser of Egypt—much as Ronald Reagan and George H. W. Bush would aid Saddam Hussein in the 1980s against the common foe of Iran."[45]

Why, in the end, did the CIA yearn for Kassem's departure? As Saddam Hussein would do later, General Kassem sought to arm Iraq against Israel and claimed that Iraq had a legal right to Kuwaiti oil. But again, it's the threat he represented to Western oil interests that seems to have sealed General Kassem's fate. On December 11, 1961, he made the fatal decision to sign Law No. 80, which confiscated from the Iraq Petroleum Company (IPC) all territories where that Western consortium had not yet drilled. Since IPC wells covered less than 1 percent of Iraqi territory, Kassem in effect deprived the oil majors of 99 percent of their concession's initial territory.[46] In his *New York Times* article, Roger Morris noted that, shortly after the Baath Party came into power in 1963: "Western corporations like Mobil, Bechtel and British Petroleum were doing business with Baghdad—for American firms, their first major involvement in Iraq."

In Indonesia: The CIA behind the Coup and the Massacres of Suharto

In Indonesia, the oil industry that appeared as early as the end of the nineteenth century was the pure creation of Royal Dutch Shell, the large Anglo-Dutch company. However, here again, the American petroleum giants gained considerable ground. In 1940, Caltex (the joint venture created in 1936 by SoCal and Texaco to commercialize the oil "to the east of Suez," in Saudi Arabia and elsewhere) discovered the two largest Indonesian petroleum fields, Duri and Minas, on the island of Sumatra. This was a thorn in the side of the Europeans and, especially, of the Japanese imperial forces.

As early as 1957 and 1958, the CIA armed secessionist movements and paramilitary rebellions from the extreme right in the petroleum zones of Sumatra and Sulawesi, where Caltex and Socony Mobil, among other

American companies, possessed enormous interests.[47] American rubber and oil investments had to be defended in the face of Mao's China, which was close by, financing its own insurrectional movements. During this troubled period, Socony Mobil's representative in Indonesia was visiting the head of the US Navy in Washington, to obtain weapons for Indonesia's chief of state.[48]

In 1965, the CIA strongly supported General Suharto when he supplanted third-world leader Sukarno, after the failed coup attempt of September 30 for which Indonesia's Communist Party had been held responsible. This party, the PKI, was a valuable ally of the Sukarno regime. It was also the third-largest Communist Party in the world, after those of China and the Soviet Union. As a result of the coup, as in Iraq, the CIA provided General Suharto with a list of four thousand to five thousand persons suspected of Communism, who were arrested and, for the most part, executed.[49] In all, between 250,000 and 500,000 Indonesians were assassinated during the purge of the PKI, which lasted from October 1965 through early 1966. American diplomatic cables and subsequently declassified CIA reports suggest that the United States was surprised by the extent of the repression.

A Motive for the Vietnam War?

What role did the appetite for crude oil play in the United States' initial engagement in Vietnam, and later, in the escalation of conflict after President Lyndon Johnson's arrival in the White House 1965? Everything indicates that it was not negligible.

On the eve of the First World War, when, having already assured his fortune, he traveled across Asia as an itinerant expert, the eminent geologist and future US president Herbert Hoover would have been one of the first to hypothesize that there were hydrocarbon fields, technically unexploitable, in Indochina. In the early 1960s, those who defended the deployment of US troops in Vietnam likely had a strategic interest in these oil resources. In 1963, in response to the Economic Club of Detroit, the worldwide automotive industry capital, President Kennedy's vice deputy secretary for political affairs, U. Alexis Johnson, used the following arguments: "What is the attraction that Southeast Asia has exerted for centuries on the great powers flanking it on all sides? Why is it desirable, and why is it important? First, it provides a lush climate, fertile soil, rich natural resources, a relatively sparse population in most areas, and room to expand. The countries of Southeast Asia produce rich exportable surpluses such as rice, rubber, teak, corn, tin, spices, oil, and many others."[50] U. Alexis Johnson was a member of the executive committee of the National Security Council from 1961 to 1964, before becoming vice ambassador to the government of South Vietnam.

After 1965 and President Lyndon Johnson's decision to send American soldiers to fight in Vietnam, many of the war's opponents claimed that in his role as a Texas senator, John Kennedy's successor had long been a faithful defender of the oil companies' interests. In fact, since the general elections of 1940, Lyndon Johnson was the best at raising money from Texas oil magnates for the benefit of the Democratic Party. In any case, nothing indicated that oil played any role in Johnson's decision to trigger the escalation of the conflict in 1965. However, six years later, Johnson's successor, President Nixon, was in turn accused by some of making the Vietnam War a war of oil. The *New York Times* discovered that geological studies had been carried out along the coasts of South Vietnam as early as 1967—the year the United States systematically began to employ massive incendiary bombs that used gasified petroleum jelly: napalm. The oil industry then increased their focus on offshore drilling and analyzing the geophysics of the landscape above the continental shelves. Two studies off the coast of Saigon were conducted by a Texan geophysics company. This company indicated that the studies were sponsored by "nine international companies" but refused to specify which companies these were.[51]

On June 10, 1971, the pro-American government of South Vietnam auctioned the concession rights covering 414,000 square kilometers of the sea around the Mekong Delta.[52] The announcement of this auction had been postponed several times since December 1970, due to the intense controversy triggered, in both the United States and Vietnam in response to introducing such an initiative at the height of the war.[53] The most fruitful concessions were obtained by four American companies (Mobil, Exxon, Marathon Oil—yet another offshoot of the Rockefeller empire—and Texas Union), and by Shell, strongly established in nearby Indonesia.[54]

Oil was certainly not the only nor the most obvious aim of American interventionism during this time—the most intense, somber period of the Cold War. For Washington, DC, though, it was a constant motivator, necessary for the optimal functioning of its war and peace machine. Even if it required, as we will now see, the leader of the "free world" to sacrifice the will of its people to please its own petroleum merchants.

THIRTEEN

Big Oil's Planetary Empire and the Rockefellers' Hegemonic Ambitions

Ten months before the Pearl Harbor attack, the ambition that led the United States to assume their role as the great superpower of the second half of the twentieth century was proclaimed in a famous editorial published by *Time* magazine. This text, titled "The American Century," was republished in many American publications, including the *Washington Post* and *Reader's Digest*.[1] Its author, the founder of *Time* itself, Henry Luce, preached that the United States had entered the war in order to defend democracy. In terms that, retrospectively, betray a worrisome assurance, he urged Americans to embody the role of the good samaritan, a role that, according to him, would be the Americans' role from then on. Head of the nation's most influential conservative publication, Luce—pure product of the East Coast's WASP elite and prominent member of Skull and Bones, the oldest of prestigious Yale University's secret societies—called upon Americans "to accept wholeheartedly our duty and our opportunity as the most powerful and vital nation in the world and in consequence to exert upon the world the full impact of our influence, for such purposes as we see fit and by such means as we see fit."*

Neoliberalism and Technology Meet

At a time when American society adhered to rigid race and class hierarchies, who were the "we" Henry Luce invoked? To understand the manner in which

* The Skull and Bones society was created in 1832. Three former US presidents were members: William Taft (1909–1913), the son of one of its founders; George H. W. Bush (1989–1993); and George W. Bush (2001–2009).

American power spread at the end of the Second World War, we must ask who had the stronger hold over the other: Uncle Sam or Big Oil? American political power deployed its empire thanks to a control over sources and flows of oil far greater than those of other nations. But simultaneously, the oil establishment was able to put its own interests above the political sovereignty of the American people.

One coincidence: At this defining moment in the age of petroleum, the turning point when worldwide production and consumption of crude oil tripled its growth rate, Hannah Arendt defined the nature of modern imperialism. The German philosopher and naturalized American citizen saw imperialism surge in the Boer War, a story of inextinguishable greed fed by a fabulous source of natural wealth: the gold mines of Transvaal, South Africa. In *The Origins of Totalitarianism*, Arendt's 1951 masterpiece, the Jewish political scientist showed how, thanks to the huge profits generated by industry, empires had escaped being states to become corporations. "For the first time, investment of power did not pave the way for investment of money, but export of power followed meekly in the train of exported money," Arendt pointed out regarding Transvaal gold. At a pivotal moment in the history of black gold and US power, Arendt (who had just become an American citizen) added a remark that illuminated the methods of the capitalist state: "The only grandeur of imperialism lies in the nation's losing battle against it. The tragedy of this half-hearted opposition was not that many national representatives could be bought by the new imperialist businessman; worse than corruption was the fact that the incorruptible were convinced that imperialism was the only way to conduct world politics."

Three years after the publication of *The Origins of Totalitarianism*, Hannah Arendt's former companion, the great and worrisome German philosopher Martin Heidegger, a former Nazi party member, produced a major text on the essence of technology suggesting it had inflated the human ego to the point of blinding it, and in so doing had seized not just nature but also humanity.[2]

The International Oil Cartel Scandal: Justice against "Realpolitik"

In 1947, Big Oil was mired in the worst scandal since the dismantling of the Rockefeller empire: The American Senate discovered that SoCal and Texaco had overcharged for the Saudi oil that they sold to the US Navy. The proponents of new US imperialism had little reverence for political power. They shamelessly hoped to reverse a traditional constraint: In the old British Empire, the

Anglo-Iranian Oil Company and its majority shareholder, the British Crown, used to sell Persian Gulf oil to the Royal Navy at cost.

In Washington, a Senate committee summoned these two American oil companies to explain themselves. The senators recalled the unwavering wartime support SoCal and Texaco had received in Saudi Arabia. In April 1948, the Senate published its brutal conclusion: "The oil companies have shown a singular lack of good faith, an avaricious desire for enormous profits, while at the same time they constantly sought the cloak of United States protection and financial assistance to preserve their vast concessions."[3] The administration responsible for implementing the Marshall Plan discovered the so-called phantom cargo scheme through which cheap oil from the Persian Gulf was invoiced at the higher price of US crude. Clashing with Aramco, James Moffett (who had not obtained the remuneration that he had requested in his meeting with Roosevelt on the subject of Saudi Arabia) publicly broke the law of silence: While the Royal Navy had paid 20 cents a barrel for Persian oil, the American government had to pay $1.48, the renegade of Standard Oil revealed.[4]

Little by little, senior officials began to notice the unusual arrangements between public authorities and the major American oil companies. In the State Department, some voices denounced the weighty, ever-growing menace of the black gold industry. In a 1950 memoir, the adviser to George McGhee (the influential oil diplomat intimately connected to the Texan networks of black gold) noted that the Iraq Petroleum Company, Aramco, and the Anglo-Iranian Oil Company each held concessions extending more than 250,000 square kilometers, equivalent to the size of many American states. This advisor, Richard Funkhouser, noted with false candor that "to have had Texas-Oklahoma-Louisiana oil fields controlled by one company would have had obvious disadvantages" for the interest of American citizens.[5]

The US government's objections to the trusts, which (in appearance) were corrected forty years earlier when John Rockefeller's original Standard Oil Company had been split, resurfaced in December 1949. The Federal Trade Commission (FTC) launched an investigation on the practices of the oil majors, at the same time questioning Washington's policies in respect to the aforesaid. The result of these investigations was delivered in a report whose title alone was explosive: "The International Petroleum Cartel."[6] For the first time, this report showed in detail the principle and contents of the secret agreements of Achnacarry and the Red Line, signed in 1928. The FTC established that the world's seven most powerful oil companies (the five American major companies, as well as Shell and the Anglo-Iranian Oil Company) had jointly agreed to control all the major oil regions, all refineries and pipelines outside of the United States; that they shared technical patents; and finally, that they had agreed to maintain artificially high crude oil prices.

Frightened by a potential scandal, the Truman administration at first chose to keep the report secret and only agreed to publish an expurgated version in August 1952, after press leaks. The charge this report contained could not have been worse for the White House. And the trap was full of irony. This occurred in the middle of the oil blockade against Iran, and the chief American diplomat, Dean Acheson, was just about to broker an agreement between the oil companies to ensure there would be a sufficient amount of crude oil on the market despite the Abadan blockade. In addition, the American army called for Big Oil's full collaboration in supplying the US troops engaged in the Korean War. In spite of the Truman administration's reluctance to act, the US Department of Justice was determined to conduct an antitrust trial. Their objectives could not have been more ambitious. They aimed to end, all at once: Big Oil's hegemonic control outside the United States; American production quotas; the secret quotas that controlled oil sales in foreign markets; the limitation of American imports and exports; and the exclusion of independent American oil companies outside of the United States.[7]

Secretary of State Dean Acheson did not intend to let this happen. In April 1952, he wrote to the Department of Justice's antitrust division. He began by stating for the record that he certainly did not mean to interfere in judicial affairs. Then he warned: To go after the American oil companies would destabilize the Middle East and hinder the objectives of American foreign policy.[8] Truman's attorney general, James McGranery, disregarded him and provoked an unprecedented *casus belli*, a "case for war." He announced the appointment of a grand jury, which, in the course of the following months, summoned the heads of no fewer than twenty-one oil companies, among which were, of course, the five major American oil companies, as well as the Anglo-Iranian Oil Company, Shell, and the Compagnie Française des Pétroles.

This was open war between the Department of Justice and American diplomacy. Here were two opposing visions of the state, one based on the principle of equity, the other on pragmatism. These visions faced off over two reports submitted in January 1953 at the National Security Council, in the last few days of Harry Truman's presidency, in a kind of blueprint of the continual struggle at the core of the most powerful of democracies. The argument defended by American diplomacy, supported by the Departments of Defense and the Interior, held that: "American oil operations are, for all practical purposes, instruments of our foreign policy."[9] It argued that, in the midst of the Cold War, an antitrust trial against the oil companies might foster the belief that "capitalism is synonymous with predatory exploitation."[10]

London and Paris lent their support for this argument. The British foreign secretary, Anthony Eden, mocked the "witch-hunters," and Paris sent Washington a formal protest against the American Department of Justice's initiative.[11] Some noted that Radio Baku, the Soviet radio station disseminating propaganda in the Middle East, increased their derisive references to the "international oil cartel." The Department of Justice's reply was scathing. "The world petroleum cartel is an authoritarian, dominating power over a great and vital world industry, in private hands. National security considerations dictated that the most expeditious method be employed to uncover the cartel's acts and effects and put an end to them . . . A decision at this time to terminate the pending investigation would be regarded by the world as a confession that our abhorrence of monopoly and restrictive cartel activities does not extend to the world's most important single industry."[12]

The White House Bows Down before the American Oil Empire

Which side would be chosen by the American president, who, like some of his predecessors and (we will see) most of his successors, had close ties to the black gold industry? In his youth, Harry Truman took oil-prospecting risks and incurred great financial loss by selling his shares just before his former partners discovered an important oil field. The democratic leader who succeeded Franklin Roosevelt liked to say that he could have become a petroleum billionaire rather than president of the United States.[13] Truman certainly had ambivalent feelings about Big Oil. (In 1942, the president presided over the Senate commission that brought to light the ties that persisted between Jersey Standard and IG Farben, well after Hitler's arrival in power.) Truman adopted a position of pragmatism. On January 12, 1953, eight days before handing the Oval Office to General Eisenhower, the president decided to transform the lawsuit to a simple civil suit, which considerably weakened the case (petroleum companies would, for example, be able to challenge every subpoena). The lead lawyer for the Department of Justice was summoned to the White House to hear the fateful news: "On Sunday evening President Truman sent for me and we met in the living quarters of the White House. His purpose was to tell me and the person representing the State Department who was present two things: first, that he reached his decision with great reluctance and he was constrained to take that decision . . . solely on the assurance of the chief of staff General Omar Bradley that national security called for that decision; and second, that he wished the civil action to be vigorously prosecuted."[14]

The White House once and for all agreed to connect the fate of the American empire to that of its petroleum industry. US political powers would now allow this inseparable couple to pursue their course without hindrance, flowing from the fields of crude oil to the reservoirs of the largest American machines—Cadillacs, B-52s, Chase Manhattan Bank—along the slopes of least resistance.

President Truman's wish for a vigorous civil procedure was ignored by his Republican successor. Shortly after Dwight Eisenhower's ascension to the presidency, the White House produced a confidential memorandum that stated very clearly: "It will be assumed that the enforcement of the Antitrust laws of the United States against the Western oil companies operating in the Near East may be deemed secondary to the national security interest."[15]

On January 14, 1954, the big American companies obtained the National Security Council's explicit guarantee that they would never face antitrust litigation for their participation in the new consortium that was preparing to take control of Iranian oil.[16] Within this consortium, the State Department and its principal negotiator, Herbert Hoover Jr., worked to advantageously position American Big Oil. In order for the war machine's inner workings to function well, the sword of American justice was laid at the feet of private companies. In 1956, the responsibility of the judiciary action against the oil companies was removed from the Department of Justice's purview and entrusted to diplomats. Incidentally, the oil companies' defense was assured by Sullivan & Cromwell, the powerful firm that counted among its associates John Foster Dulles and Allen Dulles.[17] Respectively, these two brothers directed the State Department and the CIA under Eisenhower. At the end of his second term, during his farewell speech on January 17, 1961, General Eisenhower surprised US citizens by warning: "In the councils of government, we must guard against the acquisition of unwarranted influence, whether sought or unsought, by the military-industrial complex. The potential for the disastrous rise of misplaced power exists and will persist."[18] This famous warning was interpreted as referring to the armament industries, but it applied at least as much to Big Oil, the indispensable partner of American domination.

Peak Empire for American Oil

The carte blanche granted in 1953 by the Eisenhower administration launched the apex of American companies' control of world oil production. They were able to consolidate their control for two complementary reasons. The first reason was on a human scale. It resided in the shrewdness of the oilmen and, even more, in the trepidation of those who had an insatiable need for black gold: the political and military leaders of developed consumer countries and (for the moment) those of the producer countries. Starting in the 1950s, an intensification of joint shareholdings between oil majors in the Middle East elevated cartelization to a pinnacle of sophistication. Secret agreements established in 1928 to counter overproduction were broadened and revised, in order to strengthen the control of a burgeoning market.

The second reason, by far the strongest, originated within the physical nature of the Earth itself. The abundance of new oil fields around the Persian Gulf greatly surpassed even the greedy appetite for oil that had seized humanity after the Second World War, during the Thirty Glorious Years. The oil majors were certainly politically able to consolidate their cartel, despite the fact that it was exposed by the American administration in 1952. Yet the push to deepen their secret pacts seemed to come from beyond their own volition or capacity. The alliance of private companies, which aimed to ward off potential grievances and thwart the ambitions of the oil-producing countries, seemed somehow required by the abundance of black gold itself: More than ever, their goal was to prevent overproduction and price wars that would cause them to lose profits. Mother Nature's generosity revealed itself to be so enormous that it rendered inoperative, and even futile, the sacrosanct principle of free competition. The allocation of scarce resources is the raison d'être of the market, the spontaneous balance between supply and demand. Henceforth, the scale of US production confirmed that oil was anything but rare. Behind the powerful oil companies, the black, fluid energy source imposed itself as the fundamental master of their game.

In Iran, despite a two-year boycott, the crisis affected neither oil prices nor production. The global course of crude oil remained remarkably stable. On May 15, 1951, just a few days after the British oil embargo was imposed on Iran, Washington made it known that American companies would not come to Tehran's rescue; they would respect the embargo. Saudi, Iraqi, and Kuwaiti extractions were simply accelerated and the total oil production of the Persian Gulf remained unchanged.

After the fall of Mossadegh in August 1953, it was necessary to explain to the other Gulf countries that their extractions had to decrease in order to reintroduce Iranian crude oil to the marketplace. Howard W. Page, a Jersey Standard vice president responsible for the company's Persian Gulf interests, at a time when it was preparing to integrate into the new consortium that controlled Iranian exports, spoke of how King Ibn Saud was convinced to accept a slowdown in Aramco's growth. Page's tortuous arguments, as presented to the Saudi Arabian sovereign, could hardly appear to have been in good faith or even truthful. Page said: Ibn Saud "was told that the Aramco partners [weren't going into Iran] because we wanted more oil anywhere, because we have adequate oil in the Aramco concession, but we were doing it as a political matter at the request of our government."[19] As is often the case when one follows the dialogue between the American petroleum companies and the House of Saud, the argument of the menace of political "chaos" and of Communism was fully operative.[20] The Bedouin king, now almost senile (he died on November 9, 1953), was conciliatory. According to Aramco's vice president, the old Leopard of the Desert merely replied, almost docilely: "Yes, but in no case should you lift [in Iran] more than you are obligated to lift."[21]

The Cartel Mechanism

With the 1954 creation of Iranian Oil Participants, the official name of the consortium that took control of Iranian oil after Mossadegh's fall, the major American oil companies reached the zenith of their world dominion. It was the outcome of a process that begin in 1928 with the Achnacarry and Red Line Agreements, as well as the arrival of American—and French—oil companies in Iraq; continued with the 1934 creation of the Kuwait Oil Company; and was spectacularly reinforced when Jersey Standard and Socony entered Aramco's capital in 1948.

The organization of the Western petroleum consortium in Iran exemplified the most successful and subtle form of crude oil production's cartelization. The statutes of Iranian Oil Participants allowed for a covert arrangement that imposed secret quotas on Iranian production. The shah himself even ignored the existence of this arrangement until it was revealed to him after the first oil shortage, in 1974, during the course of an American Senate hearing on "the multinational oil corporations and foreign policy" of Uncle Sam. In order to avoid a hypothetical congestion in the world market, the consortium members each year secretly agreed how high or low to set Iranian production levels, without harming any of the interested parties (apart from the Iranian government, of course). A complex calculation, referred to as the "aggregate programmed quantity," established a compromise between the companies that wished to open their valves fully and those that, on the contrary, called for a low extraction level.[22] The three largest producers from Aramco—Jersey Standard, SoCal, and Texaco—were almost always among the latter, no doubt because of the opportune promise they had made to the Saudi king not to extract more than they were "obligated" to in Iran (and certainly because it was not in their best interests to flood the market with oil).[23] Several internal Aramco documents, revealed in the 1974 American Senate hearing concerning the multinationals, highlighted a direct link between the "aggregate programmed quantity" limiting Iranian production and the annual fixed production objectives jointly determined by Aramco's partners.[24]

The control exerted on Iraq, the third-largest petroleum-producing country in the Persian Gulf, completed the arrangement that mastered black gold control in the Middle East (and the world). After revealing the existence of the "aggregate programmed quantity," the US Senate asked Jersey Standard's vice president, Howard Page, what would have happened if the Iraqi extractions suddenly went too high. Page said the following: "I admit we would have been in on a tough problem, and we would have had to lower our liftings from the [Iranian] Consortium down to the minimum we could possibly take there and meet the agreement."[25] But, in fact, it was always the opposite situation that prevailed. Whenever one of his cartel partners asked Howard Page, "Can you swallow this amount of oil?" the leader of Standard Oil's first subsidiary invariably responded: "Of course, with Iraq down."[26] The major oil companies deemed it superfluous to preserve good relations with the intractable political regimes that succeeded in Baghdad, and strove instead to be faithful allies of Riyadh and Tehran. Again, as from the outset in 1928, Iraq and its oil served as an adjusting tool for large Western companies to keep control of Middle Eastern oil. A role, as we shall see, that persisted.

The alliance of oil majors had been organized through a planetary web of joint shareholders. American economist John M. Blair, the kingpin of the 1952 report on the international oil cartel, who continued his investigations throughout his career, referred to the establishment, beginning in the 1950s, of an "oligopolistic interdependence" even more effective than the system outlined in 1928 by the secret agreement of Achnacarry. According to Blair, the organization that emerged after the war allowed the large companies to agree on the essentials—the pace at which the industry, as a whole, should grow—without requiring them to operate cohesively. Blair had shown that the total production of Western companies in the major "third world" oil countries had grown each year from 1950 until 1972, the eve of the first oil shortage, at the incredibly regular rate of 9.55 percent, despite the 1960 creation of the Organization of the Petroleum Exporting Countries (OPEC), and in spite of the significantly different (and seemingly erratic) evolution of oil extraction in each of these countries and the supply disruptions engendered by a succession of political crises in Iran, Iraq, and Nigeria, and also in Indonesia and Libya.[27*] Another indication of the efficiency of the oligopoly that shared world oil production during the Thirty Glorious Years: From 1945 until the first oil shortage in 1973, the price of a barrel of oil remained almost constant. If the competition had competed freely, the "posted price" of Arabian Light oil (in other words, the price posted by the oil majors) would not have remained steady at just under $2 for almost three decades; it would have been less stable. And, above all, it would have been, on average, lower. An internal analysis of Jersey Standard warned as soon 1950: "It appears that in the future, the Middle Eastern crude oil accessible to Jersey Standard may significantly exceed its needs."[28]

What was the degree of solidarity between the oil majors and the omnipresent constellation of independent American petroleum companies in the United States? We can simply note that Aramco's Arabian Light costs very little to extract, much less than oil drawn from Texan wells. Surveying the Aramco partners' quibbling as they tried to find a way to "support" their oil prices, an American Department of Justice lawyer in charge of the antitrust trial against the oil giants evoked the "unusual and unexpected spectacle of a customer complaining that the price they paid was too low."[29]

The abundance of black gold in the Middle East went well beyond the Western oilmen's wildest dreams. The crude oil there cost so little to extract, and the major companies had collaborated to fix their prices so effectively (in particular by limiting production), that the business had become clearly extraordinary. In 1952, the Jersey Standard president congratulated himself in front of shareholders: "We're very fortunate in having an extremely sizable and potentially important stake in the Arab world with a relatively very small investment."[30] The internal reports of the time of the American majors describe "[v]ery remarkable profits."[31]

That was an understatement. In the Middle East, in this blessed era for Big Oil, it took less than two years to recoup capital investments (wells, pipelines, etc.). The profit rate was five times higher than that of American industry in general! No other industry has ever been as profitable as black gold, not even the African mines. Starting in 1963, the consortium that exploited Iranian petroleum reached an absolute record, with a profit rate of over 100 percent: Every year, the profits garnered exceeded the shareholders' total investment. This staggering profitability level (increased, without a doubt, by the ingenuity through which Iranian production was secretly rationed) was a payoff for the lean years spent awaiting the fall of Mossadegh.

---

* See *infra*, chapter 16.

The Middle Eastern oil controlled by the cartel of major oil corporations released unprecedented power. This power escaped the Western democratic "powers," which was called upon to serve it continuously, instead.

The Hidden Empire of the Seven Sisters

On the eve of the first oil crisis of 1973, the major Anglo-Saxon companies controlled at least 91 percent of Middle Eastern petroleum production, and 77 percent of the production of the "free world" outside the United States.[32] This uncompromising leadership, remarkably stable from 1954 to 1970, earned Jersey Standard, Mobil, SoCal, Texaco, Gulf Oil, BP, and Shell the nickname "Sette Sorelle": the Seven Sisters. This acrimonious name was made famous by industrialist and Italian politician Enrico Mattei, called "the engineer," who founded ENI (Ente Nazionale Idrocarburi), the transalpine national oil group. Unlike the French oil company, who knew how to preserve its role as "eighth sister" and "youngest child," ENI never managed to draw up even a stool at the banquet table of the world petroleum empire.*

At this table, the large US companies had, of course, been able to extract the best oil shares. In the United States, in 1970, Jersey Standard, Texaco, SoCal, Gulf Oil, and Mobil enjoyed "only" 34 percent of the production, but possessed 42 percent of reserves.[33] They shared up to 80 percent of the control of main pipelines, the arteries of the "first world" economy.[34] That year, the American territories' production reached its zenith: more than 10 million barrels per day (Mb/d). It was enormous. The United States had become more than ever the number one producer, extracting as much crude oil as the combined production of Iran and Saudi Arabia (3.8 Mb/d each) and Venezuela (3.7 Mb/d). The ground beneath America supplied more than a quarter of the world's production outside the Communist bloc.[35] But the American nation's control of the streams of oil was trifling in comparison to the amount controlled by US companies: Thanks to their concessions on the five continents, the five Yankee Sisters alone ruled almost half of world production![36]

The world of oil corporations was headed by a close coterie, almost exclusively composed of WASPs. Among the leaders of the Seven Sisters, Jews were rare (even more so since they were unwelcome in the Persian Gulf), not to mention other minorities. Intermarriages between oil families were common currency (for example, between the children of Howard Page, the vice president of Jersey Standard responsible for the Middle East, and Sir William

* See *infra*, chapter 16.

Fraser, baron of Strathalmond, the unyielding patron of the Anglo-Iranian Oil Company).

We do not know the exact number of shares that the Rockefeller family held in each of the Seven Sisters during or after the Thirty Glorious Years. The only in-depth analysis goes back to 1938, when the Securities and Exchange Commission, the watchdog of Wall Street, undertook to unravel the details of holdings, trusts, and indirect holdings of the richest US family. It was revealed that the Rockefellers remained the principal shareholders and retained operational control of the four main offspring of John D. Rockefeller's original empire, officially dissolved twenty-seven years earlier, in 1911: Standard Oil of New Jersey—which became Exxon (20.2 percent share); Standard Oil of New York—which became Mobil (16.3 percent); Standard Oil of California—which became Chevron (11.3 percent); and finally, Standard Oil of Indiana—which became Amoco (12.3 percent), absent from the Middle East but a key player in the United States.[37]

From 1945 to the early 1970s, the five major American oil firms and Shell never ceased to increase the interlinking of their capital, through cross-shareholdings held by New York's most powerful banks, fabulously enriched by their oil investments.[38] These banks included Morgan Guaranty, Chemical Bank, Bank of America, First National City Bank, and Chase Manhattan Bank. These last two were the most important commercial banks in the United States. They were the keystones of the empire of the Rockefeller dynasty. Beginning in 1960, Chase Manhattan Bank was chaired by its number one shareholder: David Rockefeller, only forty-five years old, the grandson of John D. Rockefeller.[39] Inescapable in many industrial sectors, Chase was particularly pervasive in the activities of the oil majors. As David Rockefeller himself said, "oil had become so central to Chase's profitability."[40] As for Chase's primary rival, First National City Bank, it was essentially "the oil bank" developed by James Stillman at the end of the nineteenth century, thanks to capital brought in by William Rockefeller, the brother of Standard Oil's founder. From 1952 until 1967, First National City Bank was chaired by another radiant heir of the blossoming American capitalist aristocracy, a Yale University alumnus and 1924 Paris Olympics gold medalist in rowing: James Stillman Rockefeller, cousin of David Rockefeller and grandson of both William Rockefeller and James Stillman.

Without the control exerted on petroleum production, which Washington obliged and for which Wall Street capital provided the lifeblood, the Thirty Glorious Years would not have happened. Without this new energy afforded by Big Oil—which the entire industrial system seemed able to draw upon without limit—Western power would not have been able develop as it had. According to the industry's past and present defenders, these forces helped the cartel, established by the corporate heirs to the original Standard Oil, establish legitimacy. Oil was so readily available that the Seven Sisters had no other choice than to cooperate, certainly to avoid dangerous levels of overproduction, but also to carry out their mission: to guarantee the continued phenomenal flow of affordable energy, which was indispensable if capitalist democracy was to accomplish its triumphant ideals.*

In spite of such a posteriori justification, at the time, the Western petroleum oligarchy did everything possible to maintain secrecy. The very existence of such an oligarchy only flickered to the surface briefly thanks to the determination of a few US civil servants and congressmen, with the FTC report in 1952, and in 1974, when a Senate subcommittee on multinational corporations was convened.

The Senate subcommittee obtained direct testimony about the persistence of a secretive, centralized organization of the Seven Sisters cartel, which lasted at least until 1971. The American major players were then seeking to form a united front in the face of the Arab producers' growing resolve. A certain George Henry Schuler, the head of a small oil company associated with BP in Libya, testified before the Senate subcommittee that on January 30, 1971, he had been invited to a meeting of what he called the "London Policy Group."[41] Schuler testified that this group had joined forces to negotiate against OPEC. After his meeting with the London Policy Group, Schuler was invited on the same day to speak via telephone to another group, convened on the premises of Mobil in New York. This second group was led by John K. Jamieson, then president of Jersey Standard. Described by Schuler as the "meeting of the chiefs," the New York group usually took on the problems that were "too big for the London Policy Group."[42] Jersey Standard, the eldest of the Seven Sisters, the first heir to Standard Oil, maintained its preeminence within the cartel of oil majors sixty years after the John D. Rockefeller empire had been dismantled. Defended on that day by George Henry Schuler, the Western companies' willingness to cooperate started to wither, basically because, as we shall see, a new, largely unexpected factor emerged: the end of the petroleum surplus.

---

* At the forefront of these defenders today is Daniel Yergin, author in 1991 of the history of the oil that remains the essential reference: *The Prize*, awarded the Pulitzer Prize. Yergin subsequently became the vice chairman of IHS, the key economic intelligence firm in the domain of petroleum and energy in general. IHS often acts as the oracle of Big Oil, repeatedly taking the position of industry; according to critics of Daniel Yergin, this state of affairs indicates the existence of a conflict of interests at the heart of his approach as a historian.

The Rockefeller Paradox: Promoting Charity Alongside Economic Neoliberalism

In the twentieth century, there was no family more suspected of diabolical conduct than the Rockefeller family—not even the Rothschilds (who also owed an important part of their fortune to oil). Portly, with aquiline nose and stately bearing, David Rockefeller was a confident man. He spent his whole life as the favorite target of conspiracy theorists of all stripes, some of whom denounced him as the architect of a clandestine global government that aimed to enslave the masses to benefit a plutocratic elite, through sophisticated manipulation techniques.* The charge was unfair, in more ways than one. Deeply marked by the ethical rigor of the Baptist faith, the Rockefellers were without doubt the most influential philanthropists in history. Created in 1913 with the spiritual mission "to promote the well-being of mankind throughout the world," the Rockefeller Foundation has funded the work of dozens of Nobel Prize winners. In the course of the twentieth century, researchers paid by the Rockefeller Foundation developed the first antibiotic, and discovered the yellow fever vaccine and the chemistry of viruses, DNA, blood groups, tumor biology, cell biology, methadone, and the antiretrovirals. Extreme wealth sometimes also perpetuates extreme errors. Over the course of thirty years, and despite numerous studies showing they were wrong, a team from the Rockefeller Institute of New York sacrificed hundreds of thousands of laboratory animals and prescribed many inefficient sprays to cure polio.[43] The Rockefellers have funded a tremendous number of academic centers, mainly in the United States and Europe: the University of Chicago, founded in 1890 by John D. Rockefeller; Switzerland's International Center for Genetic Epistemology in 1955; the School of Medicine of the University of Lyon; and the London School of Economics. By providing their subsidies to the Emergency Rescue Committee during the Second World War, the family helped hundreds of European artists and scientists escape the Nazis: for example, Hannah Arendt, Claude Lévi- Strauss, Thomas Mann, Max Ophüls, and even Leo Szilard, one of the fathers of nuclear fission and the atomic bomb. In 1940, in Marseilles, calling on Varian Fry, an American journalist commissioned by the Emergency Rescue Committee, French member of the resistance Stéphane Hessel just missed the chance to enable Walter Benjamin's escape.

In the 1950s, the Rockefeller Foundation founded virus research laboratories in Latin America, Africa, and Asia. As we will see, it also played a founding role in the Green Revolution, which gave a large number of

* Some went so far as to claim he used mind-control techniques inherited from aliens.

developing nations that were US allies, such as India, access to modern agricultural techniques, putting an end to their endemic famine and sparking tremendous population growth.

Whether by serving on innumerable executive councils related to their interests, or across the powerful American and international institutions on which they have left a profound mark, the second and third generations of Rockefellers have invariably defended the same ideology: neoliberalism—the pure, hard economic liberalism of the neoclassical school. At that time it indeed amounted to a defense. In the West, the period following the Second World War ushered in an era of unprecedented abundance (first and foremost, energy abundance). The US national debt was not yet a burden for the government; on the contrary, it was a tool that spurred economic growth, like a springboard: the famous "multiplier effect" of the great guru of the time, the English economist John Maynard Keynes. The government became Keynesian: it regulated, taxed, redistributed, and frequently intervened in the "free and fair" competition between large private firms. In the United States, as elsewhere, political leaders only stopped interfering with large firms when those firms were involved in outposts too precious and too powerful to be subjected to public scrutiny, especially those outposts established by American oil companies in the Middle East.

Diametrically opposed to this Keynesian approach, David Rockefeller paid tribute in his memoirs to the influence of two illustrious economics professors whose teachings he subscribed to in the 1930s, during his studies at Harvard and then at the London School of Economics (LSE): Joseph Schumpeter and, especially, Friedrich Hayek, the eminent authority on neoclassical economic thought. Hayek spent most of his career in two bastions of academia funded by the Rockefeller Foundation: After teaching for nineteen years at the LSE, he obtained a position at the University of Chicago in 1950. Hayek's theses did not become prevalent until the late 1970s. Nonetheless, they exerted a certain influence, as early as the period of the Thirty Glorious Years, on a number of major lawmakers on both sides of the Atlantic: in France, for example, the political scientist Raymond Aron and, especially, Jacques Rueff, General de Gaulle's economic adviser in the 1960s.

The "Chicago School of Economics" emerged from the lineage of Hayek's teaching. It emerged during the period, in the bosom of the University of Chicago that John D. Rockefeller had founded, with its champion Milton Friedman, the hero of the neoliberalism and monetarism that triumphed at the end of the twentieth century. Proudly biting the hand that fed him, in 1967 Professor Friedman denounced the subtle ambiguity of the Rockefeller's ideological stance. Exposing the duplicity with which Big Oil exerted its influence on the political affairs of America and the world, Milton Friedman, then fifty-seven years old, wrote in *Newsweek* magazine: "Few U.S. industries sing the praises of free enterprise more loudly than the oil industry. Yet few industries rely so heavily on special governmental favors."[44] In effect, thanks to the golden gimmick and (as we will see) thanks to the influence the petroleum industry equally exerted on both the Democratic and Republic Parties during the 1960s, the average tax levied on the five major American oil companies did not equate to even 5 percent of their income.[45]

Was the champion of neoliberalism the devil's advocate, or the court jester? In his memoirs, David Rockefeller distanced himself from Milton Friedman, when Friedman, having become the leader of the Chicago School, secured its shift to the neoliberalism that followed the petroleum crisis of 1973. Rockefeller outlined his objection to Friedman's "cavalier dismissal of corporate social responsibility."[46] However, at ease with a dreadful ambiguity, David Rockefeller congratulated Friedman on the role he had played in Chile as head of the "group of young economists who for the most part had trained at the University of Chicago," called to Chile after the military coup d'état mounted with the CIA's help in 1973: "Despite my own abhorrence of the excesses committed during the Pinochet years, the economic side of the story is a more constructive one. . . . [Milton Friedman and his colleagues] counseled the general to free Chile's economy from the restraints and distortions it had labored under for many years."[47]

The contradiction Milton Friedman displayed is apparent. Its source had been anticipated as early as 1951 by Hannah Arendt in *The Origins of Totalitarianism*. In a chapter titled "The Political Emancipation of the Bourgeoisie," the political scientist, writing about the individualist trend that Hayek and Friedman later championed in the twentieth century, wrote that as long as we are motivated by individual interests, the thirst for power will be "the fundamental passion" that guides us.[48] The Rockefellers' ethical compass rested on the same principles as the game of poker, so dearly loved by Americans—a game in which each player initially has the same chance to accumulate chips, but the player who wins the most chips in the first hands monopolizes the game and becomes more and more difficult to beat as the game advances.

The Rockefellers, Patrons of Pivotal Global Institutions

Some of the most powerful American and international institutions bore the fingerprints of the Rockefeller dynasty. When, in 1946, the United Nations announced their intention to establish headquarters in New York, John D. Rockefeller Jr. tasked his son Nelson, David Rockefeller's older brother, with purchasing for $8.5 million 17 acres of land on the banks of the Hudson River, where the UN headquarters were erected. David Rockefeller later recognized his father's initiative and declared that his father, the son of Standard Oil's founder, "was a man of peace, who believed deeply in the ultimate benefits of peaceful cooperation and continuous dialogue among the peoples and the nations of the world."[49] Initially, two of David Rockefeller's other brothers, John D. III and Laurance, had proposed that the United Nations be built on the immense family property of Kykuit, a hundred kilometers north of New York City, near Woodstock.[50] In New York, on the banks of the Hudson, the design of the UN complex, in which Le Corbusier and Oscar Niemeyer were involved, was headed by Wallace Harrison, Rockefeller's personal architect. Early in his career, Harrison had participated in designing Rockefeller Center, constructed in the heart of Manhattan starting in 1930. He then drafted two of its extensions, the Time-Life Building (which beginning in 1959 housed Henry Luce's media empire), and the Exxon Tower, completed in 1971, which displayed Standard Oil of New Jersey's new official company name.

The Rockefeller family's influence on the genesis of the United Nations was not limited to the organization's headquarters. In 1939, the Council on Foreign Relations, Washington's vital geopolitical think tank, collaborated with the State Department to create a commission on postwar studies. The existence of this commission, known as War and Peace Studies, "is strictly confidential," wrote one of its authors, "because the whole plan would be 'ditched' if it became generally known that the State Department is working in collaboration with any outside group."[51] The Rockefeller Foundation offered a gift of $350,000—a very substantial sum at the time—to finance the whole project. Several members of the War and Peace Studies commission, founders of the international monetary postwar order, had been involved in the Dumbarton Oaks Conference in 1944, which had outlined the provisions of the Bretton Woods agreements. Some members also had participated in the preparations for the June 1945 international conference held in San Francisco, which established the United Nations.

The Dulles brothers were eminent members of War and Peace Studies commission. In wartime, while working at Sullivan & Cromwell, his Manhattan law firm, John Foster Dulles also figured among the experts who conceived State Department proposals that led to the creation of the United Nations.[52] His brother, Allen Dulles, was at the heart of the industrial and financial establishment that helped develop the intelligence network that became the foundation of the CIA, created in 1947; in 1953 (with the fall of Mossadegh as first triumph), he became the omnipotent director of the agency. The Dulles brothers had close ties to the Rockefeller heirs. David Rockefeller had been friends with the Dulles brothers since the 1930s. At the time, the young heir often frequented John Foster Dulles' New York home to woo his future wife, Peggy McGrath, who resided there.[53] In 1941, in room 3603 of Rockefeller Center, Allen Dulles was appointed head of the New York branch of the Office of Strategic Services, the CIA's predecessor, then closely nurtured by Britain's MI6.[54]

American diplomacy's financial arm, the World Bank, created in 1945 by the Bretton Woods agreements and situated one street away from the White House, was chaired for two decades by senior leaders of Chase Manhattan Bank: John McCloy, from 1947 to 1949, followed by two former vice presidents of Rockefeller's bank, Eugene R. Black, from 1949 to 1963, and George D. Woods, from 1963 to 1968.

The Council on Foreign Relations (CFR), where the UN statutes were largely conceived, was itself one of the main crucibles of the Rockefeller network. Most major American politicians at some point spent time in the CFR, including Henry Kissinger, Dick Cheney, Jimmy Carter, Madeleine Albright, and Alan Greenspan. Established in Manhattan, the CFR was, par excellence, the institution where the American political, financial, and media establishments met, and where they dreamed up their vision for the world. Founded in 1922, the CFR had been conceived in the image of Chatham House, created in London a year earlier—a prestigious forum reserved for the highest intellectual luminaries and policies of the British Empire. John D. Rockefeller Jr. had financially supported the CFR since its inception, notably by participating in the acquisition of its first building. And it was the widow of Harold Pratt, a former director of Standard Oil of New Jersey, who in 1944 donated the New York residence where the CFR still meets today, two blocks from Central Park in the Upper East Side. As early as 1946, David Rockefeller participated in a CFR study group on the "reconstruction in Western Europe," which strongly influenced the development of the Marshall Plan. In 1949, at age thirty-four, he became one of the CFR's directors, at that point the youngest person to have held that position. Then, in 1970, he rose to become president of the CFR, taking the place of John McCloy, whom he had succeeded as the head of the Chase Manhattan Bank ten years earlier.

Another prestigious sanctum of the Rockefeller sphere, this time secret, was the Bilderberg Group. Whether it was a place where—as, ironically, David Rockefeller himself described it—"omnipotent international bankers plotted with unscrupulous government officials," or simply the most select forum reserved for North America's and Western Europe's economic, political, and media elite, the Bilderberg Group met each year in strict confidentiality. Its dozens of members were hand-picked by a committee whose essential pillar was, for decades, David Rockefeller.[55] The man who became one of the most decorated men in the world was among the nine Americans invited to the first meeting of the Bilderberg Group in May 1954.[56] This meeting in Holland's Bilderberg Palace was initiated by Queen Beatrix's father, Prince Berhnard of the Netherlands, who was unique in that he had briefly served in an SS brigade before becoming a Royal Air Force hero, and, later, founding the World Wildlife Fund.[57]

Without any compensation, Big Oil demanded and received Washington's support for its global expansion strategy. Big Oil violated antitrust laws and overcharged the US Navy. But the White House, in order to avoid scandal and avoid the idea that capitalism was "synonymous with predatory exploitation", chose to pardon all this, on behalf of the nation's geostrategic interest. Not only was a vital part of this interest—control of world crude oil access—unconditionally assigned to the Seven Sisters cartel, but also, by extension, the Rockefellers' hegemonic desires heavily influenced many key international institutions, and often their driving ideologies. It was, however, Washington's policy game, radiating from the White House, that granted the Standard Oil heirs and other American black gold tycoons their heaviest influence.

FOURTEEN

Big Oil Asserts Itself: A Matrix of Political Power in Washington

During the 1950s and 1960s, as oil imposed itself as the mother of all raw materials, the networks of finance, heavy industry, the armaments industry, politics, and the secret service were intricately linked, with oil as their cornerstone. The unequal profits garnered after 1945 by the black gold industry, and the fundamental role Big Oil played in the powerful US economy, allowed Big Oil to tighten its grip on Washington. Through an intense game of revolving doors, in which powerful figures rotated through business leadership and political office, as well as through political financing and espionage, the American oil industry deployed an influence as omnipresent as it was discreet throughout the 1950s.

John McCloy, Grand Vizier of the American Establishment

The intertwining of political and industrial networks surrounding black gold manifested itself in the course of a powerful New York business lawyer named John McCloy. With his stark, brilliant gaze and sturdy physique, he was portrayed in 1962 by the *New Yorker* as the undisputed leader of the American establishment. McCloy approached the Rockefeller clan for the first time in 1912, during his summer vacation from school. An ambitious law student, only seventeen years old, he knocked on the massive door of the family's impressive home in Mount Desert Island, Maine, and offered his tutoring services.[1]

By the 1930s, McCloy had become a business lawyer responsible for sensitive international cases, in particular in Germany (in 1936, he was a guest in Adolf Hitler's box during the Berlin Olympic Games).[2] In April 1941, eight months before Pearl Harbor, President Roosevelt elevated him to the highly strategic position of assistant secretary of war. After the Allied victory, Nelson Rockefeller invited him to become a partner in his family's law firm (renamed, for the occasion, Milbank, Tweed, Hadley & McCloy), of which Chase Manhattan Bank was the principal client.

After McCloy's 1947 to 1949 ascension to presidency of the World Bank, Truman appointed him US high commissioner for Germany, where he served from 1950 to 1953. Yielding without balking to the pressure from the new West German ally, he granted forgiveness to many German industrial commanders in Nuremberg for their roles within the Nazi regime. Among them were several directors of IG Farben, the chemical giant that had supplied Zyklon B, and with which Wall Street and Standard Oil of New Jersey had been unwilling to sever ties. Among the IG Farben leaders McCloy pardoned was Fritz ter Meer, the Nazi representative at the heart of the German chemical industry, who had been directly involved in the "Final Solution," supervising the construction of a chemical plant at the Auschwitz extermination camp. In 1956, the restrictions on convicted criminals at Nuremberg having been lifted, Fritz ter Meer was able to assume the presidency of Bayer AG, IG Farben's former subsidiary.[3] The industrialist Alfred Krupp, sentenced at the Nuremberg trials to twelve years of imprisonment and the confiscation of his property, was also released in 1951, along with ten other leaders of the gigantic conglomerate Krupp had inherited, and which was returned to him.

The *Washington Post* published a drawing showing McCloy unlocking Krupp's cell, with Stalin in the background, taking a photograph to immortalize the event.[4] On February 15, 1951, President Roosevelt's widow, Eleanor, wrote McCloy a letter asking: "Why are we freeing so many Nazis?" McCloy replied that, in his view, "There was certainly a reasonable doubt that [Alfred Krupp] was responsible for the policies of the Krupp company."[5] Accused by some of having been bribed by Krupp's American lawyer and having liberated him because Germany was a US ally in the Cold War, the high commissioner replied: "There's not a goddamn word of truth in the charge that Krupp's release was inspired by the outbreak of the Korean War. No lawyer told me what to do, and it wasn't political. It was a matter of my conscience."[6] As a kind of American proconsul in West Germany, John McCloy made other controversial decisions on behalf of Cold War imperatives. At the beginning of 1950, he approved the CIA recruitment of a former Nazi general, Reinhard Gehlen, to serve as head of West German counterespionage.[7] He was implicated in the refusal to extradite to France the SS criminal Klaus Barbie, who after the war infiltrated the Bavarian Communist Party in the service of American counterespionage (which then organized his escape to Bolivia in March 1951).[8] Ten years later, having ceded the presidency of the Chase Manhattan Bank to David Rockefeller in 1960, McCloy became an important adviser to President Kennedy on both security matters and oil policy. This role did not prevent him from providing legal counsel to each of the Seven Sisters, including in the antitrust trial that had been ongoing since 1953, in the face of increasing pressure and the accrued grievances of the OPEC countries. McCloy confided to journalist and historian Anthony Sampson that regarding the directors of the Seven Sisters, "My job was to keep 'em out of jail."[9]

Henry Kissinger, Adviser in the Rockefellers' Service

Rockefeller influence also appears in Henry Kissinger's ascension. The young Jewish German refugee entered the United States just before the war, enrolled in Harvard University, where he was mentored by Fritz Krämer, also from Germany, a senior Pentagon advisor who had studied international policy in the Prussian geostrategic school of thought: The only thing that counted in his eyes was the political and, if necessary, military power of the state.[10]

After graduating from Harvard, in 1955 Kissinger directed a nuclear weapons Special Studies Project at the Council on Foreign Relations. This group included David Rockefeller, who noticed Henry and suggested that his elder brother, Nelson Rockefeller, recruit Kissinger for another Special Studies Project the following year; this one, funded directly by the Rockefeller brothers, significantly impacted US foreign policy.[11] Nelson Rockefeller then claimed an important place within the US intelligence apparatus. Starting in 1940 and during much of the Second World War, he served as the coordinator of the Office of Inter-American Affairs, a powerful propaganda office in Latin America. From December 1954 to December 1955, as assistant to President Eisenhower, he advised the administration regarding psychological warfare in the context of the Cold War. Part of his role was to represent the president within the Office of Operations Coordination, the White House committee responsible, in particular, for supervising clandestine operations. CIA director Allen Dulles also sat on this committee.[12]

The Special Studies Project, established in 1956 by the ambitious Nelson Rockefeller, and led by Kissinger, was responsible for geostrategic matter and counted among its members Edward Teller, the father of the hydrogen bomb; press magnate Henry Luce; and two other Rockefeller brothers, John D. III and Laurance. Also included was Dean Rusk, the Rockefeller Foundation president, who later left to become chief diplomat for Democratic Presidents Kennedy and Johnson. The work of the Special Studies Project, which continued until 1960, aimed to define the future American geopolitical strategies. The commission published a report in December 1957, two months after Sputnik's launch, and called for a massive intensification of military investments, including nuclear weapons, to counter the Soviet threat. This report was widely echoed across the United States.[13] It deeply influenced President Eisenhower's State of the Union address delivered the following month, in January 1958.

Elected governor of New York that same year, Nelson Rockefeller once again called upon Kissinger, this time as his personal adviser on foreign affairs, and then as advisor in three unsuccessful runs for the position of Republican Party candidate, in the presidential primaries of 1960, 1964, and 1968. Kissinger remained with Nelson Rockefeller until Richard Nixon was elected president in 1968 and appointed Kissinger to head the National Security Council. Nelson Rockefeller then paid Kissinger $50,000 to settle any remaining accounts.[14] The two men's paths crossed again in 1974, this time at the White House, where Kissinger again served as chief American diplomat, while Rockefeller became Gerald Ford's vice president.*

The Protégés and Protectors of the American Oil Industrialists: Lyndon Johnson and Richard Nixon

John McCloy worked for the Seven Sisters oil companies, and Henry Kissinger owed his political clout in Washington to the heir to Standard Oil, the largest of those companies. Beginning in 1940, oil money (through its tax shelters and its infusion into political careers) incessantly increased its hold on the political life of the United States. But the nebula of Big Oil was changing. It was no longer solely held in the grip of the oil majors but was influenced by a close-knit throng of independent petroleum companies, large and small, often from Texas: producers, service providers, and equipment suppliers, supported by small cadres of executives and staff. Differing by culture and ambition from the East Coast's elite internationalist companies, these companies—simply called "independents"—embarked on a path to promote their turbulent, insatiable interests and exert incomparable political influence on the United States and, a few decades later, on the entire world. This special interest group had three successive champions during the 1950s and 1960s. Lyndon Johnson was its first staunch ally; the second, with whom relations proved stormy, was Richard Nixon. The third was George H. W. Bush.

* See *infra*, chapter 20.

Prior to becoming John F. Kennedy's vice president in 1961, Lyndon Johnson had served for twelve years as a Democratic senator from Texas, and was one of the most ardent Texas petroleum industry defenders in the US Congress. As early as Roosevelt's 1940 reelection campaign, Johnson made himself indispensable to the Democratic Party by soliciting campaign funds from the richest Texas oilmen. Throughout his political career, he counted among his greatest sponsors the brothers George and Herman Brown, founders of Brown & Root, the industrial Texas conglomerate that specialized in infrastructure and naval tanker construction, absorbed in 1962 by the Halliburton group, the world petroleum infrastructure leader that included Aramco among its major clients.[15] It was during Lyndon Johnson's tenure as senator that the Texan cities of Houston and Dallas emerged as key pillars of the American military industrial complex.*

Johnson also obtained, from John Foster Dulles, a seat at the table for independent American petroleum, in the consortium that shared the Iranian oil after the coup d'état against Mossadegh.[16] From the time he became Senate majority leader in 1955, Johnson was an unwavering defender of the "oil depletion allowance." Instituted in 1913, this old, opportune tax deduction was designed to compensate the oil companies, protecting them against the ebb and flow of capital "depletion," in other words, the exhaustion of their crude oil reserves. Thanks to this strong, generous abatement, maintained at 27.5 percent of the oil companies' revenue, the more oil a field produced, the greater the benefit to its owner. For the independent petroleum companies, the political struggle in Washington to defend and maintain the "depletion abatement" became a question of survival during the Great Depression, when the black gold rush of the giant oil field of East Texas shook the entire industry in 1931. Elected in Texas for the first time in 1937, Lyndon Johnson "stood like Horatio at the bridge for years, defending depletion against all comers" each time anyone in Washington purported to limit this provision that favored oil more than any other US industry. (In total, by the end of the 1960s, it represented some $160 billion of income loss for the US Treasury.)†[17]

As vice president, Lyndon Johnson was relegated to an uneasy silence when President Kennedy railed against the oil companies in an effort to end the "depletion abatement," as well as the no less lucrative golden gimmick. John Kennedy's younger brother, Bobby, sent Department of Justice officials to audit several independent petroleum companies, to put them in their place. Having just become president after John Kennedy's death, Lyndon

---

* See *infra*, chapter 18.

† Publius Horatius Coclès, legendary Roman hero.

Johnson abruptly but unsurprisingly abandoned the Kennedy brothers' battle against Big Oil.[18]

Politically rooted in California oil money and protected for many years by powerful Texan oil allies, Richard Nixon also was considered a friend of the petroleum industry throughout his career. Starting when he entered politics in 1946, this Pacific war veteran supported the "depletion abatement."[19] From the outset, he was connected to representatives of the major oil field service companies, from the powerful Chandler family of Los Angeles, whose railways prospered in large part by transporting oil, to the East Coast financial establishment royalty, in particular Prescott Bush, the father of George H. W. Bush.[20] In September 1952, two months before the presidential election, while he was running on the Republican ticket as Eisenhower's vice-presidential candidate, the revelation of a slush fund supplying his campaign nearly cost the young Nixon his career. Several oilmen, including Herbert Hoover Jr., had fueled this slush fund. In extremis, Nixon gave a televised speech, still considered an exemplar of its genre, during which he successfully appeased millions of US viewers by promising that he would keep only one gift he'd received: a small black and white dog that his children called Checkers. Sixteen years later, in 1968, during the campaign that led to the Republican Party nomination in the race for the White House, Nixon reiterated his promise: He continued to defend the highly controversial "depletion abatement." "Tricky Dick," as he was nicknamed, then received $60,000 dollars from the boss of ARCO, an American oil major; an envelope with $215,000 (the existence of which was not proved until later) paid by the Mellon family, principle Gulf Oil shareholders; and no less than $750,000 given by John M. King, a powerful independent oil baron.[21] But, as we will see, once he reached the presidency in January 1969, Nixon bitterly disappointed the oil industry.

The Bush Family, from Father to Son, Stand at the Heart of Financial and Political Networks

As his career began, George Herbert Walker Bush stood at the exact confluence of American financial, political, and intelligence powers, via the oil industry that became its nexus in 1945. At that time, the future forty-first president of the United States was a novice oilman supported by powerful familial interests, both industrial and political; furthermore, he interested the CIA.

Between the First and Second World Wars, Prescott Bush, George H. W.'s respected and feared father, like his friends Allen Dulles and John McCloy, was one of the brightest agents of the financial global Wall Street empire, on which the American intelligence machine relied at the beginning of the Cold War. At the end of the Second World War, at twenty-one years of age, his son, George H. W., appeared poised to become his successor. To the already unsurpassed American cocktail of economic and strategic intelligence ("The business of America is business," said US president Calvin Coolidge during the Roaring Twenties), the young George H. W. Bush seemed to have been led quite naturally to add the powerful catalyst of oil.

Prescott Bush played an essential role in his son's emerging career in petroleum, probably in the secret service, and finally in his political achievements. In 1924, before the age of thirty, Prescott became a powerful Manhattan business banker, just five years after his return from the First World War, where he served as an intelligence officer. Athletic, a heavy drinker, imposing because of both his size (6.3 feet) and his booming, authoritarian voice, Prescott Bush belonged to Yale's macabre, exclusively WASP secret society, Skull and Bones—as did his son and his grandson, the forty-third president of the United States.* Prescott was himself the heir of a pious, industrious, wealthy East Coast family. His father, Samuel Bush, ran a major Ohio steel factory, the Buckeye Steel Castings company, chaired by none other than Frank, the lame duck of the Rockefeller family, younger brother of the founder of Standard Oil. Samuel Bush succeeded Frank Rockefeller in 1908 and became an influential industrialist, presiding until his retirement in 1927 over a company that counted among its main clients the railway networks controlled by the Morgans, the Rockefellers, and railway baron Edward Henry Harriman.[22] During the First World War, the American government hired Samuel Bush to organize the production of small arms and ammunition. The main firm he had to supervise—the one that, after having done business with Germany, took the lion's share in US contracts—was the Remington Arms company, controlled by Percy Rockefeller, John D. Rockefeller's nephew, heir to the greatest share of National City Bank.[23]

In 1926, Prescott Bush was named vice president of a very prestigious Wall Street bank, W. A. Harriman & Co., directed by his own father-in-law, George Herbert Walker (later, George Herbert Walker Bush would bear his name). Renamed Brown Brothers Harriman in 1931, by the end of the 1930s the firm imposed itself as one of the biggest investment banks in the world. As a partner of Brown Brothers Harriman, Prescott Bush oversaw the firm's German clientele. He was, in 1924, one of the directors of the European Union Banking Corporation (UBC), a New York–based bank founded by his

* Beginning in 1856, the Bonesmen (members of Skull and Bones) met secretly in The Tomb, a heavy stone fortress on Yale's campus.

father-in-law to manage the interests of the Thyssen family, the most powerful German industrialist family. Like IG Farben, the Thyssen empire played a key role in financing Adolf Hitler's rise to power and arming the Reich. A 1941 survey by an Office of Alien Property Commission concluded that UBC was actually a front controlled by the Thyssen family.[24] Prescott Bush remained director of the UBC after the United States entered the war in December 1941. During the Autumn of 1942, the United States seized UBC's capital on behalf of the Trading with the Enemy Act. The Roosevelt administration had evidence that in the United States, Brown Brothers Harriman was a smoke-screen for financial interests decisive in the Third Reich's emergence. Yet the investigation stopped there. In Germany, after the war, the American army seized Nazi documents implicating Brown Brothers Harriman, as well as Prescott Bush personally. Robert T. Crowley, second-in-command in the CIA's Cold War covert operations, notably responsible for relations with Wall Street, declared: "The file was damning."[25] Crowley accused John McCloy, the high commissioner for Germany, as well as Allen Dulles, future head of the CIA, of stifling the scandal.[26] Brown Brothers Harriman was one of the major clients of the Sullivan & Cromwell international law firm that employed Allen Dulles and his elder brother John Foster Dulles.* Like many on Wall Street, before the war Allen Dulles had conducted business with industries that financed the Nazis, and after the war was himself at the heart of a campaign to recruit former Nazis to the CIA.[27] Allen Dulles made the CIA prosper by calling on those who ran some of America's most powerful firms. Prescott Bush was clearly one of these men.[28] Continuing his international banking activities after the war, Prescott Bush managed the investments of the House of Saud and the House of Sabah, the ruling family of Kuwait.[29] While Prescott Bush served as a Republican senator for Connecticut, he still found time, in 1955, during a golf game along Hong Kong's China Sea, to debrief naval secret agent William Corson after a failed assassination attempt on Chinese prime minister Chou En-lai during the Bandung Conference.[30]

With the Rockefeller and Dulles brothers, Prescott Bush was counted among the great Republican Party leaders, who fared better than Democrats at courting General Eisenhower, to win victory in the presidential election of 1952. Prescott Bush also played a key role in Richard Nixon's fledgling

* A document dated January 1933 shows that Foster Dulles then intervened on behalf of Brown Brothers Harriman, to ask Berlin for clarifications regarding a steel company of Silesia that Poland wanted to nationalize. During the war, this company used prisoners, and in particular deportees, from Auschwitz. The existence of a possible link with UBC at the time of its seizure has not been established. (See Duncan Campbell and Ben Aris, "How Bush's Grandfather Helped Hitler's Rise to Power," *The Guardian*, September 25, 2004).

political career, opening the doors to the East Coast establishment and providing unwavering support during the slush fund scandal of 1952.[31] Before Nixon was narrowly defeated by John F. Kennedy in 1960, Connecticut's Senator Bush conducted an intense campaign in favor of his protégé. His son, George H. W. Bush, the hugely ambitious young oilman who dreamed of surpassing his father's success, also participated in that campaign.

George H. W. Bush: Thrust into the Black Gold Business—and the Intrigues of the CIA

In 1948, at age twenty-four and with his Yale diploma in hand, George H. W. Bush renounced the aristocratic comfort he had enjoyed in his native New England. He migrated to a small rural West Texas village, Odessa, accompanied by his wife Barbara and their first-born, George W., only two years old. Henry Niel Mallon, a calm, discreet man who hired George H. W. Bush and sent him to the far reaches of the desert, was close to the Bush family, so close that George liked to call him his "favorite uncle."[32] A Yale alumnus and member of Skull and Bones, Mallon headed Dresser Industries, one of the US oil industry's largest equipment and service providers, after the 1928 buyout by W. A. Harriman & Co., the bank Prescott Bush directed. For two generations, the latter occupied a seat on Dresser's directorial board, presenting himself as Niel Mallon's "chief adviser."[33] Three former CIA officers indicated that Dresser had frequently provided cover for the operations of the agency.[34] The firm's strategy was to control the maximum allowable number of key industry patents, whether to accelerate the pumping of crude oil or to compress natural gas to enable its transport by pipeline.[35] In the years preceding the Second World War, Dresser began to purchase a number of armament firms. Much later, in 1998, the firm became one of the pillars of America's military-industrial complex when it merged with its main competitor, Halliburton, chaired by Richard "Dick" Cheney, who also served as George H. W. Bush's defense secretary (from March 1989 to January 1993), and as vice president to George W. Bush (2001 to 2009).

When, in 1948, George H. W. Bush arrived in West Texas, this desolate region was the destination of the ultimate black gold rush for classic "conventional" crude oil in US territory. While the little city of Odessa blossomed with oil industry employees, its more fashionable sister city, Midland, 30 kilometers away, saw the mass arrival of a "swarm of young Ivy Leaguers [who] created a most unlikely outpost of the working rich," a report of the time proclaimed.[36]

The new fields drilled in this zone were much deeper than any previously drilled in the United States: They needed more capital, technology, and equipment.[37] After the 1940s, on the road that led to Midland, the metropolis of Dallas strengthened its position as the center of two highly valuable extraction activities that attracted tremendous investment: not only petroleum but also uranium, in which Texas was rich. Many local oil magnates invested in these activities, attracted by the profits guaranteed by the nuclear arms race.

After a few weeks sweeping warehouses and repainting the drilling equipment from Dresser Industries, George H. W. Bush quickly received a sensitive mission. During the fall of 1948, while Washington sought to convince Yugoslavia's Marshal Tito to break from Moscow, the Yale graduate was put in charge of showing an envoy from Belgrade around the Dresser facility.[38]

Five years later, on March 27, 1953, George H. W. Bush cofounded his own oil company, Zapata Petroleum, of which he became vice president. The name, chosen by gringos, led one to wonder whether this was a particularly ironic tribute to Emiliano Zapata, the Mexican revolutionary. Exactly two weeks after the enactment of the statutes of Zapata Petroleum, on April 10, Niel Mallon, the head of Dresser Industries, wrote to CIA boss Allen Dulles about their upcoming meeting at Washington, DC's Carlton Hotel: "I have invited a close personal friend, Prescott Bush. We want to talk about our Pilot Project in the Caribbean and have you listen in."[39] We do not know if this was a Zapata Petroleum matter, but the events that followed indicated that it was quite possible.

A part of Zapata Petroleum's initial million dollar capital had been contributed by George H. W. Bush's father, Prescott, as well as by his maternal grandfather, the powerful banker George Herbert Walker, who died the same year. The Bush clan's network allowed them to raise funds from the *Washington Post*'s proprietor, Eugene Meyer, another Skull and Bones member, who was the first president of the World Bank before the éminence grise of the Seven Sisters, John McCloy.[40] Among those associated within Zapata Petroleum were the brothers Hugh and Bill Liedtke, who, after purchasing shareholdings controlled by George H. W. Bush in 1959, merged the company in 1963 with another oil company from Oil City, Pennsylvania. The result of this merger, the Pennzoil group, became one of the most important independent oil companies in the United States.

In 1954, George H. W. Bush became president of Zapata Offshore, an offshoot of Zapata Petroleum that specialized in offshore drilling. The growth of offshore drilling, as technical as it was costly, marked an essential crossroads in oil's history. In 1956, when Bush set up shop in Houston, on the Gulf of

Mexico, offshore drilling already filled nearly the whole horizon of the US petroleum industry's future. In spite of the new wells drilled in West Texas, it had become more and more complicated to replenish the reserves of the oldest petroleum fields located on land. Many, often the easiest to access and the most prolific—whether in Pennsylvania, Oklahoma, California, or the Houston, Texas, area—had provided for more than half a century. And the American economy now sucked at their breast more voraciously than ever.

Perhaps oil exploration was not Zapata Offshore's first objective. Its profits (unlike those of Pennzoil, the other offshoot of Zapata Petroleum) never materialized. Starting in 1953, George H. W. Bush's principal Zapata Offshore collaborator was even younger than he was. Only twenty-seven years old, Thomas Devine had left the CIA to join the company.[41] He resurfaced in the intelligence agency in June 1963, in disconcerting circumstances, as we will see.

Although it has never been publicly acknowledged, George H. W. Bush was, more than likely, an estimable American secret service asset well before President Gerald Ford appointed him CIA director in 1976. In March 1958, Eisenhower was forced to suspend weapons sales to Cuba when General Fulgencio Batista's corruption and brutality, in the face of the rebellion led by Fidel Castro, upset the press and American public opinion. Also in 1958, a few months before the Castro revolution succeeded, Zapata Offshore secured a drilling platform near an isolated island south of the Bahamas, forty miles north of Cuba.

The island belonged to Howard Hughes. The famous Hollywood producer, industrial oil magnate, and aviation multimillionaire had for a long time been known to render services to the CIA.[42] Scorpion, the drilling platform Zapata Offshore used, partly belonged to Gulf Oil; Kermit Roosevelt, the man who led the coup d'état in Iran, sat on its board. CIA director Allen Dulles had for a long time been Gulf Oil's advisor for Latin American operations.[43]

Zapata Offshore seems to have played a part in the US maneuvers that culminated, on April 17, 1961, in the Bay of Pigs disaster (the failed invasion of Cuba by anti-Castro activists who were supported by the US Army and trained by the CIA). Testifying thirty years later, John Sherwood, a former agency officer involved in the colossal and foul anti-Castro operation, said: "Bush was like hundreds of other businessmen who provided the nuts-and-bolts assistance such operations require. He was no spy. None of these guys were. What they mainly helped us with was to give us a place to park people that was discreet."[44] Another official involved in the operation testified: "George Bush would be given a list of names of Cuban oil workers we would want placed in jobs. . . . The oil platforms he dealt in were perfect for training the Cubans in raids on their homeland."[45] The US Army brigadier general Russell

Bowen later spoke of having worked under the orders of George H. W. Bush, the future American president: "Bush, in fact, did work directly with the anti-Castro Cuban groups in Miami before and after the Bay of Pigs invasion, using his company, Zapata Oil [sic], as a corporate cover for his activities on behalf of the agency."[46] The oil company named Zapata served as a cover for the CIA, and George H. W. Bush was one of those valuable businessmen, opportunistic or patriotic, who the agency frequently dealt with: This assertion was supported by Robert Crowley, who worked for George H. W. Bush when the latter became CIA director in 1976, as well as by William Corson, the master spy Prescott Bush met in Hong Kong in 1955. Corson led covert missions for four American presidents (and was the first lead intelligence officer to denounce some of the monstrous acts of the American army in Vietnam).[47]

George H. W. Bush and the Intersection of Oil and Intelligence Work in Cuba, Kuwait, and Mexico

Before Fidel Castro nationalized the country, the largest amount of capital invested in the island of Cuba was American capital (and not only Lucky Luciano and the Mafia's casinos). Brown Brothers Harriman, Prescott Bush's prestigious investment bank, sustained tremendous losses as a result of Cuba's nationalization, including a sugar beet plantation of at least 80,000 hectares that belonged to the Punta Alegre company.[48] Among the Yankee businesses who were deprived of tens of millions of dollars' worth of goods and profits following Castro's victorious revolution in January 1959 were a significant number of oilmen. Jersey Standard owned a $35 million refinery in Cuba.[49] CIA director Allen Dulles reacted by creating the Cuban Task Force in December 1959, headed by Tracy Barnes, one of Dulles's partners in armaments during the Office of Strategic Services (OSS) period. Barnes, authorizing officer of the 1954 Guatemalan coup, was also bound by marriage to the Rockefeller family.[50] To counter Castro, this psychological warfare specialist acted under the direct authority of Vice President Richard Nixon. In his memoirs, Fabian Escalante, the leader of Cuban counterespionage, affirmed that in order to "bring together the necessary funds" for operations that culminated, three months after Kennedy's inauguration, in the Bay of Pigs landing (Operation Zapata), Nixon "had assembled an important group of businessmen headed by George Bush and Jack Crichton."[51]

The magician of Big Oil, Everette DeGolyer, launched Jack Crichton's career in black gold after Crichton returned from the war. Crichton was one of those Yankee oilmen who lost considerable holdings after Castro's revolution. In August 1953, Crichton became vice president of Empire Trust, a New York investment firm in the process of developing Cuba's oil production through the Cuban-Venezuelan Oil Voting Trust (CVOVT), a Havana company founded in 1950. The world of oil and finance was decidedly small: The president of Empire Trust was the best man at John McCloy's wedding.[52] The trust was presented as "something very like a private CIA" by the historian of the American establishment, Stephen Birmingham.[53] As for the CVOVT, its mission went much further than that of an ordinary investment company: "to assure continuity of management and stability of policy for shareholders of twenty-four oil companies in South America."[54]

In November 1956, the CVOVT acquired a huge concession from the Batista regime, covering some 60,000 square kilometers, or more than half the total area of Cuba.[55] In the late 1950s, the first drilling project entrusted by the CVOVT to Standard Oil of Indiana produced very little, but its output grew rapidly, which whet Wall Street's appetite. The *New York Times* called it "highly encouraging."[56] Oil prospecting was well underway when the Castro revolution took full control of the island in January 1959 and quickly suspended all permits.[57] Cuba never became an oil producer of any significance. (Its development remained stifled by a chronic lack of oil, imposed by an American embargo that lasted a half-century.)

Crichton was successful during his extremely active career as an oilman. He established his reputation (and probably his fortune) by participating, starting in the late 1940s, in the evaluation of the giant Kuwaiti oil field of Burgan, the largest in the world after Saudi Arabia's Ghawar.[58] George H. W. Bush also knew how to successfully conduct business in Kuwait, capturing the contract that allowed Zapata Offshore to drill the first wells off the emirate.[59] In the early 1960s, in the Persian Gulf, George H. W. Bush's company was associated with Sedco, then the world's offshore drilling leader, at the forefront of many decisive innovations in this highly technical field.[60] Bill Clements, Sedco's founder and president, became the Pentagon's second-in-command under President Nixon in 1971, a post in which he continued under Nixon's successor, President Gerald Ford.* Then, in 1979, he became the first Republican governor Texas elected in a century.

In 1960, without advising his shareholders, George H. W. Bush also associated his company with a Mexican company, Perforaciones Marinas del Gulf (Permargo). His main partner's name was Jorge Diaz Serrano. In 1976, Serrano

* In 1984, Sedco was absorbed by Schlumberger, then sold in 1999 to Transocean Ltd., the owner of the Deepwater Horizon platform, at the site of the April 2010 Gulf of Mexico oil spill.

became the head of Mexico's national oil company, Pemex, created in 1938 when Mexico was the world's first producer to liberate itself from the major Anglo-Saxon oil companies and nationalize its petroleum. John Sherwood, one of the CIA officers who claimed that at the time Bush collaborated with the agency, affirmed: "He never did any spying, he simply helped his government arrange to place people with oil companies he did business with. The major breakthrough was when we were able, through Bush, to place people in Pemex, the big Mexican national oil operation."[61] In 1983, Jorge Diaz Serrano was removed from the presidency of Pemex by the Mexican parliament, and served five years in prison for having diverted tens of millions of dollars.

When, in 1988, an American journalist attempted to clarify the exact nature of the relationships between Zapata Offshore and Permargo, the Securities and Exchange Commission, constable of Wall Street, found that the 1960–1966 archives of George H. W. Bush's former company had been "inadvertently destroyed" sometime after Bush's accession to the US vice presidency.[62] According to the US Navy's master spy, Lieutenant Colonel William Corson, the CIA had recruited Serrano to obtain Permargo's logistical assistance in the fight against the Castro regime.[63] The method Bush may have used to "buy" Serrano was to sell him one of his drilling platforms, Nola-I, at a reduced price. "It was mighty generous of Bush to sell us the rig because we were taking his place. . . . We replaced Zapata," Serrano told the press after his conviction.[64]

Big Oil and Lee Harvey Oswald's Guide

Finally, among all the clues and testimonies linking George H. W. Bush to the CIA's covert operations at the juncture of petroleum, financial, political, and military influence, what can we make of the relationship between the future forty-first American president and a petrogeologist Russian immigrant who guided Lee Harvey Oswald, the man accused of murdering President Kennedy in Dallas on November 22, 1963?* On September 5, 1976, George de Mohrenschildt, the sixty-five-year-old man of Russian origin living in Dallas, appealed to the new director of the CIA, George H. W. Bush. He complained of being followed everywhere and of being wiretapped, he believed, by the FBI. "I tried to write, stupidly and unsuccessfully, about Lee H. Oswald and must have angered a lot of people. . . . Could you do something to remove this

* Most of the details that follow concerning George de Mohrenschildt and his connections with George H. W. Bush are drawn from the study on the Bush family, harsh but nevertheless thorough, published by Russ Baker, *Family of Secrets*, op. cit. (including chapter 5, "Oswald's Friend").

net around us?" The CIA director reported to his subordinates, in writing but in vague terms, that he actually had known George de Mohrenschildt since the early 1940s. Then he responded to de Mohrenschildt, to say he was sorry, but he could do nothing for him, that his fears were unfounded, and that he was probably the subject of simple journalistic harassment. Two months later, the wife of this Russian aristocrat was committed to a psychiatric hospital. On March 29, 1977, two months after Bush left the directorship of the CIA, George de Mohrenschildt was found dead, shot in the mouth by a bullet from a hunting rifle. Suicide, concluded the police.

Looking backward: It was the summer of 1962, a year and a half before Kennedy's assassination. On his return from the USSR in July, Lee Harvey Oswald, a pro-Communist young American with an unstable character, moved to Dallas, where he quickly made the acquaintance of George de Mohrenschildt. This sympathetic businessman, who specialized in raw materials investments, immediately took Oswald under his wing. He helped him to find employment and housing, and even helped him with his stormy marriage to Marina, his Soviet wife.

The de Mohrenschildt family was deeply rooted in petroleum. At the beginning of the twentieth century, in tsarist Russia, his father and one of his uncles were the head of operations of the Nobel brothers' oil company in Baku, on the shores of the Caspian Sea. In 1915, the tsar sent one of George de Mohrenschildt's other uncles to Washington to argue for the United States' entry into the First World War.[65] In the aftermath of the Bolshevik Revolution, Ludwig Nobel's son, Emanuel, resolved to sell the Baku wells long operated by the de Mohrenschildts. On May 13, 1919, Standard Oil of New Jersey acquired these wells for $11.5 million, paid by the elder John D. Rockefeller himself.[66] Eleven months later, Baku fell into the hands of the Red Army. The de Mohrenschildt family escaped and, as early as 1920, George's elder brother, Dimitri, was admitted to Yale University. The following year Lenin launched his New Economic Policy; Harriman's bank, directed by George Herbert Walker (grandfather of George H. W. Bush), was one of the companies vying to relaunch the Russian petroleum industry.

George de Mohrenschildt for a long time took refuge in Poland, along with part of his family. In May 1938, twenty-seven years old, he landed in the United States and moved into a luxurious Park Avenue apartment in Manhattan. In the course of the Second World War, while his elder brother Dimitri served in the OSS, George de Mohrenschildt worked with a French intelligence officer in an operation aimed at preventing the Third Reich from procuring petroleum products from the United States. He traveled to Venezuela, where he worked after the war for Pantepec Oil, a company involved in many of the CIA's operations in Latin America.[67]

In 1950, George de Mohrenschildt invested in the West Texas black gold rush. In the dusty cowboy city of Abilene, he started a small petroleum investment firm with his nephew by marriage, Eddy Hooker. Hooker was a close friend of George H. W. Bush: They were roommates in an upscale private Massachusetts high school and both served as pilots in the Pacific during the war, and then attended Yale. George H. W. Bush was Hooker's best man and, after Hooker's death in 1967, was his daughter's guardian until she married the heir of a wealthy Cuban farming family exiled by Castro.[68]

Starting in 1950, established as a petroleum geology consultant, George de Mohrenschildt worked for CVOVT, the company that six years later acquired prospecting rights on over half of the Cuban island. The Russian geologist then crossed paths with Jack Crichton, who later said of him: "I liked George. He was a nice guy."[69] De Mohrenschildt moved to Dallas in 1952. There he was admitted as a member of the powerful Dallas Petroleum Club, to which belonged all the heavyweights of the local industry, such as Neil Mallon ("favorite uncle" of George H. W. Bush) and Everette DeGolyer. He was in contact with DeGolyer's son-in-law, George McGhee, the diligent American petroleum diplomat.

Responsible for investigating Kennedy's assassination, the Warren Commission was not surprised by the impressive contacts George de Mohrenschildt maintained at the time with the Texan petroleum elite. His address book included H. L. Hunt; George Brown, cofounder of Brown & Root; and even Jean de Menil, naturalized American and grand patron of the contemporary French art world, director of the Schlumberger Group's Houston subsidiary, and the husband of its heiress, Dominique Schlumberger. De Mohrenschildt was sent on many missions to Haiti by another local petroleum baron, Clint Murchison (an influential political kingpin when it came to tax statues benefitting Big Oil). In 1958, another Texan oil magnate, John Mecon, sent de Mohrenschildt to Yugoslavia. Mecon's foundation was later identified as having transferred funds for the CIA.[70]

George de Mohrenschildt was the person in closest contact with Lee Harvey Oswald between June 1962 and the tragic day of November 22, 1963. It was to de Mohrenschildt that Oswald addressed the famous photo in which President Kennedy's presumed assassin stands, pressing a rifle against his hip with one hand and holding several newspapers in the other. De Mohrenschildt was the witness interrogated longest by the Warren Commission, on which sat Allen Dulles, the head of the CIA fired abruptly by Kennedy following the bloody Bay of Pigs debacle; John McCloy, the grand vizier of Standard Oil; and even future president Gerald Ford. In a report others deemed insufficient, the Warren Commission drew no firm conclusions about the role George de Mohrenschildt played regarding Oswald.

In 1985, a journalist accidentally unearthed an FBI memo under the heading "Assassination of President John F. Kennedy," dated November 29, 1963, a week after the assassination, written by J. Edgar Hoover himself.* In this memo, the legendary FBI director assessed the hypothesis that there had been a project to "undertake an unauthorized raid against Cuba," conducted by anti-Castro militants, who would have been emboldened by Kennedy's killing. Hoover cited two sources that indicated that such an assumption was unfounded. One of these sources was a certain "Mr. George Bush of the Central Intelligence Agency." It must have been George H. W. Bush.† It was elsewhere confirmed that George H. W. Bush had been in Dallas on the eve of President Kennedy's death: He spoke at a Sheraton hotel meeting of the American Association of Oilwell Drilling Contractors (never in the course of his long public service career did the future forty-first American president mention this fact).[71] American journalist Russ Baker reported that, among other coincidences, on April 26, 1963, five months before Kennedy's death, George de Mohrenschildt met, on the New York premises of the Train, Cabot and Associates investment bank, a CIA agent who very likely was Thomas Devine, George H. W. Bush's close collaborator in Zapata Offshore.[72]

Mr. Bush Goes to Washington

George H. W. Bush began his political career in the Autumn of 1962, when he took charge of Republican Party finances in Houston, then in the process of becoming the planet's petroleum capital and already home to one of the world's largest concentrations of black gold millionaires. That same year,

---

* Journalist Joseph McBride found Hoover's memo while conducting research for a book about American director Frank Capra, author of *Mr. Smith Goes to Washington*. (See "The Man Who Wasn't There: 'George Bush' C.I.A. Operative," *The Nation*, July 16, 1988.)

† The denials submitted by the CIA during the 1988 presidential campaign, as well as those from vice- and future president George Herbert Walker Bush's press secretary, are unconvincing. The CIA advanced that a certain George William Bush, who it proved unable to locate, had been employed in 1963 at CIA headquarters in Langley, Virginia. The man was a night shift employee in document analysis, a low-ranking civil servant in his probationary period. Located without difficulty by the press, he stated that he had never been the recipient of any governmental interagency reports. In any case, his very subordinate function did not place him in any position to receive them. (See Russ Baker, *Family of Secrets*, op. cit., p. 9–12.)

George H. W.'s father, Prescott Bush, decided not to run for reelection as a Connecticut senator, ending his own political career. In 1964, still inexperienced but ambitious and confident, George H. W. Bush lost the Texas senate race, while Democratic presidential candidate Lyndon Johnson won by a landslide over Republican Barry Goldwater. Goldwater had managed to win the Republican nomination over Nelson Rockefeller. Prescott Bush had by this time dropped his old ally Rockefeller, whom he had previously supported, during Rockefeller's previous unsuccessful campaign in 1960. He claimed this was because of the Standard Oil heir's divorce and 1963 marriage to his former mistress. The Bush clan was in the process of switching its old political alliances, gravitating around the oil majors and Wall Street, toward independent oilmen and the cluster of new industrial powers emerging alongside what some called the "Texas Raj."

George H. W. Bush resigned the Zapata Offshore presidency in 1966; he was elected to the House of Representatives the following year, representing the district of West Houston, the exclusive neighborhood located a good distance from Houston's gigantic refineries. Prescott Bush moved heaven and earth so his son was appointed to the committee responsible for drafting tax laws. Thanks to the direct intervention and support of future president Gerald Ford, then the Republican Party leader in the House of Representatives, George H. W. Bush became the youngest representative elected in over sixty years to join this very coveted and influential committee, guarantor of the famous "depletion abatement" for oil companies, which George H. W. of course defended zealously.

In 1968, with the support of many oilmen and thanks in particular to his own vow to defend the "depletion abatement," Nixon became the Republican candidate in the race for the White House, at the expense of Nelson Rockefeller, who unsuccessfully sought his party's nomination for the third time. Bill Liedtke, Pennzoil's founder and George H. W. Bush's former Zapata Petroleum associate, was by far the Nixon campaign's most prolific regional financial director, with the help of his many ultra rich Texan friends.[73]

John Connally, the Democratic governor of Texas who, at John F. Kennedy's side, was seriously injured in Dallas on November 22, 1963, may even have sided against his party in the name of defending oilmen's interests. In January 1972, before a congressional commission, Democratic senator Fred Harris accused Connally of discreetly joining the effort to convince Texan millionaires to fund Nixon's campaign because the Republican candidate, unlike his Democratic adversary, supported the "depletion abatement." Fred Harris clarified that it was not "idle" to ask if this was the reason Connally then became Nixon's secretary of the treasury.[74] For his part, Prescott Bush once again set his networks in motion, this time to ensure that his son, aged forty-four (a tender age to serve on Capitol Hill), was nothing less than the Republican Party vice-presidential candidate. The long list of industry leaders who wrote to encourage Nixon to make George H. W. Bush his vice president included, unsurprisingly, Bill Liedtke and Niel Mallon of Dresser Industries. Also appearing on this list, among other prestigious names, were the heads of Chase Manhattan Bank, National Distillers and Chemical Corporation, and, of course, Brown Brothers Harriman.[75]

Despite all these personal requests, instead of George H. W. Bush, Nixon chose for his running mate the combative and charismatic governor of Maryland, Spiro T. Agnew. Prescott Bush never forgave the American president, his former protégé: "I fear that Nixon has made a serious error here," he wrote.[76]

As president, Nixon was judged as having betrayed Big Oil's interests. In September 1969, only eight months after his arrival at the White House, the Republican president yielded to a Democrat-dominated Congress: He agreed to lower the "depletion abatement" from 27.5 percent to "only" 22 percent of the American oil companies' incomes. An emissary of Nixon sent to ease tensions in the small petroleum town of Midland, Texas, George H. W. Bush's home base, said he had momentarily feared he would be lynched.[77] That year, 1969, brought the first signs of upheaval in the oil economy. US black gold production had more and more difficulty keeping up with the growing, insatiable domestic demand for oil. The "favoritism-ridden and expensive oil import quota system," as *New York Times* described it, was under fire by critics to whom Nixon once again yielded. The quotas had been introduced by Eisenhower, who since 1959 had limited oil importation in the name of national security.[78] Nixon's alleged inability to demonstrate, in the face of OPEC, the firmness that David Rockefeller, John McCloy, and many oil bosses demanded of him, lost him the support of Big Oil. By the time the ambiguous Watergate scandal exploded in 1973, Big Oil's support had already been handed off to the Republican party's new champion of Big Oil: George H. W. Bush.

Greed, pride, egocentrism: The psychological mechanisms that animate those who seek unbridled power are beside the point. Rather, it's the resources indispensable to the expression of power that drives those who seek it to obey the weightiest demands of the material world. The material challenges imposed by the need to access oil supplied the necessary force to leverage the rise of multiple political figures central to contemporary history. The end of the 1960s saw two opposing factions, two world visions, at the heart of the

Republican Party: that of the "traders" (representing the traditional industrial establishment) and that of the "warriors" (politically more to the right, from the conservative Southern states and, in particular, Texas). These factions were personified by two oilmen with drastically different perspectives. Nelson Rockefeller represented the internationalist ethic of the oil majors and their large Wall Street banks, whose strategies, global by necessity, could never be unilaterally pro-American. Rockefeller's influence on the American right was profoundly deep in the 1960s; the traders, who counted themselves in the "progressive, liberal" wing of the Republican Party, willingly called themselves Rockefeller Republicans. George H. W. Bush, the embodiment of East Coast high society radically transformed into a Southerner, was above all the custodian of independent American oil interests and the other vast US industries that revolved around black gold. Because of the wealth surrounding his electoral base, his own ethics rested on a deliberately jingoistic viewpoint: In the end, the only thing that mattered was the primacy of American power. In the movement he embodied, black gold restored the political shield and regenerated the pride of the reactionary South, until then considered backward by the United States.

However, in all cases, the institutions mobilized by the traders, and more and more by the warriors—Exxon, Chase Manhattan Bank, Brown & Root, Dresser Industries, Bechtel, or Halliburton—imposed themselves as the key vehicles required for Big Oil to assert itself as vital to American power. They were the only colossal bodies able to metabolize the most efficient energy source in existence, for good or for ill, and they continued to disperse energy to the four corners of the world, after having concentrated it in places of power. There were places of primary power: the petroleum drilling fields. There were places of central power: petroleum company headquarters—with their physical capital (drills, pumps, pipes, refineries)—and the banks that leaned on them. And, finally, there were places of ultimate power: Washington, for example, where, beginning in the 1950s, significant political power stemmed from the power conferred by controlling crude oil sources.

At that time, a single factor constantly hindered the sound functions of such a system: political instability. In response, the massive thermo-industrial and political system had a single response: to reimpose the status quo, no matter what.

FIFTEEN

Saudi Arabia and Gabon: Nations Spoiled by Oil

Among the nations beginning to emerge during the decolonization period, those that held the most extensive oil resources saw their destinies engulfed by their exploiters: industrial powers, petroleum companies, and local potentates. Within this triumvirate, the oil companies exerted a decisive power. They tried to ensure that these young nations did not have the slightest chance to escape their clutches.

The Kingdom of Saudi Arabia, for example, was built by and for the major American oil companies; and the small West African state of Gabon could not overstep its primary, vital role in keeping France operating as a second-tier industrial power.

The Myth of the US-Saudi Special Relationship

What were the original ingredients of the so-called special relationship that, for more than half a century, has linked the fortunes of American power to the Kingdom of Saudi Arabia? What were the bases of the most stable alliance of the modern era, between two vastly disparate regimes and nations, which revealed its ambivalent nature on the fateful day of September 11, 2001? What was the nature of the covenant between Aramco, the Americans' "largest single overseas private enterprise," and the House of Saud, the Saudi royal family?[1]

The ultimate Western power had always taken great care not to appear colonialist. In Saudi Arabia, more than anywhere else, the Americans had carefully crafted the impression that they were respectful guests. Anything but colonizers. Early on, Aramco's directors of political affairs undertook a sophisticated public relations campaign.* In the 1950s, each US employee who arrived in Dhahran received a manual in which we can discern the beginnings of the special relationship myth: "The Arabs have not had the advantage of the education and prosperity of America. The great majority are unlettered and untrained, poor and poorly clad. They are, however, very proud of their religion, their traditions, their country and of being Arabs. They are friendly, soft-spoken and polite. Americans are also justifiably proud people, and no surrender of this pride and self-respect is required on either side. All that is required is respect and decent consideration for the other fellow as a human being. Americans are guests in Saudi Arabia."[2] One Aramco director pushed artistic license to its limits when he described the company as "unique," characterizing it as an "informal and egalitarian organization."[3]

If it ever existed, this "egalitarian" organization certainly never extended egalitarianism to its Arab employees. The latter, often local, were poorly qualified in the early 1950s: "Local workers could do rough masonry and carpentry, but practically no skilled native craftsmen worked in the area of Dhahran," said the US Army Corps of Engineers, responsible for developing Dhahran's infrastructure.[4] In this Aramco city that had sprung up from the sands bordering the Persian Gulf in the late 1940s, fountains of distilled water and some toilets were reserved for American-only use.[5] The Americans' comfortable residential area was surrounded by electrified barbed wire. In essence, the American compound in Dhahran was organized according to the strict racial segregation rules of the US South, the Jim Crow laws that lasted until 1965. The same principles had long governed oil cities from Beaumont, Texas, to Bakersfield, California, from Maracaibo, Venezuela, to Tampico, Mexico: Employees were separated and paid different wages according to race.

Commonly called "ragheads" by the Americans and "coolies" by the British, Saudi workers were not allowed to live with their families.[6] The few Americans who established friendly contact with Arab families outside the circle of the House of Saud and their court were returned to the United States.[7] In Dhahran, white American executives built spacious California-ranch-style homes, with air conditioning and swimming pools. A significantly more spartan neighborhood was reserved for Italian employees, followed by a "servant camp" for Indians, and finally, the palm huts of "Saudi camp."

Aramco's Arab employees quickly aspired to emancipation. In May 1955, Saudi workers refused to board the miserable buses the company had reserved

* It wasn't until 2007 that American historian Robert Vitalis was able to deconstruct the myth of the "special relationship," revealing previously unpublished details, in *America's Kingdom* (op. cit.).

for them, which made few stops, often many kilometers from their barracks.[8] This boycott was ineffective.* In 1956, knowing that some Saudis appreciated the cowboy films broadcast on the local television channel launched by the US Air Force, Aramco conducted interviews to ensure that the Arabs "did not identify themselves with the Indians, a possibility that had caused momentary anxiety."[9]

The situation hardly seemed to have evolved by 1964, despite the integration of the first "white collar" Saudis from the aristocracy. The year when Malcolm X made his famous pilgrimage to Mecca, the ill treatment inflicted on the Saudi employees remained commonplace in Dhahran. For example, a group of progressive parents of American students described the conditions faced by Saudi school bus drivers this way: "They are screamed at, called names, among the lesser epithets is 'Dirty Arab' and on occasion physically abused. The school knows this situation, the community knows, but it still continues."[10]

The bullying commonly inflicted on Saudi employees by Aramco's American personnel had little chance of displeasing the Saudi princes. Superimposed on a society that was brutal, viscerally racist, and possessed a tradition of slavery, this American style segregation seemed to take hold in Dhahran much more easily than it did in Iran, Venezuela, or Mexico.

The strikes and riots that beset Aramco during the 1950s and 1960s were always severely and shamelessly repressed by the king. The famous special relationship between the United States and Saudi Arabia did not unite American oilmen with Saudis as a whole but did, of course, unite them with the royal Saud family and its entourage. More than once, this special relationship turned against the people of Saudi Arabia. As early as 1945, hundreds of Saudi workers stopped work on several occasions during the overwhelming Arabic summer, to demand wage equality and an end to discrimination. After the first strike, in July, the emir representing the royal authority used one of his black slaves to assault the ringleaders.[11] In the face of the mounting protests, Aramco, in agreement with royal authorities, granted them a few concessions that were pure formality, such as the superficial improvement of a tiny, fly-infested "hospital" reserved for Saudis, which the American consul in Dhahran itself called a "shame for the company and, indirectly, for all Americans."[12]

Massive strikes were held in 1953, the tip of the as yet unrecognized, early pan-Arab nationalist offensive. Then, in May and June of 1956, many Saudi workers again demanded an end to American privileges, going so far as to

* Just seven months later in Alabama, Rosa Parks sparked the Montgomery Bus Boycott, a major event in the Civil Rights movement, by refusing to relinquish her seat in the "black" section of a segregated bus when the "white" section was full.

demand the closing of the US Air Force base in Dhahran. On June 9, workers thronged at Aramco's headquarters in the young oil city to wait for King Saud. The successor of Ibn Saud, the kingdom's founder, who had ascended to the throne in 1953, was traveling to a banquet given in his honor by Aramco's leaders. When the Royal Cadillac approached the strikers, they cheered for the sovereign as they unfurled a banner that read: "Aramco Saudi Workers Welcome the King. Death to Imperialism and Traitors."[13] Two guards broke formation and tore the sign to pieces. Two days later, a Royal decree prohibited, purely and simply, any strike or assemblage.

Exchanging Oil for Security: Bribes That Buy Stability

Looking beyond the myth, many have described the special relationship between the United States and Saudi Arabia as an exchange of oil for security. The longevity of this special relationship rests on the longevity of the House of Saud—longevity that itself depends on this dyptich of bribes for stability. From the inception of the special relationship, oil paid for the royal family's security, and the king systematically paid those clans who served as his allies. In 1943, after Roosevelt declared that Saudi Arabia was vital to US defense, monies from Saudi Arabian oil began to flow in abundance; there followed, according to Jack Philby, British renegade counselor to King Ibn Saud, an "orgy of extravagance and misarrangement, accompanied by the growth of corruption on a large scale in the highest quarters."[14]

David Rockefeller—whose Chase Manhattan Bank counted among its customers not only several members of the Saudi royal family but the Saudi central bank itself—described the House of Saud's oil operation as that of a "family economic enterprise," in which King Faisal, in order to keep his unruly clan "loyal and satisfied, distributed 20 percent of oil revenues amoung the six hundred or so members of his family before making the remainder available to the government."[15] In 1951, the London weekly magazine the *Economist* captured in more brusque terms the basis of the Saudi Arabian sovereign's strategy regarding the petroleum windfall: "Officially these funds go to 'raise the living standards' of his people; in fact all but 10 percent, which he pays mainly to tribes to keep them quiet, goes into the bottomless pockets of the king, his immediate family and entourage. Understandably enough, the Saudi family is only too ready to use its revenues in ways that preserve the medieval character of the country."[16] This strategy remains in place today, to a greater or lesser extent, in the majority of the oil-producing countries.

With such partners commanding the kingdom, it was easy for Aramco to pull off this predatory game in Saudi Arabia. Contrary to the apologist history that touts the American powers' "exceptional" benevolence compared to earlier, colonialist powers, Aramco's greed matched that of the Anglo-Iranian Oil Company in neighboring Iran. After Mossadegh took power in Tehran, a senior American official who had recently been transferred from Iran to Saudi Arabia noted that the Iranian political class was much more "mature" than that of Saudi Arabia. He added: "The number of those controlling the destiny of the country and the purse strings of the oil revenues could be counted on two hands. . . . As a result, the Aramco officials can ignore public opinion, which is neither a force nor a voice."[17]

Nothing at the time seemed to shake the Saudi royal family's confidence in Aramco's leaders. Yet the latter had plenty of things to hide: In September 1951, six months after the nationalization of Iranian oil, they burned all the "highly confidential" company documents stored in Saudi Arabia, "till the last small scrap of burned ash had gone with the wind."[18] It was an unnecessary precaution, all things considered. The Sauds were not at all prepared to question the alliance, thanks to which they had escaped the usual precarious destiny of nomadic warlords. In 1952, Abdullah Tariki, one of the first Saudi oil officials, said that, thanks to a manipulation in crude oil price, the Saudi Crown only effectively received 22 percent of the oil profit, in spite of the so-called 50/50 sharing agreement obtained by the Saudis two years previously. It was a subtle swindle of the same ilk as the one Pérez Alfonso (with whom Tariki soon founded OPEC) uncovered in Venezuela in 1943. Aramco agreed to pay compensation, and the scandal passed unnoticed, even while the Iranian crisis was in full swing.

The diplomatic delegation in Saudi Arabia was not elevated to an ambassadorial rank until 1949, sixteen years after SoCal's prospectors arrived on the Pirate Coast. Its embassy was built in Jeddah, at the edge of the Red Sea, on the other side of the Arabian Peninsula's desert, more than 1,000 kilometers from Dhahran and the oil fields. As early as 1941, Aramco's ancestor, CASOC, set up its own private diplomacy and intelligence department in Dhahran: Called the Government Relations Organization (GRO), it determined and widely implemented American policy in Saudi Arabia until the oil crisis of 1973.

In 1946, the GRO created a division of Arab affairs whose structure was modeled on the Cairo office of the intelligence agency OSS, the future CIA.[19] The GRO hired several former CIA members from positions in Saudi Arabia, Iraq, and elsewhere in the Middle East.[20] Did the GRO serve as a cover for CIA agents, or were these agents working for Aramco? These revolving-door hiring practices blurred intelligence genres in the shadow of Saudi black gold and seeded the privatized American intelligence that reached its height during the Reagan-Bush era, metamorphosing, notably, to give birth to Al Qaeda. Beginning in the late 1940s, the GRO employed agents throughout the kingdom and supplied Aramco's leadership with all varieties of analyses. In March 1947, J. Shores Childs, the first full-time US ambassador to Saudi Arabia, was concerned to see Aramco's influence extend "into almost every domain and phase of the economic life of Saudi Arabia." Overwhelmed, and imagining that political leadership could still prevail over business leadership at the edge of the Persian Gulf, he cautioned Washington in a diplomatic cable with an almost pathetic tone: "We can, of course, make a fetish of the free enterprise system and in its name avoid any attempt to exercise a control over the octopus represented by Aramco." Extending the metaphor once used to describe Standard Oil, the American ambassador to the new land of black gold continued, "The longer we delay, however, the deeper its tentacles will be spread and in the end the policy of the Government of the United States in Saudi Arabia and in the Middle East may be dominated and perhaps even dictated by that private commercial company."[21] In 1954, another American diplomat wrote that the special relationship was established "with a King who thinks like an oil company and an oil company that thinks like a King."[22]

No doubt gifted with practical sense, Colonel William Eddy, who had been the interpreter at the meeting between President Roosevelt and King Ibn Saud in February 1945, contributed to the restructuring of US intelligence efforts that led to the creation of the CIA in 1947.[23] That same year, he abandoned his US government functions, officially to protest President Truman's support of Zionism.[24] It was then that he joined Aramco. In reality, Colonel Eddy's long career in Aramco's leadership seemed coupled with a role as master spy at the heart of the CIA.[25] This special kind of double agent exemplified President Coolidge's view that "the business of the American people is business." Colonel Eddy imposed himself as King Ibn Saud's privileged interlocutor in all the sovereign's negotiations with Aramco, while pretending to have cut ties with the US government, which was unlikely.[26]

Reacting to a 1952 American report on the "international oil cartel" and speaking on Aramco's behalf, Colonel Eddy mocked the attitude of BP in Iran: "We do not teach political democracy, nor try to replace prime ministers, nor monarchs, like the Anglo-Iranian Oil Company. . . . We admit the King can kick us out any minute. Nor would we invoke the World Court, nor ask for warships."[27] A few months before this oath of neutrality, Aramco pressured Riyadh and Washington to force the departure of Saudi Arabia's finance minister, Abdallah Sulayman.[28] The latter, a small, brilliant man full of energy, one of the king's closest advisers, had raised objections to the American oil companies since one of his subordinates, Abdullah Tariki, had discovered the deception of the 50/50 agreement. Sulayman's account led King Ibn Saud to accuse Aramco of insulting and duping his regime. Despite the GRO's retaliatory smear campaign, presenting (with some evidence) Saudi Arabia's finance minister as a corrupt alcoholic, the king did not yield and retained Sulayman in his position.* Little by little, the ambiguous balance of mutual dependence between Saudi Arabia and Big Oil was perfected.

The United States Builds Saudi Arabia's Modern Infrastructure

Eddy concluded his 1952 profession of good faith, stating: "We do not aspire to be an East India Company. We prefer banks and Bechtel."[29] Again, the assertion of this CIA agent, Aramco's consultant, appeared as ambiguous as it was unconvincing. Bechtel, one of the largest American public works companies, embodied the confluence of interests between Washington, Big Oil, and the Saudi Crown.

Indispensable in creating petroleum, military, and nuclear infrastructure, Bechtel was co-directed during the Second World War by John McCone, who succeeded Allen Dulles as CIA director in 1961.† Bechtel was Aramco's exclusive industrial partner in development. The company, which had managed the contract for the Trans-Arabian Pipeline since 1947, had also built most of the first modern infrastructures of the young Saudi kingdom (roads, embassies, ports, airports, power plants, luxurious modern palaces, and prisons). Starting in the 1950s, the US firm often cooperated with the first Saudi construction company, the Saudi BinLadin Group. The cooperation between these two companies retained remarkable stability: Two years after the September 11, 2001, attacks (the masterpiece of Osama bin Laden, one of fifty-three sons of the BinLaden Group's founder), the BinLaden Group and Bechtel both remained associated with the capital of a San Francisco investment firm, the Fremont Group (formerly Bechtel Investments).[30]

* The Intercontinental, the first hotel in Dhahran's neighboring town, Damman, was funded to the tune of $11 million by the Department of Finance on a property owned by Minister Abdallah Sulayman (see Robert Vitalis, *America's Kingdom*, op. cit., p. 132). Such a practice, however, was not unusual in Saudi Arabia.

† In 1956, Bechtel was responsible for the construction of the first private American nuclear reactor, Dresden-1.

Bechtel's directors knew how to make the concessions necessary to keep their relationship with the House of Saud firm for half a century. In 1947, for example, the company agreed to build the first railway line in the country, linking Riyadh to Dhahran, the Aramco city. For several months, Bechtel and Aramco challenged the relevance and profitability of the project. Bechtel eventually gave in to the insistence of the king, who wanted the train, a symbol of progress. Vainly seeking a Truman administration loan to finance it, one of Bechtel's vice-chairmen declared: "What great difference does ten million dollars make so long as the United States' best friend gets what he is convinced is best for his country?"[31] As if by a miracle, Bechtel lowered the estimated cost of building the rail line by $20 million, and Aramco gave the king the necessary funds. The softness of the desert ground meant that thousands of tons of concrete had to be delivered to build the railway. When there wasn't enough concrete, the Americans dumped torrents of crude oil on the sand, forming a crust on which they could install the tracks.[32] On October 25, 1951, Emir Saud, the heir to the throne, hammered in the final spike, one made of gold.[33] Like his father, he dreamed that black gold would make possible a fantastic irrigation network, routing the water from the Tigris and the Euphrates, far to the north, to the desolate central desert plains. But this dream was physically impossible: Neither unlimited wealth nor Saudi oil was enough to transform the arid land to verdant country.

Although the Truman administration refused to assist Aramco this time, the American government, in addition to implementing the golden gimmick and granting other fiscal largesse, took decisive measures to ensure Aramco's definitive presence in Saudi Arabia, as well to ensure Saudi Arabia's emergence as a modern state. As early as 1944, the American military helped Aramco build its first important refinery, between Dhahran and the Ras Tanura terminal, with the support of James Forrestal, who was Texaco's former lawyer and the secretary of the navy.[34]

It was also the American military that enlarged the first airfield built by the American oil companies in Dhahran. Although its construction was planned postwar, the aerodrome held obvious strategic interest. A US Navy memo, written two months before its construction was endorsed by the Truman administration, noted: "The mere existence of an American military airfield in Dhahran would contribute to the preservation of the political integrity of Saudi Arabia and to the maintenance of our interest in the oil fields."[35] This airfield housed US Air Force equipment until 1962, as well as Aramco's equipment. In order to safeguard the Saudis, Washington conferred upon this airfield the status of full-time military base, the most important in the region—the only one capable of accommodating B-29s, the largest American bombers of their time, at the end of the 1940s. To preserve appearances, the airfield's property returned to Saudi ownership in 1948: The American military paid Riyadh rent in order to use runways and hangars the US itself had financed and built.[36] The airfield became a symbol of the long, coinciding destiny of the Americans and the Saudis. When it was inaugurated as the Dhahran International Airport in 1961, its civil air terminal was built by the leading Saudi construction company, the BinLaden Group.[37] The airport plans were drawn up by a Japanese-American architect of international repute, Minoru Yamasaki, who also, in 1962, designed Towers 1 and 2 of Manhattan's World Trade Center, which were destroyed on September 11, 2001. Minoru Yamasaki's conception of the World Trade Center was inspired by the great mosque of Mecca, whose modernization was overseen by the BinLaden Group.[38]

The US government also played a direct role in the development of Saudi Arabia's modern governing institutions. In 1951, Washington financed the organization of the Saudi Arabian Monetary Authority, established the following year in Riyadh. This agency was responsible for issuing recommendations to stabilize Saudi currency, backed by the US dollar, and reforming the rudimentary management and extremely opaque budget of the Crown.[39] During the 1950s, the American administration also guided the creation of the Saudi Ministry of Defense and took charge of numerous military training missions.[40] Finally, in 1963, "the Saudis expressed a sense of their own limitations," and the House of Saud, anxious in the face of Egyptian president Nasser's progressive propaganda, appealed to American military engineers to establish a national television station.[41] According to US Army Corps of Engineers historians, "[t]he Saudis understood the potential of television in a society that was largely preliterate."[42] The Ministry of Information's main function "was to control the flow of public information through official supervision of mass media such as radio, television, film, and press."[43]

Despite its preeminent position, though, the American government chose to abandon its primary role in piloting American interests in Saudi Arabia. Historian Irvine H. Anderson described American diplomacy in Saudi Arabia and the other Persian Gulf oil countries from 1945 to the beginning of the 1970s this way: The Americans' "position was clearly to stay out of the oil business, but to reinforce things that the companies did. It worked out very nicely that way. So frequently when the companies would come in to ask for an 'OK' on a particular policy, the official response was 'no objection.' That was the code word for the fact that we support you."[44] This attitude perfectly matched Hannah Arendt's characterization of modern imperialism, in which "export of power followed meekly in the train of exported money."[45]

Washington's passivity regarding Aramco was often justified as a tactic that gave American diplomacy free rein to support the young state of Israel. But the explanation is a little lacking. Riyadh was officially deeply hostile to Zionism. But, secretly, the king admitted as early as 1957 that "Israel, as a nation, is now a historical fact and must be accepted as such."[46] Following this and despite the cries of members of Congress accusing the White House of duplicity with respect to the Jewish State, from 1954 to 1958, the Eisenhower administration validated the first deliveries of modern heavy weaponry to the Saudi kingdom, which, up to that point, they had lacked: eighteen Walker M-41 assault tanks, nine B-264 bombers, and sixteen F-86F Sabre5 fighter jets, the original elements of an arsenal destined to become as sophisticated as it was overabundant.[47] These weapons sales were the first fruits of the reciprocal dependence between Riyadh and Washington—a relationship much like that of faithful vassals to their overlord, in this case Aramco, though that reality seemed to escape both parties.

Big Oil Perpetuates Absolute Monarchy in Saudi Arabia

The White House remained at a distance when, in August 1952, Saudi Arabia sent a regiment of eighty men to occupy a portion of the Buraimi oasis, located beyond its eastern border. The territory was shared by the emirate of Abu Dhabi and the sultanate of Oman, allies of the British Crown. But it was claimed by the Saudi Crown, and Aramco suspected it contained oil.[48] Turki bin Abdullah al-Otaishan, who controlled the Saudi regiment, was the emir of Ras Tanura, the small peninsula sheltering Aramco's giant oil port; his family company, the Al-Otaishan Group, became a vital player in Saudi Arabia's petroleum industry. The Saudis attempted to bribe Sheikh Zayed al-Nahyan, prince of Abu Dhabi, offering him £30 million to stop his elder brother, who directed the emirate, from authorizing "Standard Oil" to prospect in the Buraimi oasis, but Sheikh al-Nahyan refused.[49] Aramco's division of Arab affairs, acting as the House of Saud's intelligence agency, dispatched men on the spot and unearthed the fact that in the past, Buraimi oasis residents had sometimes paid a tribute to Riyadh. The American oil company provided Washington with evidence to substantiate Saudi Arabia's claims, which eventually failed in 1955 thanks to the pugnacity of British diplomats.[50] During the three-year dispute, Aramco even provided Saudi soldiers with transport to the oasis.[51]

In 1960, the Saudi royal family and inner circles of Saudi Arabian power were divided over the question of whether or not the regime should evolve into a constitutional monarchy with a modern parliament. This was at the height of pan-Arab nationalism, of which Egyptian president Nasser was the public face. There were two opposing camps: The reformers had the ear of the King Saud, while the partisans of an absolute monarchy stood under the banner of Prince Faisal, the king's half-brother and future successor. Kim Philby, the British, pro-Soviet double agent, son of King Ibn Saud's former adviser, mentioned in the *Economist* that the eventuality of a "Saudi revolution" was "imminent."[52]

Prince Talal, twenty-first son of Ibn Saud, one of the primary supporters of an evolution toward democracy, had taken in 1958 the unprecedented step of creating a national council to draft a constitution. The plan was to greatly limit the king's powers and initiate some electoral processes. In 1960, Prince Talal naively pushed the American embassy to help it keep King Saud on the side of those "determined to carry on the fight for reform as the only means of saving the future of the [royal] family."[53] An embassy attaché responded that it was American policy to avoid interfering in Saudi Arabia's internal affairs. Talal asked why, then, had Colonel William Eddy, who was always allied with Prince Faisal, champion of the status quo, increased his visits to the American Embassy and King Saud bin Abdulaziz? The attaché replied that he knew nothing about it, and in any case, Colonel Eddy worked for Aramco. Talal said he knew that Eddy now worked for Allen Dulles, the director of the CIA.[54] He was quite likely correct.[55]

That summer, American intelligence alerted Aramco's Government Relations Office: Saudi officers were preparing a coup and intended to impose democratic reform on Saudi Arabia. The conspirators planned to assassinate Prince Faisal, as well as several Ministry of Defense directors. They let the Americans know that, once in power, their republic would ally with Aramco and not with Egyptian president Nasser. Yet it was precisely Aramco that exposed the plot. The GRO's liaison officer, William Palmer, was the one who in 1951 had burned, as a precaution, all incriminating Aramco documents after Iran's nationalization of oil.[56] Never before, nor afterward even until today, would Saudi Arabia come so close to escaping dictatorship.

In early 1961, the Saudi king clearly signified his preference for the reformist camp when he named Prince Talal the Minister of Finance. Just thirty years old, Talal formally proposed, in September, that Saudi Arabia adopt a constitution. Doubtless pressured by Faisal, the Saudi king then called for the resignation of Talal, their younger half-brother. After a speech against the autocracy, delivered at a press conference in Beirut, Lebanon, on August 15, 1962, Talal, the leader of the "free princes movement," was forced into exile:

His passport and assets were confiscated. He went to Egypt, where Nasser welcomed him with open arms.

Two years later, Prince Faisal surrounded the King Saud's palace and dethroned him. Faisal was appointed regent on March 4, and took the throne on November 2, 1964. Neither Aramco nor American president Lyndon Johnson had the slightest objection. After abdicating, King Saud lived in exile in Switzerland. He was erased from Saudi Arabia's official history. On the American side, the sovereign king who had been deposed for committing the error of allowing himself to be tempted by reform was presented as incompetent or, as David Rockefeller wrote in his memoirs, "ineffectual."[57] The very authoritarian King Faisal remained Washington's ally until a mentally unbalanced nephew assassinated him in 1975. A few months before his ascension to the throne, Faisal confided to the American ambassador to Saudi Arabia: "After Allah we trust in America."[58]

The Oil of Gabon, Fuel of the "Françafrique"

While the American government advanced discreetly in Saudi Arabia, or took great care to erase its traces in Iran, Indonesia, and elsewhere, the French were less cautious. In Washington (in Congress as well as deep at the core of diplomacy and justice), many voices constantly denounced the oil empire's influence on policy, but the democratic problem posed by this empire left the French elite practically indifferent, first under the Fourth Republic and then under General de Gaulle's presidency—a stance that more or less remains today. Therefore, Paris's behavior toward the oil-producing countries of its former colonies took a ridiculous if not tragic turn, filled with brutality and vulgar force.

Of course, the issues here were different, especially in their magnitude, from those emerging at the same time between the United States and Saudi Arabia. Unlike Washington or London, Paris was never able to become a full-fledged player in black gold. Through a succession of public petroleum companies, France piloted a hybrid form of postcolonial imperialism: Their state capitalism, seeking at all costs to obtain and maintain the precarious energy independence of a democratic European regime, rode roughshod over the emerging democracies of their former African colonies, rich in crude oil and other raw materials.

Proclaimed on August 17, 1960, Gabon's independence from France was an independence under supervision. That same day, the cooperative agreements signed with France got down to the brass tacks of the French neocolonial game. Article 4 of these agreements stipulated: "The Gabonese Republic will stock the French Army with strategic materials. When defense needs arise, exportation to other countries will be limited or prohibited." The strategic materials referred to, of course, were recently discovered oil reserves. Gabon rapidly proved to be one of the richest states in Africa, not only in oil but also in uranium, thorium, lithium, and beryl. Article 5 defined the terms and conditions under which Paris would control Gabon's underground resources: "The French Republic will be kept informed of programs and projects concerning exportation of raw materials and strategic products outside the territory of the Gabonese Republic. Concerning these same products, the Gabon Republic reserves priority for their sale to the community of States, after satisfying domestic consumption and supply needs."* Initially, Paris imposed the same stranglehold on the raw materials of all other member states of the "community" of its ex-colonies in sub-Saharan Africa, which obtained independence beginning January 1, 1960. But it was only in Gabon and neighboring Cameroon, where, among other strategic resources, hydrocarbon reserves were important, that the Gaullist regime, in the 1960s, worked carefully to maintain this stranglehold.

On February 12, 1962, Léon M'ba became Gabon's first president after his party carried the legislative elections with 99.57 percent of the votes. Léon M'ba, a small, round, vehement, ambitious fifty-nine year old, was one of those new African leaders that Paris considered "tamed"—faithful to the former colonizer, like Félix Houphouët-Boigny in the Ivory Coast or Leopold Sedar Senghor in Senegal. Before independence, when Léon M'ba was still vice president of the local French government council, he had requested that Gabon not become an independent state but remain a simple French territory. The request was rejected, to M'ba's great disappointment: "Independence, like the rest of the world," Paris replied, according to French journalist Pierre Péan.[59]

As president, Léon M'ba, called "the old man" or "the boss," upped his abuse of power and harassed his opponents personally. On February 17, 1964, he was overthrown by a group of Gabonese officers, without a shot fired. His cabinet director and future successor, the young Albert-Bernard Bongo, was flogged by the military.[60]† The next day, the perpetrators gave the presidency to Jean-Hilaire Aubame, renowned for his integrity among the Gabonese population and much more popular than Léon M'ba.

---

* Under the terms of the original text of the Constitution of the Fifth French Republic, the Franco-African "community" brought together most of the continent's former French colonies.

† He became Omar Bongo after his conversion to Islam in 1973.

But Charles de Gaulle would not hear of it. France had already lost Algeria and feared losing control of the Congo-Brazzaville, two other strategic sources of black gold. There could be no chance of letting Gabon—where petroleum production had just begun and appeared to confirm its promise—escape. Worse, Paris strongly suspected that the United States had supported the Gabon coup. It must be said that the American ambassador in Gabon, Charles Darlington, was a former director of the Mobil group, first in Iran and then in Iraq.[61] Charles de Gaulle approved an intervention plan, concocted in Paris on February 18 by Maurice Robert, head of French secret services in Africa, and by the indispensable Jacques Foccart, the "Monsieur Afrique" of the Élysée.

Also present was Pierre Guillaumat, a polytechnic engineer who served in France's resistance intelligence services, who would keep the upper hand after the Liberation and until the end of the 1970s on France's energy policy, both for oil and nuclear power. After serving as first minister of defense of the young Fifth Republic, the loyal Gaullist became the head of the Union Générale des Pétroles (UGP)—one of the main branches of the future French public oil company Elf Aquitaine. Guy Ponsaillé, UGP's personnel director, also figured within the group of French conspirators. It was he who best knew the situation on the ground. Only one person represented Gabon, and he, too, was French: Claude Theraroz, an advisor within Paris's Gabonese embassy, had the full confidence of Léon M'ba. There was only one problem. Gabon's vice president, Paul-Marie Yembit, had taken refuge in his native province and could not be located. Therefore, he could not sign the request for the French army's intervention. His letter was post-dated.[62]

The next day, one hour before sunrise, French paratroopers landed in Gabon's capital, Libreville. The same evening, after combat that resulted in the deaths of sixteen Gabonese and two French soldiers, the Libreville situation was "normalized." Frightened and humiliated, Léon M'ba regained the presidential palace, guarded by the French military. The "old man" was chaperoned by Guy Ponsaillé. The Gaullist oilman became political adviser to Gabon's president, who had been reinstated by the French army and entrusted with representing all French enterprise in Gabon.[63] From 1959 to 1969, throughout his tenure as France's president, Charles de Gaulle personally watched over the activities of the "Françafrique" network supervised by his personal adviser, Jacques Foccart. But he was mostly satisfied to validate, after the fact, the activities of the oilmen, the "Françafrique" network's principal kingpins. Most of the time, this attitude of passive and tacit agreement, oriented toward France's higher interests, was symptomatic of Paris's supporting role for its former French colonies, and for oil companies in particular. This attitude strongly resembled

Washington's practice of offering "no objection" to the major American oil companies in the Middle East.

On March 1, 1964, Léon M'ba and France faced protests in Libreville. The Gabonese president imprisoned 150 of his opponents. The opposition, although deprived of its leaders through the imprisonments, garnered enough votes in the April election to crush M'ba's candidates in the capital.

The repression was bolstered. The elite French spies operating in Africa started working for Gabon.[64] Bob Maloubier, an ex-combat-swimmer, resistance hero, and veteran of the First Indochina War, organized and trained Gabon's presidential guard. This guard was directly funded by French companies installed in Gabon, and most of all of course by the UGP.[65] The Gabonese counterinsurgency service, specializing in eliminating the opposition, was supported by Pierre Debizet, future patron of the Civic Action Service (known as the SAC), the secret militia responsible for doing the dirty work of the Gaullist regime. Jean Tropel, ace director of the French secret service's operations branch, facilitated communication with UGP's internal intelligence service; he was a kind of counterpart to Colonel William Eddy in Saudi Arabia.

One Frenchman was marching to a different drummer. The new ambassador to Libreville, François Simon de Quirielle "acted like he was serving in Vienna," journalist Pierre Péan quipped ironically. Early in 1965, he visited the presidential palace, where Leon M'ba, who had become paranoid, remained most of the time. De Quirielle suggested that the Gabonese president withdraw the French soldiers who protected his palace. The "old man" threw a fit of rage when de Quirielle compared the French garrison to an "occupation troop."[66] M'ba expelled the ambassador on the spot and called Jacques Foccart, to whom he announced that he would immediately fly to Paris and would not return to Gabon until Paris appointed a new ambassador![67] Foccart then chose a much more appropriate candidate for the position: Maurice Delauney.

Trained at the colonial school, Delauney was a prisoner in 1940, at age twenty, and first met Africans when he supervised them in factories and mines controlled by the Germans in Lorraine.[68] At the end of the 1950s, under the authority of future French prime minister Pierre Messmer, Delauney built his reputation leading the bloody repression of Cameroon's independent movement. Gabon's neighbor, Cameroon, was similarly rich in raw materials, in particular petroleum.* It was only after Delauney's appointment as the new French ambassador in Libreville that Leon M'ba consented to return. As the Gabonese president ceremonially

---

* Elf Aquitaine started oil production in Cameroon in 1977, five years after the last signs of rebellion against President Ahidjo's regime. But the country, whose crude oil extractions were limited in any case, saw its extractions begin to decline in 1985.

presented the new ambassador and his credentials, he declared that the French, "far from being of foreigners in Gabon, consider it their adopted home."[69]

Leon M'ba had been struck by cancer and, every two months, had to visit the Claude-Bernard hospital in Paris. Effectively, Maurice Delauney then became leader of the state. Here is how the Gaullist henchman recalled his discussions with Leon M'ba's vice president, Paul-Marie Yembit, during M'ba's long absences: "He [Yembit] was almost illiterate and completely unaware of how to lead a country that was in the middle of a revolution. . . . from time to time, Mr. Yembit summoned me and I saw accumulated on his table a stack of files; he asked my opinion of them and whether he should—or should not—affix his signature to the bottom of the documents. In all circumstances, President Léon M'ba had instructed me to remain a careful and vigilant advisor. I spent many hours with the Vice President who, to show that he was friendly, constantly served, in large glasses, an excellent champagne. You had to be in good shape!"[70] At the end of these grueling encounters, France's ambassador made a visit to the regime's rising star: the schemer Albert-Bernard Bongo. Only twenty-nine years old, on September 24, 1965, he was named deputy minister to the president. During that time, in Paris, the Pompidou government created a public oil division that brought together the General Petroleum Union and other oil companies belonging to the state, to form a single entity, which in 1976 was renamed Elf Aquitaine.

Albert-Bernard Bongo, freemason and former post office employee in Brazzaville, had already maneuvered between Paris and Libreville for years. On November 12, 1966, he was appointed vice-president in place of Paul-Marie Yembit, shortly before Leon M'ba, at sixty-four, was hospitalized in Paris, this time in critical condition. Jacques Foccart besieged the Gabonese president's hospital room. The Gabonese constitution did not provide that the vice president would automatically succeed the president in the event of the latter's death. Foccart intended to speedily sweep all obstacles out of the way: Charles de Gaulle's "Monsieur Afrique" wanted, at any price, to obtain a constitutional amendment from M'ba. He also wanted Léon M'ba to run for office in new elections, with Bongo as his vice president. But the "old man" did not readily embrace this, and from the foot of his hospital bed, Maurice Delauney tried to convince him. Foccart and Bongo waited in an adjoining room. Delauney persisted for quite some time. The doctors became concerned. Leon M'ba eventually gave in. He still had to undergo a grueling photo session that was used to make election posters on which M'ba was pictured seated and Bongo standing.

On February 20, 1967, the Gabonese constitution was amended, as Paris wished. Under the name of the president of the Republic of Gabon appears not the signature of Léon M'ba but that of Albert Bongo.[71] On March 19, with Bongo by his side, M'ba was reelected, obtaining 99.5 percent of the votes. He took his oath of office not in Libreville but at the Gabon Embassy in Paris. A month later, Bongo effected a governmental reorganization through which he assumed the responsibility for planning, development, and information. Léon M'ba died on November 28 at the Claude-Bernard hospital. His body was sent home to Libreville. A French ship blasted its horn several times in his honor. A military parade was held, during which the French military marched beside the Gabonese.[72]

Almost immediately, at the beginning of his presidency, Albert Bongo canceled the constitutional modification effected a year earlier with Léon M'ba, that allowed the vice president to succeed the president. He also eliminated the former political parties and created his own unique party, the Gabon Democratic Party (PDG), which he called "the renovation party." He explained: "Multi-partyism, it may satisfy the democratic ideal of the old countries of the Western world who no longer have to worry about building a nation, who find themselves already engaged in the process of modern social and technological transformations, but it can only represent, in the context of young African nations, elements of disorder and stagnation."[73]

Once again, Delauney and Bongo reorganized the presidential guard. It was mainly composed of former elite French army fighters. In this "White Guard," the only Africans admitted were members of President Bongo's tribe, the Batékés. Maurice Delauney never lost contact with Gabon. When he left the embassy in 1979, he became the head of the uranium mines company in Franceville, Gabon, where he remained the president-director until 1989. Bob Maloubier, the creator of the widely feared presidential guard, preferred to join the French public oil conglomerate Elf-Erap group in 1967. In July of that year, he was posted in Nigeria when, in Biafra, the oil province of this former British colony, war broke out.

The subjugation of Saudi Arabia and Gabon (and many other countries in the Middle East, Africa, and Latin America) by the industrial powers who extracted crude oil from their soils soon reversed itself. During the period that followed the first oil crisis, Washington elevated the House of Saud to effectively become an almost invincible force in its economy and geopolitical calculations. As for Gabon, it acquired a primary role in the hidden financing of French politics by African oil, thanks to the greed of the parties who rose to power in Paris, thus allowing Omar Bongo to create his own dynasty.

But while these protectorates arose because of black gold, and concentrated and spread their power thanks to their Western allies, new forces of emancipation emerged, which undermined the foundations of the former petroleum empire.

SIXTEEN

Cartel Against Cartel: OPEC's Painful Emergence

OPEC, the Organization of the Petroleum Exporting Countries, came into being during the second week of September 1960. Founded by Iran, Iraq, Kuwait, Saudi Arabia, and Venezuela, its goal was to regain control of the petroleum that had been dominated by Western companies. This reclamation of control over their own native resources took time to succeed, more than ten years, in fact—ironically the same amount of time it takes for a new oil field to reach its optimal production. At its creation, OPEC did not seem like a threat to the status quo. The Seven Sisters cartel organization had reached its highest refinement and the apex of its secretive power. This power was reinforced by a strategic, decisive factor that weakened the first delicate alliance of export countries: an unequalled abundance of available crude oil.

The Peak of Abundance: The Birth of the African and Asian "Elephants"

The first years of OPEC's existence coincided with the heyday of petroleum exploration. The years 1964, 1965, and 1966 established, successively, the three record highs for new oil field discoveries.[1] The world's history of oil prospecting may be reduced to a bell-shaped curve over more than a century, from the first drilling project in Pennsylvania in 1859, up to today. The peak of this curve occurred in the mid-1960s: Never before had so much oil been discovered . . . and never again.

During this prosperous era, new reserves, which the industry kept accumulating, by far surpassed the world economy's needs, launching a phase of rapid, unprecedented growth, fed from the overabundance of its own petroleum. It was a prosperous time, but politically delicate for the Western oil companies, who were constantly threatened by plummeting prices. There was only one logical solution for these companies: to put tourniquets on the main arteries of global production, without informing the foreign nations where the oil originated. The problem was that each of these young nations, on the contrary, wished to see their production, and therefore profit, grow as quickly as possible.

New "elephants" (the oil companies' nickname for larger fields of oil and natural gas) had been emerging all around the globe, as never before. First, on the shores of the Persian Gulf, fields were discovered in Qatar and Abu Dhabi, small emirates organized around humble fishing villages that Ibn Saud had not attempted to annex during the conquest of his kingdom, probably fearing the British navy that then ruled the Persian Gulf. Further north, in 1953, a giant oil geyser was discovered in Rumaila, essentially in Iraqi territory. Much closer to the sea than the oil in Kirkuk, Rumaila's reserves became the crown jewel of the Iraqi oil industry. Spilling over into Kuwait's territory, they would also become the subject of deadly fighting between these adjacent countries.

In the years following its 1949 creation, Mao Zedong's People's Republic of China faced a severe fuel shortage. Tanks and fighter planes were often grounded due to lack of fuel, as were the Beijing buses, which ran on natural gas. Even during the winter, at temperatures dipping to -30°C (-22°F), several drilling teams searched for oil in Manchuria, where the Japanese had come up empty handed in the 1930s. On September 26, 1959, finally, the Chinese discovered their "elephant": the giant field named Daqing, meaning "grand celebration." Tens of thousands of workers, mainly from the Red Army, worked to develop the field as quickly as possible. In December 1963, Prime Minister Chou En-lai announced that from then on, China would be self-sufficient when it came to petroleum. It remained that way until their economic liftoff in 1992. The discovery of Daqing was considered decisive for China's industrial development. In 1965, Mao Zedong adopted a new political slogan: "Learn the lesson of Daqing." Written at the entrance to Daqing's petroleum museum, in both Mandarin and English was this: "Petroleum has a compact relationship with a country's political, economic, and military strength."[2]

Thanks to the oil majors and their more or less rigorously tutelary governments, other "elephants" arose in Africa, on the "right side" (in other words, the more easily controlled Western side) of the Suez Canal. Beginning in 1956 in French Algeria, while in the North, the French army strengthened its repression against the National Liberation Front's (FLN's) independent combatants and the population, ten years of exploration paid off for the French oil companies in the South. There, deep in the Sahara, they discovered impressive hydrocarbon geysers: Hassi Messaoud for oil, Hassi R'mel for gas. The discovery owed much to the tenacity of the head of the French energy program, engineer Pierre Guillaumat. Guillaumat was obsessed with France's energy independence in the face of the large US and British companies, an obsession gained from the experience of his father, General Adolphe Guillaumat, during the First World War. The grand master of France's public petroleum industry and nuclear program tried to the bitter end to defend his country's access to the Algerian desert oil fields.

In 1957, on the other side of the Sahara, at the mouth of the great Niger river, after a long, vain search, Royal Dutch Shell accessed Nigeria's immense petroleum reserves, only three years before the independence of the British Crown's ultimate colony. Finally, in 1959, a Jersey Standard drill released an excellent quality crude oil in Libya, under the sands where the black fluid had eluded the German general Erwin Rommel. At the end of the 1960s, Libyan crude oil production rivaled Saudi Arabia's, providing Western Europe with a quarter of its fuel. Quickly, Libya became one of the world's largest petroleum providers. This new, major oil supplier became a wild card, landing in the middle of the game the Seven Sisters had secretively, meticulously arranged for three decades.

Enrico Mattei Defies the Seven Sisters

It was the Soviet adversary, though, who first threatened to disrupt the beautiful mechanics of the cartel of American, British, and French oil companies. The threat appeared via a maverick to whom the Sette Sorelle (Seven Sisters) owed their famous nickname: the engineer and politician Enrico Mattei, founder in 1953 of the small national Italian company ENI. Frustrated that ENI had been denied a place as the youngest child in the Seven Sisters cartel, Mattei established his reputation in 1957 when he went directly to the shah of Iran and signed an accord that made the oil majors boil with rage. For the first time, a Western petroleum company had accepted a mere 25 percent crude oil profit, instead of the famous 50 percent. With his determined gait and slicked back hair, Enrico Mattei, who had risen from a modest background through the ranks of Mussolini's resistance during the war, escalated the situation when he traveled to Moscow in 1959 to negotiate the purchase of crude oil from the USSR. For the Seven Sisters, the price to which the Soviets agreed was dangerously low, 60 cents lower than that of a barrel of Middle Eastern crude oil, which then sold for just under $2. In the midst of the Cold War, Mattei's strategy earned him many enemies. Here is how, during a March 1961 television interview, Mattei, then fifty-five years old, described what is was like to jockey for a place in the oil arena. This popular hellion of Christian democracy in Italy, a country that was poor and devastated after the war, compared the oil majors to dogs. He, who the Italian people simply and respectfully called "the engineer," said that while returning from a day of hunting, he saw a cat, "lean, hungry and low," approaching his dogs' dish: "[The cat] was very afraid, and it went forward very slowly. It watched the dogs closely, meowing, then put its paw on the edge of the bowl. [One of the dogs] threw it three or four meters, and broke its spine. This episode left a deep impression on me. We here [the Italians], we have been like a cat."[3]

The Italian oil condottiere attracted venomous criticism in the United States; the hatred it inspired in Paris and Algiers was no less potent.[4] In 1960, Paris offered ENI a concession in French Algeria, but Mattei refused, publicly arguing that he would not be the accomplice to a colonial enterprise. He would do business with the National Liberation Front, to negotiate ENI's future access to Algerian hydrocarbons, once Algeria achieved independence. He was soon added to the black list of the OAS, the "secret army" of military terrorists and French colonists who opposed the independence.[5]

Enrico Mattei's death in an airplane that crashed during a violent storm on October 27, 1962, on a flight from Sicily to Turin, remains tainted by suspicion. In 1970, film director Francesco Rosi began shooting a movie about the crash. Mauro di Mauro, the Italian journalist who Rosi employed to conduct the investigation in Sicily, disappeared in Palermo on September 16, 1970, under ambiguous circumstances. In 1972, Francesco Rosi's film, *The Mattei Affair*, received the Palme d'Or at the Cannes Film Festival. Paramount bought the rights to the film, a blend of fiction and documentary, but allowed only one very restricted showing before quickly removing it from circulation.[6] It was never reissued. In May 1994, a reformed mafia chief, Tommaso Buscetta, claimed: "Mattei was murdered at the request of the American Cosa Nostra, because his politics had caused harm to important American interests in the Middle East. It was most probably the oil companies who pulled the strings."[7] Following this accusation, Mattei's body was exhumed in June 1996. Later, new analysis conducted by a professor from the Polytechnic University of Turin, in particular, analysis of Enrico Mattei's broken signet ring, revealed probable traces of an explosion provoked by a small charge placed inside the instrument panel of Mattei's private jet.[8] Italian law enforcement reclassified the Mattei affair as a homicide but, ultimately, no suspects were identified.

The Birth of OPEC: "The Fat Is in the Fire"

Thanks to a surplus production of crude oil and the breach opened by Enrico Mattei, in 1960 the USSR was briefly able to inundate both Italy and India with cheap oil. This menace could have obliged the Seven Sisters to lower their barrel price. With this incursion from the USSR and despite the strict control secretly exercised by the oil majors for more than thirty years, "certain normal functions of prices, long rusty with disuse, began to reassert themselves."[9] In July 1960, Jersey Standard's new executive director, Monroe Rathbone, voiced his concerns to Wanda Jablonski, the elegant and distinguished journalist featured in the otherwise exclusively masculine world of the oil industry: "The tremendous amount of price discounting to third parties [Italy and India] is bad enough. If it now spreads to affiliates," directly controlled by the oil majors, "the fat is in the fire," the Standard Oil boss said.[10] A response was needed. On August 6, in an unusually bold stroke for the boss of the world's largest oil company, Rathbone announced an average decrease of 10 cents off Jersey Standard's "posted price" for a barrel of Arabian Light. The other major Western companies could only follow. From Tehran to Caracas, this unilateral decision was received with anger. "All hell broke loose," Howard Page, Jersey Standard's Middle East director, said later.[11] Page was called upon to play a crucial role in the aftermath of the events. He had strongly advised Rathbone against acting without warning the exporting countries about the price decline, and moreover without offering to share with them the resulting profit loss, of tens of millions of dollars. This price decline was all the more unacceptable to the exporting countries because they had only recently endured a price decline imposed by the Seven Sisters (in order to recapture pre-Suez Canal crisis price levels); in April 1959, the representatives of the Arab oil-producing countries gathered in Cairo and declared that, from then on, no price decisions could take place without prior negotiation.

In the hours following Jersey Standard's announcement, the Saudi Arabian oil minister, Abdullah Tariki, telegrammed his Venezuelan counterpart, Juan Pablo Pérez Alfonso. Pérez Alfonso had been the first to successfully establish, twelve years earlier, the famous 50/50 profit sharing. Tariki had already become the other thorn in the side of the oil majors, when in 1952, he demonstrated that Aramco did not follow through with the 50/50 sharing agreement recently established with the Kingdom of Saudi Arabia. To journalists who asked him how he intended to respond to Jersey Standard's outrageous *casus belli*, Tariki responded: "Just wait."[12]

The wait was brief. On September 9, 1960, representatives of the five oil-producing countries who held (but did not control) 80 percent of the planet's known crude oil reserves met in Bagdad: Saudi Arabia, Venezuela, Iraq, Iran, and Kuwait. After almost a week of discussions, the Organization of the Petroleum Exporting Countries, OPEC, was born. In his inaugural speech, the Venezuelan oil minister, Pérez Alfonso, reminded everyone that oil was an "exhaustible, non-renewable resource" and that "world reserves of crude oil would not continue to expand forever." Therefore, he said, "our people cannot let flow, at an accelerated rate, their only possibility to pass without delay from poverty to well-being, from ignorance to culture, from instability and fear to security and confidence."[13] These words enshrined what was perhaps the only major victory of Third-Worldism. They echoed those written in Nasser's 1954 *Philosophy of the Revolution*. The Egyptian leader had explained that oil was one of the cornerstones of Arab nationalism, a "vital nerve of civilization, without which all its means cannot possibly exist."

In its first years, OPEC was an empty shell. It had no political heft. Petroleum itself, simply via its discovery, defined the rules of the game. The overabundance of the oil-exporting countries' reserves and their production capabilities, in light of the world's growing demand, actually placed OPEC countries in competition with one another. To deal with these export countries, the Seven Sisters secretly forged a common front. The old divisiveness between OPEC's Arab countries, often a result of how the British Empire had divided up their borders, undermined the young organization. In 1961, General Kassem's Iraq, for example, nearly decided to invade Kuwait. The Western companies could easily pretend that OPEC would not change anything. And OPEC worried the Seven Sisters even less when, shortly after its creation, the political careers of its two founding fathers ended abruptly.

The Fathers of OPEC: Tariki, the "Red Sheikh," and Pérez Alfonso

The son of camel merchants, Abdullah Tariki was one of Saudi Arabia's original technocrats. Trained in United States, as were all the new modern Saudi elite, he graduated from University of Texas in 1947 and endured the racism of America's deep South. He was the victim of discrimination in bars where he was mistaken for a Mexican.[14] He married a young American white woman, who was doubtless as idealistic as he was (a "tramp," according to the verdict of an Aramco director).[15] Tariki told of having been the first Arab allowed to enter the luxurious Aramco camp in Dhahran, ordinarily reserved for Americans—recounting how he had never met such narrow-minded people.[16]

In 1960, at age forty-two, Abdullah Tariki became the first petroleum minister in Saudi Arabia's history, a quarter century after American drilling began. He was put in charge by King Saud, the reformer. In Aramco's internal reports, Tariki was presented, time and again, as a "pure Nasserist," or an "opportunist in search of celebrity."[17] One day he proclaimed: "We are the sons of the Indians who sold Manhattan. We want to change the deal."[18] To the Western press, Tariki became the "Red Sheikh."

The Saudi oil minister could not manage the Americans and their powerful allies inside House of Saud for very long. The technical skills Tariki had acquired in the United States—including in economics, law, and physics—were then exceptional among his compatriots. But they were not sufficient to pardon him and his political positions, which favored profound reforms in the kingdom. Tariki made his fatal error in 1961: He accused Prince and future King Faisal of corruption and nepotism. The latter fired him on March 9, 1962.

The Ministry of Oil was then entrusted to a young man who was considerably more malleable, a thirty-two year old lawyer, son of a judge from Mecca, who had studied at Harvard and loved New York: Sheikh Ahmed Zaki Yamani. Yamani remained minister of oil for twenty-four years. He became a global celebrity during the first oil crisis. Tariki was exiled to Beirut and did not return to Saudi Arabia until after King Faisal's assassination in 1975. The Red Sheikh never again returned to power.

OPEC's other founding father, Venezuelan Pérez Alfonso, had also studied in his future opponents' country. And it was in Washington that he took refuge, after spending nine months in a Venezuelan prison, following the 1948 coup. During his second long stay in America, he assiduously frequented the Library of Congress, where he analyzed, in minute detail, the functioning of the Railroad Commission of Texas. Ironically, it was the commission's sophisticated quota system, put in place after the crisis of 1929 to regulate the flow of black gold in the United States, that inspired the Venezuelan minister of oil to create OPEC.

Beginning in 1961, Pérez Alfonso felt frustrated by the dissension within the organization and by the desire of its member countries, absurd in his eyes, to produce as much oil as possible. He also observed the debilitating effect of oil money on his own country: corruption and naive economic policy.[19]

Venezuela's crude oil reserves appeared limited. As strange as it may seem today, Pérez Alfonso intended to make OPEC an ecological organization: the only entity able to restrict the production and increase the price of crude oil, both to limit the pollution of the planet and to preserve the richness oil offered so that many generations could enjoy it.[20] Pérez Alfonso later described himself as "an ecologist above all else."[21]

The one who wanted to "harvest the oil" to advance his nation, Pérez Alfonso resigned from his position as oil minister on January 23, 1963, profoundly disillusioned. Less than a year after Tariki, at just sixty years old, he withdrew from political life to his home in the Caracas suburbs. Behind the garden, near the ping-pong table, an old, rust-corroded auto awaited him, a 1950 Singer, an elegant little English car purchased in Mexico City during his years of exile. The motor had seized, the result of a gas shortage in the port of Caracas on the day it arrived, which meant no one drove it; the Singer was rendered unusable forever. The small man with the huge smile kept his darling nonetheless, and, consciously, allowed it to fall to ruins, like an "overgrown symbol of what he saw as the dangers of oil wealth for a nation: laziness, the spirit of not caring, the commitment to buying and consuming and wasting."[22]

The Seven Sisters Assimilate OPEC

OPEC carried so little weight at its inception that Switzerland, where the organization first chose to meet, refused to grant them appropriate diplomatic status. The Austrian government had the expediency and intuition to be more accommodating, and that is why OPEC settled permanently in Vienna beginning on September 1, 1965.

Despite its lack of political weight, unity, and effectiveness during its first years, OPEC's mere existence served as an excuse for the Seven Sisters cartel to further establish their dominion. OPEC's first secretary general, the Iranian Fouad Rouhani, noted helplessly that the oil majors had agreed among themselves to negotiate jointly with each individual producer country.[23] This way of doing business was illegal, of course, especially with regard to American law—as the proceedings of President Truman's attorney general against "the international petroleum cartel" had earlier shown. Whereas the Truman and Eisenhower administrations had consented to divert in extremis the arm of American justice, the same justice agreed at the outset of OPEC's creation to keep its sword sheathed, no matter what might occur. John McCloy, counsel for the Seven Sisters, obtained explicit assurance from three administrations' attorney generals (Robert Kennedy, when he was the attorney general for his brother John, then from the attorney general for Lyndon Johnson, and finally from that of Richard Nixon): Once OPEC existed, the large US companies as well as their "sisters" in Europe could collaborate without risking prosecution.[24]

More than the OPEC countries' lackluster resolve to band together politically, the simple overabundance of their oil wells comprised the greatest threat to the Seven Sisters. According to one Standard Oil of California executive, "Nobody could have lifted enough crude to satisfy all the governments in the Persian Gulf during this period."[25] Howard Page, Jersey Standard's negotiator in the Middle East, described the problem in the following manner: "It is like a balloon . . . push it in one place, it comes out in another. . . . If we acceded to all those demands, all of us, we would get it in the neck."[26] A Jersey Standard geologist, returning from the sultanate of Oman, east of the Arabian Peninsula, told Page: "I'm sure there's a ten billion [barrel] oilfield there." Page replied: "Well then, I'm absolutely sure that we don't want to go into it, and that settles it. I might put some money in it if I was sure that we weren't going to get some oil, but not if we *are* going to get oil, because we are liable to lose the Aramco concessions."[27]

The Western oil companies had little trouble inventing pretexts to curb extractions in OPEC countries and protecting themselves against any new price declines. The stars aligned almost too favorably. Howard Page said: "Sometimes, they made it easy to cut down by breaking an agreement, as in Iraq; then we could tell 'em to go to hell."[28] It is not trivial to remember that, with only 23.75 percent of its capital, it was in the Iraq Petroleum Company that the American majors had the smallest share of interest around the Persian Gulf. Page's admission, made public in 1974 under an injunction by the US Congress, took on a strange emphasis in light of Iraq's later history; deliberately restricting Iraqi production, as one keeps an ace up his sleeve, seems to have been a steady strategy of American petroleum in the Middle East.

There was only one country that Americans were reluctant to offend: Saudi Arabia. There was already an acute rivalry between the Saudis and Iran. Concerned about Iranian competition during the 1960s, Sheikh Yamani suspected Howard Page of favoring Iran. The shah was convinced of the opposite, and he was correct, because Big Oil's vital interests in the Persian Gulf were located in Saudi Arabia. Having difficulty financing his ambitious military and industrial programs, in 1966 the shah began to pressure the oil companies: He launched a publicity campaign against the oil majors and made his frustrations known to his former advisor, Kim Roosevelt. On November 16, 1966, Walter Levy, a highly influential intermediary in US industry, informed a top American diplomat, Eugene Rostow, of the nature of the secret arrangement between the consortium partners, put in place in Iran after Mossadegh's fall. He emphasized a delicate point: If the secret quotas were exceeded, the penalties provided for "overlifting" Saudi oil were much higher than they were for Iranian oil. The very fact that a secret agreement exists would be "political dynamite in the hands of the Iranians," warned Levy.[29]

It was the French, from the petroleum company CFP, who let the cat out of the bag to Mohammed Reza Shah. Frustrated that they were not authorized by the Seven Sisters to extract more crude oil, the small "eighth sister" provided Tehran with a copy of the secret agreement (or at least this is what the Iranian consortium's partners claimed).[30] A violent dispute erupted behind the thick curtains of black gold diplomacy. The shah complained bitterly to Washington. A diplomat from the American embassy in Tehran candidly told his higher-ups that the shah had put his finger on a case of "restraint of trade," reprehensible in the eyes of American law.[31] Jim Akins, the State Department's new "Mr. Oil," firmly put the imprudent diplomat in his place. He responded: "We were surprised to see the accusation put into a telegram which was given fairly wide distribution in the government."[32]

We will never know very much about the dealings that ensued between Washington and Tehran. Jim Akins made it known that if the shah persisted in his grievances, Iran would share Iraq's fate and would see restricted production.[33] Was this because the shah owed everything to Uncle Sam? After a few minor concessions on the part of the oil majors in Iran, everything returned to order.

The Six-Day War: The Weapon of OPEC Oil Fizzles Out

In 1967, imbedded in the Arab countries' military fiasco during the Six-Day War was yet another fiasco: the Arabs' first use of oil as a weapon. Colonel Gamal Nasser was anxious for his country, Egypt, this time actually armed by the USSR, to take revenge on Israel eleven years after the Suez Canal crisis. Starting in 1966, attacks against Israel by Syria, then Egypt's close ally, were becoming more and more frequent. On May 16, 1967, encouraged by Damascus, Nasser advanced the Egyptian troops in the Sinai Peninsula, demanding the departure of the UN troops who had controlled the canal zone since 1957. On the night of May 22 to 23, Egyptian ships blockaded the entrance to the Gulf of Aqaba, which received most of Israel's essential petroleum shipments from the Persian

Gulf. It was cause for war. The Israel Defense Forces, or Tsahal, retaliated on June 5, and the next day at dawn, the Egyptian and Syrian air fleets were completely destroyed. On June 10, following one of the most intense armored offensives in history, Israel tripled the size of its territory. Egypt lost Sinai and the Gaza Strip, Syria was cut off from the Golan Heights, and the Kingdom of Jordan lost the West Bank as well as East Jerusalem.

Meeting in Baghdad on June 6, the Arab oil-producing countries decided to implement an embargo against the United States, the United Kingdom, and West Germany, whom they accused of being Israel's allies. The unexpected debacle of Egypt and Syria obliged the Arab countries to show solidarity. Two pro-Western regimes were in a particularly delicate position: Kuwait and even more so Saudi Arabia, which since 1962 was conducting a proxy war against Egypt in North Yemen, with the active support of the United States and the United Kingdom. Not to show solidarity in the face of the Jewish State was political suicide. But the House of Saud took the precaution of sending its army to protect all the US oil installations in its territory, fearing demonstrations and acts of sabotage from its own population.

The embargo was brief. The Arab producers quickly understood that they were shooting themselves in the foot, offering their competitors an advantage. Iran and Venezuela, who had no intention of participating in the Arab boycott, increased their exports, and of course the United States did as well.

Following Sheikh Yamani's counsel, King Faisal decided to limit his embargo to Great Britain and the United States. Washington's and London's faithful ally thus could save face at little cost, since at that time these two countries imported very little Saudi oil. At the end of June, Aramco was authorized to resume its normal activities. The other Arab countries, in particular Iraq, which had initiated the embargo, had no other choice than to follow swiftly on the Saudis' heels. Physical reality once again imposed its contingencies on political ambitions: The glut of oil-production capacity deprived the embargo of any chance to succeed.

Nevertheless, the importance of Arab petroleum for the American petroleum industry seemed capable of impacting Washington's policy with respect to Israel. Two years after the Six-Day War, upon returning from a tour in the Arab world in late autumn of 1969, David Rockefeller, the head of Chase Manhattan, told his friend "Henry" (Kissinger) about his conversation with King Faisal, in which Faisal had complained for the umpteenth time about US support to Israel. A month later, on December 9, President Nixon received David Rockefeller in the White House's Oval Office. There, he found the Seven Sisters' lawyer and Rockefeller's predecessor at the head of Chase Manhattan,

John McCloy, as well as the leaders of Standard Oil of New Jersey, Mobil, and Amoco, and finally Robert Anderson, Eisenhower's former secretary of the treasury, who had become a director of Dresser Industries (the Texan firm through which George H. W. Bush had made his entry into the petroleum industry).* To these five gentlemen, responsible for enormous interests in the Arab oil-producing countries, the American president gave the speech that his secretary of state, William Rogers, was preparing to record that same evening. With an unprecedented tenor, this speech called on the Israeli army to withdraw from the occupied Palestinian territories and to make Jerusalem a "unified" city. But Israel's prime minister, Golda Meir, dismissed it unhesitatingly and made it known that, on the contrary, from now on East Jerusalem would be open to Jewish settlers. Nixon's reaction was nonexistent; the *New York Times* exposed, two weeks later, the American president's meeting with the elite players of Big Oil.[34]†

---

* Amoco's president, John Swearingen, was one of Chase Manhattan's directors. (See David Rockefeller, *Memoirs*, op. cit., p. 314.)

† The episode raised the ire of US pro-Israeli voters. David Rockefeller claimed he had been manipulated by Nixon, who he said had organized the leak, thus finding a shrewd way to let Israel alone while still making guarantees to Arab leaders. (See David Rockefeller, *Memoirs*, op. cit., p. 274–280.)

SEVENTEEN

The Leapfrog Effect: Algeria, Biafra, and Libya

Three serious warning shots announced the beginning of the decline of the Western oil empire in the 1960s: Algeria's independence, the Biafran war in Nigeria, and the accession to power in Libya of Colonel Muammar Gaddafi.

The first two of these events brought into play France's vital interests. In Algeria, Paris played a diplomatic game, with an immense stake in the oil capital held by the Rothschild bank, on the board of which sat future French prime minister Georges Pompidou. In Nigeria, the French intelligence service engineered despicable, deadly attacks—for the black gold of Biafra, and with General de Gaulle's full support. By decisively staking his claim in a Libyan desert overflowing with crude oil, Gaddafi finally created the opportunity for all OPEC cartel countries to escape, one after another, like frogs leaping out of a bag, from the confines of the Western companies' secret cartel.

For France and the Rothschild Bank, Georges Pompidou Negotiates the Future of Algerian Crude Oil

A quarter of a century after Mexico nationalized its oil, the Algerians took the first solid poke at the Western stranglehold on the planet's best black gold reserves. The petroleum that spurted out of the Saharan depths at the close of the 1950s was a central factor in the negotiation of the Évian Accords, signed on March 18, 1962, which drove Algerian independence that same year. The French desire to maintain control of Algerian hydrocarbons considerably slowed and complicated this negotiation.

Since the end of the Second World War, France wished at all costs to elude America's grip on Persian Gulf oil and free up more of that oil for its own company. The French Bureau of Petroleum Research, created by General de Gaulle's provisional government just after the war, coordinated extensive research all over France. The country was almost devoid of drilling equipment, and its machinery that was still working ran on steam power or came from Germany. Paris ordered equipment from Texas and, in 1951, activated many drills throughout the country.[1] The exploration met with little success, except for the discovery of the natural gas field in Lacq, in the south of the Aquitaine. Very hot and under extreme pressure, this gas was exceedingly toxic and corrosive, full of hydrogen sulphide. When the Lacq drill exploded in flames, it was necessary to call in a specialist from the United States, Myron Kinley, who after many attempts, managed to seal the leak by injecting cement into the base of the well. [2] Despite the danger and immense technical difficulties, the Lacq gas was exploited until it was exhausted (the last tap and the last compressor were cut off in 2013). But this gas was far from sufficient. France had no oil, so once again (as it had done in Iraq in the 1920s) it sought oil elsewhere.

The 1956 discovery of the gigantic Saharan crude oil reservoir Hassi Messaoud (the "well of luck") sparked France's hope of a resurgence in power, at the precise moment when it became clear that France was doomed to lose Algeria, where the reservoir was located. In March 1957, Charles de Gaulle made in the Sahara his last public statement preceding his return to power the following year: To him, Algerian oil was "a great opportunity for France," one that could "change our destiny."[3] So far absolute, France's dependence on foreign oil was reduced solely thanks to Algerian oil, going from 90 percent in 1960 to 60 percent in 1962, when Algeria became independent.[4]

Paris tried until the end to separate the Saharan oil of Hassi Messaoud from the "Algerian question," to make a kind of commonwealth, jointly managed (under France's authority) by its neighboring colonies, who were soon independent. This commonwealth project was developed in late 1956 by future Ivory Coast president Félix Houphouët-Boigny, who was then a key official in French prime minister Guy Mollet's socialist government. On January 10, 1957, seven months after Hassi Messaoud's discovery, a new law established the Common Organization of the Saharan Regions (OCRS). The goal of this so-called pre-postcolonial institution was "the economic expansion and social promotion of Saharan areas of the French Republic, managed by Algeria, Mauritania, the Niger, the Sudan and Chad."[5]

By August that same year, Southern Algeria's four large southern territories became the two French departments of the Sahara: the Oasis and the Saoura.

The creation of these two departments, wholly integrated in the OCRS, was an astute move that allowed France to breach the legal link between the Sahara and the Algerian territory. The OCRS's budget was devoted to creating a tarmac that led to Hassi Messaoud, constructing airfields, and establishing lines of communication: in short, to creating optimal conditions to exploit Saharan oil, more than 600 kilometers of desert south of the Algerian coast, where the war of independence was raging. After the military coup of May 13, 1958, in Algiers, directed by Gaullist general Jacques Massu, General de Gaulle rose to power on June 1, through a relatively popular coup d'état. De Gaulle immediately surrounded himself with pragmatic men determined to maintain access to the raw materials of Africa, beginning with the oil of Algeria: Jacques Foccart; Pierre Guillaumat, who had so vigorously championed Algerian oil exploration; and finally, Georges Pompidou, the future prime minister who succeeded Charles de Gaulle as the head of the French state. Pompidou had also worked at the Rothschild bank starting in 1954, becoming one of its executive directors in 1956 and remaining in that role until the Algerian coup d'état. Shortly after he left General de Gaulle's cabinet in January 1959, Pompidou resumed his position as administrator of the Francarep, the Franco-African petroleum research company, and in late February was also appointed a member of the Constitutional Council.[6] He retained these two high functions, public and private, until he was made prime minister by de Gaulle in April 1962. Earlier, while at Rothschild, Pompidou had personally made the call that linked Francarep's capital and the Rothschild bank with two other powerful business banks, Lazard and Paribas.[7] Francarep was among the first investors in Algerian oil. According to Jacques Getten, at the time secretary general of the Rothschild brothers firm, Francarep was simply the "most important" of the companies managed by the bank.[8]

The Rothschild bank had many interests in Algeria on the eve of its independence. Petroleum was not only the most significant but also the most promising. The bank was among the largest shareholders of the Compagnie Française des Pétroles (CFP). CFP exploited the oil from Hassi Messaoud, in lively competition with the various public corporations that, under the stewardship of Pierre Guillaumat, formed the Elf group in 1967.[9]

Pompidou had no executive function in the French government when de Gaulle put him in charge of the first secret contacts with the pro-independence FLN leaders of the Algerian Republic's provisional government, to negotiate the end of the war and the conditions of independence.[10] On February 20, 1961, in Lucerne, Switzerland, where he first met the FLN leaders, the atmosphere was glacial, close to the breaking point.[11] On March 5, the Rothschild bank's executive director went to Neuchâtel to continue, on France's behalf, negotiations with the FLN.[12] At 1:15 a.m. on March 20, a bomb exploded in front of the Rothschild bank headquarters at 23 Rue Laffitte in Paris; the attack was attributed to pro-French Algerian "militants."[13] On March 30, a news release announced the start of formal negotiations that a year later led to the Évian agreements, in which Pompidou was instrumental. A few days later, the Associated Press in Paris received a letter claiming that several recent attacks in the capital had been carried out by a group probably close to the OAS, the "secret army" opposing Algeria's independence. The authors claimed that "the complete liquidation of their petroleum assets by the major business banks, notably the Rothschild bank, as soon as Mister Pompidou returned from Switzerland, proved that high finance had already sacrificed the French Sahara."[14]

After Political Independence, Algeria Fought for Petroleum Independence

Twice more, from May 20 to June 17 at Évian and then from July 20 to 28 at Lugrin, official negotiations foundered precisely when it came to the status of the Western Sahara.[15] All earlier French offers to settle the "Algerian question" had failed because they denied the Algerians the right to make the Sahara an integral part of their future country.[16] One "absolutely daft" proposal was this: The Western Sahara would remain French, connected to the Mediterranean by a narrow corridor, also French, located on both sides of the pipeline that carried oil to Hassi Messaoud.[17] The French position was all the more difficult to maintain since the independence reached in 1960 by Mali, Niger, Chad, and Mauritania, which had been able to keep their entire Saharan territory. It was true that they had not discovered oil in their territories and, as in Gabon, access to other raw materials, notably uranium, was reserved for France.

Several historical analyses have concluded that France's focus on the Sahara during the Évian negotiations was above all motivated by its desire to maintain four bases to test nuclear bombs and one to test chemical weapons.[18] By maintaining military infrastructure in the desert, Paris would doubtless also be able, if necessary, to defend the Sahara's oil and gas wells.

The situation opened up during a press conference on September 5, 1961, when General de Gaulle for the first time publicly admitted that the Saharan Departments of Oasis and Saoura were, in fact, an integral part of Algerian territory. Everything moved so fast then that, for their part, the Algerians agreed that in return, they would include in the Évian agreement a continuance of previously signed contracts, as well as the existing oil code.[19] This code, adopted in November 1958, six months after de Gaulle took power, stipulated that the French state had the right to examine capital investments in Algerian oil, and in particular, to preclude foreign oil companies from investing, except in partnership with French companies.

Therefore, French investors preserved the essential: ownership and control of virtually all of crude oil production in Algeria, despite the country's new independence. The Évian agreements specified that Algeria would receive only a minor stake in just one of some fifteen oil companies then active in Algeria: the SN Repal, hitherto exclusive property of the French state.[20] The FLN had no other choice than to compromise, since the technical skills necessary to run the thirty oil deposits already in production were practically nonexistent within the Algerian population. But the Algerians, who mourned hundreds of thousands of their dead on July 5, 1962, their independence day, skillfully and without delay found a way to put an end to this vassal state.

In the aftermath of independence, a regime of joint Franco-Algerian sovereignty over Saharan oil was established. This shared rule was embodied by the Saharian Organization (OS), a binational institution that stepped in to oversee all matters related to oil operations, before the inexperienced Algerian administration could take matters directly in hand.[21] The Algerians complained that the oil companies systematically withheld information, and only presented their account information to the OS.[22] They resolved to be patient: They needed time to develop the skilled leadership necessary to run the national oil company, Sonatrach, created on December 31, 1963. Sonatrach's first director, Belaïd Abdesselam, "Mr. Oil" of Algeria, who remained minister of oil until 1977, was only thirty-four at the time of independence. But he was tough (on his wedding night, he gave his guests the speeches of Adolf Hitler, that he had translated, line by line) and was known as a devoted patriot.[23] Abdesselam, who had studied in San Francisco, played the American oil companies against the French, and little by little succeeded in shaking off the trusteeship of the former colonists. As early as 1964, he entrusted two American companies with a shared strategic task: the assessment of Algeria's national hydrocarbon reserves. One of these companies, D&M, founded by Everette DeGolyer, was recommended to Belaïd Abdesselam by an American geologist who proved to be an agent of the CIA.[24]

Having become president of Algeria in June 1965, Houari Boumediene instituted a political platform of socialist inspiration. He opened the door to the Soviet Union, still leaving the door open to American petroleum companies.[25] The French oil companies, who knew they were on probation, played less and less fairly. In 1967, the public Elf group, which directly implemented the independent energy policies sought by Guillaumat and de Gaulle, refused to support the ambitious prospecting program Algeria wished to launch.[26] The national oil company, Sonatrach, opened its first service station in the center of Algiers on May 8, 1967. French oil representatives were present. The date was chosen to honor the thousands of victims of the Algerian massacres at Setif, perpetrated by the French army on Armistice Day, May 8, 1945, and then, on Charles de Gaulle's orders, during the days that followed.[27] Relations between the Algerian and French oil companies seemed to have reached a point of no return after the Six-Day War: The Elf group appeared to practice a scorched-earth policy. One of its subsidiaries, CREPS, falsified the technical data it transmitted to Algiers: The field of Zarzaïtine was exploited under conditions that threatened to damage the reservoir irreparably.[28] The ultimate divorce took place in 1969, the year Algeria officially joined OPEC. Soviet engineers arrived at Hassi Messaoud in 1969 and contradicted the French engineers at the site, stating that production could be increased at the Algerian oil field, a task they undertook in 1971.

No one was surprised when Houari Boumediene announced the complete nationalization of the oil and gas sector on February 24, 1971. He had been assured support from US oil companies, who had formally committed to continue purchasing Algerian gas and oil, and to provide their assistance running the wells.[29] Michael Forrestal, a partner in the New York law firm Shearman & Sterling and the son of James Forrestal (the artisan who had planned and launched Saudi oil and Roosevelt's secretary of the navy), was one of the Americans assisting Sonatrach.[30] Elf floundered, packed up its bags, and went home. Not the CFP, which, despite the state's interest in the company, represented purely financial interests before all else, particularly those of the Rothschild bank. That is why Total is still present in Algeria today, where the French petroleum group collaborates with Sonatrach.

The Algerian public petroleum sector remained the essential source, fueled by foreign companies, of profound corruption in the highest military and political spheres of independent Algeria.

Biafra (1967–1970): France's Other "Dirty War," for the Oil of Nigeria

"The interest, it was the oil, the oil of Nigeria!" Françafrique's old guard, Maurice Delauney, exclaimed without shame, forty years after the Biafran war.[31] Gaullist France's behavior in the Biafran war merits special mention, even among the most hideous episodes in which a Western power sought to take oil from a poor country.

Nigeria obtained its independence from the United Kingdom in 1960. With forty million inhabitants, this vast country included more Africans than all the former French colonies combined. Shell's discoveries, and those of BP beginning in the late 1950s, revealed that Nigeria was also, by far, the African country richest in oil. In the mid-1960s, its extractions equaled those of the Gulf's main producers; in 1971, Nigeria joined OPEC, instantly becoming one of the heavyweights.

From neighboring Gabon, the French neocolonists remained on the lookout at the edge of the fabulous new private reserve of Nigerian petroleum, which British oil companies controlled. In 1963, the French secret service sent a former OAS sympathizer, Lieutenant Colonel Raymond Bichelot, to Abidjan so that he could closely follow the developing Nigerian petroleum business.[32] In 1964, Safrap, a subsidiary of one of the French public companies that soon formed the Elf group, established a magnificent bridgehead in the middle of Shell's and BP's territory: They acquired permission from the Nigerian government to prospect on a 3,000 square kilometer area, and then, the following year, to prospect on a 30,000 square kilometer tract east of the Niger Delta, in the Igbo ethnic group's territory, which was full of oil. The first French barrels in Nigeria were extracted in 1966.[33] On January 15, 1966, young Igbo officers commanded by General Johnson Aguiyi-Ironsi overthrew the first Republic of Nigeria, murdering twenty-seven of its leaders. Mostly Christians and Animists, the Igbos were influential throughout Nigeria, but some among them believed they were unjustly excluded from power. In the days following the coup, an anti-Igbo rebellion broke out in the north part of the country, which was mostly Muslim. The massacres caused thirty thousand Igbo deaths by October, driving hundreds of thousands of Igbos to take refuge in their region of origin, southeast Nigeria. General Aguiyi-Ironsi was assassinated on July 29. A military junta dominated by the Muslims from the north and led by General Yakubu Gowon took power. It wanted to reconfigure the country's territories, which would deprive the Igbos of the greater part of oil resources in their territory. On May 30, 1967, Colonel Odumegwu Ojukwu, the military governor of southeast Nigeria, proclaimed the region's independence. From then on, it was called Biafra.

In Paris, this secession was immediately perceived as a windfall by General de Gaulle, who confided to his faithful Jacques Foccart that he wished to "fragment" or compartmentalize Nigeria, to diminish its power.[34] With the support of his friends Omar Bongo in Gabon and Félix Houphouët-Boigny in the Ivory Coast, starting in November 1967, Foccart organized secret weapons deliveries via airplane, often under the guise of humanitarian assistance. The ruse was financially supported by the white racist regimes of South Africa and Rhodesia, which also sought to undermine Nigeria's increasing authority within Anglophone Africa.[35]

France's weapon deliveries intensified considerably in the summer of 1968. On May 24, the Biafran rebellion lost control of the Port-Harcourt oil terminal, its only access to the Atlantic: It therefore seemed obvious that it had lost any chance of victory. The war was atrocious. The army's Russian Mig fighter planes, flown by Egyptians, were masters of the air and multiplied the carnage on roads filled with Igbo refugees. The British secret service was very effective at helping to reduce the pockets of Biafran resistance. London's goal was to counterbalance the influence of the Soviet Union, which delivered weapons to Nigeria, and, especially, to preserve its access to oil in the face of the French.

The war killed between two and three million, a majority of them Biafran civilians, including many children starved as a result of the food blockade imposed by the Nigerian army. Biafran government posters recommended catching lizards, snakes, and rats. People ate flies and even wood. Starting in August 1968, the famine killed six thousand people per day.[36] In 1969, Biafra remained diplomatically isolated. The Foccart network, however, never interrupted arms deliveries except briefly, in March, in an attempt to force Ojukwu, the leader of the Biafran rebellion, to compromise with the regime in Lagos (Nigeria's political capital). But, against all logic, the latter refused. The clandestine arms shipments resumed, always with a green light from General de Gaulle.[37]

Following the summer of 1968, faced with inevitable military defeat, Paris tried to win a media victory, to win over the international community. Without doubt, its ultimate goal was to obtain a ceasefire that would allow France to turn its investments, made possible by the Biafran rebellion, into black gold profits. More than in any other Western country, citizens in France were shaken by the reports of many journalists, whose transport to Gabon was facilitated by Jacques Foccart.[38] In the minds of many, the gruesome images of Biafra's children, their stomachs inflated, evoked the concentration camps. The French secret service decided to make good use of the horror aroused in public opinion. It leaked to *Le Monde* and to other French newspapers that what was happening in Biafra was nothing less than a "genocide," pure and simple. The testimony of the French doctors of Doctors Without Borders in Biafra, who were on their very first mission, contributed to support the propaganda. Maurice Robert, in charge of French secret service operations in Africa, later confided: "We wanted a shocking message to heighten public awareness. We could have used the word 'massacre' or 'crush,' but 'genocide' seemed like it would have the most impact. We communicated to the press specific information on Biafra's losses, so they quickly adopted the expression 'genocide.'"[39] A term that, according to the United Nations, in no way reflected the reality: "butchering" or "carnage," certainly. "But genocide, no," wrote the French journalist Jean Guisnel.[40]

On January 11, 1970, Colonel Odumegwu Ojukwu fled Biafra for the Ivory Coast. On January 14, Biafra capitulated and ceased to exist. The French petroleum companies almost managed to preserve their essential interests. Safrap had to watch the National Nigerian Petroleum Company acquire 35 percent of its capital in 1971, and then 55 percent in 1974, when Safrap was renamed Elf Nigeria. In 1975, Elf Nigeria's production reached eighty thousand barrels per day, and continued to increase thereafter, although it lagged far behind the production of Shell and BP.[41] Several French oilmen argued that the Biafran war was fanned by France's secret service, against the will of Safrap's leaders.

Libya and Colonel Gaddafi's Ultimatum: The Wheel Turns in OPEC's Favor

When, at the end of the 1950s, King Idris of Libya chose to open the door wide to small, independent American oil companies, he overlooked the fact that he was introducing a spark that could eventually explode the Seven Sisters cartel. In 1955, the minister of Libyan oil, Fouad Kabazi, chose to auction a petroleum concession that had been divided into many pieces. "I did not want my country to begin as Iraq or as Saudi Arabia or as Kuwait. I did not want my country to be in the hands of one oil company," he said.[42] Fouad Kabazi was perfectly aware that the major American companies maintained the status quo regarding the Arab countries' petroleum production, including by purchasing concessions they did not plan to drill. The Libyan idea to sell concessions of very limited size allowed an independent company that discovered oil to put pressure on its neighboring major oil companies, telling them: "Look here; your neighbor has discovered oil. You are almost on the same structure. Come on, now; try to drill."[43] The Libyan oil was too promising to afford the major companies the luxury of not playing the game, and risking being excluded from the oil party in Tripoli.

In 1959, Standard Oil of New Jersey was the first company to extract the excellent Libyan crude. But ten years later, after Tripoli joined the OPEC cartel in 1962, there were independent American companies such as Occidental, Continental, and Marathon that controlled more than half of Libyan production.[44] It was an unprecedented situation, and the Seven Sisters measured its implications too late. For the first time outside of the United States, independent US oil companies found themselves with free reign in the face of the majors. Fueled by the greed of these independents, the lightning-fast growth of Libyan production created a gap that continued to widen throughout the 1960s, between the "posted price" of the oil majors, which remained imperturbably between $1.70 and $1.80 per barrel, and the free-market price at which the independent companies sold Libyan crude oil, which fell as low as $1.30 per barrel. However, the overwhelming market shares of the Seven Sisters allowed them to keep the upper hand. They continued to impose their law, all the more easily because the corruption of King Idris's regime initially seemed to allow them to keep a tight collar around the necks of their small independent competitors.[45]

Everything changed on September 9, 1969, with the arrival in power of twenty-seven-year-old Libyan officer Muammar Gaddafi, after a military coup that, without bloodshed, ousted King Idris the night of August 31 to September 1. This coup may have preempted by a few days another planned coup developed by the CIA. According to two former American intelligence executives, when Gaddafi's soldiers approached King Idris's guards, they believed that, according to plan, these were CIA agents, and relinquished their weapons without being asked.[46] Gaddafi himself was surprised at the ease of the operation.

Three months after the coup, at a summit of Arab leaders in Morocco, Colonel Gaddafi announced, in his own distinctive style, that he was about to shake up the established oil order: He publicly aimed a handgun in the direction of King Faisal of Saudi Arabia, the most powerful Arab despot owing allegiance to the major American oil companies. Gaddafi made it known to twenty-one companies present in Libya that he wanted a 40 cent increase in the barrel price of Libyan oil. He proved his seriousness by discussing with Moscow the potential opportunities on the other side of the Iron Curtain.

At the outset, Gaddafi was also an unexpected ally to the State Department's new "Mr. Oil" in Washington. Gifted with a smile as charming as it was enigmatic, Jim Akins was the same man who, three years earlier, had fired an American diplomat in Tehran who sided with the shah once he came to understand how the oil majors were limiting Iranian production. An accomplished

Arab scholar, raised in an Ohio Quaker family, Akins played a crucial and at first confusing role in the chain of events that began in Libya in 1970 and that led to the oil crisis of 1973. At forty-four years old (the same generation as George H. W. Bush), Akins immediately supported Gaddafi's grievances, legitimate in his view, and strongly encouraged the American petroleum companies to move quickly to increase the barrel price of Libyan oil.[47]

Jersey Standard's directors rejected Tripoli's demands and refused to increase prices by more than 5 cents per barrel. Gaddafi then skillfully targeted the independent US company that had invested the most in Libya: Occidental Petroleum (Oxy). Led by Armand Hammer, a seventy-one-year-old Jewish New Yorker of Russian descent, Oxy was too small to afford to forego Libyan oil. In May and June of 1970, Gaddafi forced the company to reduce its production by half, which in a few weeks went from eight hundred thousand to four hundred thousand barrels per day.[48] Tripoli accused Oxy of extracting crude oil too rapidly, which risked damage to one of the fields it had been authorized to drill. Armand Hammer implored Jersey Standard's directors to help him and save Oxy from bankruptcy by giving him access to concessions outside of Libya. But Hammer wanted too much, and Jersey Standard refused.[49] In August, the old tycoon had to face Gaddafi alone, in negotiations. Each evening, he boarded his private jet to escape Tripoli's scorching heat and return to his Parisian residence. He ended up conceding to Gaddafi, and on September 4, 1970, announced that Oxy would pay 30 cents more per barrel and be taxed more heavily.[50] One after another the other principal independent companies operating in Libya—Continental, Marathon, and Amerada Hess—also capitulated, under the threat that otherwise, their wells would be closed.

The oil majors refused to participate in the debacle. However, for the first time in their history, they found themselves out of step, between the independents, which sold their crude oil much more cheaply than they did, and Colonel Gaddafi, who required that they pay much more for Libyan crude oil. A grave threat weighed on them from that moment, stemming from "the whole nexus of relationships between producing governments, oil company and consumer," Shell's CEO later commented.[51] In Manhattan, from its offices at the summit of the Chase Manhattan Bank, the Seven Sisters' lawyer, John McCloy, tried to orchestrate a counterattack. Three successive American presidents had granted him that leeway, disregarding the antitrust laws.

Three days after Armand Hammer announced his agreement with Gaddafi, McCloy managed to convene a meeting of representatives of all the Seven Sisters, with Jim Akins representing the Nixon administration, at the State Department in Washington. The British companies, Shell and BP, were more inclined to play hardball. Shell proposed that the Sisters challenge Gaddafi to nationalize his industry, something the company was sure Libya would not dare to do. But Jim Akins was hostile to anything that might provoke Tripoli and asserted that US allies in the Persian Gulf would refuse to follow Gaddafi. A British oilman present during this decisive meeting recounted it this way: Akins "was hypnotised by the Saudi Arabians. He said that there was no question of Saudi Arabia following Libya. I said you must be joking and nearly walked out."[52] In this fateful instance, Jersey Standard's president, John Jamieson, openly refused to take part. SoCal and Texaco adopted the accommodating attitude advocated by Jim Akins and agreed to comply with Gaddafi's conditions, no doubt less concerned by their limited interest in Libya than by their future in Saudi Arabia.[53] At this point, the still-recalcitrant Sisters had no choice but to concede.

Gaddafi's success was a "source of embarrassment" for the other OPEC countries, confirmed an oil cartel official who witnessed these key weeks.[54] Subsequently, one after another, the "frogs" escaped, to use the image John McCloy himself used in the report he presented before the US Senate four years later.[55] Iraq, Algeria, Kuwait, and Iran in their turn demanded a price increase. Only Saudi Arabia held back from this action, as Jim Akins had said they would. And on December 9, 1970, a year and three months to the day after Gaddafi's coup, the OPEC countries met in Caracas to adopt a joint resolution in which they showed unprecedented pugnacity. This resolution, No. 120, instantly became famous in the Western consulates, because it advocated a general increase in prices, as well as the immediate opening of negotiations by each of the OPEC's member states.[56] It amounted to the second and true birth of OPEC.

These radical new circumstances blossomed into something more important: the very fruit of the spring of oil civilization. The year 1970 marked an essential turning point. After a quarter century of feverish economic growth, world demand for crude oil had finally caught up with the pace of crude oil extractions. The thirst for petroleum, always growing, especially in the West, caught the oil companies unprepared.

The wheel had turned. It would soon be the OPEC countries' turn to dictate their law. But if they had this capability, it was not because they themselves had the foresight to increase their political weight in the face of the West; the 1967 fiasco of the Six-Day War oil embargo was proof of that. Rather than any of the players' political will, it was the natural distribution of intact crude oil reserves that rotated the wheel and soon reversed the nature of the global oil market.

Among OPEC's members, those who most benefitted from the new nature of the oil market were those that were at once the richest in crude oil and the least populated: those able to extract a lot of black gold and who consumed very little themselves. These were the autocratic states of the Arab deserts and of course, first and foremost, the Royal House of Saud.

Freed of many constraints by the matrix of its black gold industry, humankind saw its material needs grow ever faster: not in calmly linear fashion, but explosively. In one generation, from 1945 to 1970, world consumption of crude oil accelerated at a breathtaking rate; the curve soared, but in 1970, for the first time, something seemed to impede its trajectory. In September 1969, Albert Allen Bartlett, a physics professor at the University of Colorado in Boulder, taught a new course that he called "Arithmetic, Population and Energy."[57] It was a course he was invited to teach countless times during the decades that followed. In it, he explained how a constant and apparently modest annual growth percentage actually led to faster and faster growth—in other words, exponential growth. "The greatest shortcoming of the human race is our inability to understand the exponential function," cautioned Bartlett. In mathematics, an exponential function is one that grows increasingly quickly—and infinitely—toward an invisible limit that it never quite reaches. Mathematicians call this limit an asymptote.

The baby boom generation awakened to a historical experience both irresistible and ensnaring: the asymptotic experience. This experience was born of the explosion of the libido, not just sexual libido but desire in the broad sense, illuminated by the sheer volume of fossil fuel energies released from the Earth by human industry; an explosion of vital energy that nothing seemed to curb. However, it was not without limits.[58]

EIGHTEEN

The Golden Childhood of the Oil-Made Man

Suddenly an unexpected phenomenon hatched: abundance. Was it first engendered by the desire for a new world, or by the material that allowed one to satisfy this desire? In the French collective memory, abundance was the postwar miracle symbolized by a GI perched on his tank, distributing chewing gum to liberated citizens—who had no idea that chewing gum was one of the innumerable products derived from petroleum. Up until the end of the 1960s, the United States reigned as masters of hydrocarbon production, as well as masters of production of petrochemicals made from naphtha and natural gas.[1] Over the next three decades, new, almost daily, wonders born of petroleum were taken for granted: high octane gas, jet aircrafts, bulldozers, tarmacs, Caterpillar equipment, artificial fertilizers, synthetic rubber, nylon, prophylactics, balloons, hard vinyl, PVC, acrylic fabric, Tergal, expanded polystyrene, polyester resin sailboats, neoprene, Plexiglas, glue, paint and polyurethane foam, Teflon, Kevlar, silicone, acetylene lamps and welding torches, polycarbonate glass for astronauts, Tupperware, Bic lighters, Spontex sponges, DAFA gloves, synthetic carpets, hula hoops, Frisbees, inflatable swimming pools, drinking straws, and lawn mowers—the budding middle class received these wonders as pure products of the fecundity of the American ethos, a fireworks offered by their creativity, the spirit of initiative and the vigor of the young, victorious nation.* No one bothered losing time considering the physical conditions that made possible this fireworks display, this new technical world.

* Manufactured in the United States for the first time in 1935, by American chemist Wallace Carothers in DuPont's laboratory, nylon was used during the war to manufacture parachutes. The Americans made an acronym out of it: "Now you lose, old Nippon."

What Fuels the Takeoff of Growth?

In 1960, American economist Walt Whitman Rostow compared the point at which an economy begins to grow to a plane's takeoff. His famous work, published that year, *The Stages of Economic Growth: A Non-Communist Manifesto*, combined economic theory with praise for capitalism, and imposed the standard explanation of the economic-growth phenomenon that was boiling at the surface of the globe. Once the "takeoff" is successful, economic growth must inexorably lead to the ultimate societal stage that the Americans were experimenting with, and the Europeans were beginning to discover: mass consumption. A brilliant, Yale-educated economist, Rostow helped design the Marshall Plan before being hired in 1955 by Nelson Rockefeller, then President Eisenhower's special assistant, to oversee strategic research on the USSR. In 1966, he became national security advisor to President Lyndon Johnson, playing a key role in the intensification of the Vietnam War.

According to Rostow, the desire to invest freely fuels the takeoff of the growth economy. And capital can be released only when the company and state adopt the ideal of progress, an ideal that cannot be dissociated from a desire for wealth. In Rostow's formulation, the preconditions for the growth of the industrial economy were strictly of a social and political nature: Only the human dimension counted in the model Rostow supplied.

Why not consider the literal nature of the fuel that facilitated the deafening takeoff of the post–Second World War growth economy? Such an iconoclastic, pragmatic vision began to emerge at the end of the 1960s, precisely as the golden age of the growth economy drew to a close, particularly in the works of three American authors: economist Nicholas Georgescu-Roegen, physicist Robert Ayres, and ecologist Howard Odum. The approaches to the economy that each developed independently converged into a new school of thought sometimes called biophysical economics. Biophysical economists emphasized the condition they believed necessary for economic growth to takeoff: the concrete possibility of radically increasing the amount of energy that flowed into the economic machine.

Since the Industrial Revolution, in every country, the correlation between energy consumption and economic growth has been extremely tight. Is it growth, driven by a desire to invest and get rich, that provoked an increase in energy demand, among other consequences? Or are growth and energy consumption reciprocally linked, moving in concert like the frigate and its supply ship? According to the fathers of biophysical economics, the correlation between the quantity of energy available to an economy and the intensity of its activity allows us to understand the metabolism of the industrial economy. For these economists, energy consumption and economic growth align and result in a kind of mutual spiral: If growth leads to an increase in energy needs, then available energy, depending on the degree of efficiency with which it can be used, may determine the limits for potential economic development.*

Robert Ayres eventually concluded that merely increasing oil production "due to discoveries [will] enable goods and services to be produced and delivered at lower cost. This is another way of saying that energy flows are 'productive.' Lower cost, in competitive markets, translates into lower prices which—thanks to price elasticity—encourages higher demand." More energy thus breeds more consumption, which creates more profit to finance the extraction of more energy: A spiral emerges, which leads to the depletion of energy sources.[2] Energy (and petroleum, primarily) is the liquid matrix of the industrial economy's entire growth phenomenon: Again, as in modern war, it is the great enabler. Without an abundance of available energy, no growth is possible. Western postwar powers spontaneously placed this parameter, overlooked by classic economic science, at the heart of their political and geopolitical strategies. And it turned out that the pace of economic growth in the 1950s and 1960s (for which the governments of the rich countries continued to sigh nostalgically) coincided with an unequalled period of oil production growth.

Energy abundance became the vital basis for the prodigious material abundance that spread through the United States, Western Europe, and Japan starting in 1945. West of Europe's Iron Curtain, between the end of the Second World War and the petroleum crisis of 1973 (the years France came to call the Thirty Glorious Years), countries did not experience a takeoff of exponential growth until Europe received a source of energy as tremendous as the one Americans celebrated during the Roaring Twenties. Europe's source turned out to be Arab and Persian oil, routed according to the strategy drawn up in 1945 by the US Navy and implemented, beginning in 1948, with the Marshall Plan. That oil restarted Europe's economic engine by increasing the flow of gasoline in its carburetor. Simultaneously, a similar scenario played out in Japan, where economic aid resulting from an agreement with the Americans, in a policy framework known as Reverse Course, helped the Nippon archipelago escape a permanent shortage of raw materials, primarily oil.

* This idea connected with the hard sciences, in particular the reflections on thermodynamics by physicist Ilya Prigogine (see infra, chapter 30), and subsequently was developed by many unorthodox social science authors, including American anthropologist Joseph Tainter, who had studied the "spiral energy-complexity" concept discussed in chapter 2. We will return to this decisive point in chapters 29 and 30.

The quantities of crude oil discovered on all continents were so great during the Thirty Glorious Years that they drove the Seven Sisters to maintain their secret pact, put in place during the Great Depression, so they continued to protect against overproduction woes and preserve their profit margins. However, these margins were scarcely noticed by consumers: black gold was so abundant that everyone could afford it during the spring of the consumer society. Gas even seemed like "a gift," in view of the glowing opportunities it provided to peasants (transformed into farmers by rototillers), motorists (now commonplace), industrialists, and bankers.

At that time, oil appeared so readily accessible that it became easy to overlook how indispensable it had become to every technical activity. Since oil was omnipresent and made the impossible possible, it was erased from daily concerns: It became sublimated, an almost virtual source of concrete positivity. Beginning in the 1950s, gasoline was available to almost everyone in the Northern Hemisphere (yet still only the military and the wealthy in some other parts of the world) a formidable number of "energy slaves": soon, on a daily basis, even the rich countries' most humble workers had access to a physical power equivalent to dozens of tireless men, for leisure as well as for work. Longed for since the dawn of the industrial era, the democratization of abundance was realized with the triumph of the petroleum industry. The era of the oil-made man was fully launched.[3]

1950 was the year of the great change. From this pivotal time, the economy almost tripled its growth rate. For a century, world production had grown at a nearly static rate of less than 2 percent per year. Five years after the Second World War ended, the economy shifted gears. Global growth became incandescent, with an unprecedented annual rate of around 5 percent per year, which did not decline until the crises of the 1970s. The value of world production and industrial production, in particular, nearly tripled between 1950 and 1970![4] US growth accelerated, but it was especially the European economy, bloodless after the war and revived by the Marshall Plan, that enjoyed the greatest boost, with annual growth rates sometimes exceeding 10 percent per year. This was a time of full employment and a flourishing "welfare state."

As far as oil was concerned, the takeoff was still more spectacular. World production accelerated steadily after Pearl Harbor. World consumption more than quadrupled between 1950 and 1970, rising from 11 to nearly 48 Mb/d (approximately 43 Mb/d outside of the USSR)! World production of concrete, in which fuel oil plays a key role, also quadrupled during the same period. Buildings were cheap, and water tower and other structures characteristic of the modern world were springing up everywhere, on both sides of the Iron Curtain and in the most populated contact zones of the developing world. Each city surrounded itself with branching roads of cement and asphalt. Another irrefutable change marked the 1950s, although at the time it was ignored: the curving trajectory of concentrated carbon dioxide rising in the atmosphere, which was systematically measured from Hawai'i's Mauna Loa volcano observatory beginning in 1958.

Western Europe's crude oil consumption multiplied tenfold in the space of twenty years, reaching 12 Mb/d in 1970. However, it did not catch up with the US consumption rate, which doubled in that same period, reaching 14.7 Mb/d in 1970.* But from 1950, American thirst for more and more oil clearly exceeded Uncle Sam's production capacity, although America was still increasing its output: In 1970, more than ever the world's premier producer, the United States still had to import 3.4 Mb/d, almost double French consumption at the time.[5] Until 1940, in the United States, oil's main function was still industrial and domestic heating. After the war, motor power rose to the top, as the number of cars, planes, ships, locomotives, construction equipment, and combines exploded.[6]

Everywhere in the industrial world, the lightning-fast rise of petroleum needs overshadowed the continued development of other fossil fuels. After the war, coal still supplied the world economy with twice as much energy as oil did. By 1970, however, the situation had fully reversed.[7] Oil use had begun to outpace coal use in 1960.[8] In 1951, the number of diesel locomotives in the United States for the first time exceeded those powered by steam.[9] Elsewhere, the transition was encouraged when, half a century after the great British precursor, several major industrial nations saw the coal mines that allowed the initial boom begin to decline, lacking intact deposits that could be extracted economically. This decline began in 1957 for France and Belgium, in 1958 for Germany (only for anthracite, the most sought-after form of coal energy), and in 1966 for Japan and South Korea.[10] In France, the production decline was quite rapid: Industrial authorities planned to close mines they were forced to subsidize when coal extractions, increasingly difficult, became too expensive.[11] What would have become of these industrial nations if they had lacked the disposable capital and strategic recourse necessary to obtain their share of energy sources from afar?

Not only was oil more effective than coal, but it was becoming less expensive. And less polluting. The coal used for heating was responsible for many fires and the famous "killer smogs" that for a century had stifled

* The United States became crude oil importers beginning in the 1920s.

poorer neighborhoods. In December 1952, for example, a dreadful cloud of pollution killed thousands of Londoners. Starting in early 1950, London, New York, and many other large industrial cities began, little by little, to ban coal. On cargo ships, heavy fuel systematically started to replace coal: The branching of the world's global cargo network continued to evolve, thanks to oil. Yet, oil no more fully replaced coal than television replaced the radio. Dethroned as technological society's first energy source, the old king of energy nevertheless continued to advance, although it advanced much more slowly than black gold. More and more, coal was used to turn the turbines of the electric power plants in major industrial centers and urban areas. Natural gas, a form of light hydrocarbon that was previously little-used except in the United States and often simply burned as flares in oil refineries, began to displace coal for heating. Starting in 1950, methane developed almost as quickly as did its elder brother, oil. In total, the fossil fuels consumed in 1970 (oil, coal, and natural gas) weighed three times as much as in 1950: Consumption of energy and global economic output had increased by the same order of magnitude.[12] The growth of each reached the same exponential rate. In 1970, the first nuclear plants had hardly broken ground in the United States, hydroelectric dams occupied a secondary role, and other renewable energy sources were nonexistent. Coal, oil, and gas: The sediment of organic materials decomposed over hundreds of millions of years was the almost exclusive energy source fueling the takeoff of mass consumption and capitalist society.

Exponential Growth and Asymptotic Experience

Nuclear energy was perceived as the logical and even necessary successor to coal and hydrocarbons. Admiral Hyman Rickover, father of the US nuclear fleet and a central figure in the development of the first atomic power plants, declared on 1957: "For more than one hundred years we have stoked ever growing numbers of machines with coal; for fifty years we have pumped gas and oil into our factories, cars, trucks, tractors, ships, planes, and homes without giving a thought to the future. Occasionally the voice of a Cassandra has been raised only to be quickly silenced when a lucky discovery revised estimates of our oil reserves upward, or a new coalfield was found in some remote spot. . . . Fossil fuels resemble capital in the bank. . . . A prudent and responsible parent will use his capital sparingly . . . in order to pass on to his children as much as possible of his inheritance. A selfish and irresponsible parent will squander it in riotous living and care not one whit how his offspring will fare." He concluded, prophetically: "I suggest that this is a good time to think soberly about our responsibilities to our descendants, those who will ring out the Fossil Fuel Age."[13]

During the 1960s, the US government conducted approximately thirty tests to discover whether the atomic bomb could drill new passes, expand the Panama Canal, or extract hydrocarbons still trapped in their bedrock. But this project, called Plowshare, did not lead to any concrete industrial development—due in part to radioactivity. The atomic bomb did not compete with TNT or diesel on public construction projects. The first civil nuclear installations proved to be as lengthy to develop as they were costly. In the United States, as in the Soviet Union and France, engineers gradually realized that the atom could play little more than a subsidiary role to oil. Despite the discovery of the theory of relativity and all the other giant discoveries by physicists, the naturally flammable liquid standby remained the main source of industrial energy—the most efficient, adaptable choice.

In the first scenes of 1964's *Doctor Strangelove*, Stanley Kubrick filmed a nuclear bomber sucking kerosene from its supply aircraft as an infant sucks milk from its mother's breast. The energy at humanity's disposal, thanks to the fantastic growth of petroleum extractions after 1945, allowed people to accomplish an astonishing variety of tasks, with new technology making things that had hitherto seemed improbable commonplace. The field of possibilities that opened next spread far beyond the direct technical consequences of the energy influx. Humankind had deployed the capacity to move around, leading us, step by step, to a cascade of societal consequences that erupted far from their original material source. The release of petroleum energy multiplied the effects of all technological activity. The intensity and expansion of the human experience was heightened, and it became easier and easier to build megacities surrounded by vast suburbs; to drill ever-deeper mines and wider tunnels and canals; to erect new bridges across broader rivers; to blast through hills, creating more direct roadways; to raise gigantic shipping terminals above the sea and the swamps; to enable one person, singlehandedly, to plow formidable expanses of fields and cut down forests with a chainsaw; to transport frozen food from afar, send satellites into space, and even transport the minerals, materials, and workers necessary to manufacture nuclear power stations and nuclear bombs. Countries were developing an unquenchable thirst for black gold, based on the fact that this natural result of biological and geological evolution cost no more than a small percentage of the value of the products that multiplied thanks to it.

Brought on by logarithmic growth, the black gold addiction intensified first, of course, in the United States. But the Keynesian multiplier effects permitted by energy abundance produced their most spectacular results in countries ravaged by the Second World War. Until Japan's defeat, Japanese oil was almost exclusively reserved for the army. In 1945, dozens of Japanese cities, including Tokyo, were razed by the petroleum-based, incendiary bombs of the United States. Until oil was widely introduced by Jersey Standard, Mobil, Shell, and Gulf Oil, at the end of the 1940s, it occupied only a marginal role in the Japanese economy (providing less energy than wood). But in 1970, petroleum imports provided more than 70 percent of the energy that the Japanese consumed. Without the influx of this cheap oil, there would have been no metamorphosis of the Japanese economy. Almost nonexistent at the beginning of the 1950s, Japanese automobile production exceeded four million vehicles by the end of the 1960s, of which nearly a million were intended for export. "The German and Japanese miracles were based on improved institutional arrangements and cheap oil," summarized American economic historian Alfred Chandler.[14]

Energy concentrated and recovered by sporadic, explosive discharges: Such is the principle of the power of animal life, according to the definition proposed at the beginning of the twentieth century by French philosopher Henri Bergson.[15] In the spring of the oil civilization, the nodes of power in human society swelled like never before under an unprecedented energy influx; the explosions discharging this influx absorbed the space around it. Soaked in oil, the largest power nodes (the US Navy, General Electric, Mitsubishi, BASF, and so on) quickly encompassed virtually the entire world. By proliferating opportunities to utilize raw materials (including energy sources), causing avalanches of chain reactions, oil upset the Earth's very nature. Humanity's footprint became brutally titanic.

By discharging the fiery energy of the thermo industry, the dazzling advent of energy abundance unleashed a consumerist libido; few people lingered over the magic at its source. When understood as vital energy, this libido cannot be disentangled from the ever-increasing stream of joules that begat its myriad fruits: Both marked the spread of the very same phenomenon.

Everywhere, in all human domains, starting with politics, the "technocrat" became king. A French intellectual related the 1959 meeting in the United States between Vice President Richard Nixon and the first secretary of the Soviet Communist Party, Premier Nikita Khrushchev, who confessed to being properly blown away by the prowess of American industry: "It was clear that they spoke like two men who could fight, but preferred to dazzle one another, and showed each other their automobiles . . . gesturing, hilarious (the great men of State are clowns gone wrong), Khrushchev waved his hand under Nixon's nose: 'In ten years, we will have surpassed you and will say, *bye bye*!' This was not philosophical discourse, and certainly not poetic; they did not communicate, they expressed nothing, but they agreed on the object, they competed in the same arena, which was that of technical power."[16]

Almost all who lived through the advent of abundance, carried by the new rivers of energy and the exponential growth of the 1950s and 1960s, had thoughts of "fullness," "freedom," "fun," and even—in unanimous consensus about the welfare state—"sharing," for this abundance appeared limitless. For Walt Whitman Rostow and many others, the immense wealth produced starting in 1950 represented progress, a promising, infinite progress spawned by humankind's desire and artistry. In the hidden gap between infinite progress and finite physical conditions, the realization of that progress created an illusory experience for humanity: an asymptotic experience by which consumerist desire, the true and only necessity of progress, purported that it could grow human activity exponentially toward a limit it continually approached but never reached.

Under the exciting effects of heat, the atoms in iron increasingly agitate, and more closely mimic a ceaselessly evolving magnetic polarity. In the same manner, social turbulence appears to accelerate, the phenomena of prevailing trends are exacerbated, desires proliferate, and competition intensifies as the abundance of energy expands. This analogy could be more profound than it seems; human life was activated, excited, and could inflate hope and ambition above and beyond any prior expectations.[17]

This was the golden age of the oil-made man, his innocent childhood. Paragon of American success, the great industrialist Henry Kaiser, who was behind the Second World War's Liberty ships and Tokyo's immolation, invoked the "liberty of abundance."[18] The fantasy of progress and limitless exponential growth were expressed everywhere, magnified by the pride of scientists and mendacity of publicists. In 1945, American engineer Vannevar Bush wrote a report for President Truman called *Science, the Endless Frontier*. This wise mind, Freemason, authorizing officer of the "Manhattan Project" and US postwar science policy, confidently preached the dogma of his time: There was no physical limit that science could not defeat. He wrote: "Advances in science will . . . bring higher standards of living, will lead to the prevention or cure of diseases, will promote conservation of our limited national resources, and will assure means of defense against aggression."[19] On one point at least, the postwar dogma proclaimed by Vannevar Bush was consistently debunked in the years to come: In the great celebration of abundance, scientific advances encouraged each generation to consume more of the "limited resources," not less. Very few grasped the warning given a little later by iconoclastic American economist Kenneth Boulding: "Anyone who believes exponential growth can go on forever in a finite world is either a madman or an economist."*

The illusion of an asymptotic experience was a focal point for the "American dream," defined in 1931 by American essayist James Truslow Adams: "But there has been also the American dream, that dream of a land in which life should be better and richer and fuller for every man, with opportunity for each according to ability or achievement. . . . It is not a dream of motor cars and high wages merely." Adams, however, later deemed it necessary to add: "The American dream, that has lured tens of millions of all nations to our shores in the past century has not been a dream of material plenty, though that has doubtlessly counted heavily."[20] After the war, this benevolent dream, emanating from the United States, exceeded the expectations of many (except for those subjected to the segregation that was also rampant around the oil wells, all over the globe, controlled by American capitalists). As a slogan for America's new way of living, a New York advertiser would have been able to say, with full irony, "And you know what happiness is? It's the smell of a new car, it's freedom from fear, it's a billboard on the side of the road that screams with reassurance that whatever you're doing, it's ok, you are ok."†

Cars, the "Gothic Cathedrals" of Modernity

Petroleum allowed the fulfillment of phenomenal bursts of individual potential, across all realms where man-made machines discharged great power. People could manage ever more machines fed by oil—energy slaves. The most notable among those machines, of course, was the car.

Right up to the end of the Second World War in Europe and to some extent for decades in the United States, following the plummet of the precocious asymptotic experience of the 1920s, the automobile remained the prerogative of the rich (as well as the police and gangsters). As the 1950s

* In "The Economics of the Coming Spaceship Earth" (1966), Boulding notably wrote: "The closed economy of the future might similarly be called the 'spaceman' economy, in which the earth has become a single spaceship, without unlimited reservoirs of anything, either for extraction or for pollution, and in which, therefore, man must find his place in a cyclical ecological system."

† From the American television series *Mad Men*, created by Matthew Weiner and produced by Warner Brothers (2007). *Mad Men* portrayed with panache the state of mind of white America at the beginning of the 1960s, through the history of a fictitious New York advertising agency.

began, perceptions of automobiles markedly changed, and their new mark on society was highly visible in movies of the time. Popular heroes no longer struggled with motor vehicles just to survive (as in Humphrey Bogart's truck in *They Drive by Night*, by Raoul Walsh in 1940; or Yves Montand in a truck carrying 400 kilos of nitroglycerin meant to extinguish an oil well fire in *The Wages of Fear* by Henri-Georges Clouzot in 1953); instead, they lived through them. As early as 1955 in Nicholas Ray's *Rebel without a Cause*, the car became the extension of the ego, of life, of death, as the protagonist (played by James Dean) raced, pedal to the metal, to the cliff's edge. Dean inspired the American youth of nascent "suburbia," a world of spacious yet narrow-minded white suburbs designed for the car, an antidote to the sad overcrowding of urban industrial centers. The automobile equated with escape—whether in *Rebel without a Cause* or in Jack Kerouac's 1957 book, *On the Road,* the key novel of the Beat Generation.

And then there were Elvis Presley's pink Cadillacs—so sexy! The King possessed three, including the mythical Fleetwood Series 60, 1955 Cadillac he refers to in "Baby, Let's Play House," his first hit on the American charts. After all, the expression "rock & roll," though it often evokes the sexual act, can also mean to roll, to leave. The fins on the back of the Cadillac were inspired by the Lightning P-38, the futuristic double fuselage aircraft of the Second World War.[21] In front, round headlights were outlined by eyebrows of chrome. With extravagant curves, ostentatious comfort, high octane ratings, and vibration: Everything combined to confer, on the big American cylinders of the 1950s, an erotic charge full of gaiety and confidence, which will remain forever unsurpassed. The Cadillac's forthright, arresting beauty marks the historic origin point of man's love for his car, a sign of the blooming consumer society. The automobile was the vector of an irresistible call to dodge the dullness of everyday life. It embodied the desire for freedom. This desire was profound, even unreasonable: When the Ford company introduced the seat belt in 1955, "Everybody was opposed to it: you could not get people to use seat belts," according to one of Ford's directors at the time, Robert Strange McNamara, the future Pentagon boss.[22] New roads offered freedom of mobility to everyone, and beginning in the 1960s, distant California became the most populous American state. In the southern United States, Elvis Presley's disciples were the most ardent preachers of automobile worship, elevating it to the rank of quasi-religion. In France, those aspiring to become middle class sang along with Charles Trenet in 1959: "We're Happy on the National 7," France's highways leading to the French Riviera.

This feeling of liberation emerged alongside the construction of a society that was wholly dependent on the automobile. A vote, a salary, a car: In 1956, President Dwight Eisenhower launched the construction of a network of highways across the United States. It was the largest public works project in history. General Eisenhower prided himself that it represented a quantity of cement equivalent to "eighty Hoover dams or six sidewalks to the moon."[23] Soon this project would powerfully "change the face of America," sculpting its cities, its countryside, and its citizens' modes of living, and it would be imitated everywhere in Europe.[24] The United States' 66,000 kilometers of superhighways not only connected all major cities but also closed the distance between the offices and factories of the cities and the vast regions of suburban neighborhoods, whose population reached at least eighty-five million inhabitants between 1950 and 1970.[25] Detroit, the General Motors city, enjoyed the highest standard of living per capita of all US cities. In the image of Great America, all the wealthy nations set about to proliferate their own highway systems.

Once again, we do not really know whether industry or the state led the dance. Charles Wilson, Eisenhower's defense secretary and the man he chose to mastermind the construction of America's freeways, had chaired General Motors from 1941 to 1952. During his installation in January 1953, a senator asked Wilson whether there wasn't a conflict-of-interest risk—whether as head of the Pentagon he would be able to make a decision contrary to General Motors's interests. Charles Wilson's response is still famous today: "I cannot conceive of one because for years I thought what was good for the country was good for General Motors, and vice-versa. The difference did not exist. Our company is too big."[26]

At the heart of the "highway lobby," General Motors certainly did much to promote the development of road transport. On January 3, 1951, the US Supreme Court upheld the conviction of a primary worldwide construction company as well as a group of firms, all completely aligned with the oil industry, for having conducted in forty-two large American cities (including Los Angeles, Saint Louis, Baltimore, and Philadelphia) a "conspiracy to monopolize the transit business for their own oil, tires and buses."[27] During the 1940s, the maker of Mack trucks, General Motors, the Firestone tire manufacturer, Phillips Petroleum, and Standard Oil of California had secretly funded an urban transportation company called National City Lines in order to obtain, according to a 1946 memo from the US Department of Justice to FBI Director J. Edgar Hoover, "the elimination of electric railway cars in city transportation controlled by these companies."[28] The FBI had investigated many corruption cases of municipal councils who had deliberately allowed their networks of trolleys and electric rails to deteriorate; in Florida, for example, elected officials received new Cadillacs as compensation for their complicity.[29] Once National City Lines gained control of municipalities, the trains and trolleys were eliminated within a few months, the rails were dismantled, and train cars were abandoned or burned (in some cases, with the Mafia's help).[30] The number of electric public transport lines was reduced and ticket prices were raised, despite broad public indignation, as in Los Angeles.[31] The profits of National City Lines were reinvested not in public transport but in trucking and airlines.[32]

But this scheme only pushed the wheel of an irresistible movement a little further. This was progress. The number of motor vehicles circulating in the United States rose from 45 million in 1950 to more than 100 million in 1970. This growth was almost as rapid as in the rest of the world, where the ascension of the automobile fleet was even faster, rising during this period from fewer than 20 million to more than 150 million vehicles.

Everywhere, the car had become the most trusted mark of success and social distinction. Should the worker who aspired to the middle class purchase the largest, most eye-catching car possible? The asymptotic experience fed on the irrepressible desire of individuals to proclaim their identities by the make and model of their cars: Tell me what you drive, I will tell thee where thou hast reached the magical asymptote of progress. Volkswagen's small, humble Beetle was soon adopted by a certain intellectual elite. In 1957, Roland Barthes wrote about the luxurious and futuristic Citroën DS: "I think that cars today are almost the exact equivalent of the great Gothic cathedrals; I mean a supreme creation of an era, conceived with passion by unknown artists, and consumed in image if not in usage by a whole population which appropriates it as a purely magical object. . . . In the exhibition halls, the display car is visited with intense, loving attention. . . . It takes a quarter of an hour to broadcast the Goddess, accomplishing through this exorcism the very movement of petty bourgeois promotion."[33] The car is a temple but also an alcove, a place of intimacy that protects and conceals you from the world, while you're waiting to arrive "elsewhere." In 1967, the car chase and escape filmed by Arthur Penn in *Bonnie and Clyde* left a lasting impression, giving birth to a separate film genre dedicated to this fantasy: the road movie. At that time in the United States, when there were drive-in theatres, almost 40 percent of marriage proposals occurred in cars and, without a doubt, so did a considerable proportion of first sexual experiences.[34]

Along the roadways, the landscape was filled to the brim with signs of consumerism. Service stations, motels, commercial centers, fast foods: Again, the United States was the pioneer, where the chevron, the mark of Standard Oil of California, mated with the golden *M* of McDonald's. In industry jargon, an oil well derrick is called a "Christmas tree."

Hydrocarbons and Population Explosion: Organic Chemistry, the Modernization of Agriculture, and the Green Revolution

Beyond the edge of the road, the landscape also changed. The time of open fields had arrived. The avalanche triggered by the influx of oil at the end of the war moved to the agricultural domain. Its most direct consequence: exponential world population growth, beginning, as did so many things, in 1950. If the number of humans on Earth more than doubled during the second half of the twentieth century, it was first and foremost because, during that same period, world production of wheat, corn, and rice tripled.[35] And if these crops knew such an escalation, it was above all thanks to hydrocarbons.

It was between 1950 and the beginning of the 1970s that the fastest growth of grain crops occurred. "In 23 years, farmers expanded the grain harvest by as much as during the 11,000 preceding years, from the beginning of agriculture until 1950,"[36] wrote American agronomist Lester Brown; an important actor in this agricultural upheaval, Brown became one of the world's leading environmental figures. This harvest expansion gave greater impetus to human development than even the invention of irrigation or of writing, in Mesopotamia, at the dawn of civilization. "Since 1950 four fifths of the world grain harvest growth has come from raising land productivity, with much of the rise dependent on oil," Brown, responsible for US foreign agricultural policy at the end of the 1960s, explained.[37]

Hydrocarbons occupied a decisive role in virtually all aspects of the modernization of agriculture: mechanization; massive development of irrigation; distribution of new, high-performing seeds; or tripling, within the 1960s, the quantities of fertilizers, from thirty million to ninety million tons.[38] The ammonia synthesis that enabled the manufacture of nitrogen fertilizer was produced in tremendous quantities with a base of naphtha and natural gas: Thanks to the Haber-Bosch procedure, nearly 40 percent of protein ingested by humans was soon hydrocarbon dependent.[39]

Fertilizer's two other ingredients, potassium and phosphorus, were excavated from potash mines and phosphate rocks scattered in a small number of countries around the globe, and then routed by cargo, train, and truck.

Most of the agricultural machines and irrigation pumps operated on diesel fuel. Without abundant petroleum, it was considered impossible to achieve the very high yields of the agriculture industry—and impossible to feed an exponentially growing population. Agricultural progress is also the selection of the most productive plant varieties, but this maximum productivity cannot be achieved without abundant energy.

Modern agriculture began in the United States, but it was only in 1951 that the number of tractors (five million) exceeded the number of working farm animals. An abrupt shift followed: In only three years, from 1952 to 1955, the energy American farmers had at their disposal to work their fields quadrupled, from fifty million to two hundred million horsepower. In 1970, this power reached three hundred million horsepower.[40] Oil was the preferred fuel source that Midwestern grain farmers used to feed their corn dryers; the corn was primarily destined to feed cattle. The production of meat—and beef in particular—never stopped increasing, altering people's diets. The American harvest (especially that obtained by seeds with high yields) had a great surplus. The first agricultural exporter, the United States consolidated one of the most important factors of their powerful geopolitics: the "green weapon."

The modernization of agriculture overtook Western Europe in the 1950s, in particular through the Marshall Plan and later, in 1962, through the implementation of what became known as the European Union's Common Agricultural Policy. Productivity accelerated; it took fewer and fewer farmers to cultivate the land. There was a massive new wave of exodus from rural locales that began after the Second World War and continued until the early 1970s. It was a fractal avalanche of similar events, which shared as their first and necessary condition the advent of the abundant petroleum. In France, the agricultural labor force dropped from 38 percent in 1945 to only 10 percent in 1970, to the benefit of the industry and even more so services, which developed at full speed. In only a few years, a thousand-year-old agricultural society was swallowed up. One no longer spoke of "cultivating" a field; the term in force among technocrats and large grain producers became "exploitation": farmers exploited the earth as one exploited a mine or an oil field. With abundance, the price collapsed, it was necessary to go into debt, to modernize, or to leave for the factory. Superfluous, millions of horses were delivered to slaughterhouses. The landscape was turned upside down: The magnitude of exploitation increased, waves of land restructuring removed borders to create more open fields, tens of thousands of kilometers of fences around fields were razed to the ground. In many countrysides, the forests were soon just memories. The consumption of beef cattle, whose rearing required two to three times more vegetable proteins than pork and poultry did, was no longer a luxury. The worker who in earlier days labored to "win his bread" worked to "win his steak." Food could be transported further, and kept longer, thanks to refrigerated cargo ships and trucks. As a result, the human diet evolved once again: Gone were stored foods such as salted pork with lentils, or cabbage and pickles during the winter; long live bananas and frozen foods!

The mechanization of agriculture spread to Latin America and Southeast Asia. The "green revolution" advanced. This broad program of research and agricultural aid had been initiated at the beginning of the 1940s by the Rockefeller Foundation. In 1941, the foundation created by the world's richest family financed a research program aimed at improving Mexico's corn. Their explicit intention was to defend Rockefeller investments in the country, only three years after the nationalization of Mexican oil. The program gave such good results that it was extended to other Latin American countries, and then to India in 1956. Starting in 1960 it went global, thriving thanks to the creation of many agronomic cooperatives across the capitalist developing world—for example in India, the Philippines, and Colombia. These centers were co-financed by the Ford Foundation, the US government, and the World Bank.[41]

The green revolution responded to a mixture of philanthropic, economic, and geopolitical objectives. According to American journalist Mark Dowie, who investigated Ford Foundation and Rockefeller Foundation policies, it amounted above all to a way they could "provide food for the populace in undeveloped countries and so bring social stability and weaken the fomenting of communist insurgency."[42] Many signs indicated that this strategy was effective. In India, Mexico, and the Philippines, the technical solutions of the green revolution, associated with birth control policies, were designed by the American government and the Ford and Rockefeller Foundations, as well as by their political allies in those countries, as an alternative to socialist-inspired agrarian reforms.[43]

Between 1950 and 1970, the wheat harvest doubled in the developing world, and this was only the beginning.[44] No other policy in history had ever had a broader, deeper impact than the green revolution. As with any massive evolution, it caused many chaotic and ambivalent effects. It ended most pervasive famines, in particular in India. It was also responsible for a global population explosion. In Asia, Africa, and Latin America, it provoked a rural exodus beyond any Europe had ever seen. The oil civilization spread, swallowing everything including primal forests. Its only stopping point was the geography inaccessible to industrialized agriculture: marshes, steppes, and arid foothills. These became an inescapable universe for those repelled from society—the restless immigrant population, made superfluous, many of whom lost their reason for being.

Millions and millions of displaced farmers and their children reappeared and were absorbed in slums, where often more than half of the urban population of the southern countries piled into tight quarters.[45] Ciudad Juarez, Mexico City, Bogota, Caracas, Port-au-Prince, Lagos, Rabat, Algiers, Cairo, Nairobi, Ankara, Baghdad, Tehran, Karachi, Rangoon, Bombay, Dhaka, Jakarta, Shanghai—a planet of slums, incubating more suffering than even the Dickens, Zolas, Gorkis, and V. S. Naipauls of the world could render.[46] No proponent of agricultural modernization had wished for or predicted this tragedy. These chaotic, systemic consequences of energy abundance became the weightiest, most excruciatingly difficult outcome to undo.

A more foreseeable consequence: The green revolution also led to sharply reduced biological diversity in and around cultivated crops. It put the ecosystem and water resources under pressure. It made farmers dependent on expensive, high-performance seed, pesticides that were chemical pollutants, and equipment that ran on hydrocarbons. It perfected the ever-growing population's addiction to oil. With the upsurge in cargo freight, the green revolution disrupted the food habits of a thousand years. Nations beginning to modernize saw their urban populations explode—particularly true for oil countries such as Iraq, Saudi Arabia, Nigeria, and Venezuela—as they became dependent on agricultural products transported over long distances.

In total, the population explosion made possible by energy abundance (a necessary condition for and a limit to material abundance) multiplied oil's grip on the world, largely unwittingly created by human industriousness. In turn, each new demographic surge, each new wave of consumerism, unceasingly primed the universal oil pump.

Yet ambivalence, once again, reigned. Arriving in the city, many had greater access to health care and modern hygiene. Through pharmaceutical chemistry, oil and natural gas were able to play an important role in the democratization of access to medicines everywhere in the world. Although it constituted only a small outlet for refineries, organic chemistry, applied to hydrocarbons starting in the 1930s, allowed the sale of mass quantities of the basic materials required for medicines (amines, solvents, and the like) to be sold inexpensively to pharmaceutical laboratories after the war.* Plastic's widespread availability to hospitals led, among other things, to radical progress in the use of prophylaxis (gloves, syringes, and nonwoven disposable textiles, for instance) beginning in the 1960s.

* In 1973, the French oil company Elf Aquitaine created a subsidiary that later became Sanofi, one of the largest pharmaceutical groups in the world.

"The Sky's the Limit": Masters of Black Gold, Central Actors in Ending the Gold Standard

At the top of the social pyramid, the elite of the elite were transformed into the "Jet Set." This expression, appearing in the 1950s, referred to the class of people who had the means to travel by air as others traveled by bus. They were able to move at will in the rarefied air of high altitudes when the supersonic Concorde performed its first test flight in 1969. "The sky's the limit" of the asymptotic experience.

During the ascent to mass consumption, when what was once rare became commonplace, formal distinctions blurred. The primitive and necessary driver of all development—energy abundance, in which oil was the key—imposed the same efficient technical solutions on all quarters, determined building design and construction to a tremendous degree; a simple concrete building, made from cement likely heated with oil's flame, could look the same in Dhaka, in Kiev, in Luanda, in Sarcelles, and in Sacramento. As early as 1955, Claude Lévi-Strauss foresaw "an Asia of worker cities and urban housing projects, which it was about to become."[47] The French anthropologist saw this development as a stifling entanglement; the population explosion heralded "our future," the structuralist author wrote, the "evolution toward a finite world" for which Asia supplied the "image it anticipated."[48]

Such development could not be slowed down. In the United States, the hundreds of riots that broke out in the big cities between 1964 and 1968, mainly in the black ghettos, reinforced a broad political consensus around the need to finance the progressive policies of the Great Society desired by President Johnson.[49] Uncle Sam constantly needed more money, as the US Army's budget was responsible for almost half of US federal government expenditures.[50] From 1965 to 1968, during Operation Rolling Thunder, orchestrated by Robert Strange McNamara, the US Air Force's B-52s and fighter-jets dropped on North Vietnam more bombs than they had dropped on Europe during the Second World War.

In the United States, but also in Europe and elsewhere, the demand for the dollar continued to rise. But the cash printed remained limited because the dollar bill's value was anchored to US gold reserves, set at $35 per ounce in 1944 by the Bretton Woods agreements (with the ample participation of Rockefeller networks). The reserves did not contain enough gold to keep up with the debt that had to be incurred to follow the rhythms of the ascending global growth. The US credit crises of 1966 and 1969 aggravated tensions around housing access.

On Wall Street, the gold standard began to appear as an obsolete obstacle, useless and absurd for the purposes of financing of growth by incurring debt. At the leading edge of those who were convinced that wealth could be developed well beyond the constraints imposed by the yellow metal reserves were economists close to the black gold industry. In March 1967, David Rockefeller's Chase Manhattan Bank was one of the first economic institutions to officially propose abandoning the gold standard; in December that same year, one of Jersey Standard's principal economists, Eugene Birnbaum, followed suit.[51] President Richard Nixon heard their prayer. Four years later, on August 15, 1971, in summer's full torpor, without prior notice, he suspended the exchange between the dollar and gold, catching almost everyone off guard in his own State Department. With the stroke of a pen, the American president vanquished the only formal limit of physical reality (the amount of gold) blocking the tendency to move toward borrowing infinite money and expanding the country's debt. The unending continuation of growth made possible by the lever of debt found no other obstacles on the horizon. It seemed that this horizon could be extended indefinitely by the abundance of available energy. But, as we'll see in the next chapter, it nevertheless was threatened by a change, not economic but physical, that was no less historic for the American oil industry, one that occurred a few months before President Nixon put an end to the gold standard.

Barely a month after his unexpected financial coup de force, the American president appointed as head of international monetary affairs of the US Treasury another of Standard Oil's top financial executives: Jack F. Bennett, previously director of Esso International of New York City.[52] Bennett was put in charge of establishing the system of floating exchange rates, under orders of the man who masterminded the end of the gold standard, Paul Volcker, himself a former vice president of Chase Manhattan.[53] John Connally was appointed US secretary of the treasury; the former Texas governor was often referred to as the man of Big Oil in the heart of the Nixon administration.

Easy Rider, Promethean Aspiration, and Postmodern Descent

At the end of the 1960s, there seemed to be a symbolic pause before the fires of industry concentrated, everywhere, the grand convective movement toward similar expressions of postmodernity. The shadow of a doubt extended over this sinister and extraordinarily fertile era. Some let their hair grow, while the arrogant fins on the back of American autos receded before disappearing altogether. At the beginning of *Easy Rider* (1969), Peter Fonda, alias Captain

America, hides a wad of bills in the gas tank of his chopper, painted in the colors of the starry American flag, before hitting the road, hoping perhaps to save America from itself. Did the film's director, Dennis Hopper, have an intuition that the *ride* had become too *easy*?

But there was no turning back, surely not. In the service stations, the gas and oil merchants handed out enormous balloons as advertising. Children scrambled for them. The prodigal flow of energy given to the Western world after 1945 was metabolized by the baby boom generation, as hungrily as sugar by an anemic child. In France, Jacques Tati's 1967 film *Playtime*, in one scene, depicted a traffic jam in a roundabout as a joyful merry-go-round. The slogan "Jouissez sans entraves" ("Enjoy without chains")—voiced by Parisian students—became, through a misunderstanding, the only common rallying cry that survived the societal protests against mass consumerism and capitalism that erupted in France in May 1968. And then "each returned to his car," said French singer Claude Nougaro in Paris six years later.

The mixture of dream and horror elicited by this new world, birthed by the thermomechanical industry, reached a poignant apex in 1970 in Michelangelo Antonioni's film *Zabriskie Point*, a sort of swan song of the hippie movement. Involved in a deadly demonstration against war and racism, a white rebel steals a small pink airplane to escape Los Angeles. The young man flies over a city blistered by asphalt. Above the desert, he crosses the route of a young woman driving an adorable little 1950s Buick, already old-fashioned, toward the Valley of Death. For him, as for her, escape is impossible. As the movie ends, the young woman sees, in her imagination, the repeated explosion of a luxurious villa overlooking the desert, filled with everything modern man could need.

The bridgehead of the Texas oil industry, the city of Houston, was the ultimate manifestation of Promethean aspiration liberated by abundant energy. The city's gigantic refineries provided kerosene and liquid hydrogen, which enabled the no less gigantic Saturn V rocket to wrest itself from gravity. NASA's Manned Spacecraft Center in Houston, which helped American astronauts reach the moon on July 21, 1969, was built by Brown & Root: This Houston-based giant of petroleum services and public works also received important public contracts for the US Army in Vietnam. During the 1960s, the US government entrusted Brown & Root with another Promethean project, this one doomed to failure: Project Mohole, centered on a drill site at the bottom of the Pacific Ocean designed to reach the "Moho," the outer limits of Earth's mantle, with the far-reaching hope of someday exploiting its huge geothermal energy. When the first offshore oil platforms began to proliferate in the Gulf of Mexico, Texas's Lyndon Johnson touted Mohole as "the prelude to the future exploitation of resources at the bottom of the sea," just before the project was definitively halted for being too costly and technically infeasible.[54]

Before emerging as the leading edge of science's "endless frontier," Houston had long stagnated, despite its role as an oil terminal vital for the US economy. Born in a coastal plain, next to the first prodigious Texan drilling of the early twentieth century, the suffocatingly hot city did not host the offices of the largest oil conglomerates until the advent of air conditioners in the 1950s. Only these energy-hungry machines could convince executives of the oil companies and NASA to headquarter in Houston, often against their will.[55] Without air conditioning, Houston could not become one of the most populous metropolises (and the most congested) of the United States. In 1965, with blazing gas torches, the city inaugurated its Astrodome, a circular, covered stadium whose form was inspired as much by flying saucers as by Rome's Coliseum; it was at that time the largest air-conditioned space in the world.

Houston thus became the great node of global energy power—the laboratory of thermodynamic experience without fuel, space, or tax constraints. The largest refining and petrochemical complex on the planet, the city transformed itself in the 1960s, as city planners from Harvard University's design school described it, into a loose confederation of industrial profit centers that shared a nebulous network of infrastructure, economic and legal partnerships, and a landscape that sucked liquid money from the ground to regurgitate in solid-form monuments of glass and steel for the trusts of banks and of energy.[56] Virtually exempt from taxes since 1961, when Senator Lyndon Johnson drew NASA to Houston, the city embodied an "experimental model for almost all other developing cities in the world"; it was "the worldly city *par excellence*, . . . fully delusional in its belief that a city would be nothing more than a gigantic machine to conduct 'juicy business.'"[57]

Along another stifling coastline, Nigeria's economic capital, Lagos, arose almost at the same time as Houston, thanks to the same combination of oil and air conditioning on the world map of the modern economy. Africa's principal oil terminal was, like Houston, a chaotic generator of profits marked by a lack of organization and public services. But while Houston appeared as an opulent cultural desert, Lagos's character was that of effervescent misery, a murky, melting overflow of disappointed hopes. After the arrival in force of Western petroleum capital and the economic oil boom that succeeded Nigeria's independence in 1960, "[j]obs were plentiful, people were happy, people were free," from petroleum workers to merchants who sold scrap metal and plastic at the Jankara fair.[58] A musical movement flourished during this brief grace period: the Highlife, the mixture of Big Band, Afro-Cuban, and Calypso jazz, "symbol of progress, modernity and internationalism."[59] And then, very quickly, came the irresistible lure of oil money, the military coups, the Biafra civil war, and the establishment of a state that served the multinationals; the irrepressible misery, finally, exacerbated by the population explosion. Lagos never stopped collapsing under its own weight, like a termite mound; it was always too big for its own good but constantly regenerated by the feverish drive for black gold.

The spring of the oil civilization opened with the triumph of the Liberty ships of the Second World War and the music of Glenn Miller, Elvis Presley, and the Beatles—overflowing with energy and confidence. When this spring showed signs of waning at the end of the 1960s, as the giant freighters and supertankers articulated their technological web across the planetary body, a new musical genre was born in the heart of the desperate tumult of the Lagos port, built in 1969 in LA recordings: Afrobeat, a trance-like exercise in rejecting progress defined by and for others. This music of disillusioned resistance, created by Fela Kuti, had all the stigma common to the similar postmodern spontaneous popular movements echoing each other during the next decade throughout the homogenized urban world: reggae, hip-hop, and punk.

In a book published in 1979, French economist Jean Fourastié marveled at how the average standard of living had increased fourfold in France, in the course of what he called the Thirty Glorious Years. At the end of three decades of flamboyant postwar growth, Fourastié wrote: "The elevation of hope in life, the reduction of mortality and physical suffering, the material possibilities for the common man to access forms of information that were once inaccessible, of art, of culture, suffices, even if this common man often proves unworthy of these benefits, to make us think that the realization of the great hope of humanity in the twentieth century is a glorious era. . . . [It] did not emerge from departure from History, nor did it arrive into a future frozen into permanent prosperity and immutable happiness." He bemoaned the "few likeable but misinformed adolescents who . . . are free to criticize, even hate the 'consumer society,'" noting that they benefited from "the quality and style of daily life in France, of hygiene, of health, of social security, and all the modern means of transportation, information, communication" that society provided. "Their opinions," he concluded, "seem hasty."[60]*

The price of energetic abundance was just beginning to reveal itself.

* The major critics of consumer society were far from adolescent: Consider the works of Jean Baudrillard (the inventor of the concept), as well as Guy Debord and, on the side of the early adopters of radical ecological politics, Lewis Mumford, Jacques Ellul, Ivan Illich, Herbert Marcuse (Heidegger's former assistant, the emblematic figure of France's May 1968), and even the great mathematician Alexandre Grothendieck.

PART THREE

SUMMER

1970–1998

—◀ NINETEEN ▶—

OPEC: Scapegoat for a 1973 Oil Crisis Made in the United States?

The Arab potentates' intransigence and greed for oil have been held responsible for the historical upheaval that put a brutal end to the golden age of growth. The widespread idea that the OPEC countries were exclusively responsible for and the sole beneficiaries of the 1973 oil crisis, a stigma that has been the origin for so many stories of contemporary economic crises, is, at the very least, unfair. It is most definitely simplistic, perhaps even radically false.

In early 1971, as the American astronauts of the Apollo 14 mission played golf on the moon (with a six iron they had smuggled onboard clandestinely), wealth and world population continued to grow at an extravagant pace. The source of this roiling surge, the American oil industry—the extreme avatar of Plutous, the god of wealth—again spilled its horn of plenty on the most fertile economies in the world. One year earlier, US oil production had reached a new record. The United States was by far the largest crude oil producer in the world: With more than ten million barrels extracted from beneath its soil each day, the United States alone provided almost one-fourth of world production.[1]

In 1971, few doubted the "triumphant march of progress" or envisioned a limit to the power and abundance that was brewing, flowing everywhere on Earth, transforming the human experience. Yet fifteen years earlier, Marion King Hubbert, an American geologist with a brush haircut, had announced that between the end of the 1960s and the early 1970s, something intolerable, unstoppable, would occur: The land of black gold's horn of plenty would begin to dry up. As usual, no one heeded Cassandra.

US Oil Production Reaches a Historic Peak

On March 7, 1956, while George H. W. Bush launched offshore prospecting from the Port of Houston, a major conference in another Texas city, San Antonio, was bringing together some of the best analysts in the American petroleum industry. Hubbert, one of the most renowned geologists in Shell's American branch, was poised to take the floor when, suddenly, he was called to the phone.[2] While San Antonio's mayor completed his introductory speech, the geologist hastily left the rostrum. On the other end of the line, someone from Shell's New York office waited. The latter revealed his "considerable alarm" regarding the presentation Hubbert was about to give, about the future of the American oil industry.[3] "Tone it down," ordered the voice on the phone. "That part about reaching the peak of oil production in ten or fifteen years, it's just utterly ridiculous."

Hubbert hung up. His turn came. The scientist had no intention of changing a word of his speech. His mathematical analysis of the growth and decline of US oil fields had led him to a frightening but unavoidable conclusion: While it was growing more strongly than ever, US crude oil production was destined to peak somewhere around 1970 and then enter an irreversible decline, due to the lack of sufficient reserves. It is natural, the geologist explained: The exploitation of all finite resources will eventually exhaust them; it progresses at a certain pace until reaching its peak, when the exploitable resources are halfway depleted, and then begins to decline. Before a mystified assembly of proud Texas oilmen, the unemotional geologist drove home his message that humanity's petroleum endeavor was "a unique event in human history, a unique event in biological history. It is non-repetitive, a blip in the span of time."

Initially, Hubbert's analysis sent a shock wave through the heart of the industry. Hubbert later joked that, among Shell's directors, "I think there were some pretty red faces." His employers were unable to show that his figures were wrong. And yet his conclusion, which raised "grave policy questions with regard to the future of the petroleum industry," was redacted from his report. The prognosis itself was the recipient of incredulous sarcasm. Hubbert was forgotten during the 1960s, while the expansion of US black gold production continued even more vigorously. And thus it was that in 1971, as Hubbert had predicted, US oil production stopped growing! Production only dropped slightly in 1971, compared to 1970, but it was unprecedented since the crisis of 1929. And this time no economic crisis could explain it. In reality, during November 1970, crude oil production from the land of black gold achieved a historic record, with 10,044,000 barrels extracted on average per day.[4] The decline that followed changed the course of history.

The American "elephants," the principal, giant oil fields on which US industrial power depended, were half spent. They had reached their geological "maturity," the midpoint of their existence. One after another, their strength eroded. Oklahoma's fields were the first to decline, in 1968; Louisiana's followed, in 1970. US energy demand increased frenetically, month by month. Oil and coal shortages proliferated across the United States during the summer and the fall of 1970, to the point that in September, the American Public Power Association called for fuel rationing.[5]

The following year, American companies had to face the obvious: Texas itself, central pillar of the American energy empire, showed signs of weakness. In March 1971, the Texas Railroad Commission, responsible for regulating American crude oil production, for the first time decided to authorize setting all its valves wide open. "Texas oil fields have been like a reliable old warrior that could rise to the task when needed," Texas Railroad Commission Chairman Byron Tunnell said. "That old warrior can't rise anymore."[6]

A profound upheaval began—inexorable; neither economic nor political, but ecological. On June 4, 1971, President Richard Nixon personally addressed a special message to Congress on energy resources. In thinly veiled language, he announced the disaster that his experts and the industry had seen confirmed, month after month: "For most of our history, a plentiful supply of energy is something the American people have taken very much for granted. In the past twenty years alone, we have been able to double our consumption of energy without exhausting the supply. But the assumption that sufficient energy will always be readily available has been brought sharply into question within the last year."[7] In slightly obscure fashion, without naming oil, the American president warned his citizens that they might soon pay considerably more for energy: "One reason we use energy so lavishly today is that the price of energy does not include all of the social costs of producing it." Nixon announced a series of measures to accelerate the development of offshore oil, produce nuclear as well as "clean energy," and revive the exploitation of shale oil. But the measures had little impact, and his message to Congress went largely ignored.

A few weeks later, Nixon's announcement of the United States' unilateral decision to end the gold standard had much greater impact. By December, he authorized an increase in crude oil imports. Starting in 1972, not only had the "old warrior" oil fields ceased to rise, but they began to cower. Only eight years later, in 1980, Texas extractions had diminished by more than a quarter, and this in spite of a constant increase in the number of wells.[8] By then, dragged by the decline of Texas, all of US production had declined by 10 percent.

Demand for petroleum by those living in rich countries increased dramatically, surpassing all predictions. And nowhere did demand grow more radically than in the United States: Between 1970 and 1973, consumption grew by 20 percent, reaching 17 Mb/d that year. An inevitable consequence, US crude oil imports more than doubled in this time![9] In April 1971, practically for the first time since the Second World War, the American trade balance was in the red. Crude oil imports never ceased to weigh heavily in this balance. From that date forward, the US trade deficit with foreign countries continually worsened. The tide had turned. Four months later, on August 15, Nixon announced the United States was suspending the trade of dollars for gold, effectively eliminating the gold standard; this unexpected measure could only lead to—and in fact resulted in—a strong devaluation of the greenback. Generally, devaluation is used to readjust a trade balance deficit. As we have seen, the Big Oil experts also played a leading role in permanently ending the gold standard. It is difficult, in such circumstances, to conclude that the twin timing of the onset of American crude oil production decline and the end of the Bretton Woods monetary system were merely coincidental.

The unexpected decline of oil extractions was a disaster for the American oil industry. From Houston to Washington, everyone recalled Marion King Hubbert's prediction and finally understood the magnitude of the threat. But it was already too late: More dependent on oil than ever, Uncle Sam appeared duty-bound to turn more and more toward OPEC, especially toward the Middle East, to appease its addiction. It may seem surprising that Big Oil allowed itself to be ensnared this way. In 1945, Harold Ickes and the US Navy had the foresight to recognize the threat and respond efficiently, helping American companies consolidate their bridgehead in Saudi Arabia. However, since then, oil had worked its way into the smallest crevices of economic life. Its flow was beyond the control of the industry's minions, bankers, generals, and Anglo-Saxon ministers who had made it stream forth. The stream was now a Leviathan that no will, even that of the most powerful governments, seemed able to tame.

The peak of US oil production in 1970 was the pivotal event in the history of oil, much more than the famous crisis that succeeded it three years later and that, in many respects, appears to be its consequence. Already, since 1965, the United States had been obliged to extract crude oil in pedal-to-the-metal fashion, to their maximum capacity. Persian Gulf petroleum, under the Seven Sisters' control, already regulated the market. But from 1971, the strategic situation began drastically deteriorating, as US production was starting to decline: The United States saw, irredeemably, their ability to control the physical flow of crude oil escape them. This ability was captured by their Bedouin allies of the House of Saud, who from then on were in possession of the biggest source of the world's crude export. The petroleum polarity of the planet was reversed. A decisive parameter, but almost completely overlooked in contemporary history, the precipitous decline unlocked by American crude oil extractions after 1970 triggered a series of unstoppable chain reactions, at once economic, political, military, and ecological. These reactions continue today.

The Tehran Agreement: The Balance of Power Shifts from Big Oil to OPEC

On the other side of the oil world, OPEC countries had also noticed the tides turning. Fearing they would be sunk in the stream of Arab-oil-producer grievances unleashed by the audacious Muammar Gaddafi, the oil majors attempted to stem the turning tide beginning in 1971. On January 11, representatives from the Seven Sisters, fifteen other powerful Western companies, and the Japanese Arabian Oil Company met in the luxurious New York offices of John McCloy's legal firm (still located today at the top of the Chase Manhattan building, at the heart of Wall Street). In the antechamber of the Rockefeller condottiere, now seventy-five years old, hung the portraits of all American presidents since Franklin Roosevelt. Jim Akins, the State Department's discreet and omnipresent Mr. Oil, was there as an observer. The Western oil majors intended to present a united front: More than ever, there was no reason to worry about antitrust laws. They could not let the OPEC countries continue to make increasingly costly demands. At the end of long debates, representatives of twenty-three companies addressed a joint letter in which they warned OPEC that they would not negotiate with any of its members individually, "on any other basis than one which reaches a settlement simultaneously with all producing governments."[10]

The tactic, doubtless inspired by a kind of panic, was doomed to fail. The shah of Iran, Washington's most secure ally among the producer countries, convinced American diplomats early on that a negotiation solely between the oil majors and the entire group of OPEC countries risked obliging moderately producing countries like Iran to align with hardliners, countries such as Libya. The Saudi Arabian oil minister, the affable and prudent Sheikh Yamani, responded similarly to an emissary sent by Standard Oil of New Jersey. With a wry smile, watchful eyes, and a pointy beard that Western journalists soon referred to as "Mephistophelean," Yamani confirmed sotto voce that the OPEC countries had developed a plan to impose (if necessary) a worldwide embargo on oil.[11]

The OPEC countries managed to reverse the ultimatum. Persian Gulf producers gave Western oil companies until February 2 to comply with their conditions. In the course of a long press conference, the shah castigated the Western countries' "economic imperialism." (He did, though, accept their generous gifts at the lavish 2,500-year celebration of the Persian monarchy a few months later, in Persepolis). In this crucial face-off, President Nixon's administration failed the petroleum companies. The American government's strategists knew that Uncle Sam could no longer count on its "old warrior": American extractions had stagnated, and it would be impossible to cope with an embargo. To risk confrontation with the oil countries of the Persian Gulf was foolish.

On February 14, 1971, an accord signed at negotiations in Tehran imposed a 30-cent-per-barrel increase on Western companies buying Persian Gulf oil. A barrel of Aramco's Arabian Light became $2.15 and its price was revised every year. The increase appeared relatively modest: Throughout the 1960s, while inflation ran rampant in the rich countries, provoked by the heat of growth, crude oil costs had remained relatively unchanged, around $1.80 a barrel, due to abundant supply. The increase decided in Tehran solidified the inversion of power, in favor of OPEC. John McCloy had lost his last battle, but his clients, the Seven Sisters, were still the ones who pumped and sold most of OPEC's black gold: Jim Akins and the American oil diplomats did not hesitate to search for new, subtler, and doubtless more clandestine ways to help the United States preserve the fruits of the global energy empire they had cultivated for half a century.

Two different tracks opened up for OPEC countries. For some nations, it was the hour of revenge. Considering themselves too long despised and deceived, they quickly nationalized their crude oil production. On February 24, 1971, Houari Boumediene's Algeria appropriated almost all of the French petroleum interests, dashing the last hopes of some in Paris. In Venezuela, a series of regulations assured the government total control of the industry beginning in 1972. No new concessions were sold. From that time forth, the Western oil companies drastically reduced their investments in Venezuela—the country of Pérez Alfonso, OPEC's co-founder—and extractions from Venezuela, which had played a vital role for the Allies during the Second World

War, fell rapidly. They never returned to the record level attained in 1970. The downward slide initiated by Venezuelan oil, coupled with the decline in US production, aggravated world petroleum market tensions and shored up the Middle East's position in the center of the chessboard.

In Iraq, the support provided by the Soviet Union to the Baath Party encouraged Baghdad to pressure the Iraq Petroleum Company (IPC), the oldest Persian Gulf consortium created by the Western majors. Apprehensive, in 1970 they consented to relinquish 20 percent of the company's control and agreed to significant investments that allowed a substantial increase in production. But the Iraqis wanted more revenue and control. Especially, they blamed the Western oil majors for deliberately restricting Iraqi production during the 1960s, to punish Baghdad. On June 1, 1972, the IPC was wholly nationalized by the Baathist regime, and Iraq resumed control of many concessions owned by the old company, founded a half-century earlier. Iraq's omnipotent vice president, Saddam Hussein, took charge of the final negotiations for compensation of the IPC's European and American former shareholders. Shortly afterward, Iraq authorized the Compagnie Française des Pétroles to continue collecting its usual share of nearly a quarter of the Iraqi crude oil production, successfully playing this ancestor of Total against her British and American elder sisters.[12] Thus began a long series of strategic choices characteristic of the pro-Arab oil policy of France, largely dependent, as was the rest of Western Europe, on the crude oil of the Persian Gulf, while Baathist Iraq was included—for a short time only—on the list of Washington's threatening regimes.

It was Sheikh Ahmed Zaki Yamani, the Saudi minister of oil, who introduced an alternative path that the Western oil majors recognized as their salvation in the face of the wave of nationalizations. In 1962, Yamani had been appointed one of Aramco's directors, without, however, having any power over the intricate strategies of the American company that controlled the "greatest single prize in all history." In order not to endanger the alliance sealed in 1945 between the desert kingdom and the world's primary economic power, in early March 1969, Yamani proposed transforming Aramco into a full partnership between Riyadh and the American oil majors. This partnership would be "indissoluble, like a Catholic marriage," Yamani explained in the columns of the *Financial Times*.[13]

After the Tehran agreement of 1971, signed on Valentine's Day, twenty-six years to the day after the meeting between President Roosevelt and King Ibn Saud on the Great Bitter Lake, the American oil majors could no longer ignore Sheikh Yamani's offer. In June, several Arab Gulf countries adopted the method of negotiation initiated by the Saudi oil minister; they called for an immediate transfer of 20 percent of the ownership of the companies that exploited their oil, a portion they later demanded be raised to 51 percent.

In January 1972, Yamani negotiated with Standard Oil of New Jersey, just four months before it changed its name to Exxon.[14] The Harvard educated Saudi oil minister had the power to insist on his conciliatory middle track. Indeed, between 1970 and 1973, while American production began to wane, Saudi extractions doubled, to reach 8 Mb/d. Aramco increased their outlay and opened wide the valves that had been closed since the 1960s, when the market was still glutted. For the numerous directors of the Arabian American Oil Company, this was a bitter pill: Saudi oil threatened to escape at the precise moment it became clear that it was going to take the place of US oil as the keystone of the global energy system.

Aramco's American board of directors rapidly accepted the principle of letting the Saudi crown having 20 percent of its shares. But when, in February 1972, Yamani went to San Francisco to discuss the terms of agreement, SoCal's directors let him know that its shares were simply not for sale. On April 18, Nixon's treasury secretary, John Connally (the former Democratic governor of Texas, injured during the Kennedy assassination, was edging toward the Republican Party, becoming the close rival of its rising star, George H. W. Bush), brandished the threat that Washington would support American companies menaced by expropriation.[15]

On October 5, Yamani traveled to New York to sign a "general agreement" that also involved the princes of Kuwait, Qatar, and Abu Dhabi: Western companies would concede 25 percent of their shares within three years, then gradually move to 51 percent before 1983. The discussions between Yamani and the American Aramco partners continued for weeks in the sheikh's villa in the hills of Beirut.[16] After complex negotiations, a compromise was eventually signed in December, in Riyadh, but the terms remained largely secret.[17]

Such a compromise was made possible by the fact that the House of Saud never questioned the core precept: Aramco's US shareholders would have priority access to Saudi oil reserves. The Sisters—Exxon, Mobil, SoCal, and Texaco—had little to lose and much to gain with the Saudi Crown partaking in Aramco's capital. They could hope to benefit from the operation and increase profits, since from this point forward the Royal Family would take charge of an increasingly important part of investment costs necessary for the planned augmentation of production. The American majors partnered within Aramco planned to radically increase extraction capacities in Saudi Arabia, hoping to exceed thirteen Mb/d in 1976—a prediction that would prove to be geologically impossible—and claiming that they would achieve twenty Mb/d around the year 2000. Also, as observed in the US Senate's 1975 investigative report, "Looking ahead, the Aramco partners could expect that an Aramco producing at 20 million barrels per day would so dominate the world of oil that their market shares would not only be protected but would be increased as well, as long as they could buy back the increasing Saudi participation share of the Aramco oil."[18] As we will see, the Saudi petroleum fields never reached the outstanding levels Aramco's US shareholders hoped for. But the surge in crude oil prices that followed this agreement solidified the positions of both US shareholders of Aramco and the House of Saud, and facilitated their reconciliations.

For the time being, the price of crude oil was still the unsettled parameter in play: For what price could the Western oil companies purchase Saudi oil? Confident that he had an advantage, Sheikh Yamani prolonged the haggling for months. On September 17, 1973, in San Francisco, Aramco's US shareholders capitulated once again, agreeing to buy Saudi oil for 93 percent of the "posted price" by the Western oil majors; this initiated a new phase of rising prices before the impending oil crisis. The time when the Seven Sisters' cartel could present a united front had ended: The competition between Aramco shareholders became a negotiating lever for Yamani, who played on the nervousness of the most fragile among them, Mobil.[19]

The Saudi Arabian oil minister also benefitted from the increasing pressure exerted by OPEC's most radical members. For twenty-four hours on May 15, Libya symbolically halted their exports, creating a precedent that caused the Western world great concern. And on September 1, 1973, the fourth anniversary of his coup, Colonel Gaddafi announced the nationalization of 51 percent of all the subsidiaries of the Western oil majors in Libya. In vain, in front of the television cameras, President Nixon strove to put the Libyans on notice, recalling the fate met by the first Iranian prime minister, Mohammad Mossadegh, twenty years earlier. A *New York Times* journalist asked: "Could it really be that the President of the US has not yet grasped the predominant fact of life in the energy picture over the coming decade, that the problem is not whether oil will find markets, but whether markets will find oil?"[20] The Nixon administration, as we will see, had in fact grasped the evolving situation perfectly, and for quite some time.

The prophetic *New York Times* article appeared on October 7, 1973—just one day after Egypt and Syria unleashed a surprise attack against the state of Israel. Launched in the middle of the month of Ramadan, on the eve of the Jewish festival of atonement, the Yom Kippur War began almost exactly three years after the beginning of the unexpected decline in US oil production.

The Yom Kippur War Puts Washington and Riyadh Face to Face

The greatest mechanized battle since the Second World War, the Yom Kippur War started when Egypt and Syria invaded the territories that Israel had occupied since the Six-Day War in June 1967. From the south, the Egyptian army crossed the Suez Canal, which had been closed for six years, while Syrian tanks forged through the plateau of the Golan Heights in the north. Prior to the Israeli victory and cease-fire on October 25, 1973, more than 500 aircraft and 2,700 tanks were destroyed—an average of one airplane every hour and a tank every fifteen minutes. The war took place at a time when, in Washington, Richard Nixon was facing the worst of the Watergate scandal. Nixon largely left his national security advisor, Henry Kissinger, appointed only two weeks earlier as secretary of state, to handle the Middle Eastern crisis.

The massive military supply replenishments provided almost immediately by the Soviet Union to its Arab clients, on the one hand, and by the United States to Israel, on the other, made the conflict one of the hottest episodes in the Cold War. The air bridge installed by the American army on October 13 allowed them to deliver 1,000 tons of equipment to Israel each day, in particular the formidable Maverick antitank missiles that played a critical role in determining the conflict's outcome.[21*] On Wednesday, October 17, one day after learning that he was the co-recipient of the Nobel Peace Prize for the (precarious) cease-fire he negotiated in Vietnam, Kissinger announced "we are pouring in arms at a rate about 30% quicker than [the Soviets] do."[22] The following day, Israeli general Ariel Sharon made a decisive breakthrough, passing through the Suez Canal with nearly three hundred tanks. The role that the United States chose to play in the 1973 Israeli victory was considered an extreme test of its most stable and most necessary alliance in the region: the one with Saudi Arabia. This alliance emerged from the war stronger, more secret, and more important than ever.

By the time OPEC welcomed representatives of Exxon and other oil majors at its Vienna headquarters on October 8, the organization's member countries had long since proclaimed their intention to discuss substantial increases in the price of crude oil. The Yom Kippur War that had broken out two days earlier sharpened, without a doubt, the determination of cartel members from Arab governments. OPEC's grievances, however, had nothing to do with the fate of Israel. The shah of Iran, nearly indifferent to the Palestinian cause and

* The United States had become Israel's main weapons supplier, taking France's place in 1967, at the end of the Six-Days war.

linked to Israel by many interests, never stopped hammering home that the price inflation of goods the OPEC countries imported far exceeded the tariff increase granted for crude oil. The Western companies, however, once again revealed that they did not intend to yield. While OPEC delegates were prepared to impose a price of $5.00 per barrel, Exxon's representative responded that he could not consent to go above $3.70. On October 16, in Kuwait, the Gulf's six exporting countries (the five Arab countries, plus Iran) reconvened to impose a price of $5.12 for a barrel of Arabian Light. The following day, as the Yom Kippur War clearly turned in the Jewish State's favor, the Arab ministers of OPEC (without their Iranian counterpart) discussed the implementation of an earlier threat: using oil as a weapon to sever US support to the Israeli army. Iraqi Baathists close to the USSR forthrightly demanded the nationalization of all American oil in the Middle East.[23] Finally, the Arab ministers agreed to reduce their production by 5 percent every month, until Israel withdrew. On October 18, Saudi Arabia applied more pressure by announcing an immediate 10 percent reduction. On Riyadh radio, King Faisal announced in person that deliveries to the United States would be interrupted until the latter ceased to help Tsahal, the Israel Defense Forces. Three days later, while Israeli troops had Cairo in their sights, the Jewish State accepted a cease-fire negotiated by the United States and the USSR, officially announced a day later by the UN Security Council. Henry Kissinger's intense diplomatic efforts then led to the progressive disengagement of the opponents, starting in January 1974. On March 17, the Arab producers (except Libya) lifted an embargo that, as we will see, already had been largely mitigated by late December 1973.

In the days following Saudi Arabia's announcement of its production slowdown and boycott against the United States, the Arab principalities in the Persian Gulf, as well as Algeria, Iraq, and Libya, followed suit. The effects of the embargo on exports were felt immediately in Europe and Japan (which relied on OPEC for nearly 80 percent of its oil). The contrast between the impact of this embargo and the wet firecracker of the oil embargo during the Six-Day War was striking: It wasn't that the Arab countries were more united; it was that the oil market had changed. The boycott was felt in the United States mostly due to the lack of Saudi Arabian oil. The other Arab countries were minor suppliers compared to Aramco. Moreover, many observers in Lebanon and in Caribbean refineries made it known that, despite their hardline stance, Iraq and Libya, greedy for currency, did not fully respect the boycott against the United States, and even took advantage of it to increase their deliveries. Colonel Boumediene's Algeria (where the American oilmen rendered many services) proved to be equally prudent.[24]

During the embargo, Saudi Arabia (which Henry Kissinger had visited no less than eleven times) reduced its production by approximately 20 percent.[25] Despite the decline of their own extractions, the United States was infinitely less dependent on Middle Eastern oil than were Europe and Japan: Less than 15 percent of US needs were supplied by Arab countries in the Persian Gulf and elsewhere. Arab oilmen themselves questioned the effectiveness of the boycott against the United States as early as October: It was impossible for them to prevent petroleum they sold to a European country from being transferred across the Atlantic.[26] Saudi oil shipments were organized by the four largest American oil companies. Frank Jungers, Aramco's president, said: "We were given day-to-day instructions of what the boycott order was and who could buy what. . . . So we set up a system that determined where the oil actually ended up, every barrel. . . . This was under threat of complete Nationalization. . . . We had no choice."[27]

The Saudi's apparent intransigence had limitations. The embargo initially included the US Navy's Sixth Fleet, which patrolled the Mediterranean, and Riyadh soon extended it to the entire American military. Deputy Secretary of Defense Bill Clements picked up the phone and called the American Aramco leaders: "Find a way to get fuel to Vietnam. Our kids are dying out there fighting Communists."[28]* Frank Jungers explained the situation to King Faisal. Caught between his hatred of Zionists and his fear of Communists, and probably more worried still about alienating his valuable American protector, the Bedouin sovereign responded, while pulling on his great black robe: "God help you if you get caught, or if it becomes a public issue."[29] And thus, Faisal bin Abdulaziz Al Saud consented to allow the most powerful army of the world to operate normally, secretly breaking the Arab front behind the boycott that he himself had initiated. An unknown quantity of Saudi crude oil left the Ras Tanura terminal for Singapore, theoretically destined for countries not under embargo. There, the oil was transferred to vessels intended for American ports.[30] However, Frank Jungers talked and, shortly thereafter, a leak within Aramco got back to King Faisal. According to the American diplomatic cables from January and February, 1974, after threatening to interrupt the secret deliveries of fuel to the US Navy, Sheikh Yamani transmitted to Washington a Byzantine warning that betrayed the House of Saud's duplicity: "Jungers' stupidity should not be allowed to provoke Saudi Arabia into taking action which can [cause] severe harm to the US."[31]

* Texan Bill Clements founded Sedco, a leader in offshore drilling, and was regarded as a key player in Washington's oil lobby.

Some in Washington Predict the First Phase of the Oil Crisis

It is doubtful that Washington was surprised by the power ploy of the oil embargo.[32] Starting in 1972, the Arab OPEC nations amped up their threats, and almost all pointed to the United States as their "main enemy," as the State Department's Mr. Oil, Jim Akins, wrote in the April 1973 *Foreign Affairs* article "Oil Crisis: This Time the Wolf Is Here."[33] On May 23, 1973, King Faisal made a surprise appearance at a meeting in a Geneva hotel between Yamani and representatives of Exxon, Mobil, SoCal, and Texaco. He complained of the "inability of the government of the United States to bring positive support to Saudi Arabia" and warned that if the American oil companies did not help to change American policy concerning Israel, they "may lose everything."[34] Communicated to Washington by worried Aramco directors, the threat nonetheless failed to convince Texas oil man Bill Clements, who was number two at the Pentagon. Despite the notorious anti-Semitism of the old American ally's sovereign (Faisal often shared with his visitors the anti-Semitic *The Protocols of the Elders of Zion*), "Some believe that His Majesty is calling wolf where no wolf exists except in his imagination," the emissaries who Aramco sent to Washington remarked spitefully.[35] The king repeated his threats in the July columns of the *Washington Post* and the *Christian Science Monitor* and in September's *Newsweek* magazine, less than a month before the war broke out.[36] A further twist: On October 17, the day the embargo was announced, Kissinger and Nixon received a delegation of ministers from Saudi Arabia, Kuwait, Algeria, and Morocco in the Oval Office. It seems that there was no reference to the oil embargo at this meeting. "Was Kissinger to be criticized for a deficit of attention when it came to the economic facts of life?" a biographer of the master American diplomat wondered.[37]

The price increase decided on the eve of the previous day by the meeting of OPEC countries was no surprise to Washington. In fact, this increase was not illogical, nor even disproportionate. In 1970, the US Department of Commerce published a study that was closely analyzed by OPEC: It confirmed that the profit rate for oil companies in the Middle East was by far superior to those of any other extraction industry.[38] The Arab producers also had been directly affected by the end of the Bretton Woods monetary system in 1971, when the United States gave up the gold standard. The rapid depreciation of US currency that followed threatened to siphon money from OPEC's Arab countries. All petroleum contracts were indeed tied to the value of the dollar, while gold constituted the foundation of the whole *hawala* monetary exchange system in force in the Islamic world. In 1971, it took ten barrels of oil to buy an ounce of gold; in July 1973, after a new devaluation of the dollar in February, it took thirty-four. "What is the point of producing more oil and selling it for an unguaranteed paper currency?" exclaimed Kuwait's minister of oil.[39] The oil crisis brought a return to near parity, so that in 1974, twelve barrels of crude oil were equivalent to one ounce of gold.[40]

In the years preceding the crisis, the need for petroleum progressed at a new, wild tempo in North America, Europe, and Japan, systematically surpassing all predictions. This feverishness, combined with the beginning of the American crude oil production decline, produced outbursts of shortage well before the Arab producers' embargo. The first intermittent supply difficulties arose in the United States in 1970: In September, the president of the New York State Public Service Commission stated "that it might be necessary to resort to rationing for the first time since World War II."[41] The problems become more and more frequent in the course of the following years. Significant heating-fuel shortages were reported on the East Coast when, in December 1971, Nixon authorized the first important increase of crude oil import quotas.[42] The president definitively suspended these import quotas, of great consequence to Texan oilmen, on April 18, 1973. Six months before the Yom Kippur War, Nixon warned Congress in a nineteen-page message: "In the years immediately ahead, we must face up to the possibility of occasional energy shortages and some increases in energy prices. . . . Clearly we are facing a vitally important energy challenge. If present trends continue unchecked, we could face a genuine energy crisis."[43] Pete Peterson, Nixon's recently appointed secretary of commerce said, at the annual meeting of the American Petroleum Institute: "Popeye is running out of cheap spinach."[44]

The Arab crude oil price was likely to appreciate further, since it was still below the price of a declining US oil output. An output that was becoming incapable of coping with a frenetic demand that grew faster than all estimations. During 1973, independent petroleum companies scrambled to buy OPEC oil at prices well above the "posted price" for Arabian Light, which rose to $2.90 per barrel starting in June.[45] It was then that Aramco opened their valves wide, allowing them to compensate for the decreasing American production, which dropped by a spectacular 0.5 Mb/d between 1973 and 1974.[46] In September 1973, in the industrial state of Illinois, oil sold for around $5.20 per barrel.[47] This context made OPEC's October 16 price increase to $5.12 per barrel of Arabian Light, enabled by Iran and without any link to Israel or the war, appear less radical. Well before the 1973 oil crisis, the Nixon administration's Mr. Oil himself considered the barrel-price increase to be a natural and necessary phenomenon. In spring 1972, Jim Akins publicly stated that the price of oil was "too low" and that "it was necessary to increase it" because resources were "limited" in the face of the "alarming increase of the demand": There were "not enough deposits and not enough money to find new sources if the price of petroleum remained this low," he worried, when American production began to decline.[48] Akins repeated his analysis on June 2, 1972, this time before an Arab petroleum congress gathered at the Palace of Nations in Algiers. Speaking on behalf of America, he said: "Our main concern is supply security. We do not recommend the price increase, but we recognize that it is inevitable."[49] Before an audience as amazed as it was delighted, he explicitly referred to the eventuality of $5.00 barrel, at a time when no OPEC country had yet considered such a price point.[50] Some in the United States blamed Akins for his position and his predictions, which risked becoming self-fulfilling. But he persisted, and as he wrote in the April 1973 issue of the highly influential *Foreign Policy*: The $5.00 barrel price could be reached well before the end of the decade (as soon as August, four months later, the price was negotiated at more than $4.20 per barrel in Illinois).[51] Akins specified that such a price corresponded to a "level equivalent to the cost of other sources of energy."[52]

Let us summarize. The Arab countries' oil embargo had followed intermittent shortages in the United States, a problem that had continued since the 1970 peak of US crude oil production. This embargo does not appear to have been rigorously enforced by the Saudis, neither by the king nor by the US Aramco shareholders responsible for its implementation. The barrel-price increase that OPEC instituted on October 16, 1973, appears logical and measured, first, in light of inflation's delayed effect on barrel price and the dollar's depreciation with respect to gold; then because of the persistent gap between the prices of crude oil in the Persian Gulf and in the United States; and finally and especially because of the strong upward pressure exerted on price by the decline in American production, in a context where oil demand exploded. This price increase was announced (encouraged?) by Jim Akins, Washington's key oil figure. It is unlikely that the point of view regularly affirmed by Akins was deemed inappropriate by the American oil companies: For a decade, Akins had been part of the oil diplomacy's elite; he had become America's ambassador to Saudi Arabia in September 1973, just before the war, and remained so until 1976. It would have been impossible to obtain this position without the go-ahead from Aramco's shareholders. At this stage, the 1973 oil crisis was essentially a consequence of the United States reaching peak oil production.

Did the Shah of Iran Drop a "Financial Atomic Bomb" at the Suggestion of Kissinger?

The turn of events that followed was as surprising as it was radical. When representatives of OPEC's six Persian Gulf countries met again on December 22, 1973, in Tehran, the Yom Kippur War was thoroughly over and the diplomatic ballet led by Henry Kissinger seemed ready to produce results. But Arab producers continued to restrict their exports. Shortly before the OPEC meeting, oil was auctioned in Tehran for an unprecedented $17.34 a barrel. In Saudi Arabia, King Faisal did not want to hear about any new price increases. His minister of oil, Sheik Yamani, said shortly before his arrival in Tehran that, if it set such a high base price, OPEC would "ruin the existing economic structures of the industrial countries, as well as the structures of the developing countries."[53]

It was the shah of Iran, potentate of the only non-Arab Persian Gulf country and Washington's most substantial ally in the region, who threw oil on the fire. Behind the heavily guarded doors of his minister of finance, he wanted $14 a barrel as the official OPEC reference price: a dizzyingly high proposal. The shah instantly obtained the support of the most radical Arab oil exporters: Iraq, Algeria, and Lybia—hitherto minor, impotent players compared to Tehran and Riyadh. The Saudis had no recourse. Faced with the alternative between bargaining and the dissolution of OPEC, Sheikh Yamani chose to negotiate with the shah. "It was one of the critical moments of my life; one of the few decisions I took reluctantly," he later said.[54]

On December 23, before negotiations were officially completed, the shah alone announced, from his sumptuous palace in Niavaran, north of Tehran, the sharp increase in the price of crude oil that many Western petroleum countries feared. The "posted price" for a barrel of Persian Gulf oil was $11.65. The price of black gold had doubled overnight! The American press christened this forceful blow the "Christmas Eve Massacre." Caught short, Yamani offered the West, on Christmas Day, a consolation gift: Arab producers would renounce their plan to reduce production by 5 percent in January, and would instead promise to increase extractions by 10 percent. But this gift was not enough to reassure the rich countries: With the Christmas Eve Massacre, the true oil crisis began.*

Why did Shah Mohammad Reza Pahlavi, then fifty-four years old, show himself as suddenly more radical than the most radical of OPEC's Arab

* A reference to the Saturday Night Massacre, two months earlier, when, on October 20, Richard Nixon drove out several of Watergate's main investigators.

countries? Why did he, who had just ordered $2 billion of American military equipment, he whose security depended on the White House's good will, seem ready to risk antagonizing Washington? At no time did the question of Israel enter the equation. "Greed" was the simple motive diplomats and Western oil spokespeople, shocked and dismayed, suggested to the press; greed would remain the leading explanation of the Persian sovereign's historic coup de force, and the other OPEC leaders who accepted it, willingly or not.

Consider the justifications offered by the shah himself. They were a mixture of economic arguments, as rational as they were unexpected, and fanatic curses against a West that the shah usually regarded as a model. During the press conference he gave in his Niavaran Palace, before a select audience of international economic journalists, Mohammad Reza Pahlavi said that the new price corresponded to the minimum cost necessary for the development of energy alternatives.[55] This argument echoed one put forward eight months earlier in Jim Akins's *Foreign Affairs* article. The shah evoked a "new concept": He repeated that, from now on, the price of crude "should be the minimum that you would have to pay to get shale, for example, or the liquefaction of gas or coal."[56*] Speaking in English and occasionally in French or Farsi, the Persian ruler suddenly exclaimed: "The industrial world will have to realize that the era of their terrific progress and even more terrific income and wealth based on cheap oil is finished." Then he seemed to explode: "They will have to find new sources of energy, tighten their belts. If you want to live as well as now, you'll have to work for it." He continued, presumably referring to the violence of certain radical leftist movements then in full swing in the West: "Even all the children of well-to-do parents who have plenty to eat, have cars, and are running around as terrorists throwing bombs here and there—they will have to work, too."[57]

In Europe, already grappling with OPEC invoices that doubled in October, the shah's new doubling of the black gold price was equivalent to the dropping of a "financial atomic bomb," according to French economist Robert Lattes. In New York, Exxon's directors had a very different reaction. A company spokesman said on December 24, in the *New York Times*, that the flare-ups in OPEC's crude oil prices "weren't unexpected" and that they "underscore[d] something that hasn't been emphasized strongly enough—the days of cheap energy are ended." This anonymous spokesman of the eldest of the Seven Sisters continued: OPEC's price increase could well provide a stimulus for the development of synthetic energy sources in this country, "because it definitely eliminates the differential between the costs of synthetics and natural petroleum."[58]

---

* The shah probably refers here to oil extracted from bituminous shale.

This rationale was surprisingly identical to that outlined by the shah in Tehran. It was rich in meaning. Also on December 24, a dispatch announced that, "spurred by sharp price hikes for Middle East oil," the Western companies were intensifying their research on the new oil frontier: the North Sea.[59] Twenty days later, in Colorado, an auction for the exploitation of oil shale began, for the first time since its prohibition forty-four years earlier, during the Great Depression. Oil from shale aroused new curiosity: "With foreign oil zooming up in price and domestic oil apparently running short, and with the Federal Government committed to long term self sufficiency in energy, there is a new interest," noticed the *New York Times*.[60] More widely, in the days that preceded the crucial meeting of Persian Gulf producer countries in Tehran, the oil majors, with Exxon at their lead, had indicated that they were preparing to strongly increase their investments as of 1974, thanks to a boom in profits caused by the new crude oil price. Their purpose: to increase production as quickly as possible.[61]

How to interpret such a coincidence? It is here that the story must make a detour to examine a conspiracy theory, or at least the hypothesis of a ploy. The central role in the Christmas Eve Massacre, assumed by the shah, the United States' constable in the Middle East, raises suspicions, and some observers have accused Iran of serving American interests, albeit without evidence.[62] The shah's incendiary role was not revealed. But in January 2001, twenty-seven years after the fact, Sheikh Yamani, known until then for his measured behavior, his discretion, and his exquisite diplomacy, issued a charge capable of upending the meaning and scope of one of the cardinal events in twentieth-century history. Unjustly, this dramatic move went almost unnoticed. The ex-minister of Saudi oil, long retired from his weighty responsibilities, told two journalists of the English weekly the *Observer*: "I am 100% sure that the Americans were behind the increase in the price of oil."[63] In early 1974, according to Yamani, when the shah called for even higher price increases, the king of Saudi Arabia asked his oil minister to go to Tehran to probe the intentions of the Persian ruler: Was he not worried about the anger of their common American ally? When Yamani raised the question with the shah, the latter, according to the Saudi minister, responded: "Why are you against the augmentation of the price of oil? Is this not what they want? Ask Henry Kissinger, it is he who wants a higher price." (Yamani reaffirmed those remarks in 2010.)[64]

What credit should we give these comments attributed to the shah twenty-one years after his death, by the other key OPEC player of the era? Above all, why on Earth did Henry Kissinger encourage the shah to fan the flames of the price of crude oil, and this at the worst time for the Western economy? Sheikh Yamani explained Kissinger's motive to the *Observer*: "The oil companies were in real trouble at that time, they had borrowed a lot of money and they needed a high oil price to save them."[65] The argument was not entirely correct: The oil companies' level of indebtedness was moderate then; however, as we will see in the next chapter, it was clear that they would have encountered serious difficulties financing new projects if the barrel price had remained low. Whether or not we have confidence in Yamani's accusations, there is a strong historical connection in the chain of events that includes the decline of American oil beginning in 1971, the 1973 oil crisis, and the Western oil majors' colossal investment and undertaking to relaunch Western extractions and, among other things, embark on the conquest of offshore oil deposits, in 1974. The significance of this connection remains to be clarified, but in all likelihood, thus far it has been largely underestimated.

Did Henry Kissinger—master of diplomacy and American intelligence, and Nelson Rockefeller's former protégé—secretly maneuver to foster the escalation of oil prices in 1973, which he himself called "a weapon of political blackmail"?[66]

Nelson, David Rockefeller's brother, head of Chase Manhattan, said he was "outraged" by the OPEC countries' willingness to inflate oil prices.[67] The probability that some in Washington had seen this escalation coming from afar, or even that they had encouraged it, proved nothing. For dizzying as it was, the scandalous plot described by Sheikh Yamani late in the game remains unverifiable. However, without having to suffer from the blame, the Western petroleum industry quickly recognized it benefitted in many ways from this inflation in crude oil's value. This inflation, in fact, allowed Big Oil to delay its inevitable decline by several decades.

TWENTY

Oil Money: After Neocolonialism, a Perilous Symbiosis

Peak and decline of American extractions, liberation of the OPEC countries, and the continued expansion of crude oil consumption: In less than four years, between 1970 and 1973, the oil market went from being ruled by demand to being governed by supply. The chain of events that unraveled over the next four years, from 1974 to 1978, was no less brutal. By the time the Yom Kippur War broke out in 1973, oil provided almost half (46 percent) of all energy produced by humankind; it had usurped coal as the primary energy source just ten years earlier.[1] The flare-up of the barrel price that followed beginning in 1974 applied disproportionate pressure on all arteries of the world economy, forcing a brutal historical transformation. The torrent of "petrodollars" paid to the OPEC countries radically changed the rapport between those countries and the industrial powers, which still used nearly three-quarters of the oil produced in the world.

The embargo that Arab countries imposed in October of 1973 and the shah's move to quadruple crude oil prices between October and the Christmas Eve Massacre caused two years of violent recession, which spread like a prairie fire through the capitalist world. OPEC's leaders were held responsible for a historic economic drama. This spectacular drama overshadowed a new, long-term geological reality: the depletion of the United States' principle oil zones. However, the phenomenon appears clearly in the statistics of the time, since between 1973 and 1974 the natural decline of American extractions (almost half a million barrels per day) was almost double the reduction in oil production mandated by OPEC's political decisions.*

* American production: 10,946 Mb/d in 1973; 10,461 Mb/d in 1974. OPEC production: 29,932 Mb/d in 1973; 29,667 Mb/d in 1974 (according to BP, *Statistical Review of World Energy Workbook 2013*).

The Oil Crisis Gives Birth to Unemployment, Massive Debt, and a Housing Market Crash

The oil crisis landed a brutal blow on what was still the highest rate of crude oil demand the planet had ever known. World production rose from 53 Mb/d in 1972 to 58 Mb/d the following year: a 9 percent increase! This was such a phenomenal upsurge, in such a brief period, that it is hard to imagine how an industry as unwieldy as the petroleum industry could have accommodated the continued proliferation of consumption, with or without the political surprises of late 1973.

The oil crisis created two long-lasting effects that the world economy never got rid of: debt and mass unemployment. And despite sober efforts and the emergence of substitute energy sources (mainly natural gas and coal, both still fossil fuels, and nuclear power), the crisis did not call into question the central role occupied by oil. On the contrary, the years following the 1973 crisis tightened the bonds of enslavement to the economic empires and geopolitics of black gold. At the end of December 1973, one in five service stations in the United States ran dry.[2] There and elsewhere, fights broke out between exasperated motorists. In the Midwest, the "yellow" truckers (nonstrikers) who refused to join a strike against rationing were the targets of Molotov cocktails and sometimes bullets. Soon, many rich countries reduced their speed limits.[3] A severe affront to the American way of life, the length of the NASCAR race track was shortened. But despite the fact that the boycott was targeted at America, the United States was less hard hit than Europe and Japan, countries almost exclusively dependent on Arab oil, where up to one-quarter of the normal supply was affected by the embargo. Several European countries provisionally suspended Sunday driving. The oil crisis fanned the flames of inflation, already in full force, starting with the price of coal. In early 1974, Great Britain was forced to declare a three-day workweek in most industries, in order to save energy; Ted Heath's conservative government was faced with harsh strikes in the coal mines; the BBC stopped its programs at 10:30 p.m., all in order to save energy. France soon reinstituted daylight saving time, which had been discontinued after the occupation.

The embargo revealed the depth of developed societies' petroleum dependence, as well as the breadth of its implications. In the United States and Great Britain, there were problematic delays in plastic medical equipment deliveries.[4] On the BBC, English priests debated the question of whether it was moral for family members to bathe together.[5] Japan had toilet tissue and detergent shortages starting in October 1973, which lasted for several weeks and contributed to a collective psychosis in which Japanese housewives haunted store aisles to stock their cabinets, which aggravated the shortage.[6] In West Germany, the sugar beet industry successfully obtained an exception to reorder fuel: Even a twenty-four-hour shortage in factories would have left the sugar to crystallize in tubes, threatening to block the entire production.[7]

It was an exceedingly violent blow to the capitalist economy. In 1974, the United States and Japan endured an immediate decline in GDP, followed in 1975 by declines in France, West Germany, and Italy. It was indeed a "crisis," succeeding three decades of superabundant growth, often greater than 5 percent per year. Between 1974 and 1975, the Western production of steel dropped at least 15 percent.[8] And as in the 1930s, mass unemployment reappeared, especially in Europe, but this time it did not go away. Until then, "residual" or "frictional" unemployment had not exceeded 2 percent of the French working population, but in January 1976, it reached almost 5 percent, hitting young people especially hard. In France as in many other Western countries, its progression would never really be reversed. As the shah of Iran had announced, energy was no longer a given. Efforts naturally focused on productivity, to reduce overall wages and maintain profits in order to survive in a world where growth was more sparing and selective and where there was drastically stiffened competition. Productivity-gains research became the central concern of industry bosses in rich countries. This preoccupation was one for which many employees, in particular in the industry, were forced to pick up the tab. To counter a recession that deprived everybody, many governments resorted to fiscal borrowing: For the first time since the end of the war, massive public debt reappeared and rapidly entrenched itself. Inflation skyrocketed everywhere, often exceeding 10 percent per year. For the first time since the war, the miraculous progress of purchasing power that had become almost commonplace, ceased.

The general price increase and slowdown in activity caused by the sharp cost increase in the mother of all raw materials led to serious difficulties for the reimbursement of debts incurred in the course of the earlier petroleum expansion. Italy, weighed down by its energy bill and economically much more fragile than other developed countries, seemed doomed to bankruptcy at the beginning of 1974 (a time so frightening that the Bank of Italy's governor implored David Rockefeller, successfully, to grant the bank an emergency loan of a quarter of a billion dollars), and then again in 1976, pushing the Christian democracy to launch severe economic measures that formed the backdrop for Italy's socially and politically turbulent Years of Lead.[9]

The increase in oil prices also unlocked an explosion of debt in developing countries during the 1970s, especially where the Green Revolution triggered a strong need for petroleum products.[10] Between 1973 and 1974, the trade deficit of developing countries multiplied fourfold. In an unfortunate twist of fate, 1974 was marked by horrendous droughts, in particular in Africa where famine erupted as a result. Nations had to borrow in order to import foodstuffs from the United States and Europe, all the while faced with escalating oil bills. India, for example, saw its foreign exchange reserves melt to nothing. The brutal entrenchment of debt and a trade deficit endemic since 1974 were the first symptoms that, in the course of the next decade, led the International Monetary Fund (IMF) and the World Bank to impose their drastic remedies: structural adjustment programs.*

During 1974, in the United States, the interest rates charged by the Federal Reserve to compensate for galloping inflation reached at least 10 percent. The financial shock that followed was radical. There were numerous bankruptcies, large and small. In 1975, the City of New York itself narrowly escaped one. During the 1960s, the Big Apple had gone heavily into debt to finance an ambitious social program characteristic of the golden age of the welfare state and Keynesianism. The city had built its many projects, large building developments intended for low-income residents—a program spearheaded by Mayor John Lindsay, a liberal Republican (or Rockefeller Republican) who took office in 1966, and was personally supported and financed by David Rockefeller.†

The heart of the capitalist economy, New York had the means for such programs then: With a population comparable to Sweden's, the city's budget was equivalent to that of India's government. But at the end of 1974, New York City's outstanding debt began to saturate the credit markets. No one wanted it. To avoid bankruptcy, the city's creditors, represented by the three largest Wall Street banks—two of which were intimately tied to Big Oil—put the planet's financial capital under trusteeship in 1975. The city's government was assumed de facto by the leaders of three banks: Chase Manhattan, Morgan Guaranty, and The First National City Bank (now Citibank); the first was chaired by David Rockefeller, the last by Walter Wriston, protégé and successor of James Stillman Rockefeller (cousin of David and Nelson).[11]

On May 26, an agreement allowed bankers to exchange a billion dollars in "rotten" municipal loans for new debt securities that were guaranteed directly by city taxes and received an "A" rating. To this effect, on June 10, a day before

* It was through these programs that the IMF and World Bank granted loans to developing nations on the condition that recipients adopted a host of free-market measures, from prioritizing international trade and encouraging privatization to removing regulations restraining industry—measures frequently blamed for the escalation of poverty and inequality in the developing world.

† In his memoir, Rockefeller said that Lindsay had transformed into a "populist" soon after being elected. (See David Rockefeller, *Memoirs*, op. cit., p. 392.)

New York found itself in arrears, the Municipal Assistance Corporation (or MAC) was established to manage the Big Apple's debt and regulate spending.* But this was not enough. In October, after imposing profound budget cuts on New York's mayor, freezing wages, and eliminating twenty-six thousand municipal jobs, the triumvirate that included David Rockefeller traveled to Washington to seek unprecedented assistance, in order to save not only the municipality of New York but also their own firms from otherwise inevitable default of payment. As the first bank of the United States, Chase Manhattan in effect found itself on the front lines, facing the collapse of the nascent speculative mortgage securities market that had been in full swing since 1970, and into which David Rockefeller had ventured on the advice of one of the principal partners at Lehman Brothers bank. Battered by the whirlwind, the very survival of David Rockefeller's bank was threatened.[12] The holder of housing projects and important social reform networks, New York itself had made an inordinate use of these mortgage credits.[13] Unofficially appointed the triumvirate's spokesperson by Rockefeller, Pat Patterson, president of Morgan Guaranty, alluded to a drift toward a no man's land capable of "economic downpull of general economic activity" in an economy already in recession.[14]

In mid-October, the world's three most powerful financiers, Patterson, Wriston, and Rockefeller, were received in the Oval Office by President Gerald Ford (scarcely two months after Watergate led Richard Nixon to resign). Ford did not want to hear about a bailout.† Standing by President Ford's side, the new US vice president showed cautious understanding.[15] He was Nelson Rockefeller, David's elder brother. After several weeks of negotiations and entreaties, the president and vice president eventually grew fearful: The White House pressed the US Congress to open a $2.3 billion line of credit to the city, far exceeding the norms at that time. Wall Street was saved, but shortly afterward, New York entered a long downward spiral of impoverishment and violence (the Martin Scorsese film *Taxi Driver* touched on these themes the following year). Pat Patterson welcomed the loan, however: "There were a lot of people who would just as soon have seen New York go bankrupt. They thought it was a good thing to clean it out and get rid of the labor contracts."[16] As for Chase Manhattan and its primary shareholder, David Rockefeller, they fluctuated a little but did not sink. The trust of the bank devoted to mortgages and property loans, Chase Manhattan Mortgage and Realty Trust, went bankrupt in 1979, and Chase lost its leading position in the profits of its old cousin

* Instantly nicknamed "Big Mac."

† A few days later, Gerald Ford's decision was represented in a headline by the *Daily News* (the largest American tabloid), which remained in New York's collective memory, as: "Ford to City: Drop Dead."

and rival, Citibank. Nevertheless, not once did Chase display operational loss during this period, without a doubt largely thanks to oil money.*

The harsh sanctioning of the Big Apple's accounts by Wall Street's major banks, in retrospect, could appear as a trial run of neoliberal policies yet to come. In addition, the interdependence between the eruption of crude oil prices, inflationary pressure, soaring interest rates, financial crises, the market crash for loans and in particular mortgage credits, and nationalization of private loss did not lack common traits, as we will later see, with the crisis of 2008, the so-called subprime crisis.

Nuclear Energy Grows, but Oil Continues Its Reign in a Car-Centric Society

Oil, by far the main overall energy source in 1973, was only the second source of electricity: On a global level, oil-fired electrical plants provided a little less current than coal did, but more than hydroelectric dams.[17] The oil crisis facilitated the gradual emergence of natural gas and gave a boost to coal. But it also marked the rapid emergence of civilian nuclear power. Starting in the late 1940s, the development of the first uranium nuclear reactors remained an ancillary phenomenon, in large part entwined with efforts to develop atomic weapons. France (with Israel), the United Kingdom (in India), and the United States (with, notably, Pakistan and Iran in the political context of the "Atoms for Peace" program initiated by Eisenhower in 1953) more or less secretively helped a few select developing countries take the first steps in their own nuclear programs. However, at the end of the 1960s, only a handful of nuclear-powered electrical plants were active in industrialized countries: in the United States, the Soviet Union, Great Britain, Canada, Japan, France, and Switzerland.

With Herculean industrial efforts, the energy generated in the world thanks to the atom quadrupled between 1973 and 1980, to reach the equivalent of 160 million tons of oil; this represented a modest 5 percent of crude oil consumption in the same year.[18] In January 1975, in a State of the Union speech, President Gerald Ford put forth a grandiose plan to construct two hundred nuclear power plants over the following ten years. A little more than sixty were eventually built, making the United States by far the world's leading producer

* Chase Manhattan president until his retirement in 1981, Rockefeller noted in his memoirs: "Fortunately, the income that we generated from other sources, especially from our international loans and operations during that same period, enabled us to cover the heavy losses we had sustained." (David Rockefeller, *Memoirs*, op. cit., p. 316–317.)

of nuclear energy. Among the countries most dependent on Arab oil, France and Japan made the most radical choice in favor of the atom. In 1973, in each of these countries, fuel oil was the main energy source feeding the electric power plants' turbines.[19]

On March 5, 1974, French prime minister Pierre Messmer (confronted with the continued decline in coal production) accelerated a nuclear power program that was as spectacular as it was audacious. The urgency was such that, despite its experimental reactors, France chose to build plants whose pressurized water technology was patented by the American giant Westinghouse. Messmer, who unapologetically defended France's access to Cameroon's oil and other raw materials in the late 1950s, decided to begin the construction of thirteen power plants without troubling to consult Parliament. But in France, there was broad political support for atomic energy, even from the Communist Party: In fifteen years, Messmer and his successors, on the right and the left, made France one of the countries relying the most on nuclear energy for electricity. In spite of the long crisis that was beginning, the harvest of the Thirty Glorious Years, irrigated by Arab oil, offered France the opportunity to finance its costly, ambitious nuclear program without worrying. The steps of the grand energy staircase sometimes operate with unexpected permutations: The development of nuclear power led to the decline of domestic heating oil, and this fuel became suddenly available for diesel engines. This gave French car manufacturers the opportunity to systematically focus on diesel car production starting in the 1980s. Responsible for deadly air pollution, diesel fuel was, at that time, presented as less noxious than gasoline fuels, which were then judged as the main culprit of the environmental woe that rightly or wrongly preoccupied a majority of the Western public in the 1980s: acid rain.* For OPEC, too, the history of oil and that of nuclear energy became entangled. The influx of oil money sharpened the ambitions of the oil countries' leaders at the same time as it sharpened the appetite of Western atomic industry leaders. In 1974, the year Iranian production reached its historical record (6 Mb/d), the shah repeated his statements presenting nuclear power as an indispensable energy form. "Petroleum is a noble material, much too valuable to burn. . . . We envision producing, as soon as possible, 23,000 megawatts of electricity using nuclear plants," he said, justifying his ambition to build "twenty-three atomic power plants" in his country "as soon as possible."[20] The Americans

* In 2012, the World Health Organization estimated the number of premature deaths in France caused by harmful particles disseminated primarily by motor vehicles to be forty-two thousand, a much greater number than deaths by road accidents in the same time frame; since the 1970s, these vehicles have become more numerous but much less polluting.

provided their expertise. In 1975, the German groups Siemens and AEG won a contract of more than $4 billion to build in Bushehr the first Iranian nuclear power plant, whose construction was interrupted by the Islamic Revolution in 1979. France was responsible for the enrichment of uranium. In 1976, US president Gerald Ford approved a plan of several billion dollars that allowed the construction of a plutonium production plant. Henry Kissinger, reappointed by Ford as secretary of state, indicated three decades later in the *Washington Post*: "I don't think the issue of proliferation came up."[21] This project, too, was interrupted by the 1979 Islamic Revolution.

In Iraq, the existing regime's brutality did not scare French industrialists away from the atom, nor their government from Iraq's guardianship. Prime Minister Jacques Chirac went to Baghdad in September 1974, on the invitation of the man he then described as his "personal friend," Saddam Hussein.[22] Iraq's vice president and de facto dictator then visited France in September 1975, where he was invited to visit Cadarache, the French atomic energy commission's research center. This was his one and only official visit to the West.[23] In its wake, a contract secretly negotiated by Jacques Chirac allowed Iraq to begin construction of the Osirak nuclear power plant.[24] Barely underway, the plant was bombed and destroyed by Israeli aviation on June 7, 1981. According to a current interpretation, the Osirak plant, as well as numerous weapons that France had delivered (often paid for with great difficulty) helped France preserve its privileged access to Iraqi crude oil after the 1972 nationalization.[25]

Finally, the Pakistani nuclear program doubtless owes much to Saudi oil money. Thanks to the American "Atoms for Peace" policy, Pakistan's nuclear program inaugurated its first nuclear power plant in Karachi in 1972. In 1969, the French atomic energy commission and British Nuclear Fuels signed agreements accompanying the construction of plutonium production facilities. After the surprise test of India's first nuclear bomb in 1974, British Nuclear Fuels withdrew from the Pakistani project. Pressured by Henry Kissinger, in 1978 France renounced its own cooperation with Pakistan's nuclear military program.[26] According to several testimonies, since the early 1970s Saudi Arabia had been involved in financing the manufacture of Pakistan's future atomic bomb, after Pakistani prime minister Zulfikar Ali Bhutto convinced King Faisal of the need to offset the threat of Israel's and India's nuclear programs.[27] The Saudi royal family insisted that the funds were only used to conduct research and expressly not to test a bomb.[28] Was this in order to avoid embarrassing their American protector?

In spite of the emergence of civilian nuclear power, the expansion of natural gas and the resurgence of coal, oil was not ousted. With the 1976 recession scarcely ended, crude oil consumption in rich countries accelerated strongly (simultaneous with increased steel production). Millions were unemployed and many dreams of social emancipation had died—yet, overall, the economy had adapted to the new barrel price. The law of supply and demand (and OPEC's dictates) set those barrel prices at around $12, approximately the price level enacted by the shah during the Christmas Eve Massacre in December 1973. Many technical innovations, beginning with rapid advancements in electronics, caused energy efficiency and capital investments in the productivity of industry and services to leap forward.

In all industrial sectors, engineers were working to find (when possible) profitable processes capable of reducing oil costs. For example, the Lafarge Group, the giant French producer of construction materials, went on a quest to find combustible waste from other industries in order to alleviate the huge fuel needs of its cement kilns. But the energy savings realized by these tactics were incapable of weaning companies away from oil because they were part of an industrial economy doomed not to a technical revolution but, on the contrary, to resuming expansion as quickly as possible. Each amazing refinement in the production process permitted more extensive and ostentatious modes of consumption, in which renewed hunger for energy created a "rebound effect."*

In fact, apart from the factory directors, there was little concern for saving energy. In November 1973, President Nixon tried to initiate a program to make the United States self-sufficient in terms of energy production before the end of the decade. Very quickly, Nixon's Project Independence was met with sarcasm. When the oil embargo ended nearly a year later, Nixon's successor, Gerald Ford, abandoned the smaller cars and mass transit proposed during the embargo. The new American president reiterated the White House's support of the automotive industry, proclaiming: "I'm a Michigander, and my name is Ford."[29] In 1975, according to *Fortune* magazine, five of the seven largest American companies were oil companies (led by Exxon, in which the family of Nelson Rockefeller, vice president of the United States, figured among the primary shareholders) and the other two were auto industry giants, General Motors and Ford. In Paris, a motorway network was built along most of the once romantic banks of the Seine. The car-centric world denounced by the earliest environmental activists triumphed more than ever before. In France, as in other industrial nations, the state was in debt; the growing manna from

* As explained in chapter 2, the "rebound effect" is the process by which total energy consumption increases when the energy efficiency of an instrument of consumption improves. If, for example, the consumption of a car is halved, a motorist is encouraged to drive twice as far.

gas taxes encouraged dependence on cars. The public treasuries of consumer countries became more and more addicted to black gold. André Giraud, minister of industry under Giscard d'Estaing, quipped, "Oil is a raw material that has a strong diplomatic and military effect, a significant fiscal effect, and an incidentally calorific power."[30]

Alaska and the North Sea: The Western Oil Majors Push to Develop Extreme Oil

For the Western petroleum industry, the strategy that was emerging even before the Yom Kippur War became an absolute priority that the new barrel price would make possible: The focus was on the urgent need to drill new fields that could compensate for the decline in American production and help America escape OPEC's control. But the time of easily extracted petroleum had already begun to wane. For a long time, the Seven Sisters had set their sights on two promising major oil zones, which had the invaluable advantage of lying within Anglo-Saxon territory. There was, though, a caveat: Each of the areas was located in a terribly hostile environment: the first in the extreme north of Alaska, the other under the waters of the very stormy North Sea. The hydrocarbon resources of Alaska and the North Sea, which geologists had long envisioned to have vast reserves, could not be extracted at the prices in place before the 1973 crisis: The costs of the necessary infrastructure were simply too high. Beginning in the 1970s, almost all sources of easily extractable oil were already being exploited: The black gold industry had to cope with the irreversible completion of its golden age. Simultaneously confronted by the loss of control of extractions in OPEC countries and the explosion of consumer appetite, the Western oil majors had to make perilous tactical changes to save themselves and at the same time save the growth economy.

A short United Press International dispatch published in the *New York Times* on March 30, 1973, six months before the oil crisis, revealed the extent to which Big Oil might view the oil shock as a gift from the heavens. The article gave an account of a report by Chase Manhattan Bank (chaired by David Rockefeller) that stated "the international oil industry probably will not be able to find the trillion dollars needed to fill the world's energy needs by 1985." Chase Manhattan insisted: "It is doubtful *under present economic conditions* that the industry will be able to raise that amount. In that event, the industry would be unable to provide all the petroleum the world's markets require, because a large part of the oil and natural gas resources which would be necessary are as yet undiscovered."[31]* Perhaps this is what helped convince Sheikh Yamani of Henry Kissinger's role in the escalation of oil prices inflicted by the shah of Iran.

Even before the end of the embargo, Big Oil's profits soared. In the last quarter of 1973, Exxon's profits increased by 80 percent, and that year it held the record for the highest profit ever made by any company in the world: $2.5 billion. 1974 was better yet, and then, in 1975, Exxon overtook General Motors as the planet's most profitable company. This insolent opulence in the midst of an economic recession in the United States raised the anger of the automobile-driving American citizens: Not only had the oil majors practically organized the embargo for the Arab producers, but in addition they were its primary beneficiaries. Far from keeping a low profile, the Seven Sisters defended their high prices, explaining that the market needed it. Starting in February 1974, before the embargo on the United States had been officially lifted, Texaco, Gulf Oil, Shell, and Mobil offered themselves many pages of advertising in the American press to say that, hitherto, their profits had been too low to ensure ongoing funding of their productions. This was a very unpopular argument. It was nonetheless defended regularly in major American newspapers, in particular in the *Wall Street Journal* and the *Washington Post*, where analysts explained how the new barrel prices were a blessing because they allowed for the development of new resources. Some of these articles were clearly based on the conclusions of the Club of Rome's second report, published in November 1974. The authors of this report (which did not have the lasting impact of its first report, *The Limits to Growth*, released two years earlier) vigorously cautioned against returning to the pre-oil-crisis barrel price levels: "Such an extremely low level of oil prices would hinder the development of alternative resources until the total world oil reserves are almost completely exhausted around the year 2000."[32]

Henry Kissinger pushed this logic further. In a confidential report completed in December 1974, the secretary of state and national security advisor concluded that US oil and gas reserves were sufficient to meet US demand "for another two or three decades," but only "if prices are raised sufficiently." This diagnosis was perfectly aligned with what the shah of Iran had said in Tehran one year earlier. Moreover, Kissinger believed that US reserves of oil from bituminous shale were "sufficient for the better part of the next century," adding that "their full-speed exploitation could be limited by environmental factors and water supply."[33] In February 1975, Kissinger proposed to the European allies that they agree on a "base price" below which oil could not be sold. This

* Author's emphasis.

proposal amounted to safeguarding investments in the new reserves, most of all in Alaska and in the North Sea, as well as in synthetic fuels and nuclear and other alternative energies. A *New York Times* editorial judged that Kissinger's proposal constituted an "essential step" to reduce dependence on OPEC. Once more, Washington had positioned itself as the guardian of Big Oil.[34] The initiative didn't amount to much, but that didn't matter, since after growth resumed in 1976, crude oil prices showed no signs of declining.

Now that they had the necessary funds, the Western oil majors could proceed with their Herculean undertaking, which had been impossible previously. Five years before the first oil crisis, in June 1968, an American company short of fresh black gold reserves, Atlantic Richfield (ARCO), a distant descendant of Standard Oil, fought for survival in a prospecting campaign in northern Alaska, finally locating oil in a previously unknown, inaccessible place: Prudhoe Bay. In 1969, the Halliburton group, frontrunner in oil infrastructure, began building a base camp. The Prudhoe Bay field rapidly proved to be a new little Texas, capable of challenging the rank of the "old warrior" of the US industry, which was on the brink of collapse. Short of money, Atlantic Richfield had to appeal to its partner, the giant Exxon, to take advantage of its discovery. As had been true throughout its history, Exxon's size alone guaranteed that it was associated with all the best opportunities in American territory. Exxon became the majority partner. BP, desperately seeking an entry into the United States to lessen its dependence on OPEC, found abundant oil around Prudhoe Bay in 1969, after nine years of prospecting in vain beyond the North American polar circle. But American petroleum was not prepared to let the British company grab their market shares without taking a stand. BP, which had so far never been able to acquire a single oil corporation of significant size in the United States, was authorized, under the terms of the Byzantine negotiations, to take partial control of Standard Oil of Ohio, the original cell of the Rockefeller empire. Big Oil's interests continued to intertwine. Only, nobody had ever yet sought to pump oil at such a grand scale at a latitude of 70 degrees north, in a region that was so cold in the winter, the crude oil would freeze in a few moments if the slightest malfunction interrupted its flow. In 1972, the governor of Alaska made it known that the oil discovered four years earlier unfortunately remained too expensive to extract.[35]

That reality changed after the 1973 oil crisis. Oil workers and engineers arrived en masse. The greatest challenge was to route the crude oil up to an ice-free port at Valdez, in southern Alaska, on the other side of the largest, least populous state of the Union. At 1,300 kilometers long, the Trans-Alaska Pipeline was one of the industry's masterworks. Unlike conventional oil pipelines, it was not buried in the Earth. Constructed starting in 1974 and inaugurated in 1977, it was mounted on tens of thousands of refrigerated pillars, so that the crude oil circulating in the tube would not melt and collapse the frozen soil below it. Atlantic Richfield estimated that the cost of transporting crude oil from Alaska would rise to $6 a barrel, approximately double its pre-1973 price.[36]

The development of hydrocarbons trapped beneath the North Sea had begun as early as 1959, when Jersey Standard and Shell discovered significant natural gas reserves in the Netherlands. The Groningen field, Western Europe's largest natural gas reserve, made a fortune for the Dutch, while provoking a decline in their industry: The national currency, the florin, appreciated so much that Dutch products gradually lost their competitiveness in the export markets. This "Dutch disease," as it was baptized by the *Economist*, confirmed for the first time in a northern country—and what's more, in the cradle of capitalism—how an oil windfall could transform into a wound and atrophy the rest of a country's economy.[37] Encouraged by this first North Sea discovery, Great Britain, Norway, and Denmark clamored to grant prospecting permits. When the prospecting boundaries of the North Sea were drawn up in 1964, Downing Street's experts were so impatient that the British Crown contented itself with claiming 35 percent of the area, renouncing the greater portion to which they could have been entitled. The oil flowed for the first time in December 1969, from the giant Norwegian field Ekofisk. Fortunately for Great Britain, another giant field, baptized Forties, was discovered by BP ten months later in British waters.

Starting in 1971, the United Kingdom withdrew almost all troops stationed "east of Suez." That year, the Royal Navy transferred its strategic base on the island-state of Bahrain, across from Saudi Arabia's Ras Tanura port, to the US Navy, leaving the United States free rein in the Persian Gulf once and for all. The agony of the British Empire no doubt partly explains why, in its own territorial waters of the North Sea, London showed itself to be so generous and helpful to the oil majors. The United Kingdom was selling off its concessions and called for a very small share of the profits, compared to profits claimed from the Seven Sisters (and especially BP) by OPEC countries. Worse, the majority of the best concessions were controlled not by BP but by American companies, starting with Exxon, to the point that Parliament opened a public inquiry whose report, published in 1973, pushed the government to clamp down. The Western oil majors consented to very heavy investment. Their leaders' fury was great, beginning with that of Lord Eric Drake, head of BP, for whom the Crown remained the primary shareholder, but whose autonomy was often compared to Frankenstein's.

In 1975, when the first oil rigs started production, the Labor government of a United Kingdom in crisis decided to create a national company, the British National Oil Company (Britoil), with the right to buy 51 percent of the oil extracted in British waters of the North Sea. Does the irony of the situation need to be pointed out? The prime minister, Harold Wilson of the Labor party, battled to impose on his country's hydrocarbon reserves a sovereignty that the United Kingdom had denied to Iraq and Iran by military rule throughout a half-century. Under Margaret Thatcher, Britoil eventually was privatized and then absorbed by BP.

The industrial challenge posed in the North Sea was no less than the challenge of Alaska. The barrel price of Brent (the name of a Scottish field in the Shetland region) cost ten times more to extract than the Arabian Light of Saudi Arabia.[38] For the first time, it was necessary to anchor a fleet of oil platforms in the continental plateau midway between Scotland and southern Norway, in water that was 80-meters deep, and in the middle of a sea where the roughest swells in the world often rose. The ratio between the energy invested to make crude oil flow and the energy deposited in the oil terminals for future use was deteriorating in comparison to the old days of the Trans-Arabian Pipeline. As early as 1975, a network of pipelines capable of carrying up to seven hundred thousand barrels per day was deployed at the bottom of the North Sea. It arrived in the small Scottish port of Aberdeen, which almost overnight was transformed to a rainy Dhahran. The oil money helped the Scots obtain, without firing a shot, a broad political autonomy denied to the Catholic independence movement in miserable Northern Ireland, which at that time was living through one of the most violent periods of its history. On the other side of the North Sea, the Norwegians, surprised by their unexpected good fortune, took this opportunity to improve their lagging technology, counseled by one of the early Iraqi geologists, Farouk al-Kasim, born in Basra, educated at the Imperial College of London, and married to a Norwegian. Meanwhile, an Iranian al-Kasim participated in the direct supervision of foreign oil at the Norwegian Ministry of Petroleum and Energy.[39] The Texan oilmen, often annoyed by the Norwegians' determination to control the management of their oil and to ensure, in particular, that they neither extracted nor exhausted it too quickly, nicknamed the Norwegians the "blue-eyed Arabs."

Alaska and the North Sea would remain almost the last large production zones for conventional oil production on Earth.* With each new discovery, the industry now had to resort to a combination of increasingly delicate

* With the possible exception of ultradeep offshore fields off Brazil's coast, which in 2017 remained, however, quite far from the production levels achieved in Alaska and in the North Sea.

techniques: offshore drilling, horizontal wells, "forcing" crude extraction by injecting liquid under pressure, and so on. The fountains of oil gushing spontaneously from the ground, as in Baku, Spindletop, or Kirkuk, belonged to the memory of an already almost mythical time.

Several other challenging oil zones, not comparable with Alaska and the North Sea, attracted oil companies during the 1970s. Western companies, Americans in particular, continued to drill widely in Vietnam up to the fall of Saigon in April 1975. Hanoi's Communist regime encouraged them to resume their operations as early as the following year, but the results were never what they had hoped for: This part of the China Sea was not so rich in oil, after all.[40] In Canada, the exploitation of Alberta's tar sands began in 1973 and grew slowly starting in 1978. But the production of heavy petroleum remained too costly to attract the massive investments necessary to give it meaningful life at that time. Finally, it would be a pity to omit the misadventures of the head of Elf Aquitaine, Pierre Guillaumat, who in 1977, at the end of his impeccable career, fell for a hoax and funneled more than 600 million francs to the inventors of the Avions Renifleurs (or "Sniffer Aircraft"); its creators had falsely claimed the aircraft could detect oil underwater. Without a doubt, the master engineer of the French mine corps had lost sight of the fact that in physics, even more than in politics, miracles are suspicious.

If for a time they alleviated the wealthy Western countries' dependence on OPEC, the reserves of the North Sea and Alaska did not carry enough weight to shift the center of gravity of the oil world. As soon as its pipeline opened in 1977, Alaska reached almost its maximum production capacity, 2 Mb/d, which continued for some years after overall US production declined in 1985.

The decline of the Alaska fields began only three years later, in 1988. North Sea production rose, reaching almost 6 Mb/d in the 1990s, almost equivalent to the total remaining oil production of the United States. But the pragmatic Norwegians never forgot Marion King Hubbert's lesson. They made their calculations in the late 1970s, and estimated that production would peak around the year 2000; it in fact occurred in 2001 for Norway and as early as 1999 off the coast of the United Kingdom.

Washington Channels Persian Gulf Oil Money to Wall Street

During the 1970s, no energy strategist lost sight of the lesson of American peak oil, especially not in Washington. It was obvious that until the end, the Middle East's fantastic black gold reserves would sustain activity in this desert region at the center of the world energy stage: The last drop of conventional oil will probably be extracted somewhere around the Persian Gulf, likely in Saudi Arabia. Trivial and lucrative, the Western priority from this point on amounted to making as many friends as possible around the Persian Gulf, selling these allies weapons and other gadgets, and helping them invest their savings. Whatever the undesirable consequences of these choices might be, could they outweigh the need to maintain privileged access to the primordial and ultimate source of crude oil? After the crisis of 1973 and the death of their colonialist oligopoly, the Seven Sisters, the banks, and the Western governments behind them, quickly established the conditions for a fruitful symbiosis with their OPEC partners and allies.* A symbiosis in which the sumptuous, poisonous fruits were the "petrodollars" destined for the coffers of Wall Street banks.

The anguish that gripped the financial community immediately after the oil crisis manifested almost every day in the columns of the Western economic press, which brandished the threat that Saudi princes would soon be able to buy all listed shares. However, many immediately perceived the excellent luck that "this second great Arab incursion in western history" represented.[41] Jersey Standard's top executive from 1955 to 1971, Jack F. Bennett, was subsequently in charge of the American treasury's international monetary affairs, and then was appointed undersecretary of the treasury by Nixon on May 9, 1974, taking charge of the whole of US monetary affairs. At age fifty, Bennett succeeded his boss, Paul Volcker, who helped him establish the monetary system of floating exchange rates that followed Bretton Woods. During the following weeks, Bennett (also a distinguished member of the Bilderberg Group and CFR), with Henry Kissinger, fine-tuned a plan that helped retake control of the monetary flow from industrial countries to OPEC countries.[42] This plan came to fruition on June 8, 1974, only three months after the embargo's official end, with the ratification of a broad program of economic and military assistance to Saudi Arabia. The financial component of the agreement was summed up in one phrase. The word "oil" did not appear in the text even once.[43] However, it was certainly intended to encourage Saudi Arabia to reinvest their oil profits, generated by Aramco, in large American companies and banks, including Exxon, which remained one of Aramco's primary shareholders, next to the House of Saud. Arriving in person to sign the agreement in the Blair House residence located next to the White House, the prince and future king of Saudi Arabia, Fahd (eleventh son of Ibn Saud) announced that this agreement constituted a "new and glorious chapter in relations between Saudi Arabia and the United

* The concept of a symbiotic "bilateral monopoly," describing the situation born after the 1973 oil crisis, was revealed in 1978 by John M. Blair, *The Control of Oil*, op. cit., p. 293.

States."[44] Kissinger described it as a "milestone."[45] Jack Bennett, his mission accomplished, returned to his mothership: In 1975, he became Exxon's chief financial officer and vice president, and remained so until his retirement in 1989.

Shortly after, a young, brilliant banker and Illinois native, David C. Mulford, was appointed head of the investment council group of the Saudi Arabian monetary agency responsible for managing the Saudi treasury. Mulford, who then also directed the New York branch of White, Weld and Company, a large Boston-based investment bank associated with the Credit Suisse, had worked in the White House under President Johnson from 1965 to 1966.[46] He remained the investment advisor to Saudi Arabia until 1983. It was a daunting task: One day, as his meeting with an Asian finance minister dragged on, he realized that he was running late in handling that day's investments—$300 million.[47]

The formidable Saudi fortune, increased dramatically after 1973, was mostly redirected to the United States, in the form of US treasury bills or certificates of deposit in the private banks.[48] As early as 1974, the money that flowed from Saudi Arabia to the United States, through financial organizations such as Chase Manhattan and Citibank, was at least twice what America had paid for Saudi oil![49] That year, the Saudis invested $5 billion of their $26 billion oil profits on American soil. In 1976, the total direct investment of Saudi money in the United States reached $60 billion. In 1979, Saudi Arabia became the first holder of US treasury bills: The Saudis financed the United States' oil debt.[50] An American businessman went into ecstasy in *Newsweek*: The Saudis "are going to fuel our industry and keep our economy afloat. I say make the place the goddam 51st state."[51] With the gold standard barely abolished, black gold had become the dollar's most robust base.

Aramco Changes Hands: The Brief Duel between Americans and Saudis

At a time when the money men of Saudi Arabia and the United States were celebrating their new arrangements with pomp and circumstance, the two countries' petroleum leaders and engineers were at odds. The Saudi Crown had become Aramco's majority shareholder in 1974, holding 60 percent of the company's shares. In America, the "old warrior" Texas oil industry was on its knees, and despite Nixon's Project Independence, Ford's launch of the colossal nuclear program, and the production of the wells in North Alaska, Americans still had to import almost half their crude oil in 1979, versus a quarter in 1970. A fifth of this imported oil came from Saudi Arabia.

The situation could not have been more perilous for Uncle Sam. Thanks to the pugnacity of Senate investigators in Washington and two of the most prominent American journalists of the time, we have a few snippets of information that suggest the intensity of the secret drama that played between Washington and Riyadh, between Houston and Dhahran, during these years that were so crucial to the future path of American energy. The drama began three months before the 1973 oil crisis, in the middle of Aramco's nationalization negotiations, and reached its climax on the eve of the second oil crisis (see chapter 21). Its first indication appeared in a dispatch from H. J. Johnston, one of Aramco's leaders in Saudi Arabia, sent by order of Standard Oil of California (SoCal), on July 25, 1973, and exposed by the American Senate investigation of the oil multinationals directly after the crisis.[52] In this dispatch, Johnston revealed he had admitted to Sheikh Yamani that the amount of the "true reserves" in Saudi Arabia was in fact two and a half times higher than reported: "In the past we have given [Riyadh] ultra conservative numbers as proven reserves of 90 billion barrels." In a dry report, which did not mention the response that this revelation provoked from the Saudi oil minister, Johnston wrote to his superiors in San Francisco that he and Yamani "finally came to the position that our true reserves were 245 billion based on the method that is commonly accepted to determining these figures"—nothing less than a third of world reserves listed at the time! The Saudis later were accused of artificially inflating the official estimate of their reserves. But this dispatch showed that on the contrary, the American owners of Aramco had lied for a long time, deliberately undervaluing their assessment. Big Oil's strategies left nothing to chance: It was striking to see American petroleum announce an extremely low estimation of Saudi oil reserves during a period of abundance, in which the Western majors sought to curb the development of extractions, only to see them later admit to a much more ample (and truthful) estimation at the precise moment when production growth had become an imperative.

At the end of 1973, when Big Oil was forced to relinquish control of Aramco, the company's American engineers received orders to pump all the oil that they could, beyond any prudence and to the point of risking irreversible damage to the Saudi's megafields. In January 1974 (while the Saudi oil embargo was still formally in force), a *Washington Post* journalist claimed to have had access to internal, ultraconfidential technical reports from Exxon, Mobil, SoCal, and Texaco. According to the paper, these documents showed that Aramco engineers had been complaining since 1973 to their directors of excessive production that created a state of "severe technical difficulties," as well as "erratic production" and "huge pressure drops" in the pumps that promised to make it impossible to meet the "20 Mb/d projection" that Aramco's American engineers had dangled in front of the Saudis.[53] The Senate committee that investigated the oil companies then subpoenaed several Aramco leaders. All vehemently denied the reality of the catastrophe, except SoCal's chief engineer in charge of petroleum reserves, William M. Messick, who confirmed the report.[54] Three years later, on February 14, 1977, a press dispatch from *Petroleum Intelligence Weekly*, a highly respected industry information agency, reported a new unexpected, important drop in production, of more than half a million barrels per day.[55] Why? "Bad weather" had slowed the loading of supertankers at the Ras Tanura terminal, the agency related. Senate investigators contacted the meteorological service of the US Navy base at Ras Tanura, who responded that not the slightest storm had occurred in six months.[56] After reading other inconsistencies in the reports of the two largest American intelligence agencies, the NSA and the CIA, Senate investigators queried W. Jones McQuinn, the recipient of Johnston's July 1973 dispatch. To their great surprise, this Aramco director confirmed that they had risked damaging the oil fields with excess pumping, and that the Saudis (who were Aramco's true bosses) had ordered American oil companies to take adequate conservation measures to save their oil reserves.[57] A new series of subpoenas was issued, and investigators sought access to thousands of pages of Aramco's internal documents. Sheikh Yamani strongly opposed their release. A diplomatic crisis threatened to erupt when Seymour Hersh, of the *New York Times*, exposed part of the scandal.[58] But the thick veil of oil diplomacy held firm. Aramco's American and Saudi Arabian partners kept their dirty linen in the family and cautiously agreed to secret negotiations that led to Aramco's complete nationalization in 1980. The final Senate report, published in April 1979, was stripped of everything that could tarnish the special relationship between Washington and Riyadh. The elected members of the Senate commission who voted on this censorship included Jacob Javits, New York State senator and Nelson Rockefeller's political ally, and Joe Biden, Barack Obama's future vice president, then a senator of the small state of Delaware, where Aramco was registered.

The issue of peak oil and the depletion of crude oil reserves was a major concern for the CIA. It surfaced in a secret report dated March 1977, in which analysts of the intelligence agency delighted in announcing, very prematurely, the beginning of the decline of Soviet oil extractions "not later than the early 1980s."[59] This hasty prognosis at the very least proved, by the wealth of details it concealed (the number of wells, pace of extraction, rate of decline, state of infrastructure, and so on), the CIA's careful attention to the state of the world crude oil reserves.*

Oil Money Metabolizes in the Cold War: George H. W. Bush, the CIA, and the Safari Club

The United States showed no more than a fleeting propensity to flex its muscles at OPEC. In the embargo's last weeks, every industrialized country, and in particular France, attempted to establish its own bilateral supply agreements with the producer countries. In this general free-for-all atmosphere, Henry Kissinger convened a conference in Washington on February 11, 1974, to try to unite a common front under the American banner. A new institution, the International Energy Agency (IEA), was born on November 18, created under the auspices of the rich countries' Organization for Economic Co-operation and Development (OECD). It was supposed to stand against OPEC: a kind of anti-cartel of the nations that consumed the most oil. Among other things, to avoid alienating the Arab producers, France refused to join the IEA, which Michel Jobert, Georges Pompidou's foreign affairs minister, presented as a US war machine directed against the Arab countries and OPEC. The IEA did nevertheless set up its headquarters in Paris, but the war machine remained confined to an OECD annex in the fifteenth arrondissement and, outside of its functions in the management of strategic stocks, was, with rare exceptions, locked in the role of a vital analysis center. Early in 1975, the Pentagon reported that it had prepared an invasion plan to take control of oil fields in the Persian Gulf.[60] This was a purely symbolic approach to pressuring the Arab producers: Henry Kissinger specified from the outset that an invasion was only an extreme last resort.[61] Instead, the United States became, from afar, the primary arms supplier of the OPEC countries in the Middle East, starting, of course, with Iran and Saudi Arabia, two countries that Washington was about to use to accomplish some of it's murkiest jobs of the Cold War.

After President Richard Nixon, engulfed by the Watergate scandal, resigned on August 9, 1974, the White House for the first time became the nucleus of those who were not yet called the "neoconservatives," future promoters and architects of the two invasions of Iraq under the administrations

* The territory of the former Soviet Union oil production maintained a record production above 12 Mb/d in the 1980s, before collapsing along with the Soviet empire in the '90s. It returned as of 2006 to its record 1980s extraction level, thanks to the development of multiple new extraction projects, particularly around the Caspian Sea. The production of the Russian Federation, however, has not reached the peaks (more than 11 Mb/d) achieved by communist Russia in 1986–1988.

of oil presidents Bush, father and son. Nixon's successor, Gerald Ford, was the only American president who was not elected to that high office.* After having considered appointing George H. W. Bush, the new head of the Republican Party, as vice president, Ford finally appealed to Nelson Rockefeller. The latter, thrice an unsuccessful candidate for the Republican Party's nomination in the presidential elections (in 1960, 1964, and 1968), ex-governor of New York, embodied the liberal political establishment within the "grand old party." After Ford and Rockefeller, Henry Kissinger was the authorizing foreign policy officer. Gerald Ford's cabinet chief, Donald Rumsfeld, was forty-three years old, a former Princeton University wrestler and football team captain and a former US Navy instructor pilot. Hostile to the all-powerful Henry Kissinger's relaxed policy with respect to the USSR, Rumsfeld managed to convince Ford to thoroughly overhaul the government on November 4, 1975.[62] In what the press dubbed the "Halloween Massacre," Rumsfeld became defense secretary for the first time in his long political career. He bequeathed the position of President Ford's closest advisor to his assistant and protégé, a certain Dick Cheney, then thirty-four years old. George H. W. Bush was appointed head of the CIA. The rich oilman who twelve years earlier had advocated for the use of the nuclear bomb in Vietnam had since firmly established himself as a leader of the US right wing.[63] Although Bush was not known to have any intelligence experience—we have seen that the reality was probably quite different—President Ford did not hesitate to entrust him with "more power than any director since the creation of the CIA," according to a *New York Times* account that ran after his appointment on January 30, 1976.[64] Upon arriving in his position, George H. W. Bush formed a group of analysts responsible for toughening the CIA, then considered too lenient. To head the group, called Team B, Bush appointed another young man, thirty-two year old Yale political science professor Paul Wolfowitz, who had been recommended by another future celebrity of the US neoconservative camp, Richard Perle.[65]

In Texas, at the beginning of this same year, 1976 (perhaps in March), Jim Bath, an aircraft merchant close to the Bush family, received a phone call from Salem bin Laden.[66] The latter, a wealthy, thirty-year-old Saudi Arabian, was the heir of the construction group that bore his name, founded by his father, Mohammed bin Laden, a one-eyed mason who had become a multibillionaire and was the intimate friend of King Faisal of Saudi Arabia. Jim Bath was a former US Air Force ace. He was a near neighbor of the young George W. Bush and had become his close friend four years earlier, accompanying Bush

* Gerald Ford was appointed vice president after Richard Nixon's first vice president, Spiro Agnew, resigned on October 10, 1973, swept away by the Watergate scandal.

on training flights as the Bush clan's heir struggled to become a pilot, performing his military service far from Vietnam, in a Houston air base dubbed a "champagne unit." Bath was perfectly integrated into the petroleum network. He went duck hunting with George H. W. Bush's friend and business partner James Baker—a former Ford administration member, a business lawyer with Texas's second most powerful law firm (Andrews Kurth) and future heir to his father's place in the number one firm, Baker Botts.[67] Jim Bath even participated in several strategic meetings during Prescott Bush's unsuccessful campaign in the 1970 senatorial elections.[68] A short time after they met, on July 8, 1976, Salem bin Laden designated Jim Bath (who had no particular experience in finance) as the representative of his US interests. Bath then had the good fortune to be engaged by Khalid bin Mahfouz, the twenty-six-year-old heir of the leading Saudi bank in charge of managing the Royal fortune. Khalid bin Mahfouz and Salem bin Laden visited Houston frequently. A playboy who sang and played piano, Salem bin Laden hosted great parties for the world petroleum industry capital's businessmen in a faux château called Houston's Versailles, that Jim Bath nicknamed the "Big House."

In Texas, the Saudis were welcomed with open arms. According to his lawyer, Khalid bin Mahfouz together with the Texas Commerce Bank, in which James Baker was a major shareholder, helped to finance the construction of Houston's tallest building, completed in 1982 and today called JP Morgan Chase Tower.[69] Jim Bath, whose company almost exclusively employed former military pilots, was responsible for the management of the Bin Laden–Houston company, specializing in naval equipment. Among other investments, Salem bin Laden's holding company founded a small Texas airline, Bin Laden Aviation.[70] When journalist Craig Unger asked him—in the course of a 2002 investigation that he published in book form in 2004—whether he had been a CIA agent, Jim Bath, then retired, equivocated.[71]*

Whatever we may think of the synchronicity between George H. W.'s arrival at the head of the CIA and the business links forged by a Texan military pilot close to the Bush family with heirs of two of the most eminent Saudi families, early 1976 marked a turning point in the relations between the Kingdom of Saudi Arabia and American espionage. This turning point was the creation of the Safari Club, a secret group of intelligence services united to fight Communism—outsourcing to Riyadh a cryptic key role in financing

* Khalid bin Mahfouz attempted to oppose the publication of Craig Unger's investigation (*House of Bush, House of Saud: The Secret Relationship between the World's Two Most Powerful Dynasties*), which largely influenced *Fahrenheit 9/11*, the Michael Moore documentary that won the Palme d'Or at the 2004 Cannes Film Festival.

the Cold War, from South Africa to the foothills of the Himalayas, everywhere the funds of exceedingly wealthy Middle Eastern princes could be useful.

After Watergate and the Pinochet coup in Chile, the US Congress tried to closely supervise the White House's secret operations. In particular, a Senate amendment blocked the secret aid President Ford wanted to send to Angola to counter the USSR's and Cuba's massive support of the civil war that had begun in the newly independent Portuguese colony. Angola's oil resources—suspected as early as the 1920s and prospected beginning in the 1950s—became both the prize and the amplifier of this war that dragged on for nearly thirty years. In a top-secret message about a "recent congressional action," sent to Saudi Arabia on December 21, 1975, Henry Kissinger called for $30 million for aid to Angola.[72] After this date, according to Prince Turki, who was responsible for Saudi intelligence from 1977 to 2001, "Saudi Arabia's activities in Africa expanded dramatically—at the time when American activities in the rest of the world were shrinking because of Watergate and the restrictions placed on the government, particularly CIA activity."[73] In 1975, the amount of aid given by the United States to its ally countries was surpassed by Saudi Arabian aid, which reached $5.7 billion, or not less than 13.8 percent of the kingdom's GDP.[74]

Saudi interference extended beyond Africa. It penetrated "deep into Asia," the *Washington Post* noted in 1977.[75] In his memoirs, Kissinger recalled without regret "a helpful Saudi footprint placed so unobtrusively that one gust of wind could erase its traces."[76] At this critical juncture of the Cold War, Saudi aid was dispensed through an original initiative of Count Alexandre de Marenches, the head of the French secret service. Starting in 1976, de Marenches organized a group involving Saudi Arabia, Iran, Egypt, and Morocco—the Safari Club, which counterattacked everywhere the Soviet ogre and the Libyan fox advanced. In his memoir, the master French spy saluted the effectiveness of its US counterpart—the one who was thought to have little experience, George H. W. Bush, who he showed around Paris in March 1976.[77] During the 1970s and 1980s, oil money made the Saudi Arabian Crown a still more autonomous geostrategic power; Riyadh provided financial support to Syria, Jordan, North Yemen, Angola, Zaire, the Sudan, Somalia, Djibouti, Uganda, Mali, Nigeria, and Guinea. In Asia, the Saudis rained dollars down on South Korea, Malaysia, Taiwan, the Philippines and, of course, their Muslim allies of Bangladesh and especially Pakistan (including, as we have seen, by bringing Islamabad the necessary funds for their nuclear program).[78]

Saudi Arabia served as a covert funding source for counterinsurgency operations encouraged or organized by the American and French governments. But an embarrassing gap, obviously tolerated, soon appeared between the Saudi's secret aid to American allies everywhere in the developing world and aid that the oil princes chose to give anti-Zionist groups that Washington considered terrorists, beginning with Yassar Arafat's Palestine Liberation Organization (PLO).[79] It was a gap that never stopped expanding and one day gave birth to a bitter surprise for the West and their Middle East allies: Salafi terrorism, one of three major monstrosities that unleashed United States hegemonic policy on the Persian Gulf at the turn of the 1980s.

Arms and Oil Infrastructure: The Birth of a Symbiosis between Washington and Its Persian Gulf Allies

Starting with the 1973 oil crisis, the West's best allies in the Persian Gulf made themselves indispensable by miraculously maximizing sales for weapon manufacturers. Weapons manufacturers were one of the rare industries (along with oil) that improved despite the crisis, which encouraged their supervising governments to disregard the consequences of the manna amassed. As we have seen, President Ford and Henry Kissinger were ready to sell a plutonium plant to Iran in 1976. That year, nearly twenty thousand American military trainers lived in Iran, to the point that a Senate commission asked whether Iran was capable of fighting a war without Uncle Sam's direct involvement.[80] After the Halloween Massacre, the strategic link formed between the Ford administration and the US's Middle East policeman, the shah of Iran was so tight that Tehran offered, with Secretary of Defense Donald Rumsfeld as an intermediary, to finance the development of a new version of the American firm Northrop's famous F-18 fighter jet.[81]

Saudi Arabia, with only eight million inhabitants, in the late 1970s became the best customer for sophisticated weaponry manufactured in the United States. In 1976 alone, American arms sales to Riyadh reached $5.8 billion, double their total between 1950 and 1970.[82] France also benefitted from the boom and remained the kingdom's number two supplier.[83] But the construction of the kingdom's military infrastructure was contracted almost exclusively to the Americans. Between 1973 and 1976, the amount of the contracts for military buildings multiplied fivefold and exceeded $10 billion. The US Army Corps of Engineers oversaw the construction of much military, naval, telecommunications, and other vital modern infrastructure. In particular, the US Army Corps designed the Saudi Arabian National Guard headquarters near Riyadh, as well as its ultrasophisticated camp at the Al-Hasa oasis, erected in the vicinity of the primary oil fields, near Aramco's headquarters.[84] With almost seventy thousand men, the Saudi National Guard was the kingdom's elite army corps. It was specifically responsible for protecting the House of Saud, as well their petroleum facilities. In February 1975, the US Department of Defense contracted the training of this praetorian guard to the Vinnell Corporation. Based in Los Angeles, this mercenary firm employed US veterans of the special forces in Vietnam. Vinnell was often presented as a smokescreen for the CIA, and it was the first private US company officially authorized to train foreign troops.[85] The two American giants of the building and oil infrastructures, the Texan group Brown & Root (subsidiary of the Halliburton firm) and the Californian group Bechtel, were the leading beneficiaries of the enormous, lucrative construction contracts linked to the Saudi Arabian army and the exploitation of its black gold. Nixon's secretary of the treasury, George Shultz, left Washington in 1974 to become one of Bechtel's lead executives until 1982, when he was appointed President Reagan's secretary of state.

The billions of oil dollars generated by the oil crisis worked to calcify the powerful geostrategic and energy spine that extended from Houston to Dhahran through Washington and Riyadh. John Perkins, a former economic adviser who worked for the NSA, the secretive US electronic-information intelligence agency, summed up the effects of the oil embargo this way: "What had initially appeared to be so negative would end up offering many gifts to the engineering and construction businesses and would help pave the road to global empire."[86]

After Democratic president Jimmy Carter arrived in the White House in January 1977, the US Congress, largely aligned with the Israeli cause, attempted to put the brakes on arms sales to Arab countries. But in May 1978, when American legislators tried to block the sale of F-15 fighters in Riyadh, Sheikh Yamani threatened to increase the barrel price and slow the kingdom's oil investments.[87] The White House pressured Congress, which gave in a few days later, approving the sale of sixty F-15s to Saudi Arabia (for good measure, fifteen others were sold to Israel, along with seventy-five F-16s). Bill Quandt, a close collaborator of Zbigniew Brzezinski, Carter's national security advisor, warned the president: "The Saudi role will be crucial, unreliable, and unpredictable."[88] In 1979, the Saudi Arabian order for American weapons was ten times higher than those of Israel or any other US ally combined.[89]

Even within the administration of Jimmy Carter, the humble peanut farmer, the financial networks that gravitated around Big Oil were never far away. Carter himself, his vice president, his treasury and defense secretaries, and his secretary of state (among others) were all members of the "Trilateral Commission." This exclusive international think tank, founded in July 1973, which also included Henry Kissinger, had been designed as a global extension of the Bilderberg Group and the Council on Foreign Relations by the omnipresent David Rockefeller. Before joining the Carter administration, Zbigniew Brzezinski was the first director of the Trilateral Commission.[90] This commission (including some advisors, such as American political science professor Samuel Huntington, who were worried about the uncontrollable effects of unchecked democracy) embodied for its detractors the supremacy of the new technocratic order.[91]

Beginning in the 1970s, at least thirty thousand of the best Saudi students left each year to educate themselves in American universities—a supersonic social change for the humble tribes that had become the filthy rich of the Arabian desert.[92] The US-Saudi industrial and military symbiosis that emerged from the 1973 oil crisis and replaced the former pseudo-colonial regime was primed to produce its first schizophrenic outburst.

TWENTY-ONE

The Second Oil Crisis: A Deadly Vortex of Power Is Unleashed Around the Persian Gulf

The amazing wealth that carried the Persian Gulf countries after 1973 unleashed a hellish maelstrom of violence in the Gulf, magnified by the war machines complacently sold by their northern oil clients and patrons. In 1979, three decades of conflagration began. The primary focus was Iraq—a nation handicapped at birth, that possessed one tenth of the planet's oil reserves, and whose destiny had been unsettled since the First World War by its leaders' negligence and the cynical strategies of oil imperialists. This conflagration closely or indirectly implicated all Western powers that consumed oil, near and far. It was marked by Washington's most flagrant denials and by a sordid cruelty that served to reinforce and safeguard US domination of the world's most important petroleum sources.

This appalling history opened with a series of spectacular setbacks in the allegiances the United States had patiently built around the Gulf. They were tied to the crises that brought all of the Middle East to a white heat at a decisive moment in the Cold War. Throughout 1978, in Iran, increasingly serious problems swayed the shah's regime. Beginning in November, crude oil exports stopped abruptly. The second oil crisis had begun; its consequences were more profound than those of the first. On January 1, 1979, after months of often violent protests and their bloody repressions, Shah Mohammad Reza Pahlavi announced that he was preparing to leave Iran and take a "vacation." After two weeks, he abandoned the Peacock Throne and entered permanent exile, ousted by Ayatollah Khomeini's Islamic revolution. The "big pillar" on which Washington relied in the Middle East had just collapsed. Eleven months later, on November 4, four hundred young Iranian Shiite militants took US embassy personnel hostage in Tehran. They intended to make "Great Satan,"

America, pay. They accused America of having murdered Iranian democracy twenty-five years earlier and then having armed and trained the Savak, the barbaric political police who served the shah. Sixteen days later, on November 20, in Saudi Arabia's Holy City of Mecca, Sunni terrorists took hundreds of pilgrims hostage. They demanded the immediate removal of the Saudi royal family, who they accused of selling themselves to America. A month later, the night of December 24 to 25, Soviet troops entered Afghanistan; the Red Army was soon capable of stationing themselves 600 kilometers from the Strait of Hormuz. Funding the resistance against the USSR, with the United States' blessing, the House of Saud believed this was an opportunity to reinstate its honor in the eyes of a new generation of Muslim radicals, among whom emerged a certain Osama bin Laden. In September 1980, Saddam Hussein's Iraq attacked Khomeini's Iran. A very long war began, a war of trenches dug between the two countries' oil fields and terminals, where Saddam Hussein introduced chemical weapons manufactured with materials supplied by its American and European allies.

The American hegemonic strategy concerning the new central global energy structure had just birthed its three worst monsters: Iran's Islamic Revolution, the Salafi jihadism that matured in Saudi Arabia, and the military adventurism of Iraqi leader Saddam Hussein.

Iranian Revolution: Washington Loses Its "Big Pillar" in the Persian Gulf

It was a strange, sad fate that met the last shah of Iran. Put on the throne in 1941 by the British who had come to oust his father, then reinstated in 1953 by the Americans and the British in the name of protecting Western petroleum interests, during the 1960s and 1970s Mohammad Reza Pahlavi had tried in vain to win his people's affection through a policy of rapid Western modernization, which eventually worked against him.

Launched in 1963, the shah's White Revolution (a nonviolent revolution) was supposed to make Iran one of the five most advanced countries in the world. It enacted agrarian reform that alienated large landowners. It gave women the right to vote, angering religious leaders (already appalled by the regime's links with Israel). It launched ambitious industrial investment programs, which never managed to put an end to daily electricity outages, stop the development of slums, or alleviate traffic jams in Tehran. The White Revolution was intended to pacify the country. It offered freedom of expression at the same time that the shah's omnipresent secret police, the Savak (with whom one in every three Iranians collaborated), trained with Washington's help and tortured the shah's opponents, real or imagined, or made them disappear.[1] During the 1977 New Year's Eve banquet at the White House, US president Jimmy Carter complimented the Iranian people's "respect," "admiration," and "love" for their sovereign, whose library included the works of Gandhi and Voltaire.[2] Yet a few months earlier, when Washington had managed to convince the shah to open its prison doors to the Red Cross, the latter examined more than three thousand political prisoners who had been beaten, burned with cigarettes and chemical products, or sodomized with bottles and boiled eggs.[3]

During the 1970s, the oil boom generated enormous inflation and, in Iran as elsewhere, widened the gap between the lower classes and the rich. The presence of hundreds of thousands of foreign workers exacerbated tensions. Riots and attacks were perpetrated by opponents with growing frequency. On August 19, 1978, nearly four hundred people were burned alive in an arson attack on a cinema in the oil city of Abadan. The Savak were immediately accused. The arsonists were in fact Islamist revolutionaries; their attack remained the most murderous terrorist act until September 11, 2001.[4] The boiling point came on December 11, 1978, during the great feast of Ashura, which celebrates the martyrdom of Imam Hussein, the Shiite Muslims' symbol of resistance against tyranny. From a small pavilion of the Neauphle-le-Château near Paris, to the great embarrassment of French president Valéry Giscard d'Estaing, Ayatollah Khomeini Rouhollah, leader of the religious opposition, called for revolution: "Let them kill five thousand, ten thousand, twenty thousand. We will prove that blood is more powerful than the sword."[5] Enormous protests followed. Perhaps ten million people poured into the streets. Overwhelmed, the regime collapsed all at once, and this time its Western allies, themselves caught short, did not attempt to prevent the disaster. On January 16, 1979, the shah fled to Cairo, while everywhere in the world, the lines at the service stations got longer and longer. The fallen sovereign was allowed into Morocco, then into the Bahamas, and finally, into Mexico. The shah was doomed to a somber, wandering existence: He embarrassed Jimmy Carter, who feared alienating Iran's new rulers and feared reprisals against his diplomats and citizens. Despite the protestations of David Rockefeller, Henry Kissinger, and John McCloy, the shah was admitted into America only when he proved to be seriously ill.[6] Mohammad Reza Pahlavi was hospitalized in New York on October 23. He was then treated in Texas and Panama, before returning to Egypt, where he died of leukemia on July 27, 1980, a diagnosis that was made by French doctors years earlier but that had remained secret.

A huge crowd awaited the Air France Boeing 747 that landed in Tehran on February 1, 1979. The hieratic religious leader who descended, Ayatollah Khomeini, was a robust ascetic, sixty-seven years old, with a diet of onions, garlic, and yogurt. Exiled by the shah in 1964, he had lived in Iraq for fourteen years before being expelled by a Saddam Hussein made nervous by the virulence of the Shia militancy in his own country. The ayatollah was quickly elevated to the rank of quasi-deity by the Islamic Republic he established. As recounted by an Iranian historian exiled in the United States, "what began as an authentic and anti-dictatorial popular revolution based on a broad coalition of all anti-shah forces was soon transformed into an Islamic fundamentalist power-grab."[7] The Islamists inflicted a ruthless repression, first against the shah's faithful but very quickly against the revolutionary militants of any other faction. In the years to follow, there were thousands of public hangings of political prisoners: In the field of repression, Khomeini's regime clearly outstripped the shah's. The strongest opposition to the Islamic revolution was the People's Mujahedin of Iran. At first allied with Khomeini, the Mujahedin turned against the new regime starting in 1981, and later led a guerrilla war supported by Saddam Hussein's Iraq.

The Second Oil Shock and Neoliberalism's Political Victory

The second oil crisis began even before the shah's departure. To its opponents, the oil industry was both a symbol and a key target. Successful strikes, beginning in November 1978, led to important blockages. No fewer than thirty tankers waited to fill their tanks around the Kharg Island terminal, off the coast of Iran.

The twenty thousand executives and expatriate foremen who managed the numerous facilities were intimidated; they, for the most part, quickly left the country. On December 25, 1978, at the peak of winter demand, Iranian exports were completely halted. In London, the crude oil price instantly leaped more than 10 percent. Production in Iran, the world's second highest exporter after Saudi Arabia, remained as weak as it was erratic during the first three months of political turmoil in 1979. Panic seized the power brokers, who saw many of their supply contracts become obsolete. OPEC's rise in power had broken the old vertical structures and the cohesiveness of the Seven Sisters' cooperative. There were many more actors in the market; each tried to get the most oil and build reserves to protect themselves. The average barrel price, which had increased somewhat since 1973, rose from $14 in 1978 to more than $31, on average, the following year. It remained above $27 until 1985.

The lines and fights in the service stations returned, people siphoned gas out of other people's tanks, and vehicles ran out of gas and were abandoned on the roadside. One day in Los Angeles, a gas station attendant discovered that he had been delivered a tank . . . of water.[8] The chronic shortages continued until the summer of 1979. The bulk of Iranian production, however, finally returned to the world market. In addition, the fall in Iran's exports was almost totally offset by the end of 1978, in particular by Saudi Arabia. The United States, still much less dependent than Europe and Japan on Middle East oil, did not face a decrease in supply. Those primarily responsible for the chaos were motorists and industrialists who sought at all costs to hoard fuel, concluded an April 1979 US Treasury study that the White House tried to bury: It was too embarrassing.[9] Growth returned after the fissure of 1974–1975, and the thirst for petroleum, increased by the fear of shortage, constituted the essential cause of the soaring prices in 1979. The opportunism of a number of the market's actors cashing in on the panic (beginning with many producers, large and small) only made things worse. The supposed solidarity of the consumer countries' International Energy Agency accomplished nothing, other than the implementation of rationing measures that often increased tensions: It was every country for itself. But it was simpler and infinitely less politically risky to blame the usual suspects, Iran and OPEC. During a televised speech broadcast on July 15, 1979, President Jimmy Carter acknowledged dramatically that his country, king of oil up to this point, was vulnerable: "This intolerable dependence on foreign oil threatens our economic independence and the very security of our nation. The energy crisis is real. It is worldwide." Eight years after the beginning of the US crude oil production decline—which Alaskan oil production, started in 1977, could not fully offset—the US president, live in front of his fellow citizens, exposed the precarious position of American progress. For the most part, public opinion did not forgive the Democratic president for this admission of weakness in the American way of life. Carter knew he was putting his reelection at risk when he announced that "from now on, every new addition to our demand for energy will be met from our own production and our own conservation. The generation-long growth in our dependence on foreign oil will be stopped dead in its tracks right now." His successor in the White House, Ronald Reagan, strongly proclaimed that "America is back" but fared no better than Carter at keeping this promise.

The Iranian Revolution triggered a diplomatic and political crisis in the United States, which was a deciding factor in Ronald Reagan's overwhelming victory in the November 4, 1980, presidential elections. This crisis had begun exactly one year earlier. On November 4, 1979, in retaliation against the United States after they permitted the shah of Iran entry into the United States, Islamic militants invaded the American embassy in Tehran. For Medhi Barzagan, a liberal democrat who in February was appointed head of the provisional government with Khomeini's support, things had gone too far. As soon as he learned of the hostage situation, Barzagan—who during Mossadegh's time had been the first president of Iran's national oil company—resigned. For 444 days, Khomeini's young supporters kept 52 Americans hostage; among other vexations, they threatened to boil the feet of the embassy's chief of security in oil.[10] In the United States, where people feared a return of gas shortages, public opinion of Iran plummeted. Jimmy Carter was deemed incapable of resolving the crisis.* The president quickly banned Iranian oil imports, before breaking off all diplomatic relations with Iran in early April. But the commando operation that he authorized three weeks later to try to release the hostages failed miserably: Three helicopters out of eight crashed in a cloud of very fine sand, before a fourth accidentally collided with a supply plane.

Ronald Reagan, the Republican Party's candidate for the presidency, seized the opportunity to take advantage of the crisis. Fearing that the Carter administration would succeed in rescuing the hostages just before the November election, creating an October surprise that would swing the election, the campaign team of Reagan and his future vice president George H. W. Bush began its own parallel negotiations with Khomeini's regime. Gary Sick, Carter's strategic advisor for the Middle East, later accused Reagan of having struck a bargain with Tehran: The hostages would be held captive until after the US presidential election, and in exchange Iran would receive from Israel the spare parts it badly needed for its newly begun war against Iraq.[11] Several solid journalistic investigations supported this accusation.[12]

Whatever occurred, the American hostages were released on January 20, 1981, exactly five minutes after President Reagan was sworn into office.

The panic caused by the Iranian Revolution raised a new tsunami of inflation that was violently unleashed on the world economy, and whose consequences were even greater than what took place in 1973. Once again, the sharp, unexpected increase in the price of crude oil instantly affected transportation, construction, and agriculture—confirming oil's ubiquity. Blast furnaces

* Moreover, Jimmy Carter was ridiculed over another scandal involving another oil power hostile to the United States: Muammar Gaddafi's Libya. Beginning in 1979, Billy Carter, the president's brother, became a foreign agent of the Libyan government in exchange for $200,000. "Billygate" compelled Jimmy Carter to publicly separate himself from his brother in August 1980, three months before the presidential election.

used more coal, which reclaimed the ground that heavy fuel had taken.[13] The frenzied Japanese economy, particularly dependent on Iranian crude, was hit hard. In France, as everywhere in Europe, unemployment rates rose to a new level. In Great Britain, BP, which was still buying 40 percent of its oil from Iran, had to invoke a case of "force majeure" to interrupt its supply contracts with other companies and escape bankruptcy. The conservative government of Margaret Thatcher, who was elected on May 4, 1979, faced inflation rates of 18 percent the following year. The time of draconian monetarist policies advocated by economist Milton Friedman, David Rockefeller's protégé, had arrived. The Bank of England's interest rate was around 16 percent in 1980. The impact on the economy was brutal. In her first year and a half in office, the "iron lady" saw the number of unemployed double, approaching three million by the end of 1981. There were violent riots of the city's workers.

Appointed by President Carter in August 1979, three months before his departure, Paul Volcker, the new chief of the Federal Reserve, administered the same shock treatment to the American economy. Jimmy Carter had initially offered the position to David Rockefeller; Chase Manhattan's president politely declined the offer and "strongly" recommended that Carter appeal to Volcker (who had been a Chase vice president in the 1960s).[14] To stop the spiral of inflation that endangered the profitability and stability of all banks, the Federal Reserve increased its benchmark rate to 20 percent in 1980 and 1981. The following year, 1982, the American economy experienced a 2 percent recession, much more severe than the recession of 1974.

The chain reaction did not stop there. Because of interest rates imposed by the Fed, the central US bank, the price of American currency rose sharply. However, as the debts of many countries are often denominated in dollars and underwritten with variable rates, developing country debt exploded once again, even more than it had during the first crisis. In 1982, the first major victim was Mexico, which had launched an ambitious modernization of its industry after the first oil crisis, financed by its oil revenue. The World Bank, chaired by Robert McNamara, had multiplied the amount of its loans fourfold, building on frenetic growth and the strong oil export increase from Mexico. After the Volcker plan was enacted, the World Bank imprudently continued to encourage Mexico to accrue further debt. On February 19, the peso was devalued by 30 percent. Capital fled the country. And on August 16, the central bank of Mexico's director announced that after having paid colossal sums of money during the first months of the year, his country was no longer able to meet its payments. Extreme panic followed. After experiencing 8.7 percent growth in 1981, Mexico registered a recession of 4.2 percent in

1983. The country's development stopped cold. There was never a question of accepting a moratorium: With the Mexico crisis, the IMF inaugurated a series of its brutal neoliberal structural adjustment programs, formalizing a logic inaugurated in Chile in 1973 by Milton Friedman and colleagues from the University of Chicago. James Baker, the powerful Texan business lawyer who became Reagan's secretary of the treasury (after first serving as chief of staff during the Republican president's first term), intervened decisively in October 1985. In exchange for facilitating the repayment of their loans, he required fifteen heavily indebted developing countries to stop all public development. Among the four largest countries involved, three were major petroleum producers who, having too quickly adopted lavish lifestyles, had been surprised by the crisis: Mexico, Venezuela, and Nigeria. This was the advent of the "Washington consensus" based on the ideas of Milton Friedman—a purge of public spending through privatization and deep cuts in social programs callously administered to many developing countries that were heavily indebted as a result of the two oil crises.

The Birth of Wahhabi Terrorism, against the "Hypocrite" Saudi Princes

"Your attention, O Muslims! Allah Akbar, Allah Akbar! The Mahdi appeared!" The voice that sounded in the Great Mosque of Mecca at dawn on November 20, 1979, New Year's Day of the Islamic year 1400, was not that of the sheikh who was supposed to lead the first dawn prayer for the fifty thousand pilgrims gathered on the last day of the Hadj.[15] More than four hundred insurgents claiming to be "Muslim brothers" had seized control of the most sacred place in Islam. Some were descendants of the Ikwan warriors massacred by Ibn Saud's automatic weapons half a century earlier in the Battle of Sabilla. Very quickly, they sealed off the mosque's gates, slaughtered several Saudi police officers, and prevented the pilgrims from leaving. Among the terrorists, the one who claimed to be the Mahdi—the equivalent of the messiah in the Muslim tradition—was named Mohammed Abdullah al-Qahtani. But the real leader of the fundamentalist group, the one who in the course of a prison stay had met with al-Qhatani and revealed to him his divine election, was Juhayman al-Otaybi. He was a former corporal of the Saudi Arabian National Guard, armed and trained by the United States, responsible for the protection of the royal family and its petroleum fields. In Saudi Arabia, al-Otaybi and his disciples were already known for their fiery sermons; the echo of these, without a doubt, reached the ears of the young Osama bin Laden.[16] They verbally attacked the royal family and went as far as shredding bank notes to eradicate the king's image. For the first time, the House of Saud's duplicity was openly denounced. The rebels showed supreme contempt for Mecca's protectors: The king was also the imam of the Wahhabis, the rigorous Muslim movement at the very foundation of Saudi Arabia; the rebels in Mecca believed themselves to be purifiers. For them, the House of Saud was made of *munafiqun*, Muslim hypocrites whose real allegiance was to the Christians. The Great Mosque's loudspeakers blasted forth the claims of al-Otaybi and his fighters: They demanded nothing less than the royal family's departure and a refund of the money they were accused of taking from the people, the expulsion of Western civilian and military experts, and finally, the end of oil exports to the United States. However, just before the terrorists cut the telephone lines, an employee of the bin Laden family's firm had time to give an alert, and it was the BinLaden Group that told the king about the takeover and provided architectural plans for the Great Mosque, which they had renovated. During the siege, Salem bin Laden—the heir of the BinLaden Group who had chosen someone close to the Bush clan to act as his fiduciary in the United States—paraded by the Mosque many times, shouldering a machine gun, faithful to his flamboyant playboy image.[17]

The Koran prohibits any act of violence inside a holy place. The House of Saud pressured the first ulema of the kingdom to obtain a fatwa authorizing them to deviate from this rule. This ulema, Abdul Aziz bin Baz—an old cleric of the blind-faith tradition, convinced that the sun revolved around the Earth—had been Otaybi's teacher. Yet he quickly yielded.[18] At midday, the sultan prince ordered a general artillery attack and made the assault. He met defeat. His men descended on long ropes from an assault helicopter, reaching the middle of the central courtyard. All were massacred. Before giving up, the Saudi authorities considered sending dogs loaded with explosives or flooding the basement and electrocuting everyone (hostages included) with high-voltage power cables.[19] Vinnell Corporation, the American company of mercenaries who trained the Saudi Arabian National Guard, was called to the rescue.[20] But the director of Saudi intelligence and son of the late King Faisal, Turki Al Saud, knew that it was political suicide to more openly call for US assistance. Prince Turki, who was almost hit by a sniper bullet soon after his arrival at the mosque, consulted Alexandre de Marenches, the leader of the French secret service, with whom he had created the Safari Club. De Marenches advocated using gas. Three of de Marenches's elite French gendarmerie commandos were dispatched to Saudi Arabia, after they converted to Islam for this purpose. A toxic, nonlethal gas was injected into the mosque, but this operation also failed. The Saudi soldiers then drilled holes in the walls, and threw grenades into the rooms located below, killing insurgents but also many of the hostages. Finally, after two weeks of fighting, the remaining insurgents, pushed to their limits, surrendered. On January 9, 1980, Otaybi and sixty-two of his surviving supporters were decapitated. Never had the royal kingdom seen so many executed at once. A year and a half later, Washington received another clear warning about the threat now constituted by Sunni Islamic terrorism for the best US allies in the Arab world: On October 6, 1981, Egyptian president Anwar Sadat, the man who had signed the 1978 peace agreements with Israel at Camp David, was shot by a group of militant Muslims, probably members of the Islamic jihad.*

Afghanistan, Iran, and the Carter Doctrine

Between the terrorists' surrender of the Mosque and their execution, the Red Army entered Afghanistan on Christmas Eve, 1979. In the long term, this event not only contributed to the fall of the USSR but also shook up many political issues in the Arab and Muslim world. Oil money was the economic key to this reconfiguring. Osama bin Laden, who later assessed the Mosque's hostage-takers as good Muslims, spoke of his reaction to the announcement of the Afghanistan invasion by Soviet soldiers: "It enraged me, and I rushed to the scene. I arrived a few days later, at the end of 1979."[21] There are doubts about whether or not he took this precipitous trip. Be that as it may, Osama bin Laden, then twenty-two years old, was fascinated by the stories of miracles that beset the unequal battle of pious Afghans facing modern, brutal, unholy Russian troops: helicopters that got caught in ropes, clouds of birds announcing the arrival of fighter jets, corpses of martyrs escaping putrefaction.[22†]

As early as 1980, Jeddah, Saudi Arabia's second largest city, became the center for the recruitment of Islamists who volunteered to fight for Afghanistan. Saudi Prince Turki helped these religious militants, from the four corners of

---

* A fatwa approving the assassination had been issued by Egyptian imam Omar Abdel Rahman, nicknamed the "blind sheikh," who was condemned by the United States for his role in the World Trade Center bombing of February 26, 1993.

† Doubtless the mistrust of modern technology was a symptom of the trauma caused by the social change that was quickly and dizzyingly imposed on the Saudis by the oil industry's arrival—and also an outgrowth of the explosive hatred expressed by the most powerful conservative Islamic movements. One finds this mistrust in Ayatollah Khomeini, who criticized the "Satanic Progress" imported by "colonial powers." The contrast between Osama bin Laden the ascetic, who dressed in monk-like garb, and his half-brother Salem, a jet setter, is striking. Even for a revered visitor, the young Osama bin Laden scoffed at turning on the air conditioner in his Jeddah apartment. (See Lawrence Wright, *La Guerre Cachée: Al-Qaida et les Origines du Terrorisme Islamistes,* Robert Laffont, Paris, 2007, p. 100.)

the Arab world, "join the caravan" of the Afghan jihad, providing them with false papers. The Saudi princes made direct financial contributions to the jihad: This was, for them, a windfall capable of dispelling the humiliation they had suffered during the hostage-taking in Mecca. The BinLaden Group's employment office in Cairo served as an entry point to Afghanistan.[23]

In reality, the Saudis were in step with the CIA, who were present along with the Afghan mujahedin at least six months before the Soviet invasion. The US agency had firmly decided to undertake the defeat of the Red Army in a country that offered the promise of opening a passage to the Persian Gulf and the Indian Ocean, but whose occupation did not in itself offer the slightest strategic advantage, and which in addition was populated by invincible combatants who had already pushed back the tsar's troops during the British Empire's "Great Game." In 1998, Zbigniew Brzezinski revealed to *Le Nouvel Observateur* that the CIA had arranged to bait the USSR in Afghanistan: "We did not push the Russians to intervene, but we knowingly increased the probability that they will do so."[24] When the French weekly asked Brzezinski, formerly President Carter's national security adviser, whether he regretted promoting the full outbreak of Islamic terrorism and Al Qaeda, Brzezinski answered unhesitatingly: "What is most important regarding the history of the world? The Taliban or the fall of the Soviet empire?" His interview was published three years before the September 11 attacks. For Brzezinski, the issue of the Red Army's intervention in Afghanistan was more than "offering the USSR its Vietnam war."[25] The shadow that followed the Soviet empire's involvement in the Persian Gulf from then on inspired a new strategic direction that Brzezinski proposed to President Carter: the Carter Doctrine. That doctrine more clearly than ever positioned Arabian and Persian oil among the American empire's primary strategic priorities. Before alerting his fellow citizens about the precariousness of America's energy supplies, Jimmy Carter presented the doctrine that bore his name, both before Congress and, on January 23, 1980, live on television during his annual State of the Union address: "The region which is now threatened by Soviet troops in Afghanistan is of great strategic importance: It contains more than two-thirds of the world's exportable oil," he said. "Let our position be absolutely clear: An attempt by any outside force to gain control of the Persian Gulf region will be regarded as an assault on the vital interests of the United States of America, and such an assault will be repelled by any means necessary, including military force."

The geostrategic challenge around the Persian Gulf strengthened from empire to empire. The Carter Doctrine was strikingly similar to the 1903 British declaration cautioning the tsar of Russia, the Prussian kaiser, and the Ottoman sultan against any incursion in the Persian Gulf.[26] The Carter Doctrine was an important milestone in the history of American presence in the Gulf. How does one effectively defend a region so far away? The Pentagon, which previously had no real base in the area, accelerated the establishment of a "rapid-reaction force," decreed by Carter in 1977. This force had at its disposal one hundred thousand men ready to be deployed at any time in the Gulf, thanks to prepositioned equipment in Bahrain, Oman, Saudi Arabia, and the Diego Garcia atoll, located in the Indian Ocean, yielded de facto by the United Kingdom to the United States.*

During this time, Iran's Islamic Revolution never ceased increasing direct pressure on the "vital interests" of the United States. Beginning in January 1980, a month after the hostage situation at Mecca's mosque, Tehran encouraged the large Shiite population in the Saudi port of Qatif to rebel against the royal authority; Qatif was located in the heart of Saudi Arabia's oil complex, midway between Dhahran, Aramco's headquarters, and the giant terminal Ras Tanura. Riots followed, aimed, for example, at an agency of the Saudi British bank. On the shortwave, Radio Tehran took issue with the "amoral" behavior of many Saudi princes and continuously disseminated messages such as the following: "The ruling regime in Saudi Arabia wears Muslim clothing, but inwardly represents the US body, mind and terrorism."[27]

In October 1981, aiming explicitly at the threat Iran posed to Saudi Arabia, the new president, Ronald Reagan, established the "Reagan Corollary to the Carter Doctrine," declaring that the American army would from then on defend its vital interests not only against its external enemies in the Persian Gulf but also against threats from the inside. The same month, the Republican president had to use all his political might to get the Senate, largely aligned with Israel's cause, to approve the sale of five airborne early warning and control system surveillance planes (AWACS) to Saudi Arabia. Vice President George H. W. Bush and Reagan's cabinet chief, James Baker, weighed in favor of this plan. As soon as Caspar Weinberger became Reagan's secretary of defense, he added his support. Weinberger was the former vice president of Bechtel, the construction group engaged in multiple contracts in Saudi Arabia, whose president, George Shultz, became US secretary of state in 1982. The AWACS sales contract, at $5.5 billion, included a telecommunications infrastructure worthy of NATO, whose installation was notably awarded to the Bechtel group.[28] According to a *Washington Post* investigation whose publication was delayed until after the decisive congressional vote, this contract was the visible

* The population of this British-owned island was thoroughly expelled when the US Navy established a base there in 1971.

part of a secret plan between the White House and Riyadh: In exchange for access to the most advanced US military technology, Saudi Arabia had committed to building a network of air and naval bases that would be available to the American army in the event of a threat. This secret plan "was creating a new theater of war" for the US Army, reported the *Washington Post* investigation's author, Scott Armstrong.[29] During the next decade, the Saudis bought $200 billion worth of American military equipment and, with the help of US military and civilian engineers, constructed new deepwater ports, as well as a dozen important airfields.

On January 1, 1983, Reagan transformed the rapid reaction force into "central command"—CENTCOM in Pentagon jargon. The name signaled the high priority the Reagan administration gave to the Persian Gulf region. It also alluded to the "central region" between Asia and Europe to which it gave access; this was area that John Halford Mackinder termed the "geographical pivot of history" in his 1904 description of natural seats of power.[30] In keeping with the Americans' talent for strong images and their taste for simple ideas, the American army placed, on CENTCOM's crest, a bald eagle extending its wings across the Persian Gulf.

TWENTY-TWO

The Long Iran-Iraq War: A Lose-Lose Game Orchestrated by the Reagan Administration?

Once Washington, with Saudi Arabia's help, put the Red Army in check in Afghanistan, the war that Saddam Hussein waged against Iran in September 1980 served the United States as an instrument to curb the Iranian Revolution and its dangerous influence on the Shiite populations scattered elsewhere around the Persian Gulf and subject to Sunni rulers. The alliance between Washington and Baghdad was forged by Zbigniew Brzezinski. Carter's advisor considered Iraq ripe to be extracted, as Egypt before it was, from the sphere of Soviet influence: Baghdad must succeed Tehran as a main pillar of stability in the Persian Gulf.[1] But Iraq's return to the sphere of American influence was primarily the work of Brzezinski's predecessor, Henry Kissinger. Of course, this return became as tragic as it was crucial in defining the relationships of power at the center of the world's most strategic region. Such an outcome is understandable only by recalling the dual aim of US security policy in the Middle East, in place for more than half a century, which Henry Kissinger himself described in 2012 as: "preventing any power in the region from emerging as a hegemony; ensuring the free flow of energy resources."[2]

All Around the Gulf, Black Gold's Power Overflows, and the Storm Rises

To understand the way that the dual aim exposed by Kissinger long after the coup applied to the key role of Iraq, the most populated Arab country in the Middle East, we need to look backward. In 1963, as we have seen, the CIA's influence seems to have weighed heavily in the coup that drove out General Kassem, for the first time opening the doors of power to the Baath Party and allowing the young Saddam Hussein to begin his bloody career. In July 1968, a new coup chased non-Baath-Party members out of the Iraqi government and allowed General Ahmed Hassan al-Bakr, the Baathist leader, access to the presidency. In 1971, he appointed his cousin, Saddam Hussein, like him a native of Tikrit, as vice president. The number two Iraqi military intelligence member, Hussein, asserted in his memoirs: "For the 1968 coup you must look to Washington."[3] It is plausible that Washington strongly preferred the Baath Party to a more versatile democratic regime.

If such was the case, the Baath Party seemed unaware of it. In 1972, Baghdad nationalized its petroleum and, worse, signed a treaty of friendship and cooperation with Moscow. Iraq did not become a satellite of the Soviet Union, but rather a hard and fast client: In the mid-1970s, the Russians provided the bulk of the arms that Iraq bought with its extensive oil profits. The rest was provided by the French, who had negotiated privileged access to Iraqi oil. But Baghdad did not completely close the door to Americans, particularly not to the Texan company Brown & Root, the builders of huge oil facilities.

Henry Kissinger and Richard Nixon nevertheless decided to punish Iraq. "To keep Iraq from achieving hegemony in the Persian Gulf, we had either to build up American power or to strengthen local forces," Kissinger later explained; America chose the second solution because the first would have been an "enormous, perhaps insuperable" task.[4] In 1972, Kissinger established with the shah of Iran a plan to "keep Iraq occupied by supporting the Kurdish rebellion."[5] Since the Ottoman Empire's implosion following the First World War, the Kurdish communities had been divided, mainly between modern Turkey, Syria, Iran, and Iraq; they were regularly persecuted by the regimes of these countries. Exploiting Iraq's congenital fragility, Kissinger encouraged the shah to support Iraq's Kurds against Baghdad's central authority.[6] Having well understood the "rules" of the game, the independent Kurdish leader Moustapha Barzaki assured readers, in the June 1973 *International Herald Tribune*, "We are ready to do what goes with American policy in this area if America will protect us from the wolves. If support were strong enough, we could control the Kirkuk field and give it to an American company to operate. It is in our area, and the nationalization was an act against the Kurds."[7] But for Kissinger there was no question of permitting the Kurds to operate freely; rather, their combat was to be used as "merely an instrument to dissuade Iraq from any international adventurism."[8] Thanks to the sophisticated weaponry that the shah supplied, the Kurds frequently raided Iraq's oil installations. The shah, however, took care to keep a tight rein on the Kurdish combatants, the Peshmerga, never providing them with more than three days of ammunition.[9] In 1975, 45,000 Kurdish combatants, supported by two Iranian divisions, were keeping 80,000 Iraqi soldiers "busy." Kissinger wanted to snatch Iraq from the influence of its Soviet sponsors. The shah desired a share of the Shatt-al-Arab, the common estuary of the Tigris and Euphrates rivers, control of which had been granted exclusively to Iraq in 1937. Iraq granted that share and the two countries agreed to restore neighborly relations when they signed the Algiers Agreements on March 6, 1975, during a meeting of OPEC. Eight hours later, the shah interrupted delivery of aid, including food, to the Iraqi Kurdish rebels; Saddam Hussein then deported 200,000 to 300,000 of these Kurds to the country's south.[10] Not a single Kurdish rebel received political asylum from the United States.[11]

After the Algiers Agreements, business with Iraq resumed. Many articles, almost exclusively devoted to Iraq's social projects, flourished in the pages of European and American newspapers.[12] Saddam Hussein, who modeled himself on Stalin (eventually visiting, one by one, the "little father of the peoples'" fifteen villas on the shores of the Black Sea), mercilessly executed many Communist militants in 1978, before banning the Communist Party the following year.[13]

In July 1979, Saddam Hussein replaced his parent and protector, President al-Bakr, and formally assumed an absolute power that he held for a long time. The bloody dawn of Iran's Islamic Shia revolution constituted a deadly threat to Iraq's Sunni-dominated Baath Party. Iraq, as designed by the British at the time of its creation in 1921, presented the particularity of being dominated by the Sunni minority who primarily occupied the center of the country, while the oil fields were located in the Kurdish north and in the predominantly Shiite south. Saddam Hussein decided to attack religious and militant Shiites in Iraq. Accused of being on Khomeini's payroll, many Shiites were tortured in the Abu Ghraib prison and elsewhere: Some had their limbs severed by a circular saw, while others had their intestines filled with gasoline through the rectum; many hundred were hanged.[14] The victims included Ayatollah Mohammed Baqir al-Sadr. In the aftermath of an April 1, 1980, attack that failed to kill Prime Minister Tariq Aziz, Muhammed Baqir al-Sadr was arrested and tortured along with several members of his family, then killed by a nail driven through his skull.[15] His body was then burned and dragged by a tractor, along with his sister's body, through the streets of Najaf, one of the Shiites' holy cities.* Twenty Baath Party leaders were assassinated in April 1980; forty thousand Shiites were banished to Iran.[16]

* Muhammed Baqir al-Sadr's son-in-law, Muqtada al-Sadr, became the scourge of the American army after the 2003 invasion, in which he headed the Mahdi Army. "Long live Muhammed Baqir al-Sadr," a witness cried out, at the hanging of Saddam Hussein on December 30, 2006.

1980–1984: The Reagan Administration's First Escapades and Donald Rumsfeld's Trip to Baghdad

Iran accused Iraq of many attacks on border villages. Starting in August 1980, there were daily artillery fire exchanges.[17] On September 17, Saddam Hussein sent his Parliament a message declaring that "the frequent and flagrant violations of Iraqi sovereignty . . . have rendered the 1975 Algiers Agreement null and void." Five days later, he sent troops and thousands of tanks across Iran's border on a 650 kilometer front. Moscow condemned the invasion; Washington was very discreet, although a year earlier the United States had registered Iraq on the list of terrorist states because of the support Baghdad provided to the Palestinians. Prince Fahd, future king of Saudi Arabia, affirmed that he gave Saddam Hussein the green light on behalf of the American president Jimmy Carter!* If true, President Reagan continued Carter's policies: Even though the US Congress refused to consider lending military aid to Saddam Hussein due to the latter's hostility toward Israel, the White House secretly authorized Kuwait, Saudi Arabia, and Egypt to transfer US arms to Iraq.[18] Vice President George H. W. Bush was then one of Saddam Hussein's strongest supporters.[19]

The Iraqi president promised a "lightning war." Its main objective was to take hold of Khuzestan, in southwest Iran: a province also called Arabistan, whose important Shiite population, not Persian but Arab, wanted to escape Ayatollah Khomeini's regime. This province also happened to be the heart of the Iranian oil industry, in particular the centers of Abadan and Ahwaz. By taking control of the Khuzestan, Iraq could instantly double its oil production capacity and control between one-tenth and one-fifth of the planet's reserves. This kind of annexation was more than enough to make Saddam Hussein into what Kissinger and his successors feared: the "hegemon" of the Persian Gulf.

Oil was needed immediately, a strategic essential of the "lightning war" launched by Iraq's Saddam Hussein, which lasted for eight years. The Iranians called the Iran-Iraq War the "imposed war"; when we focus on the role played by the Reagan administration and its Israeli and Saudi vassals, it seems that it was also a manipulated war, in which Washington appeared to ensure that no winner prevailed.

---

* Ronald Reagan's secretary of state, Alexander Haig, conveyed this secret, which Fahd had shared with him, in a confidential report sent to Reagan in April 1981 and uncovered by investigative journalist Robert Parry. In this same account, Haig also told Reagan that Prince Fahd had explained that Iran had received, via Israel, spare parts for American equipment. (We will return in further depth to this question of the delivery of American arms to Iran via Israel.) See also Robert Parry, "Saddam's Green Light," *Consortium for Independent Journalism*, 1996, searchable at consortiumnews.com.

A week after the invasion of Iraq began, on September 29, Iranian shells began to rain down on Iraqi oil depots in the Fao peninsula, at the mouth of the Shatt-al-Arab. In retaliation, the Iraqi army bombarded the Abadan refinery, only forty kilometers away from Fao, on the other side of the Shatt-al-Arab. Starting on October 2, Saddam Hussein's troops began a long, unsuccessful attempt to surround this heart of Iran's petroleum industry. Also nearby, the Basra refinery, one of Iraq's principal refineries, was targeted by US-made Iranian fighter jets, manned by US-trained pilots. The pipelines that departed from the Kirkuk region, in Iraqi Kurdistan, were also targets; for better or worse, they remained virtually the only feasible way to transport Iraq's crude oil. The refinery extinguished its flames to avoid being targeted and, for security, many wells on both sides of the border were closed (their reopening was a slow, delicate operation). In a few weeks, Iraqi production collapsed and Iranian extractions, already harmed by the revolution and the departure of foreign engineers, slowed down even more.

For a year and a half, Iran and Iraq were bogged down in Iranian territory, both losing equally. The Islamic Republic of Iran Army continued, as in the time of the shah, moving to US military tunes; the Iraqi army did the same.[20] With American equipment acquired during the shah's rein, the Iranian army was vulnerable, in danger of rapidly finding itself short of spare parts.[21] "If the United States withdrew their support, the Iranian forces probably could not withstand large scale hostilities for more than two weeks," predicted the author of a document seized at the US embassy in Tehran.[22] Yet the Iranian army held fast. The deliveries of US arms from Israel between 1980 and 1983, including the high-performing TOW antitank missiles (paid for primarily by oil shipments), were undoubtedly destined for use in this unexpected resistance.[23] This equipment was valued in the hundreds of millions of dollars.[24] Without even invoking the quid pro quo of the October Surprise, it's difficult to imagine that such shipments could occur without the Reagan administration's green light—or at least without its knowledge.[25]

The war's first turning point came in March 1982. The Iranian army, more numerous and by all accounts materially advantaged, went on the offensive, using very young soldiers, often adolescents, to advance in martyrdom in the minefields of Iraq. Basra quickly found itself threatened. In April, at the request of Iran's Islamic Republic, Syria closed the Kirkuk-Baniyas pipeline, further draining Iraq's supply of foreign currency. At this juncture, Ronald Reagan decided to do everything possible to save Saddam Hussein from defeat. Iraq was removed from America's list of terrorist states. Arms dealers, mainly American, British, and French, rushed to Baghdad. Iraq also enjoyed the support of the

Saudi Arabians and Kuwaitis, amounting to tens of billions of dollars. And in 1982, the CIA authorized the sale of cluster bombs to Baghdad.[26]

During one of the obscure battles that took place at the beginning of that year, an Iraqi general dumped huge quantities of oil into the marshes south of the Iraqi city of Amara. He then ignited the waters with incendiary bombs and dropped high-voltage power cables into the flaming marshes. A Lebanese Associated Press correspondent described the corpses of Iraqi civilians floating among those of Iranian soldiers: "their bellies opened like fish, . . . women and children—the people of the marshes, those who knew what a toad was, who lived with the ducks and buffaloes and fished with spears, this civilization . . . in the process of disappearing." This civilization was living on the same land that, six thousand years earlier, was the cradle of many civilizations.[27]

The first reports of Saddam Hussein's chemical weapon use appeared in 1982. He used them almost daily, according to American diplomatic correspondence in 1983.* However, on December 20, 1983, Donald Rumsfeld, designated the White House's special envoy in the Middle East, landed in Baghdad to engage in what he later called a peace mission. The former Pentagon chief under Gerald Ford, who later became CEO of Searle Laboratory, shook Saddam Hussein's hand on behalf of the Reagan administration.† Then Rumsfeld, the former navy airman with an enigmatic, carnivorous smile, joined Tariq Aziz, deputy prime minister and leader of Iraqi diplomacy, in a tête-à-tête that lasted over two hours. No one mentioned chemical weapons. The conversation focused on the Bechtel company's proposition to build a pipeline from the Euphrates oil fields in the south of Iraq to Jordan's territory in the Red Sea's Gulf of Aqaba.[28] US secretary of state George Schulz was Bechtel's former president. Wrote Rumsfeld in a secret diplomacy dispatch he sent the day after the meeting, "I raised the question of a pipeline through Jordan. [Aziz] said he was familiar with the proposal. . . . It apparently was a US company's proposal. However, he was concerned about the proximity to Israel as the pipeline would enter the gulf of Aqaba. . . . He said they are

* That year in Beirut, Hezbollah, Lebanon's Iranian-supported Shiite movement, stepped up car bomb attacks, particularly targeting soldiers and American diplomats (a fuel-nitrate mixture proved to be the most efficient of cheap explosives). A retaliatory, CIA-sponsored attack on March 8, 1985, failed to kill a Shiite religious leader in Beirut suburbs but left 80 dead and 256 wounded, including women and children. It left several babies charred in their cribs. The operation was financed by Prince Bandar, the Saudi Arabian ambassador in Washington, who to this end withdrew $3 million and deposited it in a Swiss bank account owned by the CIA. (See Bob Woodward, *Veil: The Secret Wars of the CIA, 1981–1987*, Simon & Schuster, New York, 1987, p. 455, cited in Mike Davis, *Petite Histoire de la Voiture Piégée*, La Découverte, Paris, 2012, p. 117–118.)

† A pharmaceutical and agrochemical laboratory purchased by Monsanto in 1985 following a drastic downsizing ordered by Rumsfeld, Searle is now part of the Pfizer Group.

interested but need to find the right formula. He felt that it could be done for less than two billion dollars and recognized that it would take about two years because of the planning required." Rumsfield then notes in his dispatch that he told Aziz that he "could understand that there would need to be some sort of arrangements that would give those involved confidence that it would not be easily vulnerable." He concluded, "This may be an issue to raise with Israel at the appropriate time."[29]

Four months later, on March 5, 1984, American diplomacy officially condemned Iraq's use of chemical weapons. On March 24, American secretary of state George Shultz expressed his concern to Rumsfeld: Wouldn't this condemnation risk undermining relations with Iraq?[30] On March 26, Rumsfeld returned to Baghdad to assure Iraq that the condemnation was political posturing and, without any doubt, to discuss anew Bechtel's proposed pipeline project sale.[31] This project did not materialize, notably because Saddam Hussein was too concerned about the Israeli threat since the Jewish State had reduced his Osirak reactor to dust (to the seemingly great displeasure of Vice President George H. W. Bush and James Baker, then head of Ronald Reagan's cabinet).[32] Also on March 26, a UN survey confirmed to the whole world that Saddam Hussein had made extensive use of chemical weapons.

The next day, near Iran's Kharg Island oil terminal—including military facilities built by the US Army Corps of Engineers during the shah's rein—an Iraqi fighter jet damaged a Greek oil tanker. The *Filikon L* was transporting 80,000 tons of Kuwaiti oil. But the Iraqi pilot, who flew a French-made Super Étendard, most likely believed that the crude oil came from Iran. This was the beginning of the long "tanker war," during which the US Navy offered its protection to all oil tankers, except those carrying Iran's black gold, although Iraq bore the responsibility for the extension of the conflict. By attacking the oil tankers in a war that it no longer seemed capable of winning, Saddam Hussein hoped to push Iran to retaliate by closing the Strait of Hormuz, thus likely causing the United States to directly intervene in the conflict. But, despite its ritual threats with each new attack, Tehran was not careless enough to provoke the US Navy with such a *casus belli*. More than five hundred ships and a few offshore platforms were hit or destroyed by the end of the war, principally the result of Iraqi air attacks. These ships and platforms were most often registered in neutral countries.

While the British-made Iranian frigates effected a naval blockade in Iraq, Saddam Hussein proved unable to seriously reduce Iranian exports. However, the chaos generated in the Gulf's narrow waters threatened several times to become a larger scale conflict. On April 25 and May 7, 1984, Iraqi equipment partially destroyed two Saudi oil ships carrying crude oil from Iran. Riyadh remained neutral. On June 5, following one of Iran's first attacks in Saudi Arabian waters, two Saudi F-15 pilots (the most powerful American fighter jets) slaughtered two old Iranian F-4s that had been used in Vietnam.[33]

Saddam Hussein's Chemical War, Irangate, and the Reagan Administration's Strategy

Iran had twice as many inhabitants as Iraq: Chemical weapon use was strategically logical for Iraq and the foreign governments who didn't want to see him losing the war. From 1985 to 1990, Washington allowed US companies to sell Iraq products designated as "dual-use," civilian and military: Raw chemical ingredients allowed the possibility of manufacturing either fertilizers or weapons (in particular, mustard gas and nerve gases, such as sarin gas), and American-made helicopters that had the ability to spread these chemical weapons.[34] A *Washington Post* investigation specified that these products authorized by the United States also included "deadly biological viruses, such as anthrax and bubonic plague."[35]* The list of Iraq's suppliers included: eighty-six West German, eighteen British, eighteen American, seventeen Austrian, sixteen French, twelve Italian, and eleven Swiss companies, according to the inventory transmitted by Baghdad to the United Nations in December 2002, which George W. Bush's administration attempted to censor.[36] In total, Iraq admitted importing two hundred thousand bombs capable of carrying chemical weapons between 1983 and 1989, and dropping nearly twenty thousand of these.[37] In early 1985, Iraqi radio broadcast an official communique of the Iraqi army, warning: "Swarms of insects attacked the Arab nation from the East. But we have pesticides to eradicate them."[38]

President Reagan seemed to have clearly chosen the Iraqi camp, until November 3, 1986, when a little known Lebanese magazine, *Al-Shiraa*, revealed what became known as Irangate, or the Iran-Contra Affair. Since August 1985, the United States had supplied Iran with several thousand TOW antitank missiles as well as Hawk ground-to-air missiles, first via Israel, then, beginning in 1986, directly from the United States. A portion of the proceeds from these heavy weapons sales was used to fund the Contras, a paramilitary group that attempted to topple the Sandinista government of Nicaragua, Central America. The leak may have originated with an Iranian officer, Mehdi Hashemi, opposed to the secret deals with "Great Satan" America. A group

* There is no evidence that such weapons have ever been used.

of senior White House officials, who had designed the scheme—in violation of the embargo Washington had imposed on Tehran—justified it by stating that it was intended to obtain the release of American hostages in Lebanon, a country prey to civil war, by Iranian-supported militant groups: Hezbollah and the Palestinian Islamic Jihad. The United States obtained the liberation of three hostages. For its part, Iran took control of Iraqi oil installations on the Fao peninsula, in February 1986 winning a major victory in which American missiles played a significant role.

The tortuous history of the Irangate scandal involves several pathways where petroleum and power entangle in a rather incongruous manner. From their inception, Nicaragua's counterrevolutionary forces, the Contras, benefited from the support of the Soto Cano CIA base in neighboring Honduras. It was the agency's most important base in the world.[39] In 1979, the Sandinistas, a widespread popular movement whose leaders included Marxists, left-wing priests, and nationalists, succeeded in ousting from power the Somoza family dictatorship and its "uniformed mafia," which Washington had supported since 1932. Augusto Sandino, the revolutionary leader who had fought the Somoza dictatorship during that time, was introduced to the ideologies of the extreme left by American trade unionists from the Industrial Workers of the World, employed by Standard Oil at the Tampico, Mexico fields; Sandino would also have met the expatriate writer B. Traven, grand critic of capitalism.[40]

In October 1983, some members of the US Congress were upset when they learned the Reagan administration had helped the Contras take over port facilities that belonged to Exxon, pushing the American company to interrupt its deliveries of Mexican oil that were vital for Nicaragua.[41] In October 1984, the US Congress unanimously adopted the Boland Amendment, which strictly limited the support that the White House was authorized to provide to the Contras. During Irangate, the Reagan administration violated American law with activities in both Iran and Nicaragua. As the scandal emerged in the United States, Ronald Reagan lied several times in front of the media, and found himself under threat of impeachment.[42] However, neither he nor his vice president, George H. W. Bush, were concerned about legal repercussions, unlike secretary of defense Caspar Weinberger and the operation's direct initiators, Lieutenant-Colonel Oliver North and Rear-Admiral John Poindexter, who were convicted, though never imprisoned. Caspar Weinberger was pardoned by President George H. W. Bush on December 24, 1992, just before his departure from the White House, after losing to William Jefferson Clinton. The following year, Weinberger became chair of the powerful American economic magazine *Forbes*. The only American who was incarcerated as a result of Irangate was a pacifist activist from Indiana, Bill Breeden: He spent four days in prison for stealing the street sign from a street that recently had been renamed *John Poindexter*, and demanding a ransom equivalent to the sum transferred from Tehran to the Nicaraguan Contras via the CIA.[43]

John Poindexter's government career was briefly restarted in 2002 by George W. Bush, who appointed him head of the Total Information Awareness program, created as a result of the September 11, 2001, terrorist attacks, and later renamed Terrorism Information Awareness Office, a more suitable name. The rear-admiral had to resign from his position a year later, when it became known that his office oversaw the development of a project baptized FutureMAP, which facilitated speculation on the prediction of geopolitical events in the Middle East, particularly terrorist acts, political assassinations, or even coups d'état.[44] In 2005, a week before President George W. Bush's reelection, Oliver North resurfaced in the American media as a correspondent for the pro-Republican Fox News channel; reporting from Iraq, he touted the American troops' effectiveness in the face of the insurrection.[45]

Saudi Arabia proved to be the Contras' largest funding source by far.[46] Though totally illegal, the CIA's request for Saudi support allowed it to discreetly circumvent the Boland Amendment. As the intelligence agency's former director, Vice President Bush was not frightened away from using such a mechanism. He even appears to have been the initiative's principal promoter.[47] In 1986, a legislative inquiry conducted by Senator John Kerry showed that the Contras' air shipment network had been used by drug traffickers. American investigative reporter Russ Baker noted that the legal firm responsible for the smokescreen company that Oliver North had registered in the Cayman Islands to collect Iranian money for the Contras also operated on behalf of a small aircraft company, Skyway Aircraft Leasing Ltd., belonging to Salem bin Laden and managed by Jim Bath, the Bush family's pilot friend.[48]

In 1987, after the considerable upset Irangate caused in public opinion and for Washington's Arab allies that had been funding Saddam Hussein's war, the Reagan administration chose to increase pressure on Iran. The US Navy greatly accelerated its activities in the Persian Gulf, notably protecting the ships coming from or destined for Iraq. Beginning in March 1987, taking advantage of how the Iranian war effort was losing steam, Saddam Hussein launched with impunity a campaign to eradicate the Iraqi Kurds. During the twelve months that followed, the Kurds underwent 211 days of chemical attacks.[49] In February 1988, Saddam Hussein began an operation of massacres and systematic looting, called Al-Anfal ("spoils of war"). Approximately 80 percent of Iraq's Kurdish villages were destroyed, tens of thousands of their men, women, and children died.[50] On March 16, within a few brief hours, five thousand Kurds in the village of Halabja were killed by a cocktail of deadly gas. The images reported by Iranian journalists horrified the entire world. According to the *Los Angeles Times*, the deadly gas had been released by American-made helicopters purchased by Iraq to spread pesticides.[51]

Yet when, in August 1988, US Senate members proposed the Prevention of Genocide Act severely sanctioning Iraq for its chemical weapons usage, notably by boycotting its oil, Secretary of State George Shultz indicated that he considered the initiative "premature," and Ronald Reagan threatened to veto it.[52] The top-tier legislators opposing the sanctions included Wyoming representative Dick Cheney. According to the advisor to the Senate committee that proposed the sanctions, Reagan's national security advisor, Colin Powell, had orchestrated the decision to give Saddam Hussein a blank check to gas the Kurds.[53] Yet, for a long time, the CIA accused Iran of being responsible for the Halabja massacre—an assertion that was debunked, notably by a thorough study of the Human Rights Watch group.[54] (Fifteen years later, in September 2003, after Iraq was invaded by a coalition formed by the George W. Bush administration, Secretary of State Colin Powell traveled to Halabja; before the press, he saluted the memory of the victims.) On April 17, 1988, thanks to a massive recourse to gas weaponry and almost universal indifference, the Iraqi army launched an offensive that allowed them to quickly retake the Fao peninsula, lost two years earlier.[55] The following day, Washington radically heightened pressure on Tehran: After an American frigate struck a floating mine, the US Navy launched its largest naval offensive since the Second World War, Operation Praying Mantis, in which it engaged the USS *Enterprise*, the biggest nuclear aircraft carrier in the United States and the largest military ship in the world. The Pentagon chose several Iranian oil platforms as targets for its reprisal, marks that were as easy as they were vital. On July 3, with tensions at their height in the Gulf waters, the American ship USS *Vincennes* erroneously shot down Iran Air flight 655, killing all 290 civilians on board. On the defensive, at the end of its strength, Iran accepted a cease-fire that marked the return to the 1975 Algiers Agreement borders that had been in force before the war. At a cost of several hundreds of thousands of dead (the actual record remains uncertain), the twentieth century's longest conventional war had reached its futile conclusion.

On October 2, 1989, fourteen months after the cease-fire that ended the conflict, new American president George H. W. Bush signed National Security Decision Directive No. 26. After recalling in its preamble that "access to Persian Gulf oil and the security of key friendly states in the area are vital to

U.S. national security," this secret directive stipulated that "normal relations between the United States and Iraq would serve our longer-term interest," and that the American government ". . . should pursue, and seek to facilitate, opportunities for US firms to participate in the reconstruction of the Iraqi economy, particularly in the energy area."[56]

During the eight years of a conflict in which the United States was supposedly neutral and that, in the end, had no victor, we see that the Reagan administration fanned the flames, sometimes toward Iran, sometimes toward Iraq, and often in both directions simultaneously. According to Sarkis Soghanalian, an arms salesman omnipresent at the end of the Cold War, Jordan's king Hussein, who was one of Washington's most dependable interlocutors in the Arab world, had confided to Soghanalian that he was convinced the CIA had pushed Iran and Iraq to destroy one another.[57] What could have been the Reagan administration's objective? In May 1985, three months before Irangate's first arms delivery, a memo signed by Graham Fuller, the CIA's Middle East analyst, proposed an answer: "Our tilt to Iraq was timely when Iraq was against the ropes and the Islamic revolution was on a roll. The time may now have to come to tilt back."[58] In 1987, "official anonymous sources" cited by the *New York Times* claimed that the Reagan administration had provided the two belligerents with "deliberately distorted or inaccurate intelligence data," particularly in the form of satellite photographs that had been manipulated in order to hide or exaggerate concentrations of troops. These sources said the information had been shared in an effort to "prevent either Iran or Iraq from prevailing in their conflict."[59] Did the Reagan administration—within which thrived a few of the future key players of both Bush administrations—seek to prevent the emergence of another dominant power, besides the United States, in the Persian Gulf? If yes, then Iraq, after having served as the Seven Sisters' main lever for controlling oil-production behavior, emerged as the main instrument of such a game.

In July 1988, when Iran informed the United Nations that it accepted a ceasefire, Ayatollah Khomeini, at the head of a weary nation, intoned: "Taking this decision was more deadly than taking poison." Iran's chief religious leader pointed at those countries he considered Iran's principal enemies, beyond Saddam Hussein: "God willing, we will empty our hearts' anguish at the appropriate time by taking revenge on the House of Saud and America."[60] Khomeini died less than a year later. The United States is far from having settled its accounts with Iran and its Arab enemies.

TWENTY-THREE

The Oil Countershock: The Frenzy of the Reagan Years, the Collapse of the USSR, and the BCCI Scandal

It took Iran and Iraq decades to retrieve the record production-capacity levels they achieved before the war that pitted them against each other from 1980 to 1988. Two years after the Iranian Revolution and a year after the conflict began, hysteria and chaos helped to maintain high barrel prices. However, by 1981, in Houston, Rotterdam, and Dhahran, experts had begun to realize that a total transformation of the economic situation was looming: Ten tense years after US oil reached its peak in 1970, the world economy was poised to benefit from a new petroleum surplus.[1] Euphoria once again seized capitalists, who this time concluded that there were no limits on their enterprise, that the trees of the forest were destined to rise to the sky.

The Western Oil Empire Strikes Back

The Iranian Revolution imprinted a lesson on people's consciousness, one that even those who had been oblivious to it in the aftermath of the 1973 oil crisis could not miss: It was mortally dangerous to depend too much on oil. The tectonic movements of the global energy system, unleashed after the Yom Kippur War, had been amplified and accelerated after 1979: Thousands of factories, whole industrial sectors, shifted their energy source from oil to electricity. Electricity production itself no longer relied so heavily on oil: In the early 1980s, only one-tenth of the total electrical power generated in the rich countries of the OECD still relied on oil, compared to one-quarter in 1973.[2] The success of small, energy-efficient vehicles like France's Renault 5 and its many

Japanese competitors helped subdue but did not halt the growing demand for crude oil. This was not the beginning of a global weaning: far from it. After a long hiatus, induced by the violent recession that followed the 1979 oil crisis, world oil consumption returned to precrisis levels, around 62 Mb/d, by 1986.

Begun in response to the 1973 crisis, the delicate, onerous exploitation of the first major oil fields in a distant, extreme location (Alaska) and offshore (in the North Sea, the Gulf of Mexico, and West Africa) was bearing fruit: The abundance of these new, more expensive oil sources helped ease tensions in the black gold market. An even more complex endeavor attempted in 1973 by the American industry—the development of oil shales—suddenly became pointless. Extremely capital-intensive, it did not turn a profit due to the oil surplus. After investing at least a billion dollars in its Colony project, a large-scale exploitation of oil shale in Colorado and Utah, Exxon suddenly abandoned the venture on May 2, 1982, leaving thousands of workers and their families high and dry just a week before the company's centenary celebration.[3] Most unconventional hydrocarbons were still a quarter-century away from development. During that same year, the global crude oil market started to develop a surplus and the barrel price began a long, downward slide; the ingredients of an oil countershock had appeared.

The new oil regions that opened up after the 1973 crisis eluded the OPEC countries' control. The Western oil majors resumed the upper hand. Starting in 1981, OPEC's production was surpassed by that of the rest of the capitalist world. The irony of history and the luck of geology conspired: Thanks to the North Sea, Britain, the former colonial power, became the Gulf countries' most critical new competitor. OPEC also lost part of its market share to the USSR. The monumental Soviet hydrocarbon sources (with more than 12 Mb/d, the USSR was then the world's number one producer, ahead of Saudi Arabia and the United States) allowed them to sell their oil surplus throughout the capitalist world. Moscow thus assured itself a valuable source of dollars and other convertible currencies.

OPEC's progressive market-share loss, heightened by the Iran-Iraq War, was exacerbated by a price problem. The traditional system of prices set and administered by OPEC (which remained higher than $34 per barrel in 1981) often led to higher prices than those negotiated on open markets, to which all producers had access. The organization's member countries faced a dilemma: either lower their prices or slow their extractions. During conferences held at OPEC's Vienna headquarters in March and December of 1982, after lengthy discussions, the cartel's member states for the first time established a production quota system. In order to support their prices, they set a global ceiling at 17.5 Mb/d. The organization's prime actor, Saudi Arabia, was the country most readily able to regulate the market's equilibrium, and accepted a significant reduction in its production.

But the OPEC cartel was less efficient than its predecessor, the Seven Sisters cartel, no doubt because, unlike the latter, OPEC controlled less than half the world's production. In early 1983, at the height of the crisis, world demand declined sharply. In February, the barrel price of North Sea Brent Crude dipped below $30. Some producer countries, heavily indebted as a result of surging interest rates following the second oil crisis, experienced the plummeting barrel price as another slap in the face; the former British colony of Nigeria, whose oil was in direct competition with that of the North Sea, saw its customers desert it: For a time, the country was forced to nearly halt its crude oil exports.

In March 1983, the OPEC countries met again, this time—a paradoxical symbol—in London. All member states were allocated a specific quota for the year and agreed to lower their benchmark price from $34 to $29. However, the price of crude oil on the "spot" market (for "cash market" in which an asset is immediately paid for and delivered), which was beyond OPEC's control, continued its slide. Exxon, Mobil, Texaco, and SoCal—Aramco's former partners prior to the company's 1980 nationalization by the Saudi crown—were reluctant to pay the high Saudi Arabian oil prices. The Saudi oil minister, Sheikh Yamani, then induced a "shock treatment." After a reduction from 10 Mb/d to 7 Mb/d between 1981 and 1982, Saudi Arabian extractions fell dramatically once again, to only 3.6 Mb/d in 1985. That year, on average, the crude oil price was slightly above $27. The price drop had been halted, provisionally. The complete price collapse, the oil countershock itself, occurred later, in 1986.

The "Miracle" of Reaganomics, Solidarnosc, and Russian Gas

Ronald Reagan's presidency, from January 1981 to January 1989, marked the triumph of neoliberal policies advocated by David Rockefeller, the pope of Wall Street, who in April 1981, at age sixty-five, officially retired from the presidency of Chase Manhattan Bank. The Reagan years also coincided with those of the oil countershock. The success attributed to Reagan's economic policy was itself attributed to two critical factors directly linked to the evolving price of oil. First, the neoliberal reforms (deregulation, restriction of public expenditure, and reduced government payrolls, for instance) were inspired by the brutal oil crisis of 1979. A still more direct factor was the crude oil price decline. The combination of a strong increase, followed by a precipitous drop in the price of oil, seems to have had a rubberband effect on the industrial economy. According to the critical analysis advanced by Keynesian economist Paul Krugman, winner of the Nobel Memorial Prize in Economics, "The secret of the long climb after 1982 was the economic plunge that preceded it. By the end of 1982 the U.S. economy was deeply depressed, with the worst unemployment rate since the Great Depression. So there was plenty of room to grow before the economy returned to anything like full employment."[4] The Keynesian economist Joseph Stiglitz, also a Nobel Memorial Prize winner, established a more direct link between Margaret Thatcher's economic success beginning in 1984 and the windfall delivered by the oil countershock and by the intensive exploitation of North Sea crude oil in particular: "Oil is an asset below the ground and if you take that asset below the ground and you spend it you're poor. . . . You [the UK] squandered that wealth, you took all that North Sea oil and you did very well for that period because you were living off your wealth. You mistook the success of the Thatcher era as a success based on good economic policy when it was really a success based on living off your wealth and leaving future generations impoverished."[5]

The notion of neoliberalism's triumph was, partly, window dressing. While the White House, enabled by the IMF and the World Bank it controlled, imposed strict economic limitations and conditions on the borrowing ability of numerous developing nations, the Reagan administration's own balance sheet diverged considerably from the regulations others were subject to: Between 1980 and 1988, US public debt rose from 26 percent to 41 percent of US GDP, incurred above all due to the tax relief provided for the upper classes and by a spectacular increase in military spending.[6] The "miracle" of Reaganomics was first and foremost a prosperity that stemmed from unbridled consumption—made possible, even irresistible, by the gas-price declines that occurred on the heels of the 1974 and 1981 crises (as well as by massive debt increase in the United States and other developed countries).

The drop in barrel prices during the 1980s was an economic phenomenon amplified by American geopolitical strategy. The US military spending boom under Reagan could be contextualized as the last act of the Cold War. After the USSR became Europe's primary energy supplier, Washington managed to drown the Soviets' currency sources under a stream of Saudi oil.

On December 12, 1981, Soviet general Victor Kulikov oversaw the martial law imposed on the Poles by their head of state, General Jaruzelski. Its purpose: to combat and drive underground the popular revolt movement led by the Solidarnosc (or Solidarity) Party chaired by Lech Walesa. The Western European countries' reaction was prudent and measured. But Ronald Reagan wanted to seize this opportunity to try and deal the Kremlin's power a decisive blow. On December 29, Reagan prohibited all American companies from collaborating in the construction of a gas pipeline intended to carry natural gas from Siberia to the countries of Western Europe. Deemed vital by Paris and Berlin, which sought to limit their dependence on Arab and Persian oil, this pipeline was no less important for Moscow, because it represented a valuable potential source of foreign currency. In Brussels, the European Commission proposed that the member states of the European Economic Community join Washington's boycott. Most refused, and the list of products targeted by the European embargo dwindled away.

Ronald Reagan counterattacked by threatening to cut off the American market to any European enterprise involved in the gas pipeline's construction and any that had used any US patents to further the pipeline. This formidable retaliation blocked almost all margins of maneuverability for those large European industrial groups involved in the site, including the German Mannesmann and AEG, the British Rolls-Royce, and the French Creusot-Loire. The French minister of industry, Jean-Pierre Chevènement, went so far as to threaten to commandeer the French enterprises that refused to continue to work with Moscow, before President François Mitterrand stalled for time.[7] The Thatcher government's secretary of commerce even referred to the "repugnant" embargo imposed upon Britain by its close American ally.[8] The Trans-Siberian Pipeline construction project had to be downsized, and met with a considerable delay that cost Moscow tens of billions of dollars.[9]

The Reagan administration didn't stop there. The idea of an embargo against the Russian pipeline had been suggested to the White House by Roger Robinson, then one of the vice presidents of David Rockefeller's Chase Manhattan Bank, in charge of supervising loans to Communist bloc countries. In March 1982, Roger Robinson was appointed to Reagan's National Security Council (NSC). Within the NSC, the dynamic, affable international banker hatched the American strategy that most strongly contributed to the USSR's collapse. This strategy was enshrined in secret National Security Decision Directive No. 66, dated November 29, 1982, signed by Reagan and drawn up by Robinson to "reduce the USSR's hard currency earnings."[10] The economic war launched by the United States against the Soviet Union, which began with the trans-Siberian natural gas pipeline, then found its fatal weapon: oil.

Washington Drains the USSR's Currency Sources by Flooding the Oil Market with Saudi Arabian Crude

In the early 1980s, the USSR's power was spread too thinly around the planet, like a small dab of butter on a giant slice of bread. To finance its allies from Vietnam to Angola, or to acquire new, sophisticated industrial materials, Moscow really only had one source of currency: its hydrocarbon exports. During his five years spent managing Chase Manhattan's loans to the Communist bloc, Roger Robinson had discovered a major vulnerability in the Soviet Empire: Moscow's resources of foreign convertible currencies were dangerously limited. In the early 1980s, while crude oil prices remained high, Moscow's oil proceeds were only a little over $30 billion per year, or one-third of the annual sales of a single large US petroleum company such as Exxon or General Motors.[11] Robinson, who had served as David Rockefeller's personal assistant for two and a half years, explained the USSR's mortal weakness to President Reagan: The functioning of the Soviet Empire cost Moscow between $15 billion and $17 billion per year, and each $1 decrease in barrel price caused the USSR to lose $1 billion in revenue.[12]

The Soviet economy was already gasping for breath. Their military expenditures approached 40 percent of their GDP, mainly because of the Afghan guerrillas equipped by the United States and generously financed by Saudi oil dollars. Later, Mikhail Gorbachev told his colleagues on the Communist Party's Central Committee that, aside from vodka sales and high oil prices, the Soviet economy had not experienced economic growth in twenty years.[13] Starting in September 1983, CIA director William Casey consistently asked the new Saudi sovereign, King Fahd, to lower the crude oil price.[14] Washington maintained constant diplomatic pressure. Caspar Weinberger, the Pentagon chief who had often worked with the Saudis as a Bechtel group executive, also intervened with King Fahd. He later explained the dual benefit he had sought: The Saudis "knew we wanted as low an oil price as possible. Among the benefits were our domestic economic and political situation, and a lot less money going to the Soviets. It was a win-win situation."[15] Weinberger also described the central workings of the Reagan administration's politics regarding Riyadh: "One of the reasons we were selling the Saudis all those arms was to get lower prices."[16]

King Fahd's oil minister, Sheikh Yamani, was troubled by Washington's urgent requests, since he had only recently, in early 1985, succeeded in interrupting the crude oil price decline by strongly reducing Aramco's extractions. According to several American diplomatic cables, the veteran of Saudi oil politics harshly criticized the Americans in exchanges with foreign diplomats, waving "evidence of a conspiracy by the United States Government to drive down the price of oil."[17] Washington had many levers at its disposal to influence Riyadh. Fear and hatred of Communism was one of them. The need for the US Army's logistical support at the height of the Iran-Iraq War was another. Riyadh's resistance to pay its bills to its many American suppliers while Saudi Arabia's production had halved the oil windfall enjoyed since the 1980s also played a role.[18] At the same time, collaboration between American and Saudi intelligence agencies ran at full speed, whether on the Afghan front or in the Irangate affair.

Whether it amounted to indulging the United States, recovering lost market shares, or both, the Saudis greatly increased their extractions beginning in October 1985, eight months after King Fahd's first official visit to the United States.[19] In September, Aramco abandoned the traditional pricing system administered vigorously by OPEC, a system that descended from the Seven Sisters cartel's "posted price." Riyadh then adopted "netback pricing," a price-fixing principle designed to reflect the reality of production costs. Though the change was slightly chaotic, the other OPEC members had no choice but to join in, starting in December.[20] A pretext for a variety of volume discounts designed to attract buyers, the advent of netback pricing triggered a total price war between OPEC's members, leading to a full-scale oil countershock: The black gold price completely collapsed in 1986, a culmination of five years of slow decline. Reaching a low-water mark of 3.6 Mb/d in 1985, Saudi Arabian extractions climbed back to 5.2 Mb/d as early as the following year. Throughout 1986, in successive thrusts, Aramco abruptly flooded the crude oil spot market, setting prices that were sacrificial. (The prices, though, were still considerably higher than Aramco's profitability threshold: Oil from Saudi Arabia and the Persian Gulf in general was practically less expensive to extract than water—on the order of $2 per barrel.) When Saudi Arabia opened their valves, it weakened the quota system that Yamani himself had laboriously imposed within OPEC. The result was a free fall in world prices, from $27 on average in 1985 to $14 in early 1986, dropping to $7 in July 1986.

We do not know the actual extent of Washington's influence on the revival of Saudi Arabian extractions starting in late 1985. This revival nonetheless drove the oil countershock's nadir, from 1986 to 1989, during which the barrel price remained below $20. This revival clearly had a heavy impact on the final degeneration of the Soviet economy. Mikhail Gorbachev said he was convinced that Reagan's strategy, actualized thanks to Saudi oil, had functioned perfectly. Gorbachev testified in 2010: "Reagan convinced the Saudi Arabian King to put more oil on the market to lower the price. Our main source of foreign currency thus fell by two thirds."[21]

Mr. Bush Goes to Riyadh

In 1986, the cunning strategy in President Reagan's secret National Security Decision Directive No. 66 was hampered by an obstacle as unassailable as it was unexpected: Vice President George H. W. Bush. That year, while the barrel price kept falling, the capitalist economy's engines roared with pleasure: Again, as in the blessed 1960s era, everyone could pump as much as they liked from a reservoir of abundant, inexpensive petroleum. One April day in Texas's capital city, Austin, an Exxon service station owner, sponsored by a local country music station, realized every motorist's dream; he offered his unleaded gas for free.[22] This may sound like a good thing; however, Aramco's production increase had lowered the price so much that Big Oil itself was in danger. In Houston, the oil industry was rapidly devastated. The US oil market was completely deregulated at the beginning of Reagan's first term, during the second oil crisis, with the aim of reviving extractions and forcing prices. The time of the Texas Railroad Commission's omnipotence had ended. But the deregulation now turned against American petroleum: The profitability of many of their investments was undermined by oil's plunging value. Offshore in the Gulf of Mexico, well drilling and development projects that had been planned when prices were at their pinnacle became financial sinkholes. The whole economic structure of the southern United States, with oil as its backbone, threatened to break apart. Real estate bubbles burst everywhere. In Houston, the brand new buildings left empty by the crisis were called "glass prairies." 1986 was dogged by misfortune. The decline in US extractions resumed, after a seven-year respite made possible by Alaskan oil. A sign of the times: Whereas, in the 1950s, a young George H. W. Bush's Zapata Petroleum had drilled 128 wells in Texas and met with success each time, his eldest son, George W., drilled dry well after dry well with his own oil companies (Arbusto, which merged in 1984 with another company, Spectrum 7).[23*] George Bush Junior was derided for his alleged incompetence. But it was Texas itself that was running dry when he followed his father into the oil industry.

At the peak of the American oil industry crisis, Vice President George H. W. Bush opted for a risky initiative: He decided to openly and personally move in the opposite direction from the Reagan administration's policy. In early April 1986, as the number of drilling sites in the United States collapsed, threatening to further accelerate the decline of American production, George

---

* In Spanish, *arbusto* means *bush*. Arbusto knew so little success that, derisively, George W. Bush's critics conspicuously emphasized the second syllable, *bust*: He eventually renamed the company Bush Exploration.

H. W. Bush flew to Saudi Arabia.[24] Just before his departure, he told the press that he would "be selling very hard" to the Saudis the importance "of our own domestic interest and thus the interest of national security."[25] Bush continued, "I think it is essential that we talk about stability and that we not just have a continued free fall like a parachutist jumping out without a parachute."[26] Arriving in Saudi Arabia, Bush warned Sheikh Yamani: If the price of crude oil did not go back up, the US Congress would eventually reinstitute the customs tax on imported oil, discontinued at the end of the 1970s.[27] This new American about-face left the Saudi Arabian oil minister reeling. "Poor George," a White House official said condescendingly, stressing that Bush's position was "not Administration policy" and did not fit with its free-trade doctrine.[28] On April 6, Bush and King Fahd talked into the night in the sovereign's residence in Dhahran, the kingdom's oil capital.[29] Of course, nothing is known about their conversation. But the next day, before the press, "poor" George H. W. Bush showed himself to be inflexible: "I don't know that I'm defending the [US oil] industry. What I'm doing is defending a position that I feel very, very strongly . . . Whether that's a help politically or whether it proves a detriment politically I couldn't care less."[30]

Bush's act of insubordination and his position in favor of raising the crude oil price (and therefore the pump price in the United States) were unanimously interpreted by the American press as political suicide, which deprived the vice president of any hope for the next presidential election. Or at least, almost unanimously. In an April 8 editorial, the highly influential *Washington Post* proclaimed that low prices were undermining the American petroleum industry's ability to grow its production. The American capitol's daily newspaper hammered the argument home: "Mr. Bush is struggling with a real question. A steadily increasing dependence on imported oil is, as the man suggests, not a happy prospect."[31]

After hearing Washington's wish that they lower the barrel price, the House of Saud could have been annoyed to hear the US vice president claim exactly the opposite. On the contrary, they took Bush's argument about US "national security" very seriously.[32] This was at the height of the Persian Gulf "tanker war" between Iran and Iraq, and Riyadh could not afford to refuse to bend to the will of Vice President Bush and his allies in Houston and Wall Street. In late May, while the barrel price stagnated at $14, six OPEC oil ministers met under Sheikh Yamani's aegis in Taif, near Mecca; they agreed to set a goal of $18 a barrel. Saudi Arabia's new change in direction created collateral damage: The victim was Sheikh Yamani himself. The king seemed to want the impossible; he desired both higher prices and the continuance of Aramco's production levels. On October 29, 1986, Yamani was fired. After this blow, the great authorizing officer of the 1973 embargo, the architect of Saudi Arabian oil policy for a quarter of a century, told the press he opposed the sovereign's "economically suicidal" approach.[33] In the ultimate irony, despite his departure, it was Yamani's last negotiated compromise that prevailed: Between 1986 and 1987, Saudi Arabia agreed to lower its production from 5.2 to 4.5 Mb/d. The crude oil price rose to $19 in the United States: George H. W. Bush had won, had managed to save Big Oil.

Washington and Riyadh in Sync: The Afghanistan War to the Birth of Al Qaeda

During this period, the alliance between Washington and Riyadh approached its highest degree of maturation. The US Navy resolved to assure the security of commerce in the Persian Gulf, to the detriment of Iran, the great common enemy; and the House of Saud apparently could not get too close to Uncle Sam. King Fahd assumed a lifestyle that was more like a billionaire hedonist's than one might expect from a Bedouin sheikh descended from the austere Wahhabi tradition. Aboard his personal Boeing 747, equipped with a fountain, he liked to travel to the French Riviera. There, he went from casino to casino before boarding his 147-meter yacht, with its ballroom and two swimming pools.[34] The Arab sovereign liked to indulge in black jack and roulette, games of chance forbidden by Islam, in which it was rumored he could lose millions of dollars in a few hours. The English press wrote that sometimes the custodian of the Two Holy Mosques violated the curfew imposed on the game by British law, hiring the dealers so he could continue to play the whole night in his London suite.[35]

King Fahd's nephew, His Excellency Prince Bandar, Saudi Arabia's ambassador to the United States from 1983 to 2005, then personified the extreme closeness of the relations between the Sauds and the American political elite. Closeness and even intimacy: Bandar bin Sultan bin Abdulaziz Al Saud said he considered George H. W. Bush "almost like a buddy."[36] The prince made a name for himself in Washington as early as 1981, heading up an intense lobbying campaign that had allowed Reagan to wrest, from the pro-Israeli Congress, their agreement to sell AWACS aircraft to the Saudis. Bandar, whose grandfather had fought his way to the Saudi throne with saber thrusts, confessed ironically that on his arrival in the United States he had understood nothing of the political game of American democracy. But he judged this game "exotic and exciting," and quickly developed a taste for it: "There was no blood drawn. It was physically safe, but emotionally tough."[37] A lieutenant in the Royal Saudi Air Force, Bandar received his pilot training from the US Air Force; in June 1985, one of his cousins, Prince Sultan Al Saud, had the privilege of being one of the first non-Americans to participate in a NASA space shuttle mission.* For three decades, Prince Bandar remained the smiling, modern face of Saudi lobbying in Washington. At the beginning of the Iran-Iraq War, he was sent to Iraq to serve as an intermediary between Baghdad and Washington, and it was through Bandar that Saddam Hussein informed the CIA that he was ready to accept US assistance.[38] During the Irangate years, 1984 to 1987, a period equally decisive for the war in Afghanistan, Prince Bandar met with Caspar Weinberger, the US defense secretary, no fewer than sixty-four times.[39]

The clandestine cooperation on various fronts of the Cold War, initiated at the end of the 1970s with the creation of the Safari Club, radically intensified during the Reagan administration. Saudi oil money was an essential funding source for CIA operations, whether destabilizing pro-Soviet movements in South Yemen and the Horn of Africa, or fanning the embers of the interminable civil war in Angola, a country where the oil majors were anxious to be able to safely exploit the large reserves of offshore crude oil.[40]

But it was on the Afghan front, facing the Red Army, where the understanding between Washington and Riyadh was most fertile, and which bore its first rotten fruit. On March 27, 1985, Ronald Reagan signed his National Security Decision Directive No. 166: It was no longer simply a matter of harassing the Russians but also of chasing them from Afghanistan. One year later, for the first time, the White House overtly offered frequently illiterate mujahedeen the opportunity to toy with the latest American military technology: the Stinger. These portable missiles, guided by infrared light and sent through Pakistan to Afghanistan by the hundreds, had virtually never been used in combat. Hitting their target each time, they allowed Afghanistan's primitive guerrilla forces to inflict extensive damage on the assault helicopters the Soviet Army used to control the country. For Moscow, the occupation of Afghanistan was rapidly transformed, according to Mikhail Gorbachev, into a "bloody wound."

The technological escalation of this conflict was costly. More than half of the bill was settled with oil money. In 1986, the amount the United States spent on Afghanistan reached $600 million: This was the most significant clandestine war ever organized by the CIA. Saudi Arabia pledged to invest exactly the same

---

* The rich Saudis' taste for completing their pilot training in the United States was regarded as having helped to facilitate the September 11, 2001, attacks by giving its perpetrators access to pilot training on American soil.

sum that Washington invested in the war.[41] To these two sources of funding a third was added, much more nebulous but no less important: The money from rich Arabs collected by religious charitable organizations around the Persian Gulf. "It was largely [private] Arab money that saved the system," a Pakistani general later emphasized.[42] This money was intended for four radical militant Islamic groups active in Afghanistan. One of them, Ittehad-i-Islami (Islamic Union for the Liberation of Afghanistan) was headed by a strict fundamentalist Afghan, Abdul Rasul Sayyaf. The latter had virtually no connections in Afghanistan but was powerful in Saudi Arabia; the kingdom's intelligence network had recruited him early in 1980 and made him one of their stars. It was to Sayyaf that Osama bin Laden turned when the young, wealthy Saudi decided to join the Afghan jihad. Sayyaf was, of course, deeply hostile to the infidel Americans. This was also true of Osama bin Laden's other mentor, Abdullah Azzam, the hypnotic, combatant Palestinian priest who, thanks to oil money, helped recruit the pan-Islamic army that was based in Afghanistan. The men in this small army of "Afghan Arabs," which reached more than twenty-five thousand fighters in the mid-1980s, came from the Islamic Salvation Front in Algeria, fundamentalist Sunni movements in Egypt, Syria, and Iraq, and Palestine's Hamas. Abdullah Azzam met with Islamic militants in a refugee house in Brooklyn, New York, where the CIA also sought candidates for its own "holy war" against the Soviets.[43] Privately, Afghan intellectuals were concerned when they saw the Arabs, and in particular the Saudis, use their money to win a religious hold on Afghanistan.[44] The Afghans, or at least their elite, had benefited until then from the reputation of being one of the most open, tolerant peoples of the Muslim world.

Turki Al Saud, chief of Saudi secret services, affirmed that his first meeting with Osama bin Laden took place in 1985 or 1986 in Peshawar, the Pakistani city located at the extreme end of the Khyber Pass, an entry point into Afghanistan. For Prince Turki, the son of the powerful bin Laden family represented an opportunity to recruit, train, and indoctrinate jihad combatants without going through the intermediary of the Pakistani secret service, the Inter-Services Intelligence (ISI).[45] From that point on, funding from Gulf countries continued to grow. One of Sayyaf's associates reported that the war chief had "raise[d] millions of dollars from businessmen and charities in Saudi Arabia, Kuwait and the United Arab Emirates."[46] By the end of the war, this support reached $20 million to $25 million per month. These funds made Sayyaf's group prosper, although they played a marginal role in the Afghans' fighting. Before the war ended and the last Soviet soldiers withdrew in February 1989, these oil dollars helped to establish Al Qaeda.

Al Qaeda's early funding sources included some of the largest Saudi Arabian fortunes, in particular that of Khalid bin Mahfouz, heir of the first Saudi bank that had moved to Houston in the late 1970s, along with Salem bin Laden. In 1988, at bin Mahfouz's request, the Saudi royal family's personal banker offered $270,000 to the organization Osama bin Laden founded.[47] Following numerous revelations after the September 11 attacks, bin Mahfouz's lawyer let it be known that "this donation was to assist the US-sponsored resistance to the Soviet occupation of Afghanistan and was never intended nor, to the best of Sheikh Khalid's knowledge, ever used to fund any 'extension' of that resistance movement in other countries."[48] Yet, in the years after the USSR fell, the name Khalid bin Mahfouz, manager of Saudi oil money, frequently appeared on the lists of generous donors to charitable Muslim associations sending contributions to Al Qaeda.[49]

The BCCI Scandal

To transfer the funds necessary for its secret war over to the ISI and the various guerrillas implanted in Afghanistan, the CIA had recourse to numerous accounts in a most unusual bank: the Bank of Credit and Commerce International (BCCI).[50] Founded in 1972 by Agha Hasan Abedi, a Pakistani who was well connected to the largest fortunes in the Persian Gulf, the BCCI had become the seventh largest bank on Earth in little more than a decade. On July 5, 1991, it was closed by authority of the Bank of England, during the most nebulous, phenomenal financial scandal in history. The suspect accounting methods and Ponzi schemes systematically practiced by the BCCI left an abyss in their wake, into which vanished an estimated $5 billion to $15 billion. Diverse inquiries revealed the BCCI's involvement in laundering hundreds of millions of dollars from almost all forms of organized crime, including arms trafficking, drug trafficking, prostitution, and corruption, as well as the financing of terrorist groups such as Abu Nidal and even the death squads of the Medellin cartel.[51]

The BCCI also played an early key role in the clandestine operations of the CIA and Saudi intelligence. After the 1973 oil crisis, no other bank in the world experienced such rapid growth: It was perfectly positioned to receive, deposit, and invest throughout the world, and particularly in the United States, the oil dollars of the Middle East. It soon counted powerful Saudi businessmen among its major shareholders, beginning with Kamal Adham, Turki Al Saud's predecessor as the director general of Saudi Arabian intelligence. Beginning in the early 1980s, William Casey, the CIA's director and a friend of Ronald Reagan, frequently met Hasan Abedi at Washington's Madison Hotel in order to use the BCCI's services for clandestine CIA operations in Iraq.[52] BCCI's accounts were also used for Irangate, to transfer funds from the White House to the Nicaraguan Contras, via Saudi Arabia.[53] And, in 1986, Saudi Arabia's chief banker, Khalid bin Mahfouz, invested nearly a billion dollars in order to take control of 30 percent of BCCI's shares and become its principal shareholder.[54]

The investigation report US senators John Kerry and Hank Brown published in December 1992, a year and a half after BCCI was liquidated, were a terrible source of embarrassment for the CIA.* Begun in February 1988, the investigation Kerry conducted for the Senate Foreign Relations Committee focused on the relationship between drug trafficking, on the one hand, and foreign policy and American justice on the other. In 1989, confronted with the Department of Justice's passivity, Kerry entrusted the investigation's dossier to New York's district attorney, Robert Morgenthau. Morgenthau pushed Price Waterhouse, the firm that audited BCCI's accounts, to transmit a confidential report to the Bank of England in June of 1991, less than a month before the latter decided to shut down the BCCI. This report indicated that the BCCI had "generated significant losses over the last decade and may never have been profitable in its entire history."[55] The scandal did not reach the ears of the press until several weeks after BCCI had ceased to exist. In the Senate, Kerry and Brown reopened the case. During the inquiry, Kamal Adham, BCCI's representative in the United States and the CIA's main intermediary with Saudi Arabia in the 1980s, hired George H. W. Bush's former political director, Ed Rogers, as one of his lawyers.[56] Kerry's and Brown's final conclusions were most damaging for the H. W. Bush government's Department of Justice, accused of systematically turning a blind eye as evidence of the scandal's enormity accumulated.[57] Jacques Bardu, a French customs officer tasked with searching BCCI's Paris offices in July 1991, affirmed having seized the documents of an account in the name of a certain "George Bush," with over $5 million on deposit; in the hours that followed, according to Bardu, the documents were transmitted to the budget minister, Michel Charasse, then to the Élysée, and finally, following the order of an "assistant" of François Mitterrand, to the Department of the Treasury's attaché at the American embassy in Paris.[58] If these documents existed, there would be no further questions.†

* The December 1992 version, cited here in reference, is not the definitive version of the report, which according to some sources included passages that were redacted at Henry Kissinger's request.

† When questioned by the author in December 2013, Michel Charasse said he did not recall such an episode. Jacques Bardu had since died.

Among other depravities, the American secret services were accused of using a series of BCCI accounts to finance the war in Afghanistan by laundering opium money. "If BCCI is such an embarrassment to the U.S. that forthright investigations are not being pursued it has a lot to do with the blind eye the U.S. turned to the heroin trafficking in Pakistan," a American intelligence officer confided to *Time* magazine.[59] According to American historian Alfred McCoy, "the Pakistan-Afghanistan borderlands became the world's top heroin producer, supplying 60 percent of U.S. demand. In Pakistan, the heroin-addict population went from near zero in 1979 to 1.2 million by 1985, a much steeper rise than in any other nation. . . . As the Mujaheddin guerrillas seized territory inside Afghanistan, they ordered peasants to plant opium as a revolutionary tax. Across the border in Pakistan, Afghan leaders and local syndicates under the protection of Pakistan Intelligence operated hundreds of heroin laboratories. During this decade of wide-open drug-dealing, the U.S. Drug Enforcement Agency in Islamabad failed to instigate major seizures or arrests."[60] In 1995, the CIA's former head of operations in Afghanistan, Charles Cogan, admitted that the United States had knowingly overlooked enormous heroin trafficking: "I don't think that we need to apologize for this.* Every situation has its fallout. . . . There was fallout in terms of drugs, yes. But the main objective was accomplished. The Soviets left Afghanistan."[61] BCCI oil money, the religious puritanism of its Saudi shareholders, and its clients' drug money, blended in improbable, explosive ways.

As a *Time* magazine journalist wrote in 2001, "The discovery of the CIA's dealings with B.C.C.I. raises a deeply disturbing question: Did the agency hijack the foreign policy of the U.S. and in the process involve itself in one of the most audacious criminal enterprises in history?"[62] BCCI, which invested huge sums in raw materials trading, beginning with oil, represented an irresistible attraction for the CIA, which since the end of the Vietnam War had sought to escape the overly scrupulous control of the American Congress. According to the testimony of a BCCI executive, "What [BCCI founder] Abedi had in his hand [was] magic, something [Saudi intelligence chiefs] Kamal Adham or even Prince Turki didn't have."[63] Located across the world, including in more than fifty developing nations, BCCI was, according to *Time* magazine, "a vast, stateless, multinational corporation that deploy[ed] its own intelligence agency, complete with a paramilitary wing and enforcement units, known collectively as the Black Network."[64] BCCI was liquidated in 1991; beginning in 1993, its primary shareholder, Khalid bin Mahfouz, paid nearly half a billion dollars to

* The financing of covert operations by opium money was already a familiar tactic; it was apparently pioneered in Indochina by the French intelligence agency SDECE.

reach a settlement in the New York justice system's fraud case and pay BCCI's creditors.[65] Many turning points on the trail from the Safari Club to Al Qaeda, through the BCCI, remain unknown or difficult to interpret. But it is clear that by entrusting the financing of many clandestine operations to Saudi oil money and relying on the cryptic networks financed by Riyadh, the American empire had created what became the most evident form of its own nemesis. In 1989, after the Soviets withdrew from Afghanistan and when George H. W. Bush moved into the White House, Pakistan's prime minister, Benazir Bhutto, warned the new American president: "The extremists so emboldened by the United States . . . are now exporting their terrorism to other parts of the world."[66] Bhutto, the first woman to lead a Muslim majority nation, whose 2007 assassination was claimed by Al Qaeda's head of operations in Afghanistan, warned Bush: "You are creating a veritable Frankenstein."[67]

Since 1976, when Bush had become director of the CIA, the American intelligence service had sought to harness the power of Saudi oil money, unleashed by the oil crisis. Did this tremendous, unstoppable power unseat the Americans at the end of the 1980s? Many in Washington were fearful when, in March 1988, the press revealed that Riyadh had turned to China to purchase ballistic missiles capable of transporting nuclear bombs, or when, in the course of the investigations that led to the BCCI scandal, it was discovered that the BCCI had financed the laboratories of the worrisome Dr. Abdul Qadeer Khan, the father of the Pakistani nuclear bomb—and the Reagan administration could do nothing about it.[68] Also in 1988, Aramco eliminated the last symbol of Big Oil's domain: The Arabian American Oil Company changed its name to become the Saudi Arabian Oil Company, in other words Saudi Aramco. Dr. Ali Al-Naimi, educated at Stanford University, assumed full direction of the richest company in the world. The "greatest single prize in all history" no longer belonged to Americans.

The BCCI and George W. Bush's Good Fortunes

The links between Washington and Riyadh were not, despite all that, broken. They continued to metamorphose. In November 1988, with full financial support from his oil industry friends, George H. W. Bush easily won the presidential election against his bland Democratic opponent, Michael Dukakis. Bush's Persian Gulf allies appeared to be absent from the scene, at least at first glance. George H. W. Bush's eldest son, George W., worked nearly full-time on his father's campaign, to the detriment of his own oil career, a career that was never very dazzling, apart from two providential windfalls. Two years earlier, in the middle of the oil countershock and a series of bankruptcies he had endured in Texas, George W. Bush's small, debt-ridden company, Spectrum 7, had been purchased by Dallas-based Harken Energy for $2.25 million. By the terms of the transaction, George W. received $600,000, a seat on the board of directors, and a consulting post that paid more than $50,000 per year, while leaving him ample time to work on his father's ascension to the White House. To the question of why Harken Energy was so generous to Bush Junior, Harken's founder, Phil Kendrick, responded: "His name was George Bush. That was worth the money they paid him."[69] The main shareholders in Harken—a company that *Time* magazine presented as "one of the most mysterious and eccentric outfits ever to drill for oil"—on the one hand included Alan Quasha, son of an influential friend of Ferdinand Marcos, the Philippines' corrupt pro-American dictator, and on the other hand, included Harvard University's powerful investment funds, directed by Robert G. Stone Junior, son-in-law of Godfrey Rockefeller, who was a cousin to Nelson and David Rockefeller and, incidentally, close to the Bush family.[70] Stone had been a business associate of "former" CIA agent Thomas J. Devine, George H. W. Bush's partner in Zapata Offshore.[71] He had also served beside Henry Kissinger on the board of directors of Freeport Mining, a giant precious metals mining corporation dominated by the Rockefellers and specifically associated with the CIA-supported coup that brought Suharto to power in Indonesia in 1965. Shortly after Bush Junior joined Harken Energy's board, Jackson Stephen, one of the largest independent US bankers and a generous donor to the Republican Party, drew up a rescue plan for the small Texan oil company.* To actualize it, Stephen appealed to UBS, the Swiss banking sector giant, a BCCI partner. When the rescue plan failed at the last minute, Stephen turned to one of the largest Saudi Arabian fortunes, that of real estate and finance mogul Abdullah Bakhsh, who became one of Harken Energy's largest shareholders.[72]

In spite of its powerful partners, Harken's affairs were in bad shape. Like Spectrum 7, the company drilled dry well after dry well, losing $12 million in 1989, then $40 million the following year. To hide some of its losses, Harken resorted to baroque accounting practices, which later were compared to those in the Enron scandal.[73] Yet in January 1990, twelve months after George H.W. Bush arrived in the White House, Harken found a new petroleum opportunity. The small company, which had never drilled either abroad or at sea, astounded

---

* Also on Harken's board of directors was the famous Hungarian-born billionaire George Soros, who, for one reason or another, became George W. Bush's major financial opponent in his 2004 reelection race.

the petroleum world by winning, at the giant Amoco's expense, a contract for prospecting off the coast of Bahrain, a small island-state in the Persian Gulf. Located opposite Dhahran and largely subservient to Saudi Arabia, Bahrain is home to the US Navy base in the Persian Gulf. The three key characters in Harken Energy's successful bid (the Houston oil consultant who paved the company's way in Bahrain, the American ambassador to Bahrain, and the prime minister of Bahrain himself) were all connected to the BCCI. After the scandal about Khalid bin Mahfouz's bank broke, the *Wall Street Journal* observed: "The number of BCCI-connected people who had dealings with Harken, all since George W. Bush came on board, likewise raises the question of whether they mask an effort to cozy up to a Presidential son."[74]

After the announcement of the "incredible agreement" obtained by Harken Energy, investors, reassured by the US president's son's place at the heart of the company, rushed to invest their capital even while Harken was reeling with debt.[75] On May 17, 1990, George W. Bush participated in an urgent meeting of Harken's directors: The company was likely to run out of funds in three days![76] George W. made it known that he wished to quickly sell his shares so he could buy the Dallas baseball team, the Texas Rangers. Harken's legal staff formally discouraged him from doing so: The president's son ran the risk of being prosecuted for insider trading. But George W. ignored the advice, and on June 22 he pocketed $848,560, a few weeks before the announcement of Harken's new catastrophic losses.[77] The transaction was judged legal by the Securities and Exchange Commission (SEC), the Wall Street police. And George W. Bush bought the Texas Rangers, his first step in a dazzling political ascension that would make him governor of Texas in 1995. The president of the SEC had been appointed by his father, George H. W. Bush. He was a former partner of Baker Botts, the petroleum industry's legal firm. This was also true of Robert Jordan, the lawyer who defended Bush Junior before the SEC.[78] In September 2001, after he became president of the United States, George W. Bush nominated Robert Jordan as ambassador to Saudi Arabia. Harken's drilling off the coast of Bahrain produced nothing, and the island-state remains the only Gulf country poor in black gold.

Big Oil at the Heart of the Financialization of the Economy

The Reagan years marked the financial world's transformation into a gigantic casino in which Big Oil played a large role in key developments of the period: sophisticated financial products, the arrival in force of retirement funds, aggressive speculation, the proliferation of hostile takeovers, acquisitions using leveraged buyouts, megamergers, share buybacks, and the launch into orbit of shareholder dividends and executive compensation, as well as massive layoffs.

The total deregulation of the crude oil market led the Western oil companies to the edge of financial speculation, with BP and Exxon at the vanguard. Exxon was particularly innovative, thanks to the impetus of the group's financial director, Jack Bennett (who was involved in establishing the floating exchange rate system in 1971, and then channeled Saudi petroleum money to Wall Street in the wake of the 1973 oil crisis), and the ingenuity of JP Morgan—the preferred bank of Exxon, the West's oil-industry leader.[79] Starting in 1973, central Manhattan's Exxon building was JP Morgan's headquarters—a sign of the bank's subordinate position in relation to the juggernaut of American black gold.[80] Notably, the collaboration between Exxon and JP Morgan was the origin of "credit default swaps," a financial derivative that became famous during the 2007 financial crisis. JP Morgan had invented credit default swaps to allow Exxon to raise the funds necessary to pay a record $5 billion fine levied by the state of Alaska after the Exxon-Valdez oil spill, on March 24, 1989.[81] In the year 2000, JP Morgan merged with the historic Rockefeller enterprise, the Chase Manhattan Bank. During the 1980s, the multinationals and, in particular, the oil majors had grown to such tremendous proportions that they clamored for colossal investments beyond the funding capacities of wealthy capitalist families. Capitalism had been "democratized." Or at least, it needed to access the savings of the "everyman." Expedited by computers, this opening once again magnified the power of capital and, at the same time, magnified the greed of the new small shareholders.

This democratization of financial investment went hand in hand with an aggressiveness that was exacerbated in the transactions. It was the heyday of the financial shark. During the Black Monday crash on October 19, 1987, the culmination of a fevered period of leveraged buyouts and mergers, 70 percent of the world share market was already under the control of investment and pension funds, responsible for managing the savings the middle class had accrued during the Thirty Glorious Years, especially the retirement accounts of tens of millions of Americans.[82] The year prior to the crash, 346 acquisitions of more than a million dollars occurred, versus only a dozen per year in the early 1970s.[83] The largest merger-acquisitions were produced within the American oil industry. In 1981, Mobil attempted to absorb Marathon Petroleum—both descendants of Standard Oil. Marathon counterattacked the following year by striking a deal with a "white knight," US Steel. In August 1983, Mesa Petroleum, the company of T. Boone Pickens, who created the prototype for the hostile raids of the 1980s, attacked a much larger company: one of the Seven Sisters, Gulf Oil. Mesa Petroleum, which had invested heavily in highly expensive offshore drilling in the Gulf of Mexico, was then overwhelmed by the oil countershock. But the giant Gulf Oil was more fragile yet, since it had lost control of its fabulous Kuwaiti reserves when the emirate nationalized its oil in 1975, and also as a result of the decline in its Texan production. By taking recourse in high-yield bonds (the infamous "junk bonds," then much in vogue), Pickens bought a remarkable number of largely undervalued Gulf Oil shares. However, Mesa Petroleum failed to take full control of the Texan firm. Pickens merely pocketed huge capital gains during the slaughtering of Gulf Oil, at the end of which, in March 1984, the company the Mellon family had created at the turn of the century ended up being swallowed by Chevron (SoCal's new name). Chevron thus absorbed the company that had opened its door to Persian Gulf petroleum a half-century earlier, for $50,000.

Finally, on December 31, 1984, Texaco acquired Getty Oil, the company created by J. Paul Getty, the richest man in the world since John D. Rockefeller. But there was a glitch: Getty Oil was already promised to Pennzoil, one of the largest American oil companies, as a result of a merger between a former Standard Oil subsidiary and George H. W. Bush's original oil company, Zapata Petroleum. On November 19, 1985, Texaco was ordered to pay $10.53 billion to Pennzoil, in what remains one of the most expensive civil trials in history. Preferring to avoid this more or less hostile wave of acquisitions, Exxon chose to consolidate to protect itself against potential takeovers and retain its shareholders, spending billions of dollars to purchase its own shares. The company also reduced its staff by 40 percent, in what was one of the most severe severance schemes of the 1980s.

Oil Reserve Declines: The Roots of Evil

The mergers and acquisitions movement initiated in the 1980s, which continued during the following decade, was a last resort, a kind of diversion. The American oil companies were discovering that it was more and more difficult, expensive, and uncertain to acquire new black gold reserves by prospecting. It was much simpler and cheaper to buy competitors (and their reserves).

A long illness without a cure had begun to gnaw at the black gold industry. After reaching its peak in the mid-1960s, the annual number of petroleum discoveries declined, while consumption continued to grow. These arcs intersected as the 1980s began: From then on, year after year, humanity burned more oil in the atmosphere than it discovered in the Earth. And the gap between consumption and annual discovery of conventional oil would widen. In 1979, the year of the second oil crisis, the industry reached its maximum "proven and potential" reserves: After this date, the global reservoirs of conventional oil (in other words, classic liquid oil) emptied faster than prospectors were capable of replenishing them.[84] This phenomenon was not noted in the statistics of "proven" reserves, the only accessible, and somewhat misleading, statistics outside the industry. The Western oil majors were the first victims of this trouble, a long time in the making. After being driven out of many OPEC countries, they had reinvested their enormous profits from the 1973 and 1979 crises in exploration territory in the United States and elsewhere, hoping to bring renewed vigor to the great country of black gold. But at the end of the 1980s, the result of these efforts proved disappointing to the United States. Despite the extension of offshore drilling and the construction of THUMS, artificial islands in Long Beach created for oil drilling, in 1986 California joined Texas and Oklahoma in a downward decline.*

The impasse was also visible in Alaska. In the early 1980s, when the Prudhoe Bay field that opened in 1977 had already begun to reveal its limitations, the oil companies resolved to look to the glacial waters of the Arctic Ocean. A geological structure located 100 kilometers north of Prudhoe Bay, baptized Mukluk—for the lined boots of the Inuit—raised great hope. But this drilling, undertaken in the most extreme conditions ever faced by oilmen, failed. In December 1983, the verdict came down: Mukluk was a dry well. In 1988, the Prudhoe Bay field began its long decline. The US oil reserves continued to dwindle. Harold Ickes' prophecy threatened to materialize: This time, "America's crown, symbolizing supremacy as the oil empire of the world," seemed to be genuinely "sliding down over one eye."[85] The oil majors found no more success at the other end of the Americas. When the Falklands War broke out on April 2, 1982, exploratory drilling off Argentina's south coast, begun in 1980 by Argentina's national company, YPF, as well as by Exxon, Shell, and the French company Total (the former CFP), occupied a significant place in the minds of the belligerent Argentines and Brits. From Downing Street's point of view, the strategic issue of this distant war, which ended two months later when the Royal Navy regained the archipelago, seemed otherwise insignificant. Was the Thatcher government as tempted by the prospect of beefing up the British energy empire as they were taken with

* The name of these islands, built in 1965, is an acronym of Texaco, Humble, Union Oil, Mobil, and Shell. The THUMS islands provided an opportunity to refine horizontal drilling techniques.

the North Sea?* The twenty or so drills interrupted by the conflict never were high producers. This southern Atlantic zone proved to be, once again, poor in black gold (until today, in any case).

During the 1980s, as Western crude oil stocks began to shrink, the reported tally of "proven" reserves controlled by major Persian Gulf countries leaped nearly 60 percent. This explosive figure, advanced without external validation by OPEC's five main member states, later elicited heavy suspicion—with good reason, at first glance anyway.[86] In 1984, Kuwait's House of Sabah was the first to release such staggering numbers, asserting that its crude oil reserves had increased by 37 percent from the previous year, rising from 67 billion to 92 billion barrels. OPEC made the decision, deemed sound, to index new production quotas based on the reserve amounts announced by its member countries. From then on, one-up-manship ran rampant. In 1986, the Islamic Republic of Iran announced it had just barely surpassed Kuwait, with 92.9 billion barrels, whereas it had claimed merely 59 billion a year earlier. The same year, the United Arab Emirates reported that their reserves had leaped 194 percent. Saddam Hussein had to weigh in: In the midst of a war against Iran, Iraq demonstrated a certain amount of black (gold) humor in 1987 when it claimed that its crude oil reserves had risen by 38 percent, to achieve precisely "100.0" billion barrels, a figure it subsequently maintained year after year! Last but not least, Saudi Arabia followed suit in 1988, officially claiming that its reserves had risen by 50 percent, to 255 billion barrels. This proclamation arrived the year Aramco became Saudi Aramco.

This episode exposed, for the first time, the secret and arbitrary character of OPEC's official oil reserve figures, doubtless the most strategic and closely protected economic data in the world. Yet, at the time, the impressive inflation of figures advanced by the five main Persian Gulf players (Saudi Arabia, Iran, Iraq, Kuwait, and the United Arab Emirates) went unquestioned. Oil depletion belonged to a future so nebulous and distant that the vast majority of observers were indifferent to the question of reserves. But in the 2000s, when the depletion problem became more than hypothetical, the Gulf countries commonly were accused of lying, pure and simple. Such accusations must be revised in the light of the admission made to Sheikh Yamani by Aramco's American directors in July 1973, concerning the "true" amount of Saudi Arabia's reserves at that time: 245 billion barrels. Knowing that, during the period of

* Margaret Thatcher's principal motives were undoubtedly driven by domestic policy: The looming elections the following year were the probable reason for the British response to the Argentine landing in the archipelago. The Falklands War spiked the popularity of the "iron lady," who previously had been widely reviled for brutal measures taken during the crisis sparked by the second oil shock; Thatcher triumphantly carried the legislative elections in June 1983.

overabundance that preceded America's peak petroleum levels in 1970, the Seven Sisters had systematically restricted production growth in Persian Gulf countries using a method both logical and relentless, and understanding that deliberately lowballing official reserve announcements served their objective to limit extractions, one can assume that when the other Gulf countries, such as Saudi Arabia, nationalized, they inherited oil reserve amounts that were, at the very least, conservative. Moreover, the subsequent evolution of Persian Gulf production suggests that the figures advanced by the OPEC countries beginning in the late 1980s more or less conformed to reality.

A decade after the main exporting countries nationalized their oil, the Western oil majors, which played a vital role for industrialized countries, had become minor players.They controlled only low quantities of petroleum compared to the nationalized Saudi, Iraqi, and Iranian companies. During the first oil crisis, the costly development of the first petroleum fields that required complex, extreme extraction techniques did not alter this situation. Since the advent of the oil age, the principal route to the zenith of the American dream had originated in the region where the US Army established a "central command": the Arabo-Persian Gulf—drowned in oil, transfigured by greed or ravaged by war, oscillating between hatred for and bewilderment about the world black gold had set in motion.

TWENTY-FOUR

Dear Saddam: The Gulf War, the Fate of the Iraqi People, and the Long-Term Interests of Uncle Sam

The main reason the Saudis maintained a close alliance with the "infidel Americans" was the sparse population of the immense Saudi desert, constantly vulnerable to more densely inhabited Persia and Mesopotamia. In contrast, Saddam Hussein, once Iran's enemy thanks to the support of the Arab petroleum princes and the United States, hoped to make Iraq a great revolutionary Arab power, independent (like Iran) of any external influence. However, Iraq, like Iran, had been bled dry by their eight years of war, which seem to have been well agitated by the Reagan administration. Saddam Hussein's army remained strong, well cared for by its numerous suppliers. But maintaining this army required tremendous petrodollar support, and Hussein, who expected that Iraq's long struggle against Tehran's ayatollahs at least afforded him a certain prestige among Sunni Arab potentates, was quickly disillusioned.

Saddam Hussein Turns against His Patrons

Baghdad was not compensated for having obstructed uprisings by the Shiite population, supported by Iran, against the Sunni Arab princes' regimes. Saudi Arabia and Kuwait, which had largely financed the Iraqi army precisely to this end, refused to erase any portion of the heavy debt Iraq had incurred, which reached nearly $90 billion by the end of the war. Baghdad accused Kuwait of having stolen crude oil from the Rumaila fields for years. More than a third of Iraq's oil wells were located in this giant crude reservoir, one of the richest in the world, but its borders ran a few kilometers into Kuwaiti territory. Iraq had mined these wells during the war against Iran and, according to Baghdad, Kuwait had benefited unduly from horizontally drilling under the border to siphon off Iraq's black gold. Having long claimed Kuwait as its "nineteenth province," Iraq argued that the small emirate had stolen $2.4 billion of crude oil this way. In addition, both Kuwait and the United Arab Emirates exceeded the production quotas OPEC had established for them in 1990 by nearly 40 percent. Baghdad assessed this loss at $14 billion, much more than Kuwait had loaned Iraq during the war.[1] Saddam Hussein then repeated to anyone who listened that Kuwait was provoking "another war against Iraq."[2]

In early 1990, Saddam Hussein assured President George H. W. Bush that, despite his threats, he would not attack Kuwait. And Washington exerted zero pressure to help Baghdad and Kuwait's House of Sabah, the wealthy emirate's ruling family, find a way out of this impasse. It was a dispute between two valuable United States allies. Extending the agreement sealed during the Iran-Iraq War, Bush asserted that "[n]ormal relations between the United States and Iraq would serve our longer-term interests and promote stability in both the Gulf and the Middle East."[3] In May, the American and Iraqi intelligence services continued to exchange information. In July, the White House encouraged granting new loans that favored Iraq and rejected the attempts of the secretaries of commerce and defense to continue restricting the exports of "dual-use" materials (civil and military), a practice in place since 1985. And then George H. W. Bush's old Texan compatriot, James Baker, secretary of state, reminded Bush that Saddam Hussein—who at age twenty-two had participated in the CIA-organized assassination attempt of General Kassem—was a dangerous ally. On July 19, Baker received a worrisome memo warning that Iraq had acquired chemical, bacterial, and nuclear arms, "probably proliferation related," from American companies. A week later, on July 25, Baker, seeming to discover in extremis what he described as "Iraq's extraordinary aggressive weapons proliferation efforts" implemented by Saddam Hussein, called for stronger controls on exports to Iraq.[4] In Baghdad, the same day, April Glaspie, US ambassador to Iraq, was summoned by Iraq's president; she was later accused of not having clearly warned Saddam Hussein against any military action.* In any event, the latter had already made his plans. On July 27, Prince Bandar, the Saudi Arabian ambassador in Washington, passed along Baghdad's new assurances that there

---

* According to an alleged transcript of the meeting, published in September 1990 by Baghdad, Glaspie declared: "We have no opinion on your inter-Arab conflicts, such as your dispute with Kuwait." Some believe that this document proves that Washington had implicitly given Saddam Hussein the green light. However, Tariq Aziz, the leader of Iraqi diplomacy, who attended the meeting, declared to ABC news that Saddam Hussein did not doubt in the least that the United States would react "severely" to an invasion of Kuwait.

would be no war. This was a lie. On August 2, the Iraqi army invaded Kuwait without difficulty. Kept in check for eight years by Iran, in a single day Saddam Hussein doubled his crude oil reserves: The Iraqi leader controlled a fifth of the world's petroleum, nearly as much as Saudi Arabia. The second monstrous child of US hegemonic policy in the Gulf had revolted against his mentors. Saddam Hussein's military adventurism revealed the degree to which the Arab oil princes depended on the American military. But Hussein greatly overestimated his strategic assets and, more so, underestimated President George H. W. Bush's determination to protect the United States' "vital interests" in the Persian Gulf, which the White House had explicitly committed itself to defend.*

Some initial reactions from Washington gave the impression that George H. W. Bush's strategists were vague or indecisive. One White House NSC member described the attitude that dominated the council's first meeting after the invasion: "Hey, too bad about Kuwait, but it's just a gas station, and who cares whether the sign says Sinclair or Exxon?"[5] Without a doubt, this witticism was a sign of a puzzling moment. The Bush administration "hawks," particularly Secretary of Defense Dick Cheney (approaching his fifties), quickly assessed the situation. On August 5, George H. W. Bush descended from his presidential helicopter, strode across the White House lawn, and declared in front of cameras—in a manner that suggested his syntax and delivery had been rehearsed to emphasize anger and underscore his resolution: "This will not stand, this aggression against Kuwait." No other president could have had such a strong predisposition to take up the cause for Kuwait, with his own oil company in charge of drilling the very first wells off the coast of the emirate (as Bush himself confided a few weeks later during a White House dinner).[6]

The US Army already had all the necessary plans in place. With the agreement of General Colin Powell, chairman of the joint chiefs of staff, CENTCOM's commander, General Schwarzkopf, had led several simulations of direct confrontation with Iraq barely three weeks before Saddam Hussein's troops entered Kuwait.[7] However, to implement the response against what the Pentagon presented as the "fourth army of the world," solid rear bases would be indispensable. Saudi Arabia was designated as their host. The masters of Riyadh, Mecca, and Dhahran had little empathy for their Kuwaiti neighbors: The Al Saud family had generally despised the Sabah family since the perilous trials and tribulations of young King Ibn Saud, and when Prince Bandar went to the toilet, he would say, "I've got to go to Kuwait."[8] President Bush was aware that Baghdad's objectives were not limited to Kuwait: In exchange for his withdrawal, Saddam Hussein demanded $4 billion, the forgiveness of

* According to the Carter Doctrine and its Reagan corollary.

his debt, and a readjustment of the border, which would allow him to obtain better sea access than the restricted access imposed by the British in 1921. On August 3, the Iraqi president offered to withdraw his troops, proposing to meet with Arab leaders in Jeddah, Saudi Arabia, two days later. Washington opposed such a meeting and warned its allies and Arab clients against any negotiation with Iraq. General Colin Powell himself was convinced that Saddam Hussein was perfectly aware that Saudi Arabia was too big of a fish for him.[9] However, the Bush administration attempted to convince everyone, beginning with the Saudis, that Saddam Hussein firmly intended to invade the Saudi kingdom, thereby threatening to plunge the world's economy into chaos. Also on August 3, in the aftermath of the invasion, Dick Cheney presented Prince Bandar with satellite images that seemed to indicate that one of the Iraqi armored divisions was heading toward the Saudi Arabian border.

Riyadh received the message loud and clear. That same day, King Fahd indicated that he'd made the "historic decision" to invite the Americans to protect his kingdom. The Saudis "didn't just want [Saddam] ejected from Kuwait; they wanted him destroyed," James Baker wrote. "For them, the *only* solution was an American-led war that would annihilate Saddam's military machine once and for all."[10] In the eyes of many Muslims around the world, the House of Saud had done something irreparable, by calling the largest army of infidels in the world to protect the holy land of the Prophet Muhammad. Throughout the Persian Gulf, in the business realm, most experts expected King Fahd merely to solicit modest air protection from Uncle Sam.[11] But beginning on August 8, the US Eighty-Second Airborne Division arrived in Dhahran. Every fifteen minutes, a large Lockheed C-5 Galaxy cargo plane landed on the airstrip of the Saudi petroleum capital, designed by the US Army Corps of Engineers. The Saudis hoped to limit publicity for this grandiose historic spectacle as much as possible. The few journalists who managed to make it to the spot witnessed the manifestation of the all-powerful US military, the most titanic of all human institutions, and by far the most voracious consumer of oil.[12]

Iraq, Fourth Army of the World or Phantom Threat?

With American troops on the ground in Saudi Arabia, the White House rejected any possibility of a negotiated withdrawal of Iraqi forces. From August 1990 to the end of what would become the Gulf War, in February 1991, Washington rejected no fewer than eleven peace proposals put forward not only by Iraq, but also by Jordan, Libya, Morocco, and two main suppliers of Saddam Hussein's army: the Soviet Union and France.[13] Iraq's aggression was severely punished: George H. W. Bush decided to confirm that his country was the singular hegemon at the center of planetary oil production.

From August to December of 1990, Washington assembled the widest military coalition since the Second World War. Secretary of State James Baker, former and future lawyer for Big Oil, traveled the five continents on a journey he described as "an intricate process of cajoling, extracting, threatening, and occasionally buying votes."[14] Egypt saw billions of dollars in debt erased in exchange for its participation in the coalition, unlike Yemen, from which the United States withdrew support overnight. A sizeable historical factor heavily supported the Bush administration's efforts: the fall, one year earlier, of the Berlin Wall. The United States was the only remaining superpower. On September 11, 1990, in what came to symbolize the advent of American hyper-domination, George H. W. Bush announced: "a new world order can emerge: A new era freer from the threat of terror, stronger in the pursuit of justice and more secure in the quest for peace."[15]

The US-assembled coalition that gathered to oppose Iraq consisted of thirty-one nations. It was joined without a hint of hesitation by President François Mitterrand, and was even implicitly supported by the moribund Soviet Union.* But the alliance remained largely dominated by the American military, which supplied more than half of the deployed troops. In Saudi Arabia, President Bush prepared the largest military deployment since the Vietnam War. On November 8, 1990, while repeating, "I would love to see a peaceful resolution," Bush announced that 200,000 additional American soldiers would join the 230,000 already deployed in the Gulf. The US agreement with the House of Saud was flawless. In mid-November, during the Thanksgiving holiday, George H. W. Bush and his wife Barbara went on tour with American troops amassed in the Saudi Kingdom. Recently divorced, their daughter Dorothy remained alone with her child at the White House. It was then that Prince Bandar's wife, Princess Haifa, invited Dorothy to spend the rest of the holidays in their company. This gesture moved President Bush so much that he shed a tear; the first lady of the United States nicknamed the prince ambassador of the Saud in Washington, "Bandar Bush."[16]†

* Paris committed some nineteen thousand soldiers. The French Defense Minister Jean-Pierre Chevènement, hostile to this war, who would later qualify it as "neo-colonial" (interview with the author, 2015), submitted his resignation to President Mitterrand on January 29, shortly after the offensive began.

† According to former CIA agent Robert Baer, this nickname, which garnered a lot of press, was given during one of Prince Bandar's visits to George H. W. Bush's Maine summer residence in the early 1990s. See Robert Baer, "The Fall of the House of Saud," *The Atlantic Magazine*, May 2003.

Opposing the nearly one million soldiers assembled by the US coalition, Saddam Hussein amassed between 250,000 and 500,000 men in Kuwait and southern Iraq. This was according to the Pentagon; many signs indicated that Dick Cheney had been able to supply Saudi Arabia, and the international media, with numbers that largely overestimated the forces engaged in Iraq. In August 1990, a Japanese newspaper and, in September 1990, the American network ABC submitted Russian satellite photos of Kuwait to an American military expert. This expert, Peter Zimmerman, shared the images with other experts. His verdict: "All of us agreed we couldn't see anything in the way of military activity on these snapshots," which were "astounding in their quality."[17] In fact, as *Newsweek* recounted, "all they could see, in crystal-clear details, was the U.S. buildup in Saudi Arabia."[18] The images ABC presented omitted a 30 kilometer strip of Kuwaiti territory: Perhaps the bulk of Iraqi troops were hidden there? In early January 1991, Florida's *St. Petersburg Times* obtained the missing snapshot. It did not reveal any hidden Iraqi troops. Zimmerman stated that from what he could see, the numbers of Iraqi troops in Kuwait did not reach even a fifth of the figure advanced by Dick Cheney. Once the American offensive was launched, many reporters noted the weak Iraqi resistance. Evoking a "phantom enemy," the New York daily, *Newsday*, quoted a high-ranking American officer as saying: "There was a great disinformation campaign surrounding this war."[19]

By the summer of 1990, all the pieces were in place to eradicate Iraqi troops from Kuwait. The only remaining task was to convince the American people. It was a delicate matter. On August 11, Kuwait's exiled ruling family hired the largest PR firm in the world, Hill & Knowlton. "Free Kuwait" T-shirts were distributed across the United States, and information days were organized on college campuses. But it was not enough. Washington had long supported Iraq, whose leaders appeared no more or less autocratic and anti-Semitic than Kuwait's ruling family. As for the atrocities committed by the Iraqi army in Kuwait (looting, rape, arbitrary hangings, and the like), the White House had taken no action when, two years earlier, the Iraqi Kurds had been Hussein's victims. George H. W. Bush felt the need to proclaim that, "The fight is not about oil" (although after the war, Dick Cheney admitted without hesitation that "the marriage of Iraq's military of 1 million men with 20 percent of the world's oil presented a significant threat").[20]

In early October 1990, while hundreds of thousands of American soldiers participated in Operation Desert Shield, which aimed to protect Saudi Arabia against an Iraqi threat that appeared increasingly illusory as the weeks passed, 70 percent of Americans favored the continuation of negotiations and rejected the recourse to a military option.[21] However, on October 10, a fifteen-year-old Kuwaiti's shocking testimony before the US Congress changed American public opinion. The girl, whose first name was Nayirah, did not give her family name or speak under oath. Before the sub-committee on human rights, this Kuwaiti wept as she reported that in a maternity ward in Kuwait City, she had witnessed Iraqi soldiers pluck babies from their incubators and throw them down, where they "left the babies on the cold floor to die." The story ran on a loop in the Western media. George H. W. Bush referred to it many times and did not hesitate to evoke the Holocaust, frequently comparing his former ally Saddam Hussein to Adolf Hitler. After the war, it was revealed that this story was pure fiction, wholly crafted by Hill & Knowlton, and that the young Nayirah was none other than the daughter of Kuwait's ambassador to the United States and a member of the House of Sabah.[22] But this PR stunt was enough to reverse public opinion in the polls and, in early January, a slim majority of Americans stated that they favored the war against Saddam Hussein.[23]

Instant Thunder and Desert Storm: The "Flogging" of Iraq Begins

On January 17, 1991, the war opened with a bombing campaign of colossal intensity. Its code name was Instant Thunder. For forty-six days and forty-six nights, the US Air Force committed almost half of its combat aircraft from around the world and on average dropped almost as many bombs per day on Iraq and Kuwait as it had dropped on Germany and Japan during the Second World War.[24] The famous "surgical strikes" were largely a fiction complacently relayed by CNN and other American networks, which didn't report the carpet bombing by B-52s (of which the coalition supplied no images).[25] An official report of the American Government Accountability Office revealed, after the fact, that the efficacy of America's so-called intelligent weaponry was "overstated, misleading, inconsistent with the best available data, or unverifiable."[26] Beyond its goal of destroying Saddam Hussein's war machine, this long aerial bombardment campaign aimed to destroy infrastructure vital to the Iraqi population. Instead of aiming for easily repairable electrical lines or transformers, the strikes systematically targeted Iraq's electric power plants and refineries. Iraq was thus deprived of 92 percent of its electricity production capacity and 80 percent of its fuel production capacity, according to journalist and French diplomat Éric Rouleau's account in the Council on Foreign Relations' magazine, *Foreign Affairs*. American bombs also targeted petrochemical plants;

cement factories; aluminum, textile, electrical cable, and medical equipment facilities; and even radio and television stations.[27]

Considered in the 1970s a model of development in the Middle East, Iraq was suddenly set back more than half a century. Without electricity, it became impossible to filter tap water or properly operate the sewers. "The whole of the drinking water system of Iraq [was] destroyed or near collapse" after the bombing, according to the World Health Organization. Not only did the destruction of the Iraqi electrical system greatly overstep the UN Security Council resolution, it violated the Geneva Conventions, which made it illegal to interfere with or deprive civilians of vital resources (and notably, water). The US Air Force declared after the fact that Operation Instant Thunder provided "leverage" on postwar Iraq, destroying "valuable facilities that Baghdad could not repair without foreign assistance."[28] Secretary of Defense Dick Cheney believed that all targets selected by the US military, including the Iraqi's central power plants, were "perfectly legitimate"; "If I had to do it over again, I would do exactly the same thing," insisted the future vice president of the United States.[29]

On February 15, while the bombing campaign was underway, Baghdad announced that it would comply with Resolution 660 of the UN Security Council: Iraq was ready to withdraw from Kuwait. Bush shunned the offer, calling it a "cruel hoax." On February 21, Moscow made it known that Saddam Hussein was ready to withdraw his troops within three weeks, in exchange for a day of cease-fire and the lifting of sanctions. Colin Powell responded with an untenable counterproposal: Iraq had one day to withdraw, without a cease-fire. "If, as I suspect, [the Iraqis] do not move," said Powell, "then the flogging begins."[30] According to American journalist Bob Woodward, during this time President Bush told his principal advisers: "We have to have a war." [31] On February 23, in the aftermath of the American ultimatum, Mikhail Gorbachev picked up the red phone to tell Bush that "Saddam ha[d] caved" and agreed to withdraw without conditions. [32] Another refusal: "It was too late for that," George H. W. Bush justified after the fact.[33] The US national security advisor, Brent Scowcroft, would later note, "It was essential that we destroy Iraq's offensive capability" to reach the American objective.[34] Scowcroft would highlight that this "major" objective "had not been feasible to list . . . openly as such while a peaceful solution to the crisis was possible."[35]

The "flogging" began on February 24: After the "thunder" came Desert Storm. The land offensive launched from Saudi Arabia was directed by General Schwarzkopf, nicknamed "the Bear," a son worthy of his father, the great protector of the shah of Iran and his oil. Norman Schwarzkopf Junior, a Vietnam War hero, had commanded the infantry battalion responsible (though not under his command) for the sinister My Lai massacre in 1968.[36] The Saudis were supposed to lead the operations. Formally, King Fahd was the commanding general of the "joint forces" and Prince Khaled Al Saud was commander in chief of "foreign forces." They kept up appearances. However, voluntarily or not, Saudi fighter-bomber pilots often missed their targets and, in his memoirs, General Schwarzkopf took the luxury of insulting General Khaled, writing that "his military skills are not, by far, as important as his princely origins" and adding that "the difference between him and other generals" was that Khaled Al Saud "had the authority to sign checks."[37] Two days after the offensive began, the remaining Iraqi troops in Kuwait fled along the roads that led from Kuwait-City toward Basra in any way they could, riding tanks, troop transports, and ordinary trucks, buses, and automobiles: Between twenty-five thousand and fifty thousand Iraqi soldiers were killed by infinitely superior air and ground forces during forty-eight hours of desperate retreat, along what was nicknamed the "Highway of Death." An American soldier described it as a "medieval hell."[38] But the carnage, which extended 60 kilometers, passed almost unnoticed in the Western media. Dick Cheney welcomed the "best-covered war ever" by journalists who had no other way to access the front than to be embedded in units chosen for them by the military.[39]

As they withdrew, Iraqi troops set fire to more than seven hundred of the approximately one thousand oil wells in the Kuwaiti desert. The result caused, without a doubt, the worst black tide in history, spreading more than a thousand kilometers over coasts once renowned for their pearl oysters. Smoke from the wells spread as far west as Romania, and the Chinese complained of black rain falling on the Himalayas. On February 23, Iraq denied responsibility for the fire and called for a UN inquiry, which was never held. Surely this amounted to a new stunt by Saddam Hussein. Or perhaps not? Twelve years later, a few days before Bush Junior conducted his invasion of Iraq, the American Gulf War Veterans Association reported matching testimonies of American special forces soldiers sent between Iraqi lines and the advancing American front in order to blow up Kuwaiti wells.[40] For weeks, dozens of specialized teams, provided especially by American petroleum service companies, were called in to extinguish fires of Biblical proportions.* Their working conditions were severe and hazardous. The Brazilian photographer Sebastiao Salgado, who went on-site to complete his monumental opus about labor, testified: "There I lived the apocalypse and saw the black symbol of humanity. . . . It was like fighting against the end of the world, a world drenched in black and death."[41]

* German filmmaker Werner Herzog drew on this for his most maligned film, *Lessons of Darkness* (1992).

On February 28, President George H. W. Bush declared a cease-fire. Two days later, the United States' armored Twenty-Fourth Infantry Division ravaged the Iraqi Hammurabi Division, which was in retreat; the Americans destroyed two hundred armored vehicles with virtually no loss of their own. The next day, the "fourth army of the world" surrendered from a fight in which it had never really been equipped to engage. The coalition had lost nearly two hundred soldiers, killed by the enemy; the majority were Americans. On the Iraqi side, the unverified death count reached one hundred thousand soldiers killed during a month of bombing and a week of catastrophe. This enormous figure came from the US military, which never provided an estimate of civilian victims.

Neither Bush nor Clinton Offer Relief to Iraqi Insurgents: Saddam Will Remain

Despite the coalition's overwhelming superiority, President Bush decided not to send his troops to Baghdad; he left Saddam Hussein, this new "Hitler," in power. Eight years later he justified his decision by explaining that "trying to eliminate Saddam . . . would have incurred incalculable human and political costs."[42] (The use of the adjective "incalculable" constitutes a remarkable premonition). Bush Senior affirmed that to "occupy Iraq" would have meant "unilaterally exceeding the United Nations' mandate," a mandate that, as we have seen, already had been exceeded during the war and was violated further during the years of the embargo that followed.[43]

Saddam Hussein had scarcely acknowledged his defeat when a number of Iraqi soldiers rebelled in the south of Iraq. In mid-March, the insurgents seemed to have taken control of the majority Shiite municipalities of Basra, Curbala, and Najaf. In the north, little by little, the Kurds also took control of major cities. Saddam Hussein's regime seemed like it was about to implode. But, to their dismay, the rebels did not receive the slightest support from the coalition led by the American military on behalf of the House of Saud. They could not obtain the weapons taken to Saddam Hussein's forces. Despite an influx of only vague verbal warnings, General Schwarzkopf authorized the defeated Iraqi army to use its attack helicopters, as long as those helicopters did not approach American troops.[44] He thus greatly facilitated the repression of the insurgency.[45] On March 27, the White House stated that it "had made no promises to the Shi'as or Kurds." During the war, coalition aircraft had dropped twenty-one million leaflets on the Iraqis, calling upon them to "join with your brothers and demonstrate rejection of Saddam's brutal policies."[46]

On February 15, 1991, Bush himself had called the Iraqi people "to take matters into their own hands and force Saddam Hussein, the dictator, to step aside."[47]

At the end of the first week of April, Saddam Hussein managed to crush the rebellions of the Shiites in the south and the Kurds in the north. According to different estimates, this suppression caused between twenty thousand and one hundred thousand deaths.[48] In the middle of winter, nearly a million Kurds took refuge in the mountains of Turkey and Iran. A Western humanitarian categorized it this way: "It was at the time the largest, swiftest mass exodus recorded in history."[49] Turkey, allied with the United States within NATO, looked with disfavor on the arrival of Kurdish refugees from Iraq and offered very little aid; between fifteen thousand and thirty thousand Kurds died on their escape route or in refugee camps.[50]

It was only once the Kurdish rebellion was squelched, on April 8, that the United States, Great Britain, France, and Turkey prohibited the Iraqi army from flying over northern Iraq. The coalition waited sixteen months longer, until the end of August 1992, to impose a no-fly zone over the Shiite south. For this delay, the *New York Times* noted, there was no "plausible explanation."[51] The White House's strategic advisor, Brent Scowcroft (who was Henry Kissinger's principal assistant at the time Kissinger manipulated, among other groups, the Iraqi Kurds), attempted to explain Washington's passivity with respect to Saddam Hussein and this war in a 1999 book he co-authored with George H. W. Bush. Regarding the objectives of this war, Scowcroft wrote: "The trick here was to damage his offensive capability without weakening Iraq to the point that a vacuum was created, destroying the balance between Iraq and Iran, further destabilizing the region for years."[52] Referring indirectly to the lack of assistance to the Shiite and Kurd rebels, Scowcroft emphasized: "Neither the United States nor the countries of the region wished to see the breakup of the Iraqi state. We were concerned about the long-term balance of power at the head of the Gulf."[53] For this balance, from the point of view of the main power "at the head" of the Persian Gulf, a moribund, inert Iraq constituted the most profitable long-term option.

The US strategy at the end of the Gulf War seems at first glance confusing, even contradictory. The initial UN mandate to the coalition was satisfied: The Iraqi troops had left Kuwait. President Bush did not want the American army to advance to Baghdad, yet on the evening of the victory, he wrote in his journal about Saddam Hussein, "He's got to go."[54] He did not lift a finger, however, to support the Kurdish and Shia insurgencies. Neither was Hussein's departure the object of the ongoing embargo imposed by the United Nations eight months earlier, in the aftermath of Iraq's invasion of Kuwait, and renewed on April 3,

1991, by UN Resolution 687: The embargo was intended to force Saddam Hussein to accept the destruction of his biological and chemical weapons, his military nuclear program, and his medium- and long-range missiles. Although Bush Senior had stated at the beginning of the war that Baghdad could join the "family of nations" once it respected the UN resolutions, one month after the adoption of Resolution 687 on disarmament, he declared: "My view is we don't want to lift these sanctions as long as Saddam Hussein is in power."[55] The destruction of Iraq's famous "weapons of mass destruction" (largely acquired thanks to the goodwill of their American former ally during the war against Iran) was largely accomplished by the mid-1990s, and in October 1998, the International Atomic Energy Agency found that Iraq's nuclear program no longer existed.* Sanctions against Iraq, however, lasted until the fall of Saddam Hussein's regime in May 2003, bringing to its knees the Iraqi population, with which President Bush assured the world he had "no quarrel" during his speech on the "new world order" of September 11, 1990.

Was the final, informal purpose of the sanctions against Iraq to make Saddam Hussein unpopular enough that Iraq's population would be prompted to depose the despot who governed them and had already subjected them to two wars? Regarding Iraq, there is perfect continuity between the strategies of Republican George H. W. Bush and those of Democratic President Bill Clinton, who succeeded him in the White House in January 1993. In March 1995, the Clinton administration refused to support a coup, coordinated by the local CIA base in Iraqi Kurdistan, which united the efforts of senior Sunni officers, two rival independent Kurdish movements, and Shiite opposition groups in the south of the country—in short, all forces capable of standing up against Saddam Hussein. Appalled by the Clinton administration's passivity, the head of the CIA field office, American intelligence officer Robert Baer, resigned two years later. Considered "without a doubt the best CIA agent in the Middle East," Baer, aghast, provided a detailed account of this aborted coup attempt.[56]†

If the sanctions against Iraq had not been intended to lead to an uprising against Saddam Hussein, what was their purpose? Again, the history of oil

---

* Between 1991 and 1998, nearly 3,500 United Nations inspectors visited 3,400 Iraqi sites, including 900 secret military installations. American Scott Ritter, responsible for inspections, noted that by the end of this period, "we did ascertain a 90–95 percent level of verified [Iraqi] disarmament." General Hussein Kamel Majid, Saddam Hussein's son-in-law, who for ten years was responsible for Iraq's unconventional weapons program, defected in August 1995 and declared to the United Nations and the CIA that Iraq had destroyed all its weapons of mass destruction shortly after the end of the Gulf War. In 2001, William Cohen, Bill Clinton's secretary of defense, later informed the George W. Bush administration that "Iraq no longer poses a military threat to its neighbors." (See Larry Everest, *Oil, Power and Empire: Iraq and the U.S. Global Agenda*, op. cit., p. 188–193.)

† "Best agent" in the opinion of renowned American journalist Seymour Hersh.

flows down cryptic paths, raising difficult questions, such as: To what extent had the outcome of the first war against Iraq, led by Bush Senior, made it inevitable that sooner or later, there would be a second, in the minds of men like Dick Cheney who orchestrated both conflicts? During the embargo years, the theory of an American conspiracy flourished among Iraqis. Former CIA agent Robert Baer presented this theory (without, however, mentioning whether his own experience lent it credibility): "The United States needed [Saddam Hussein] to preserve peace in the Gulf. Bogeyman, kingpin, Saddam Hussein terrorized his neighbors. The mere whisper of his name made the oil kings seek refuge in the skirts of the superpower."[57] It is always imprudent to attribute unambiguous intentions to such a broad institution as the executive power of the United States, and all the more so if one suspects these intentions have been disguised. The primary motives of the American post–Gulf War strategy have not yet been clearly decrypted and may never be. But if the goal was to oust Saddam Hussein, why leave him in power so long after they had filed down his fangs?

Half a Million Iraqi Children: A "Fair Price to Pay"?

The long UN embargo on Iraqi exports (almost exclusively petroleum), which would be succeeded by an Oil for Food program in 1996, deprived an already heavily indebted country of currency, rendering it incapable of purchasing the equipment it needed to repair its systems for electricity, drinking water, irrigation, sewage, and medical care—destroyed or rendered inoperative during the bombing in January and February, 1991. In March 1991, noting the bombing's "near-apocalyptic results," Finland's Martti Ahtisaari, deputy secretary general of the United Nations and future Nobel Peace Prize winner, noted that "Iraq has, for some time, been relegated to a pre-industrial age, but with all the disabilities of post-industrial dependency on an intensive use of energy and technology."[58] The sudden development induced by an oil windfall weakens the backbone of oil nations as surely as that of adolescents who grow too quickly.

Previously, the Iraqi population had been one of the richest and best fed among the non-Western nations. Due to the embargo, the unemployment rate rapidly reached 50 percent of the population and climbed to more than 70 percent in the industrial sector. Annual per capita income dropped from $2,700 dollars in 1989 to less than $500 in the late 1990s, making Iraq one of poorest countries on the planet.

The embargo triggered carnage at the heart of the civilian population, who were victims of a kind of unpublicized bacteriological war. In 1995, a UN report described the situation: "In Baghdad untreated sewage has now to be dumped directly into the river—which is the source of the water supply . . . and all drinking-water plants there and throughout the rest of the country are using river water with high sewage contamination."[59] In 1996, all Iraqi wastewater treatment plants remained out of commission and, even in 2002, UNICEF estimated that 40 percent of the water consumed in the city of Basra consisted of untreated wastewater.[60] Everywhere in the overheated Iraqi cities, especially in Baghdad and Basra, cholera, typhoid fever, and severe diarrhea were rampant.

Hunger also wreaked havoc in the lands that had seen the birth of agriculture. The Iraqi fields suffered from a lack of irrigation, spare parts for farm equipment, and fertilizer and pesticides—often banned because of potential civilian and military "dual purpose." The harvest fell by as much as two-thirds.[61] Before the Gulf War, Iraq, whose population had exploded during the second half of the twentieth century due to the Green Revolution, already imported almost three-quarters of its food; the United States, in particular, exported a significant portion of their rice crops to Iraq. After the war, Iraqis' access to food was strictly rationed.

American presidents Bush Senior, Bill Clinton, and Bush Junior accused Saddam Hussein's regime of being entirely responsible for this nightmarish health and food situation. Yet the Pentagon and the American intelligence agencies had issued warnings, at least as early as January 1991, about the consequences of destroying electric power stations and their effects on Iraq's drinking water.[62] What's more, during the twelve embargo years, Washington and London vetoed over 90 percent of the more than 1,500 contracts proposed to supply civil equipment to Iraq. Among the products blocked at one time or another by the American and British advisors to the United Nations were hospital fans, vaccines, truck tires, and water tanks. In 2001, a UNICEF report pointed to the obstruction of contracts, valued at $500 million, intended for water treatment. The United States opposed many contracts in the name of their potential "dual use," civilian and military, the same "dual use" from which the Reagan administration had complacently looked away during the Iran-Iraq War and whose definition, in this instance, was extended to telecommunications equipment and oil facilities.[63]

Until May 1996, Saddam Hussein refused to accept the Oil for Food program, proposed in August 1991 by the United Nations. He, too, bore a heavy responsibility in the deterioration of the health situation in his country. This program allowed Iraq to export its oil under the control of the UN Security Council, in order to obtain revenue to cover its population's humanitarian needs. Bill Clinton and, later, George Bush Junior accused Saddam Hussein of squandering the benefits of this program. In fact, the Baghdad regime received more than $2 billion in bribes from foreign companies buying oil or providing equipment to Iraq, which represented less than a tenth of the value of the foodstuffs acquired in the Oil for Food program: approximately $23 billion, which per year, per capita, represented less than half the income of a Haitian.[64]

The Oil for Food program led to a resounding scandal: Baghdad offered very advantageous petroleum contracts to reward various kinds of support; the many beneficiaries included not only Russian, French, and Chinese companies and individuals, who had the right to free or discounted barrels, but also the oil majors, American or not.[65] Trusted because it was conducted under the auspices of the United Nations, this practice of giving in-kind benefits was not uncommon: For the oil despots, black gold was always the currency that purchased political power and influence.

After it was established in December 1996, the Oil for Food program doubled food rations for Iraq's poor. Improvement was a relative term. In 1999, UNICEF estimated that a quarter of newborns were underweight, and a quarter of the children under five years old (or one million persons) suffered from chronic malnutrition.[66] The United Nations' two successive humanitarian coordinators in Iraq, Irish Denis Halliday and German Hans von Sponeck, resigned on October 1, 1998, and February 2, 2000, respectively, in protest against the Security Council sanctions, described by Halliday as "genocidal," and as "a tightening of the rope around the neck of the average Iraqi citizen," according to von Sponeck. Yet in 2003, a few weeks before the outbreak of the second invasion of Iraq, the UN executive in charge of the Oil for Food program called the distribution system established by Baghdad the "most efficient in the world."[67]

The Oil for Food program yielded good results on the oil side. It allowed Iraq to return to producing more than 2 Mb/d, its pre–Gulf War production rate, putting back on the world market a source that had been almost constantly absent since the start of the Iran-Iraq War in 1980. The UN Security Council ensured that Baghdad received only about half of the royalties. Almost a quarter were destined for Kuwait. The large oil infrastructure and services companies that had suffered losses during the invasion of the emirate also received reparations from Baghdad, not only for destroyed equipment but also, more often, for simple slowdowns in their business. Among the primary beneficiaries were several American petroleum giants: Texaco, Mobil, Bechtel, and Halliburton.[68]

The United Nations allowed many petroleum service companies to sign contracts to help Iraq repair its infrastructure and relaunch its crude oil extractions. Among these companies, the Clinton administration allowed two subsidiaries acquired in 1998 by Halliburton, the American oil services giant, to sell Iraq water-treatment and petroleum equipment for $73 million.[69] The line between authorized and unauthorized contracts was subtle: Washington eventually blocked contracts valued at close to $1 million, signed by one of these subsidiaries, to provide spare parts, compressors, and fire-fighting equipment intended to repair the Khor Al Amaya Iraqi offshore platform, severely damaged at the beginning of the war against Iran and completely destroyed by the coalition during the Gulf War.[70] Dick Cheney, who in 1995 had become Halliburton's president, committed himself to convincing the US Congress to review its policy of unilateral sanctions, which according to Cheney penalized American firms.[71] In April 1996, Cheney explained to an audience of oilmen: "The problem is that the good Lord didn't see fit to always put oil and gas resources where there are democratic governments." He added that due to the sanctions, US companies found themselves in the situation of "the bystander who gets hit when a train wreck occurs."[72] The former Pentagon chief was not concerned that his commentary would be judged as contradicting the past and future policies of his own political office.

Whatever the core agenda may have been, the destruction of Iraq's vital infrastructure during Operation Desert Storm, the subsequent obstructions to rebuilding, and the insufficiency of the fruits received from the sale of oil ultimately led to the decimation of the Iraqi population. In 1999, UNICEF estimated that between 1991 and 1998, nearly 500,000 Iraqi children under five years of age had died from disease or malnutrition as a direct result of the sanctions.[73] In 2002, Baghdad calculated the balance sheet at 1.7 million dead children.[74] These figures were already circulating when, on May 12, 1996, a journalist on *60 Minutes*, the flagship show of the American CBS network, asked President Clinton's secretary of state, Madeleine Albright: "We have heard that half a million Iraqi children have died. I mean, that's more children than died in Hiroshima. And you know, is the price worth it?" Without hesitation, Madeleine Albright responded: "I think this is a very hard choice but the price—we think the price is worth it."

TWENTY-FIVE

Planetary Harvest: The Time of Scandals

More than the Gulf War itself, the control of Iraqi borders and airspace that came out of the war strengthened Washington's strategic position in the Persian Gulf. During the 1990s, the United States' military presence reached twenty thousand troops on average. American and British aircraft carried out no fewer than 280,000 sorties above Iraq. In December 1998, during a widely criticized bombing campaign authorized by Clinton, the US Navy advanced two of its Nimitz class super–aircraft carriers, the most powerful warships in the world, to the Gulf's interior.*

After the fall of the Berlin Wall and the collapse of the Soviet empire, concentrations of troops dwindled in Europe, and strongly increased everywhere that access to raw materials was in dispute: around the water in the valleys of Jordan, the Indus, and the Nile; around the wood, stones, and other precious minerals in Colombia, Indonesia, the Philippines, Sierra Leone, and Liberia and along the Congo River; around the oil, of course, from the Gulf of Guinea, the Persian Gulf, and the former Communist republics bordering the Caspian Sea. After the death of the USSR, access to raw materials, which, against the earlier backdrop of ideological conflicts, had not been fully acknowledged, became the blatant cause for most of the conflicts, latent or not, around the planet.[1]

* The USS *Enterprise* and the USS *Carl Vinson* successfully carried out Operation Desert Fox, authorized by President Clinton in order to punish Iraq for its lack of cooperation in the inspection of its infrastructure. The operation's merits were widely criticized, notably by Paris and by several UN inspectors, in particular American Scott Ritter, who judge Baghdad's cooperation satisfactory.

"Grapes of Wrath": Al Qaeda and Wahhabism

Ironically, the bloodiest ideological fissure since the Cold War was born out of the West's control of the mother of all raw materials. In the wake of Instant Thunder's outbreak, a headline in the *Independent*, the British daily, announced that weapons had been drawn between Islam and Christianity.[2] The famous "clash of civilizations" promised two years later by Harvard professor Samuel Huntington (former consultant for the Trilateral Commission and the Carter White House National Security Council) found its most powerful catalyst in the anger of a young heir for the untold riches that Christians had lifted from the Arabian desert.[3]

A few days after Iraq invaded Kuwait, Osama bin Laden wrote to King Fahd, asking him to stop asking America for assistance.[4] On behalf of President Bush, Dick Cheney had sworn an oath to Crown Prince Abdullah that American troops would leave the kingdom as soon as the Iraqi danger was quelled, or as soon as the king asked him to do so.[5] In early September 1990, Osama bin Laden showed Prince Sultan, Saudi Arabia's minister of defense, and then Prince Turki, the head of the secret service, a detailed ten-page plan demonstrating how to fortify the kingdom's northern borders using bulldozers owned by the Saudi BinLaden Group.[6] At age thirty-three, the returned hero of the war in Afghanistan headed an international network of Islamist fighters. Bin Laden assured the princes he would be able to establish an army of one hundred thousand combatants, veterans of Afghanistan.[7] After hearing this presentation, Prince Turki burst into laughter, shocked by his interlocutor's "degree of arrogance and disdain."[8] The United States' hegemony in the Persian Gulf had managed to manifest the third avatar of its nemesis (the first being Iran's Islamic revolution, and the second being Saddam Hussien's military adventurism). Osama bin Laden's impetus for radicalization had nothing to do with the Palestinians' situation—and even less so the Bosnian Muslims' situation. Rather, he was motivated by the presence of infidel soldiers in the Holy Land, who came to defend the kingdom that the Western economic press nicknamed the "central bank of oil."

This was the departure point for Al Qaeda's founder, hitherto faithful vassal of the House of Saud and instrument, among many others, of American policy. In particular, Osama Bin Laden accused the Sauds of having spent $25 billion to help Saddam Hussein fight Iran, before paying $60 billion to foot the bill for the Gulf War.[9*] The rift was completed with the first attack that bore bin Laden's mark. On November 13, 1995, just before the midday call to prayer, a car bomb

* The Gulf War had cost $61 billion. Saudi Arabia and Kuwait had paid for half of this, to which can be added the invoices for military equipment.

exploded in close proximity to one of the most stylish shopping thoroughfares of Riyadh, in front of the Office of the Program Manager of the Saudi Arabian National Guard (OPM-SANG). Seven were dead, including five Americans, and sixty were injured. Among them were many American advisers from Vinnell, the security company close to the CIA, responsible for training the Saudi national guard.[10] On June 25, 1996, an explosion killed dozens of people, including nineteen Americans, in front of a military housing complex near Dhahran, at the heart of the Saudi Arabian oil industry. Again, this was likely one of Al Qaeda's early attacks. On August 7, 1998, bin Laden ordered the massacre of more than two hundred people (including many African Muslims) in a double car bomb attack against the American embassies in Dar-es-Salaam in Tanzania and Nairobi in Kenya; the date chosen was the eighth anniversary of the American soldiers' arrival in Saudi Arabia, those the terrorist leader called the "Christian crusaders."

The rift between the Sauds and Al Qaeda's founder concealed a more nuanced reality, one less acceptable to Riyadh. The Wahhabism promoted by Saudi Arabia remained the main crucible of fundamentalist Islamism. When Osama bin Laden was stripped of his Saudi citizenship in 1994, Saudi Arabian oil money continued to fuel a global network of Koranic schools. In the north of Pakistan, these madrasas formed the core of the Taliban movement that took power in Afghanistan in September 1996. Two years later, Nawaf Obeid, a young Saudi researcher in the social science field, noted: "With the Taliban, the United States have had the opportunity to observe a Wahhabi government without the moderating presence of the Sauds."[11]

For the Sauds, more powerful and more subject to Washington than ever, the "new world order" of justice and peace announced by Bush Senior on September 11, 1990, translated to a resurgence of conservatism brutally implemented by the government police, the *mutaween*. In order to stave off the uprising they constantly feared, Riyadh lavishly subsidized their subjects' food, housing, energy, and other needs. The citizens, often idle, were served by a humble populace of foreign Muslims who had almost no rights. Starting in September 1990, after their country refused to join the coalition against Iraq, thousands of Yemeni workers and tens of thousands of Palestinians were expelled by Riyadh on the pretext that PLO leader Yasser Arafat supported Saddam Hussein.[12] In the weeks following the House of Sabah's return to Kuwait, in March 1991, 450,000 Palestinians were driven from the emirate in a few days. Many were tortured, some were murdered.[13] On May 29, 1991, President George H. W. Bush presented an initiative to halt and reverse the destabilizing arms race in the Middle East.[14] Of the $28 billion worth of arms sold by the United States during the two years that followed, more than half were destined for Saudi Arabia.[15]

On the eve of the 1991 Gulf War, President George H. W. Bush said that the American military deployment in Saudi Arabia had the purpose, among others, of saving the American way of life.[16] Behind this ambiguous pretension persisted, once again, the same trivial reality, described in 1997 by CENTCOM's commander in chief, General J. H. Binford Peay III: "America's vital interests in the [Persian Gulf] region are compelling. . . . The unrestricted flow of petroleum resources from friendly Gulf states to refineries and processing facilities around the world drives the global economic engine."[17]

The Former USSR, a New Course of Resistance

The 1990s revealed an acceleration of the mad dance of corruption that oil could arouse in any situation. The industry had weathered the consecutive upheavals of oil crises. Big Oil had had the time to reorganize, and starting in 1994, world production exceeded the record level reached in 1979. What was beginning to be called "globalization" of the economy advertised the beginning of a new explosion in the global demand for crude oil. Finally, the USSR's disintegration in December 1991 returned the weary Russian petroleum industry to the capitalist market; it was worn out, arcane in its infrastructure but nearly as substantial a supplier as Saudi Arabia. While the new president, Boris Yeltsin, agreed to sell Russian oil to young oligarchs with ties to the west (such as Mikhail Khodorkovsky, who in 1995 headed the new petroleum giant Yukos), the ex-Soviet Empire's enormous production capacity formed the main course in a bacchanal of dirty money, where guests held hands beneath a table as large as the world.

Of this secret feast of the oil harvest, one can only discern the scraps, and sometimes count the victims, it left behind. Among the witnesses, their accusations patchy by necessity, figured ex-CIA-agent Robert Baer. Baer affirmed that upon his return to Iraq, he had fallen into the middle of a curious ballet involving the White House, a discovery that, he said, cost him his position within the intelligence agency.[18] On May 17, 1995, at his office in CIA headquarters, Baer received a request for information about an influential underworld contact in the world of oil: Roger Tamraz. This Lebanese venture capitalist, a naturalized American then sought by Interpol for fraudulent bankruptcy in Lebanon, was nevertheless a generous donor to the American Democratic Party, able to book appointments with Vice President Al Gore and even with President Bill Clinton. Tamraz's purpose: to get Washington on board with his pipeline project, which would allow oil to flow through the former Soviet

Republic of Azerbaijan via Armenia and the unstable self-proclaimed Republic of Nagorno-Karabagh.

The information request Robert Baer received came from Sheila Heslin, an energetic young woman responsible for Central Asia within the NSC. According to Baer, Heslin's requested investigation of Tamraz aimed to "charge" a troublemaker who was poaching in the middle of a hunting party. Several of the Western oil majors, in particular Amoco and BP, were indeed associated with the development of another project pipeline, the Baku-Tbilisi-Ceyhan pipeline, whose construction was launched four years later, transporting Caspian Sea crude oil from the capital of Azerbaijan to the Turkish port of Ceyhan, not via Armenia, but through Georgia. On September 20, 1994, a consortium led by BP and Amoco had signed on to a staggering investment project with Azerbaijan, estimated at $8 billion, to drill off the coast of Baku. Despite the fury of neighboring Russia and Iran, Azerbaijan had chosen to appeal to the expertise and capital of the Western companies in order to exploit these ultracomplex offshore resources: an operation rendered indispensable by the slow depletion of wells on the mainland of the world's oldest oil territory. This voracious bite Big Oil planted in the former Soviet Union sharpened many appetites. Sheila Heslin explained that Washington's objective was "in essence to break Russia's monopoly of control over the transportation of oil from the region" around the Caspian Sea.[19]

Perplexed about the NSC's intentions in respect to Tamraz, Baer questioned the man who oversaw Bill Clinton's reelection campaign, Donald Fowler, Democratic Party chairman. Fowler accused Heslin of being Amoco's "ambassador" to the White House.[20] Shortly before, the Azeri president (and ex-Soviet leader) Heydar Aliyev had said that the Clinton administration pressured him strongly to award Exxon a concession off the coast of Baku.[21] According to Baer, it was Sheila Heslin's boss at the NSC, Anthony Lake, who had maneuvered matters so that Exxon obtained this share of Azeri black gold (coveted three-quarters of a century earlier by Standard Oil of New Jersey).[22] After his departure from the CIA, Baer wrote: "How could I have been naive enough to believe that the role of the White House was to support all American companies, and not this or that in particular?"[23]

During Clinton's second term, Anthony Lake and his then deputy director of the NSC, Sandy Berger, were sanctioned by the American courts for having waited until long after their arrival at the White House to resell tens of thousands of dollars' worth of shares that Lake held in Exxon and Berger in Amoco.[24] The dance didn't stop there: In his tell-all book, Baer accused Sheila Heslin of having attempted to squelch a CIA investigation of the contacts between Al Qaeda and Iran, in fear of reprisals from Tehran against concessions obtained by Amoco in the Caspian Sea.[25]

Firms of lawyers and lobbyists in Washington jockeyed to convince the diplomats of the new oil-rich Caspian Sea republics that they could best open the doors to American power, as they had done in the past for representatives of the Persian Gulf and Gulf of Guinea states. The most active of these firms was the prestigious law firm Hogan & Hartson, who happened to be Sandy Berger's employer before his arrival at the White House.

The Carlyle Group Perfects the Symbiosis between the Bush Networks and Saudi Petrodollars

The practice of "revolving doors," in which powerful American business managers switch between working in the public and the private sector, was long established in the United States. But while some in Washington already had a keen interest in the crumbs of the former Soviet Union's "Big Oil feast," an investment company that had been heretofore unknown took what is now called "access capitalism" to a whole new level.*

The Carlyle Group was founded in 1987 by a former Carter administration policy advisor, David Rubenstein. Wall Street was then in the midst of a frenzy of acquisitions and leveraged buyouts. Rubenstein was an innovator. Starting with nothing, he established Carlyle as the leader of buying and selling the capital of companies not listed on the stock exchange. A Carlyle Group brochure explained the company's strategy, well on the way to establishing itself as one of the most powerful investment groups on the planet: focused on the "federally regulated or impacted industries such as aerospace/defense."[26] To succeed, Carlyle needed to be well connected and capable of outreach. In January 1989, Frank Carlucci became Carlyle's CEO a few days after having left his seat as President Ronald Reagan's defense secretary (a post in which he had succeeded Caspar Weinberger, Bechtel's former vice president). A former CIA deputy director, Carlucci had been called to the White House by Reagan in 1986 to direct the NSC and clean up the Irangate scandal. During his time at Princeton University, Carlucci counted among his comrades James Baker and Donald Rumsfeld (his roommate). After his Pentagon departure, Carlucci sat on the board of directors of approximately thirty companies and organizations, in addition to his Carlyle Group position.[27] The fall of the Berlin Wall

* This expression appears in 1993 in one of the first articles criticizing the Carlyle Group. (See Michael Lewis, "The Access Capitalists," *The New Republic*, October 18, 1993.)

led to a sharp reduction in the armaments business, offering Carlyle prime investment opportunities in gold that Carlucci skillfully grew.

One of the Carlyle Group associates, Frederick Malek, participated in President George H. W. Bush's victorious 1988 election campaign. Shortly after Frank Carlucci's recruitment, Frederick Malek dedicated himself to helping an unsuccessful, forty-three-year-old Texas oilman, George W. Bush, who he had recruited to be on Carlyle's board of directors. With what seemed to be characteristic candor, Bush Junior clearly illustrated "access capitalism": "When you're the president's son and you've got unlimited access combined with some credentials from a prior campaign in Washington, D.C., people tend to respect that." George W. Bush continued, with a slight shrug of the shoulders which emphasized the obvious: "Access is power. And I can find my dad and talk to him any time of the day."[28]

Carlyle brought to ultimate fruition the symbiosis between a certain US industry and the petrodollars of its Saudi friends. In February 1991 (five months before the outbreak of the scandal abruptly shut down the activities of the BCCI), the Washington-based investment firm was making a name for itself. Carlyle astonished Wall Street and the *New York Times* when it advised a thirty-five-year-old multibillionaire from the Saud clan to invest $590 million in the largest American bank, Citicorp.[29] Most of those funds were invested in more than a billion dollars' worth of non-creditworthy American mortgages. The transaction took place during the Gulf War, at a time when American troops, commanded by Bush Senior, were fighting victoriously on behalf of the Kingdom of Saudi Arabia. On the Carlyle side, Bush's friend Frederick Malek was on the move. Hogan & Hartson played the role of intermediary. The wealthy Saudi Arabian Al-Waleed Bin Talal Al Saud became Citicorp's largest individual shareholder. Having purchased the shares at the lowest price possible, he was poised to make a tremendous profit, thanks to the institution that became Citigroup in 1998, and remained one of the most important private financial institutions on the planet. Prince Al-Waleed, one of King Ibn Saud's many grandsons, closed the loop on a age-old process: Citicorp was the new name of National City Bank, nicknamed the "oil bank" of Wall Street a century earlier; it remained directly linked to the dynasty of William Rockefeller, whose older brother, John, had sent him to New York to invest the profits of the oil region before the birth of Standard Oil. Regardless of where they came from, petrodollars seemed to perpetually regenerate Wall Street. In the 1990s, the capital from the Persian Gulf invested in the United States and Europe was valued in the hundreds of billions of dollars.

In 1993, Carlyle took control of Vinnell, the security company that, among other things, was responsible for training the Saudi national guard, which had intervened during the 1979 Hajj hostage situation. The investment group was poised to become one of the biggest arms traders in the world. That same year, Carlyle successfully recruited a man who probably had Washington's most charmed political career of anyone who had not been elected: James Baker, President Ford's former undersecretary of commerce, former cabinet chief and secretary of the treasury for President Reagan, and then secretary of state and once again cabinet chief for his old Texan friend, President George H. W. Bush. After Bush's departure from the White House, Baker resumed his career as a business lawyer, from then on as a partner in Baker Botts, the legal firm his grandfather had founded in Houston, and as a principal legal advisor in the American oil capital. Baker agreed to rejoin Carlyle following an April trip to Kuwait, where he represented the interests of a budding young Houston-based company that specialized in energy trading: Enron. During this trip, he was accompanied by two of President Bush's sons, Marvin and Niel; the first had come to sell electronic barriers, the second sold equipment to combat oil pollution.[30]

Carlyle landed an even greater coup in 1995 when it recruited ex-President Bush himself among its advisers, and some time later, the former British prime minister John Major. What better representatives could a company dream up to do business with the Arab oil princes, who were still very attached to those two leaders? As an observer noted in the *Financial Times*: "[A]nd they bring in Bush and Major, who saved the Saudis' ass in the Gulf War."[31] George H. W. Bush defended himself by saying he had never negotiated a single contract himself. Like James Baker and John Major, he merely appeared and talked with the members of the Al Saud or bin Laden families. Gratitude did the rest. On May 3, 1995, for example, Vinnell picked up a $163 million contract to modernize the Saudi national guard.[32] Six months later, in Riyadh, several Vinnell employees were injured during the first attack attributed to Osama bin Laden. That same year, the symbiosis in the oil money crucible led several members of the bin Laden family (in which Osama was described as the renegade) to invest directly in Carlyle, as did two sons of Khalid bin Mahfouz.[33]

In France, Business Dances a Minuet on a Bed of Black Gold

The political elites of France, a second-order industrial power, blithely participated in a systematic minuet of dirty oil money. Black gold was not the only vector for the purchase of influence, but it was the most powerful. The 1990s teemed with political-financial scandals directly linked to oil interests.

First there was Iraq, with the French version of the Oil for Food scandal, during which the former French minister of the interior Charles Pasqua and the Total group CEO, Christophe de Margerie, were put under investigation (and vindicated, for lack of evidence).* Then there was the Karachi case, which related to contracts for the sale of frigates to Saudi Arabia and submarines to Pakistan, and included retrocommissions that served to finance the presidential campaign of Gaullist candidate Édouard Balladur in 1995. There was the case of the so-called Angolagate: the 1994 sale of $780 million worth of Russian arms to the ultraliberal, pseudo-Marxist MPLA movement of the president of Angola, José Eduardo dos Santos.† This sale occurred at a decisive moment in the Angolan president's victory over the UNITA movement, following two decades of civil war abundantly fed by oil money. It gave rise to the conviction of thirty-six people in 2009. Among these were the son of President François Mitterrand, Jean-Christophe (who in Africa was dubbed Papamadi—meaning "Daddy told me"), as well as then French minister of the interior, Charles Pasqua. The latter was sentenced to three years in prison, a minimum of one year without parole, and a fine of 100,000 euros. But Pasqua was acquitted through an appeal two years later, after repeating that this weapons sale had been endorsed by all the highest authorities of the standing French government in 1994. During his acquittal, *Le Monde* evoked the "magnanimity" of the Paris Court of Appeals, then presided over by a magistrate who subsequently asked to be transferred to Africa as legal adviser to the Gabon government.[34]

Finally, there is the case of Elf Aquitaine, France's large, public petroleum company. This sprawling scandal remains a striking example that allows a glimpse into the workings of the Françafrique, an opulent system of corruption and political financing with its roots in decolonization: the "longest running scandal of the Republic."[35]

Loïk Le Floch-Prigent, CEO of Elf Aquitaine from 1989 to 1993, was sentenced in 2003 to five years in prison and a fine of 375,000 euros, as were several of the company's other leaders, for charges including the abuse of

---

* On July 18, 2013, Paris prosecutors appealed the acquittal of the Total group's CEO on charges of bribery of foreign public officials.

Grandson of the founder of the champagne Taittinger, Christophe de Margerie, often nicknamed "Big Moustache," died on October 20, 2014, during a plane crash in a Moscow airport. As a result of his 2007 appointment as Total group's CEO, in the midst of a historic crude oil price spike, he was practically the only Big Oil boss to publicly recognize that "peak oil" was real, although he presented this—how could he do otherwise?—as a benign problem on the far horizon.

† Total was not the only active oil company in Angola during the civil war. The international oil industry's adaptive capacity is always surprising. To access offshore Angolan oil, the oilmen dealt with the regime of Luanda when it was still "Marxist," advised by Cuban military experts, in the face of an insurgency supported by apartheid South Africa.

public assets. Le Floch-Prigent never stopped defending himself, asserting that he was only operating under the mechanisms Pierre Guillaumat and the Gaullists had established in the 1950s. One of Le Floch-Prigent's predecessors at Elf Aquitaine, Albin Chalandon, affirmed that millions of francs had been distributed in envelopes destined for selected candidates during legislative and presidential elections of the 1970s and 1980s.[36] In April 1977, for example, a memo from Elf's intelligence unit indicated that Gabonese president Omar Bongo "put all his hopes in Jacques Chirac and believed he still ha[d] to be helped"; the note was transmitted a month after Chirac's election in the Paris town hall, and four years before his first presidential election bid.[37] Stamped "SECRET," this note was addressed to Pierre Guillaumat and Elf's "Mr. Africa," André Tarallo—Chirac's former classmate at l'École Nationale d'Administration. Le Floch-Prigent repeated that, at the time of his entry into office in 1989, he had asked François Mitterrand what he should do "with the money that Elf gives African leaders and that they redistribute in France." According to Le Floch-Prigent, President Mitterrand replied: "You do not change anything, but make sure that no one is wronged, and as for the socialists, ask my entourage."[38] During the course of the investigation, Éva Joly, the investigating judge, affirmed that she had learned that during 1989–1992 the equivalent of six months of profits (or 3.5 billion francs) had been diverted from the national oil company. According to Joly, all of Elf's operations included routes to supply secret funding. Hundreds of millions of francs were funneled through Swiss bank accounts to the heads of African petroleum nations; it was impossible to know where these funds traveled next. Alfred Sirven, Le Floch-Prigent's business director (who, like Le Floch-Prigent, was sentenced to five years in prison), at that time had two heavyweight contacts in the French political arena: on the right, Charles Pasqua, and on the left, Roland Dumas, minister of foreign affairs from 1988 to 1993.

The scandal broke out after the revelation that in 1992 Elf Aquitaine had provided very broad and generous financial support to the struggling textile firm of Maurice Bidermann, a close friend of Jacques Chirac. The Western textile industry was in decline, undermined by globalization—in other words, by the low cost of shipping fuel. African oil money came to the rescue of the old French industry, financing in particular Bidermann's factory in Corrèze, Chirac's home turf. In August 1993, the conservative government in place during socialist President François Mitterrand's term, fired Le Floch-Prigent to appoint in his place a close friend of Prime Minister Édouard Balladur, Philippe Jaffré. In September 1999, Elf, a giant public company, was absorbed by its "small" private competitor Total at the end of a hostile takeover that had been given the green light by socialist Prime Minister Lionel Jospin and his minister of the economy, Dominique Strauss-Kahn (who himself, a little later, was investigated during the Elf case by judge Éva Joly, before the charges were dismissed in 2001).*

Of the Elf case, the French recalled the pugnacity of Norwegian-born French magistrate Éva Joly, a small, blond, determined trial judge, who became the Ecologist Party's candidate in the 2012 presidential election. The French public also particularly remembered an episode in which Roland Dumas's former mistress, Christine Deviers-Joncour, an Elf lobbyist responsible for public relations, used an Elf credit card to purchase the socialist minister of foreign affairs a pair of elite Berluti shoes valued at 11,000 francs, or $2,000. The public may also have recalled a letter addressed to François Mitterrand by writer Françoise Sagan at the initiative of an Elf intermediary, André Guelfi, nicknamed "Dédé la Sardine," which secured a meeting between the French president and the prime minister of the Republic of Uzbekistan—a post-Soviet republic that has proven to be very poor in hydrocarbons. Without a doubt, they will have forgotten the villa in Louveciennes, purchased by Elf for 6 million francs paid in an under-the-table, lump-sum payment to President François Mitterand's golf partner, at the president's request, at a favorable rate.[39]

All of these ephemeral scandals illuminated only the visible tip of the iceberg. The Elf trial revealed that each transaction, every barrel sold in Africa, was subject to commissions and eventually, kickbacks. It was an open secret of French politics. Networks of political financing supplied by secret Elf Aquitaine funds to African leaders and partially diverted back to France—in particular by the president of Gabon, Omar Bongo—did not seem to have been seriously affected.[40] The sheer magnitude of these secret French networks will remain largely unknown. We did learn, however, that they extended to Germany: In 1991, after the German reunification, Elf executives reluctantly agreed to invest 15 billion francs in the former German Democratic Republic's enormous, dilapidated petrochemical complex in Leuna (the former jewel of IG Farben during the Nazi era). The contract was created at the request of Mitterrand, who wished to oblige his "friend," Chancellor Helmut Kohl. This led to the payment of enormous commissions, necessary to obtain the subsidies that could make the investment profitable. These commissions reached close to 300 million francs. They would have been paid in particular to the senior executives of Kohl's party, the Christian Democratic Union, via intermediaries from the French and German secret services.

* The state had sold all its shares in Total, the former CFP, between 1992 and 1998. Dominique Strauss-Khan's personal secretary, Évelyne Duval, had for a time been employed by an Elf subsidiary.

Paris's Fatal Games in Africa: Algeria, Gabon, Congo-Brazzaville

After its independence, Algeria's strategic position in France's energy independence was replaced by that of the former French colonies of the Gulf of Guinea.

In Algeria, however, the oil money that ebbed and flowed in France's direction was no less politically crucial. From September 1989 to June 1991, the government reformer Mouloud Hamrouche proved unable to elicit from François Mitterrand any measurable transparency in contracts with French companies, which had long adopted the habit of overcharging for their services in order to pay the commissions claimed by the old barons of the Algerian regime.[41] This inability appeared to play an important role in the disaggregation of Algeria's Reformist movement, a fleeting, peaceful alternative to the civil war that, throughout the 1990s, pitted the corrupt military hierarchy that fed on black gold: General Larbi Belkheir's "clan," against the combatants of the Islamic Salvation Front, which became the monstrous Islamic Armed Groups (GIA). Many GIA officers were soldiers who had defected from the Afghanistan jihad, a frightening Hydra also fed by oil money and the complacence of those in power.*

In sub-Saharan Africa, President François Mitterrand failed to anticipate the kind of hope that had been generated by the fall of the Berlin Wall. But he quickly attempted to play catch-up. On June 20, 1990, before the leaders of the former African colonies gathered in the peaceful French La Baule seaside resort, Mitterrand claimed he could show them a "path" long neglected by Paris in the name of other priorities: a "multiparty system, freedom of the press, independent judiciary, no censorship." In Africa, this speech spurred a powerful desire for emancipation. "National conferences" arose in many places, in order to develop new political alternatives. The game was often biased. Loïk Le Floch-Prigent spoke of how political affairs were for a long time engineered in Gabon, Congo-Brazzaville, and Cameroon, the three former colonies where Elf was firmly implanted: "There is an opposition that the president finances. He chooses it. As soon as this opposition becomes a little important, he integrates it into his government. And we know that the more he finances it, in reality, it is us who finance it."[42]

However, resounding false notes could be heard at the time of the La Baule summit, during a brief spring of the French-speaking citizens of the African

---

* Djamel Zitouni was the first emir of GIA who had not fought in Afghanistan, but he received solid military training there.

petroleum countries. On May 25, the unexplained circumstances surrounding the death of a Gabonese protestor, Joseph Rendjambé, provoked riots in Libreville and Port-Gentil, Elf's headquarters, where company executives were taken hostage. Paris soon dispatched more than one thousand soldiers. Le Floch-Prigent decided to close Elf facilities until Gabon's president Bongo could take control of the situation. Paris considered stopping support for the African dictator, but Bongo informed Roland Dumas that he was considering appealing to American oil companies. France ceded quickly, and the status quo prevailed, defended by France's elite troops. On September 16, 1990, Gabon held the first multiparty legislative elections in a quarter of a century, described by *Le Monde* reporters as a "distressing fiasco."[43] In the face of the massive election fraud that allowed Bongo to strengthen his grip on the nine hundred thousand Gabonese, Paris's criticism was purely symbolic. Gabon plunged back into its usual political torpor, conducive to oil exploitation.

A few months later, in Congo-Brazzaville, another aborted pseudo-democratic process, intoxicated by Françafrique oil money, uncorked an interethnic civil war that smacked of mafioso, from 1993 to 1997. A career officer trained by the French army, Denis Sassou Nguesso, was Congo-Brazzaville's president and dictator since 1979. He was the leader of the nation's single party, allegedly Marxist-Leninist; but that did not prevent him from maintaining strong relations with the Elf group. Thanks to the 150,000 barrels that Congolese wells provided each day, the French state had consolidated its most secure and significant private African oil reserves, after those of Gabon.

Following the conference in La Baule, under pressure from Paris, Sassou Nguesso accepted the multiparty system and renounced Marxist ideology in December 1990. Elf's leaders encouraged a seemingly innocuous former prime minister from Congo-Brazzaville's early independence, Pascal Lissouba, to run against Sassou Nguesso. Le Floch-Prigent recounted: "Sassou strongly feels that he will not lose the election, since he considers himself to be so good and generous, and the people love him so much that he will be elected. I warn him that he has on the contrary all chances of losing."[44] Pascal Lissouba, in fact, benefitted from support from the important Batéké ethnic group, the majority in the south of the country. Elf wanted to obtain an advantageous arrangement, and called on the president of neighboring Gabon, Omar Bongo, who happened to be both Lissouba's cousin and (only recently) Sassou Nguesso's son-in-law.

Loïk Le Floch-Prigent acted in accordance with his role: the unofficial governor of the most important provinces of the small, French neocolonial empire. Infinitely more revered there than France's ambassadors, Le Floch-Prigent possessed the resources of an authentic proconsul: virtually unlimited funds and an intelligence, security, and action network that was more well-established, according to Le Floch-Prigent, than the French secret service, from which Elf often recruited the best officers. In particular, Elf (simply called "the firm" by French agents) enjoyed a blank check from the president of the republic, François Mitterrand.[45] Le Floch-Prigent explained: "I did not need any special instructions from Paris, and I did not receive any because I knew in general what they wanted to achieve, which was enough for me."[46]

The French oil company chairman began by convincing Sassou Nguesso to accept a power-sharing agreement with his adversary, Lissouba, whatever the outcome of the elections might be. Lissouba affirmed that thanks to Elf's money, he had been able to buy about thirty Land Rovers for his campaign.[47] He won the presidential ballot hands-down on August 16, 1992, with 61 percent of the vote. Sassou Nguesso, a good sport, welcomed his opponent's "superb victory." But once he became president, Lissouba quickly broke the agreement negotiated through the efforts of Elf and Omar Bongo. He ousted Sassou Nguesso's supporters from the government, announced that he would reject coalitions of political power initially planned in Parliament, and announced that he would hold new elections. Lissouba aroused the ire of Elf's CEO, but nonetheless requested $150 million from Elf in order (according to Le Floch-Prigent's testimony) "to rig the elections."[48] Elf refused to continue to support Lissouba, but in March 1993 the wheel turned: In France, the right won an overwhelming victory in the legislature, forcing President Mitterrand to share space with Édouard Balladur's conservative government. Lissouba bypassed Le Floch-Prigent and traveled to Paris to solicit help from the new minister of cooperation, Michel Roussin (who, in the time of the Safari Club, served as an expert on the Françafrique for former secret service director Alexandre de Marenches). Shortly thereafter, Le Floch-Prigent was summoned to the Hôtel Matignon by the director of Balladur's cabinet, Nicolas Bazire, himself well-versed in the Françafrique. Elf's CEO did not give in: He continued to refuse Lissouba financial support. He affirmed later that he had warned Bazire and Mitterrand, as well as Omar Bongo, about the risk of civil war in Congo-Brazzaville.[49]

When Lissouba turned to the independent American company Occidental Petroleum (Oxy), the Balladur government didn't intervene. On April 28, 1993, Oxy's directors were quick to lend Lissouba the $150 million he asked for.* The Congolese president did not hesitate to pledge to repay this loan with the future profits from eighty-five million barrels of oil pumped by Elf, as well as by the Italian company Agip. The terms were advantageous for Oxy, which

* After a long, fruitful career as an outsider, Armand Hammer, the president of Oxy, died on December 10, 1990.

purchased the crude oil for $3 per barrel rather than its set price of $14. The $150 million allowed Lissouba to pay the salaries of civil servants on the eve of the parliamentary elections, which Lissouba's party easily won on May 20.

Serious clashes occurred in June and July, and then again in December. The regular army exploded. Partisans and members of the Lissouba's ethnic group confronted those loyal to Denis Sassou Nguesso, who in a decade of Elf-funded Marxist dictatorship had had the time to put aside enough to finance a militia. In August, Édouard Balladur fired Loïk Le Floch-Prigent, President Mitterrand's henchman, from his position at Elf. His successor, Philippe Jaffré, attempted to reconcile with President Lissouba. After a cease-fire on February 2, 1994, the Balladur government agreed to facilitate the training of the regular army, confirming its support for Lissouba.[50] Two weeks later, a "special intervention group" was ordered to prevent combat between rival militias in Brazzaville; it also guarded the track that connected the capital to the country's oil terminal, the port of Pointe-Noire, sometimes also called Elf City. The trains resumed running on March 1, and business resumed, too. In 1996, Philippe Jaffré was able to negotiate the redemption of a debt from Oxy.[51] Elf had a weighty argument: The French company had invested heavily in the Congo and, about forty miles offshore of Pointe-Noire, was preparing to activate a giant oil platform of concrete and steel, one of the largest in the world at the time.

In 1997, the civil war intensified. On June 5, while preparing for new elections planned for the following month, President Lissouba's troops attacked the well-defended Brazzaville residence of Sassou Nguesso. Four months of clashes ensued between Lissouba's militia, the Zulus, and Sassou Nguesso's fighters, the Cobras. Artillery duels devastated Congo's capital. The use of weaponry seemed beneficial to Lissouba, who had the advantage of Russian assault helicopters manned by Ukrainian mercenaries. However, after Jacques Chirac took office in 1995, and strove to undo the networks knitted together by his great opponent, Édouard Balladur, the Congolese president had lost Paris's support. In September 1997, when Lissouba traveled to Paris to ask for aid, Chirac simply refused to receive him, and indicated that since his term had ended in July, after the fighting began, Lissouba was no longer the legitimate head of state. Having claimed thousands of civilian victims, the war ended after one of its bloodiest episodes, with the victory, as sudden as it was unexpected, by Sassou Nguesso, who took control of Brazzaville and Pointe-Noire on October 16. The president had benefited during the last weeks of the conflict from pivotal support: the Angolan president Dos Santos, a former Marxist like Sassou Nguesso, who had just been victorious in his own civil war (thanks notably to Russian weapons under circumstances that became known as Angolagate).

Like Saddam Hussein in Kuwait, Pascal Lissouba had wanted to push his advantage too far. He had funded an independent political force in Cabinda, a small Angolan enclave in Congolese territory, where Elf exploited important reservoirs; it was nicknamed the "African Kuwait" because of its oil riches.[52] At the end of September 1997, Dos Santos, a former ally of the USSR and Cuba who was supported by Washington, justified sending troops to Congo-Brazzaville—financed by loans from Western oil companies and in particular by Elf, accusing Lissouba's army of incursions in Cabinda.[53] The soldiers and especially the tanks and MIG fighter planes Angola sent tipped the balance decisively in Sassou Nguesso's favor. In Washington, Dos Santos's allies strongly criticized Angola's intervention, which sanctioned returning Congo-Brazzaville to the bosom of France. When Sassou Nguesso's troops easily took the Pointe-Noire oil terminal, Paris launched no reprisals, despite the sporadic looting in the city, nor did they attempt to evacuate the six hundred French nationals, mostly Elf employees.

On October 26, 1997, a day before Sassou Nguesso proclaimed himself the new president of the Republic of the Congo, Philippe Jaffré rushed to Brazzaville, which was nearly destroyed and prey to looters, to negotiate the terms and conditions of the French public oil company's return to grace, before Paris had the time to do as much.[54] Denis Sassou Nguesso easily remained in power, having taken care to incorporate into his government several relatives of Lissouba, who was condemned to forced labor in absentia in 2001.

On May 14, 1999, a thousand refugees in Kinshasa, the capital of the neighboring Democratic Republic of Congo (DRC), crossed the Congo River to return to Brazzaville. In a vibrant televised appeal, Denis Sassou Nguesso called himself the guarantor of their security. These refugees were arrested upon their arrival at the river port of Brazzaville: 353 of them disappeared forever. According to various testimonies, they were massacred and then burned in the enclosure of the presidential palace. In January 2002, a criminal complaint of crimes against humanity was submitted in France. At noon on April 1, 2004, Jean-François Ndengue, head of the Congolese police, was arrested at his French home in the Paris suburbs. Investigating Judge Jean Gervillié immediately contacted the Quai d'Orsay to ensure that Ndengue did not have a diplomatic passport. Having obtained oral confirmation, the judge placed the leader of the Congolese police in custody. According to the *Canard Enchaîné*, a French investigative newsweekly, Sassou Nguesso then called French president Jacques Chirac from his Brazzaville palace to air his grievances: "What would you say . . . if I stopped the military attaché of the Embassy of France? Or if I threatened the interests of Total and its leaders on the spot?"[55]* Jean-François

* The Total group had absorbed the Elf group four years earlier.

Ndengue was nevertheless placed in provisional detention on April 2. He was imprisoned for twenty-one hours at Paris's La Santé prison. During the night, the Paris prosecutor ordered his release. On April 3, at 4 a.m., Ndengue exited the prison and immediately left France aboard a special plane chartered by the Congolese presidency.[56] By allowing a man who was accused of crimes against humanity to flee, President Chirac concluded the epilogue of a civil war with an unknown death toll, a war triggered by the unraveling of a democratic transition desired by President Mitterrand and sponsored by Elf.

Oil money bolsters the good fortune of the consumer countries' political leaders and rescues struggling companies (Harken, Citicorp, Bidermann, the Leuna refinery, for example), finances heads of state (Saddam Hussein, Pascal Lissouba), and undoes them when they covet oil fortunes that are not for them to take (Kuwait, Cabinda). Without a doubt, the same causes produce similar effects everywhere. The "Françafrique" petroleum network is a rough miniature of the web of relationships between Big Oil, its Washington champions, and its potentates of the Persian Gulf.

So many dubious ethical choices, so much dishonor brought about as a consequence of the same need: to maintain a political status condusive to properly supplying consumers with oil. After two prison terms, Loïk Le Floch-Prigent, former Elf CEO, continued to arrogantly defend imposing this status-quo logic on Congo-Brazzaville and elsewhere: "Revolution, always detrimental to the oil interests, was to be avoided. We try to avoid the chaos. My worry: I have 150,000 barrels per day there. I want to preserve this supply, the same way I want to keep it in the Gabon, Cameroon and Angola. . . . The mission entrusted to me and my only objective is to have a production which corresponds roughly to the national consumption (even if this oil is not that actually consumed in France), in order to ensure that France has sufficient production to withstand a shortage in the case of an oil crisis."[57]

A political status quo can always be defended, whatever the cost. At the end of the 1990s, oil companies saw the emergence of a more formidable challenge: to maintain the status quo of their production everywhere they drilled. Gabon reached a peak crude oil production in 1997; beginning in 1998, oil production in what until then had been the most reliable province of the small French petroleum empire began to collapse. The phenomenon was far from unique.

◀ TWENTY-SIX ▶

Grandeur and Decadence: The Explosion of Opulence, Misery, and the Human Footprint

Humanity has just two great forms of energy in this meek world. One flows in a virtually unlimited stream, the other is of finite stock. The first form, inexhaustible in the scale of human time, is drawn to the Earth by gravitational force; directly or through radiation, creating the matrix of "renewable" energy sources: geothermal, hydro, wind, solar, and biological. The second form, an infinitely more limited energy store, is provided by certain types of rock. Among this stock, only coal and hydrocarbons (fossil rocks, themselves fruits of the sun and gravity) are able to directly feed the thermal machinery of progress, unlike the radioactive ores. The beginning of an American crude oil production decline in 1970 was followed by the oil crisis of 1973, which began a stunning technical, economic, and societal chain reaction, the results of which, for the most part, shape our current world. Among a myriad of consequences, the first oil crisis began to create an interest in the politics of ecology, which raised the hope (a fantasy?) of the advent (a return?) of renewable energy sources that would replace coal mines and hydrocarbon wells.

But no. Humanity continues to follow the irreversible slope of least resistance. The powerful historical movement initiated during the first oil crisis has not steered us away from fossil fuel energy sources; we have on the contrary increased our dependence, again. We adopted the habit of describing this movement as a "crisis." A crisis evokes the idea of imperative change. The oil crises of the 1970s resulted in a galaxy of transformations, of economic and technological refinements. But they changed nothing in terms of the fundamental energy forms that humanity uses for manufacturing and transportation. From this point of view, the "crisis" following the oil crises is a sham. The meager growth of renewable energies and even the much more considerable growth of civil nuclear energy remain at the scale of marginal phenomena.* The hierarchy of major energy sources remain the same, and oil retains the primary position, in front of coal and natural gas. These three sources of carbon, fossil, nonrenewable energy continued to provide four-fifths of the energy produced by humanity even in 2018: exactly as in 1973; and since 1973, our energy production doubled! As the year 2000 approached, in spite of the oil crises and the emergence of the concepts of sustainable development and climate change in the early 1990s, oil production had increased by nearly a third compared to 1973.[1] Essentially, oil simply relinquished part of its market share to coal and gas.[2]

Lit by the fire of oil, the jumbo-jet of worldwide growth scarcely slowed its ascent during the long "crisis" period that began in 1974. After tripling between 1950 and 1975, the volume of the world economy doubled during the quarter-century that followed: The bloom of summer followed the impetuous gush of spring.†

Critics of Growth: Nicholas Georgescu-Roegen and the Club of Rome Report

During the three years between the 1970 peak of American oil production and the 1973 oil crisis, at the summer solstice of the oil era, economists and physicists articulated the worst systematic critiques of the growth economy.

The same year petroleum declined in the United States, an economic theory professor at Nashville's Vanderbilt University launched a radical attack. Of Romanian origin, born in 1906 at the edge of the Black Sea, Nicholas Georgescu-Roegen had earned his doctorate in statistics at the Sorbonne in 1930. A Rockefeller Foundation award allowed him to study in London and Harvard, continuing his work alongside the decorated Austrian economist Joseph Schumpeter, inventor of the concept of "creative destruction." He returned to Romania, and later slipped through the Iron Curtain into the United States, where he became an American citizen and continued his research at Harvard, and later in Nashville. The paper Georgescu-Roegen published in 1971 suggests such a reversal of the view on the economy that his disciples

* In 2013, nuclear energy represented 5.1 percent and renewable energies, 3.3 percent (13.3 percent counting wood and agrofuels) of the world's production of primary energy, according to the International Energy Agency.

† Measured in 1995 dollars, the world's gross domestic product rose from $5 trillion in 1950 to $16 trillion in 1975, and $35 trillion in 2000.

compared it to the heavenly message of Galileo, which "convinced the doctors of the Catholic Church to watch the sky through a telescope."[3] His article, only fifteen pages long, was quickly reprinted in the *Ecologist*, the first great ecology magazine, founded in 1970. It asserted that all economists, be they capitalist or Marxist, misunderstood the nature of human activity as much as the astronomers before Galileo had misinterpreted the movement of planets.

Since the Industrial Revolution, the economy had been described as a process of exchange between production and consumption, represented in virtually all textbooks as a closed circuit, replenishing itself. Nicholas Georgescu-Roegen surmised that nothing could be further from the truth than the idea of representing the economic process as an isolated, self-regulating phenomenon. Such a vision reminded him of the myth of perpetual movement.[4] This had been the original blunder of economic science. On the contrary, Georgescu-Roegen stressed that the economy was essentially an unidirectional, irreversible thermodynamic process: natural resources with value are fed into it, and waste without value is ejected from it.[5] The energy that was necessarily dissipated in the form of heat in the course of these processes was dissipated forever and, whatever happened, the stock of minerals through which humans have attained a level of what he called almost miraculous development could not be reconstituted.[6] Georgescu-Roegen deemed it urgent for the economy to comply with the second principle of thermodynamics, revealed at the dawn of the Industrial Revolution in the work of French engineer Sadi Carnot.[7] This principle of "entropy," which had been known since that time, describes the irremediable degradation over time of usable energy (low entropy) in unusable energy (high entropy).

While a few hundred kilometers from his office in Nashville the crude oil production in Texas, Oklahoma, and Louisiana began to decline, Nicholas Georgescu-Roegen stated: "Every time we produce a Cadillac, we irrevocably destroy an amount of low entropy that could otherwise be used for producing a plow or a spade. In other words, every time we produce a Cadillac, we do it at the cost of decreasing the number of human lives in the future. Economic development through industrial abundance may be a blessing for us now and for those who will be able to enjoy it in the near future, but it is definitely against the interests of the human species as a whole, if its interest is to have a life span as long as it is compatible with its dowry of low entropy."[8]

The iconoclastic economist, who had served the Soviet administration in 1944 as secretary general of the Romanian commission for armistice before moving to the West, grouped capitalism and Marxism together, guilty of the same Promethean hysteria which obscured the finite nature of raw materials. Neither the capitalists nor the Marxists were able to "even conceive that there is any real obstacle inherent in the human condition," Georgescu-Roegen claimed.[9] As the Apollo program came to an end, he noted that it could be "that the prodigious efforts to reach the moon also correspond to the more or less conscious efforts to find access to new sources of low entropy."[10] For Georgescu-Roegen, the trap set by this technology was simple: The more the modern economy fed the fires of industry, the closer it came to its demise. The more intense the growth, the briefer it would be. The more straws we place in the glass of world crude oil reserves, and the harder we suck, the more quickly the glass will be emptied.

The criticism Nicholas Georgescu-Roegen formulated was by and large ignored, except by a radical European ecologists who laid the groundwork for the ideology of degrowth. Yet the next critique, published in 1972, received instant worldwide attention. *The Limits to Growth* was commissioned by the nonprofit Club of Rome, founded in 1968 by Aurelio Peccei, former director of the Italian car manufacturer Fiat.[11] The work was partly financed by the German manufacturer Volkswagen. In one of the first attempts to forecast using computer modeling, the report was created under the direction of systems expert Dennis Meadows at the Massachusetts Institute of Technology. He and his colleagues studied the historical interactions between five functions—population growth, food production, industrialization, pollution, and the use of nonrenewable resources. The conclusion of *The Limits to Growth* was unprecedented: If the world carried on with a business-as-usual approach, the current economic, industrial system would collapse somewhere in the middle of the twenty-first century.

The report did not indicate which specific resource shortage or pollution source might be the origin of the collapse. It nevertheless pointed to the depletion of oil reserves and provided an early warning that there had been a rise in carbon gas emissions capable of impacting the climate. Forty years later, Meadows recognized that he had underestimated the impact of technology on agricultural yields and population growth, but also the magnitude of climate change and humanity's dependence on fossil fuels.[12] The comparison of the potential trends mapped out in the team's 1972 business-as-usual scenario, which it called "overshoot and collapse," and the actual trends from 1972 to the beginning of the twenty-first century show very close results.[13]

The Club of Rome report generated strong media interest at the first UN summit devoted to the environment, held in Stockholm in June 1972. The main person responsible for this tremendous publicity was none other than the president of the European Commission, who declared to the French weekly *Le Nouvel Observateur* that he had had a "terrible revelation" when he learned the conclusions of *The Limits to Growth*.[14] "I came to understand that we are rushing towards disaster," announced the Dutchman, Sicco Mansholt, previously known for having orchestrated an ultraproductivist reform of the European community's common agricultural policy.[15] "We wrote too many blank checks on the future," the leader surmised in the columns of the magazine.[16] The authors of *The Limits to Growth* had pled for "zero growth," in other words for the economy to plateau. But Mansholt went further, and called for complete degrowth: "Is it possible to maintain our rate of growth without profoundly changing our society? In lucidly studying the problem, we see that the answer is no. Then, even more than zero growth, we need negative growth. To put it bluntly: we must reduce our economic growth, our purely material growth, and replace it with the notion of another kind of growth—that of culture, happiness, well-being."[17]

An intense polemic ignited within industrial societies. *The Limits to Growth* was quickly translated into thirty languages. It became one of the first and most controversial environmental bestsellers. Its reputation even crossed the Iron Curtain. In the Soviet Union, the dissident author Alexander Solzhenitsyn wrote in 1973: "We had to be dragged along the whole of the Western bourgeois-industrial and Marxist path in order to discover, toward the close of the twentieth century, and again from progressive Western scholars, what any village graybeard in the Ukraine or Russia had understood from time immemorial . . . that a dozen worms can't go on and on gnawing the same apple forever, that if the Earth is a finite object, then its expanses and resources are finite also, and the endless, infinite progress dinned into our heads by the dreamers of the Enlightenment cannot be accomplished on it."[18] In accordance with Nicholas Georgescu-Roegen's analysis, the critique of the logic of growth for which *The Limits to Growth* authors and Sicco Mansholt argued angered the Marxists just as surely as the partisans of economic liberalism. The secretary general of the French Communist Party, Georges Marchais, foresaw a program of "misery" and "economic repression" for workers.[19] In his acceptance speech for the 1974 Nobel Memorial Prize in Economics, Friedrich Hayek, pope of neoliberalism and David Rockefeller's mentor, said that the Meadows report tarnished the "prestige of science."[20] His prize, awarded in the wake of the first oil crisis, marked the beginning of the triumph of the most vigorous and unrelenting of the ideologies promoting economic growth. In 1975, the Chicago School's ultraliberal, neoclassical policies were tested on a national scale for the first time in Chile, at the end of the coup against President Salvador Allende on September 11, 1973, hatched with the help of Henry Kissinger and the CIA. As we have seen, the experience was cautiously commended in David Rockefeller's memoirs.

Some of the highest eminences of the capitalist system did not ignore the warning outlined in the Meadows report, but on the contrary integrated it into their strategy for the future. At the time Friedrich Hayek was receiving his Nobel, Henry Kissinger was finalizing a confidential report exploring the consequences of the population explosion of poor countries on American security. The White House security adviser noted that "the U.S. is in a relatively strong position on fossil fuels compared with the rest of the industrialized world."[21] However, because "the U.S. economy will require large and increasing amounts of minerals from abroad, especially from less developed countries," Kissinger called for making population-control policies in poor countries a strategic priority of primary importance.[22]

Many people had begun to discover the risk of pursuing infinite growth based on finite resources. Very few, however, stopped to contemplate this new ouroboros, the self-consuming dragon of ancient civilizations. In France, in 1974, agronomist René Dumont, the first Ecologist party candidate in the presidential elections, for a time fascinated the press with the novelty and force of his conviction against the "continuing absurdity of industrial growth." He obtained only 1.32 percent of the votes, a score that his successors barely improved upon. In the United States, Jimmy Carter articulated an idea that was incredible coming from the president of the richest nation in the world. In his inaugural speech on January 20, 1977, he announced: "We have learned that 'more' is not necessarily 'better,' that even our great Nation has its recognized limits." Notably inspired by Admiral Hyman Rickover, key architect of the American nuclear program, Jimmy Carter intensified his warnings about the precarity of an industrial civilization based on oil during his time in office. But some of Carter's statements were premature. For example, in April 1977 he said, "We now believe that early in the 1980s the world will be demanding more oil that it can produce."

As a result, his successor, Ronald Reagan, ridiculed him. Upon taking office in 1981, Reagan removed the thirty-two solar panels that Carter had installed on the White House roof. And during the inaugural speech for his second term, on January 21, 1985, the Republican president professed that "There are no limits to growth and human progress, when men and women are free to follow their dreams." A few months earlier, in the journal *Science*, scholars in the United States had shown that on the contrary, throughout the past century, US developments in GDP, labor productivity, and prices had been strongly related to the level of energy use.[23] They introduced the concept of "energy returned on energy invested" and found that this return tended to decrease decade after decade: The extraction of oil, in particular, required more energy, to the extent that the petroleum industry was forced to install its rigs in sites that were increasingly more precarious, offshore or in the north.[24] Although this article became a classic in scientific and ecological circles—it failed to undermine the nearly universal belief preached by President Reagan.

The Illusion of Dematerialization and the Dream of an Inexhaustible Energy Source

The idea of growth without limits fed a myth that had appeared at the end of the oil crises: the dematerialization of the economy. Starting in the 1970s, world energy consumption certainly continued to increase, but for the first time, it rose significantly more slowly than economic growth. This disconnect probably initially arose with the rapid advent of computers, which opened the possibility of a fantastic intensification of economic activity. In the United States, another precursor: The ratio between the mechanical force driving the economic machine and GDP reached its historical peak in 1970, then began to decrease, precisely at the time when American crude oil production began to do the same.[25] An enigmatic concomitance.

After moving from agriculture to industry in the early 1950s, in the 1980s the workers of the rich countries shifted en masse toward the service sector, which soon employed more than two-thirds of the working population in the West and Japan. At the end of the twentieth century, each new growth point required a quarter less energy than it had needed a generation before. Many concluded that economic growth was in the process of ridding itself of its fundamental energy base. Wrong. Certainly, industry everywhere redoubled its efforts to increase its efficiency, and regained its profit margins, which were ravaged during the oil crises. But in rich countries, labor productivity never again increased as it had during the Thirty Glorious Years. The outcome was an increase in the productivity of capital, which justified recourse to increasingly radical remedies: computerization and automation beginning in 1970, mergers and concentrations starting in the 1980s, and finally, delocalization beginning in 1990. Many of these developments explained the emergence and persistence of mass unemployment throughout industrialized countries.

Electronics, managerial streamlining, delocalization: Each of these three phenomena multiplies the impact of energy necessary for growth. They "liberate" industrial tasks from the arms of workers (over and above those who are unemployed) to ensure the growth of the service sector. None of these, however, could have occurred without having, as their base, a source of cheap, malleable, abundant energy. Initially presented as "clean," in the 1990s the electronics industry was both a major source of pollutants—Silicon Valley was responsible for more chemically contaminated waters than anywhere else in the United States—and extremely energy-intensive.[26] It used a lot of oil—whether for plastics, the solvents for which it had inordinate uses, or the extraction of its raw materials and the delivery of its components and products. Company mergers and globalization were dependent on mass transport of people and goods. Between 1970 and 2000, the quantity of oil dedicated to transportation doubled, rising from six to twelve billion barrels per year![27] For the first time, in the early 1980s, the oil needs of transport exceeded those of industry. The changeover of many oil-run power stations to natural gas, the resurgence of coal, and the birth of civilian nuclear energy helped to maintain modest fuel prices, which, in turn, promoted the use of different kinds of fuel for transport. Without inexpensive fuels for cargo and container ships, there could have been no globalization of the economy.

In spite of the inventiveness of engineers, crude oil consumption never stopped growing during the "summer" of oil, rising from 48 to 73 Mb/d between 1971 and 1998. It simply increased more slowly in rich countries than it did in the rest of the world. Around the second oil crisis of 1979, first in the United States and then across the planet, oil consumption per capita eventually stabilized after a century of expansion: It reached a long, undulating plateau that persists today.[28] But total crude oil consumption continued to increase as the world's population grew. It is a crucial phenomenon: Economic growth, oil consumption, and total global energy consumption even in 2013 continued to evolve simultaneously in similar fashion, linked together in every way.[29]

Globalization is the result of the "rebound effect" discovered by British economist Stanley Jevons during the time of the earliest oil wells: To the extent that technical progress increases the efficiency with which a resource is used, the consumption of this resource is likely to grow rather than shrink. The more the mouth and stomach expand, the more the appetite grows, and vice versa. Without inexpensive diesel for ships, there would be no globalized trade. It is difficult to imagine a civil nuclear industry without a plentiful petroleum supply to extract and transport the uranium ore, and to build the plants. It is still more difficult to imagine the predominance of the service sector without an underlying framework of powerful, sophisticated energy industries. It is in the rich countries, where there are more service industry workers, that per capita energy consumption remains by far the highest.[30] Abundant energy and inexpensive petroleum seem, of necessity, doomed to remain the ground floor for "growth." Without these, almost nothing that has followed would have been feasible. Global crude oil reserves are like a huge, melting glacier that shrinks, floods, and fertilizes side valleys hitherto deserted, out of reach. The rebound effects begin to cascade. Before 1970, the entire industrial sector barely used twenty kinds of metal. In the year 2000, industry had use of sixty different metals.[31] Their extraction has become ever more difficult, in other words more energy-intensive in order to deploy large-scale technical innovations and, particularly, to meet the exploding needs of the computer and aerospace industries.[32] American automobile engine efficiency increased greatly following the oil crises, allowing cars to travel twice as far per gallon between 1980 and 2000. At the same time, the number of miles traveled each year by American motorists increased by one-fourth, after having stagnated since the 1950s.[33] In addition, newer models tended to weigh more each year: Performance gains were thus immediately undone, and cars became "overequipped."[34] In the automobile's early years, municipal leaders did all they could to pave new roads to their cities' shopping centers. Starting in 1970, in the United States and little by little in other rich countries, the relentless increase in automobile traffic (encouraged by improved engine performance) obliged circumvention of city centers, which led to the systematic establishment of supermarkets in the suburbs designed for motorists. The rebound effects led to "ratchet" effects, to new ways for humankind to imprison itself with its own technology, in a manifestly irreversible way.

The waves of energy necessary for economic production were more concentrated and swift, constantly irrigating new economic activities. Thanks to technological progress, the efficacy of these energy streams increased, allowing a dramatic increase in the creation of new wealth, while leaving enough disposable energy to feed new supplementary streams, whether they were undersea oil wells in the North or uranium mines beneath the Sahara and in Central Asia, indispensable for manufacturing still more machines, all around the globe.

On August 15, 1971, Washington's decision to abandon the gold standard, supported by Big Oil, erased all tangible limits of monetary measure. From then on, money creation became nothing more than compensation for the creation of debt. Wealth creation became something of a self-fulfilling prophecy. The wealth deployed in service to humanity's singular desire to accrue debt to enrich itself, and the obligation to repay this "debt money" with interest, shored up the economy to keep abreast of its exponential growth.*

* The documentary *Money as Debt* (2006) by Canadian Paul Grignon, cited, for example, Marriner S. Eccles, former president and governor of the US Federal Advisory Council: "If there were no debts in our money system, there wouldn't be any money."

After 1971 and the abandonment of the gold standard, the only binding limit on monetary creation that could prevent limitless debt growth was the US economy's capacity to regulate resource use according to economic growth. Beneath technical progress and productivity gains, this ultimate material limit retained the capacity to deploy the energy of humans and materials: to move, accelerate, brake, heat, cool, extract, crush, bend, deploy, synthesize, densify, spread, recycle, communicate, . . . and so on.

Regarding access to oil and other fossil fuels, limits still seemed far off. In wealthy countries, beginning with the first oil crisis and even more so following the second, private and public debt increased much faster than economic growth.[35] In the United States, in particular, the ratio between the total debt (of individuals, companies, and the nation) and GDP increased, from 150 percent in 1980 to 300 percent at the end of the millennium—more than in the aftermath of the 1929 crisis, and only the beginning of what was to come.[36*]

The desire to break down physical barriers that prevent the economic machinery from accessing infinite energy never ceases to bear down on humanity. In 1986, Mikhail Gorbachev had to abandon the age-old Soviet dream of diverting the course of Siberian rivers to irrigate the steppes of Central Asia: The project would have required energy too immeasurable even for Russia. In 1989, the Western press was passionate about the phenomena of "cold fusion." American scientists claimed to have managed to trigger a nuclear fusion in a simple container of water. But this source of free, inexhaustible energy soon revealed itself for what it was: a pipe dream. More promising was the cooperative international project launched by Gorbachev's initiative in 1985, in order to develop nuclear fusion plants. In the 2000s it gave birth to the highly ambitious, costly ITER, an international thermonuclear experimental reactor research project. *Iter* is the Latin root of "the way." But the project of harnessing the hydrogen bomb's power in a central power plant, attempted for decades by researchers such as the Soviet dissident Andrei Sakharov, may well be perpetually blocked by insurmountable limits, not intellectual but physical.

The "Big Bang" of Misery and the Ambivalence of Progress

Embarking on a path by technological means has a commensurate dark side. Abundant energy multiplies wealth but, by a powerful tendency

* The rate of total indebtedness of the United States exceeded 350 percent after the financial crisis of 2008.

toward enantiodromia, it also leads to a dramatic increase in the number of miserable people living on Earth.* The demographic explosion unleashed by the Green Revolution led, after a generation, to the "big bang of Urban Poverty."[37] In poor countries, the mechanization of agriculture made it possible to feed a rapidly growing population. But it was simultaneously the engine of the impoverishment of the people: Farms expanded and required less manpower, particularly in regions that were modernizing faster, around major urban centers. The "landless" became legion, and most had no other choice than to leave the countryside.† By the hundreds of millions, they piled into the slums. The Green Revolution caused an immense rural exodus in southern countries, unlike anything Europe ever experienced. In 1910, London had seven times more inhabitants than it did in 1800, but at the end of the millennium, the population of Dhaka or Lagos was forty times larger than in 1950.[38]

The "Washington consensus" imposed in the 1980s, under the stewardship of Big Oil champions like George Shultz and James Baker—splashed oil on this global tragedy. The structural adjustment programs of the IMF and the World Bank were especially designed to suppress the regulation of agricultural tariffs, precipitating the ruin of many small farms in the Southern Hemisphere, while in the United States, as in Europe, ultramechanized agriculture enjoyed generous public aid.[39]

The demographic explosion of the developing world was above all an explosion of a myriad of slums surrounding new urban centers. Between 1970 and 2000, the total urban population of poor countries almost tripled, from seven hundred million to two billion people.[40] In many developing country cities, more than half the population dwelled in slums.[41]

The oil countries were not much better off, often due to their leaders' negligence, as well as to the paradoxical capacity of energy abundance to create deficit. During the 1970s in Iran, hundreds of thousands of destitute, unemployed workers migrated to Tehran in vain, attracted by the mirage of

---

* *Enantiodromia* refers to movement in a reverse, or opposite, direction. Carl Jung used enantiodromia to describe the reversal of a reality, calling it "the most marvelous of all psychological laws," the law that describes the tendency of things to become their opposites. This upends notions of complementary duality, like the concept of yin and yang, and is linked to entropy. According to the poet Hölderlin, "But where the danger is, also grows the saving power"—and vice versa. Jung wrote, "Everything human is relative, because everything rests on inner polarities; for everything is a phenomenon of energy." (Carl Gustav Jung, *L'Inconscient Individuel et l'Inconscient Collectif ou Supra-individuel* (1942), *La Réalitié de l'Ame, Structure et Dynamique de l'Inconscient*, Book One, Le Livre de Poche, Paris, 2008, p. 65–70.)

† The name used by the movement of landless rural workers (Movimento dos Trabalhadores Rurais Sem Terra) of Brazil, whose creation was formalized in 1984.

the shah's colossal construction projects. At the same time in Caracas, the largest city of Venezuela, which was inundated with oil money, squatters' camps were constructed during the night, destroyed in the morning by the police, and then reconstructed the next night, until the struggle exhausted the squatters.[42] At age seventy-two, three years before his death, Juan Pablo Pérez Alfonzo, Venezuela's former oil minister and one of OPEC's founders, who had become an ecologist, in 1976 looked at the developments within his country and concluded that regarding oil, his country was "in the process of drowning in the devil's excrement."[43]

Starting in 1974, in the wake of the first oil crisis, Cairo experienced an urban boom, due particularly to money sent home by Egyptian workers in Saudi Arabia.[44] The Nasserian state was unable to heal the wounds caused by simultaneous explosions in population and Egyptian debt. In the 1980s, Lagos, the largest city in Nigeria, was growing twice as fast as the whole of the population, even though its economy was at the bottom of an abyss, pinched by its debt explosion and plummeting barrel prices during the oil countershock.[45] At the same time, the Mexican government attempted to slow the development of slums by reserving the bulk of its subsidized housing for the middle classes and the members of the powerful trade unions (especially those of oil). The effort was in vain: A decade later, no fewer than one in five Mexicans inhabited the sprawl of Mexico City, where a fifth of the population lived in slums.[46]

The shah's former minister of science and higher education and Iran's ambassador to the United Nations until 1971, Majid Rahnema was a privileged witness of the disillusionment and horror caused by the deleterious consequences of energy abundance. He deemed the crushing misery of modern slums to be chasing the "worthy poverty" of the old rugged societies toward rapid disintegration.[47] Rahnema, who had helped to instill in the shah the dream of transforming Iranian society through oil, became a travel companion of the radical environmentalist Ivan Illich. He wrote: "The spread of widespread misery and poverty is an obviously unacceptable social scandal. . . . But it is not by increasing the power of the machine to create goods and hardware products that this scandal will end, because the machine that put this effect into operation is the same one that consistently produces misery."[48] His diagnosis is questionable, but a symptom remains: In the year 2000, while the world population continued its rapid expansion and reached six billion inhabitants, nearly a billion people (one out of six) suffered from malnutrition.[49]

On the other side of the world, the summer of the oil civilization introduced the advent of the baby boomers, an opulent generation from any angle; in France, they called the 1980s "the money years." The boomers were a generation full of pride and hope, materially fulfilled like no other before it, yet jaded and often dissatisfied. While in 1973, Stevie Wonder sang "Living just enough . . . for the city," Bruce Springsteen catalyzed the spirit of the time in 1975 with "Born to Run," his most emblematic song: "In the day we sweat it out on the streets of a runaway American dream / At night we ride through the mansions of glory in suicide machines . . . / Beyond the Palace hemi-powered drones scream down the boulevard / . . . The highway's jammed with broken heroes on a last chance power drive." This period gave rise to a vague guilt, perhaps fear, that the hubris of the oilmen might one day be chastised. In 1979 (the year of the second oil crisis), *Apocalypse Now* won the Palme d'Or at the Cannes Film Festival. The film tells the story of an American patrol boat returning up the Mekong River at the end of the Vietnam War. Before the conclusion of the nightmarish trip, American soldiers, who never leave the boat except to search for mangoes, assist in the destruction of a Vietcong village by attack helicopters, go water skiing, and refuel. The first words chosen by Francis Ford Coppola when his film was presented were: "My film is not a movie, my film is not about Vietnam. It is Vietnam. It is what it was really like, it was crazy. And the way we made it was very much like the way the Americans were in Vietnam: we were in the jungle, there were too many of us. We had access to too much money, too much equipment, and little by little we went insane."[50] At the end of the assault scene in the Vietnamese village, Colonel Kilgore (played by Robert Duvall) makes the most striking comment in the film: "I love the smell of napalm in the morning. You know, one time we bombed a hill, for twelve hours. When it was all over I walked up. We didn't find one of 'em, not one stinkin' dink body. The smell, you know that gasoline smell, the whole hill. Smelled like—victory."* With *Mad Max*, also released in 1979, and even more so with *Mad Max 2* in 1981, George Miller, an obscure Australian director, offered the "post-apocalyptic" genre its greatest successes, capturing a fantasy vision of one of the great inconsequential fears of developed societies: the fight to the death for the last drops of oil. That same year, in the USSR, Andrei Konchalovsky's film *Siberiades* depicted the glorious and pitiful aspects of the illusory transfiguration of Russia by the triumphant petroleum industry.

The Russian oil industry's disintegration after the collapse of the Communist bloc caused two Communist countries in the developing world to endure the brutal experience of the oil shortage envisaged in *Mad Max*. Virtually deprived of inexpensive gas from the former USSR, North Korea was plunged into a famine that caused perhaps millions of deaths during the 1990s; the increased use of forced labor was far from sufficient to replace the energy

---

* Napalm is a derivative of oil.

slaves of agricultural machines idle due to a lack of fuel.[51] Thanks to saner politics and a better climate, Cuba's population planted crops around Havana and succeeded in keeping the specter of famine at bay.*

Everywhere in the world, the close link between energy consumption and quality of life was confirmed, up to a certain point. In the West, obesity was becoming a commonplace scourge, in particular within the working class, to the extent that fast-food portions swelled; the proportion of overweight children in the United States doubled between 1970 and the year 2000, from 15 percent to 30 percent. The Human Development Index (HDI), developed in 1990 within the United Nations, correlates life span with the duration of schooling and per capita income. In almost all countries, this index increases as energy consumption increases (allowing improved access to water, health, and education). But this parallel ascent is interrupted: Beyond a certain threshold, HDI stops growing. France, for example, has an HDI equivalent to that of the United States, although a French citizen consumes half the energy that an American does. Other indices measuring quality of life in detail provide troubling results, a sign perhaps of another enantiodromic evolution. This is the case according to the Index of Social Health published by Fordham University, which aggregates sixteen series of sociological data on the United States (average salary, poverty among children, and the frequency of violent crimes, for example). Progressing at exactly the same rate as the GDP through the 1960s, in the United States this social health index spiked in 1973 and then began a long slide, despite the continuation of economic growth and energy consumption.

A secondary effect of the pull toward a suburban universe compelled by the automobile, ghettoization of American urban centers played a major role in the deterioration of social health in the United States. In 1990, in Chicago's central district of Woodlawn, located in the vicinity of the university founded a century earlier by John Rockefeller, only 44 percent of men had jobs, and the infant mortality rate was higher than that of many developing countries.[52] In Los Angeles, the megalopolis erected on now largely defunct oil fields, seeing explosive growth and preyed upon by gangs, saw the first major political group to call for a limit to urban development: the "slow growth" movement, tinged at times with racist overtones, dominated by the homeowners of suburbia. At the end of the millennium, high on energy abundance, the global urban population was on track to become more populous than rural regions.†

* In 1999, Grupo de Agricultura Orgoenica, the Cuban institution responsible for overseeing the development of organic farming on that island, was awarded an "Alternative Nobel Prize" in Sweden, by the Right Livelihood Award Foundation.

† This shift fully materialized in 2007, according to the United Nations. In 1900, only one human in seven lived in a city.

The oil-made man still dreamed of embracing his car, embodying the attributes of power represented by the exhaust from the tailpipe; cigarette brands and energy drinks figured among the omnipresent sponsors of automobile championships. However, starting in the 1980s, American cinema took malicious pleasure in systematically depicting car crashes, destroying these big, angular, raucous, and gleaming toys. And robotically round, "biodesigned" economy cars came on the market, given to undistinguished appearance and colors, often metallic gray, which obscured them on the universal bluish-gray of asphalt. A year after the Gulf War, in 1992, Rage Against the Machine, a hitherto unknown Los Angeles metal group, caused worldwide reverberations with their song "Killing in the Name," which was received as a cry of fury against the powerful technology of industry and the military, and which became a kind of hymn for the young supporters of the rising counterculture movement. The genre of techno music was born in Detroit at the end of the 1980s, as Motown, which once had been the richest US city, found itself on the road to poverty as a result of robotic advances and global competition. Techno music evoked reproduction, amplification, iteration, modulations of the power games. In 1997, the British group Coldcut recorded one of the biggest techno hits, mixing an Indian song of the Amazon forest with the noise of chainsaws and revving engines. That year, Ted Kaczynski, known as the Unabomber, was tried for murderous attacks perpetrated in the United States against what ecological terrorist called the "industrial-technological system."

Sustainable Development, Ecological Footprint, and Global Warming

The well-founded principle of societal development that rests essentially on economic growth (itself dependent on the abundance of fossil energy) was for the first time questioned on the international scene during the Rio Earth Summit, organized by the United Nations in Rio de Janeiro in early June 1992. It was at this summit, one of the biggest media events in history, that the press first heard of sustainable development.* Radical environmentalists never stopped denouncing this concept as an oxymoron. The primary proponents of sustainable development included a Canadian oilman (Maurice Strong, secretary general of the Earth Summit) and a Swiss cement billionaire (Stephan Schmidheiny, founder of the World Business Council for Sustainable

* More than ten thousand accredited journalists attended.

Development, who in 2012 was sentenced to sixteen years in prison by an Italian justice for an asbestos scandal in his company, Eternit).[53]

Another concept attracted the media's attention during the Rio summit: that of the ecological footprint. Developed beginning in 1990 by two researchers from University of British Columbia, Vancouver, the ecological footprint measures the resources consumed in order to meet the needs of different walks of life. It implies that if all the inhabitants of the Earth consumed as much as the Americans, five planets would be needed (three would be needed for the French). In total, humanity's needs would exceed by 20 percent the Earth's biological capacity. According to the researchers, this capacity already was surpassed during the 1970s. The developed countries were by far the main cause of this situation: The richest fifth of the world population consumed four-fifths of the world's resources; this had been true throughout history, but the imbalance appeared more striking than ever. Since the beginning of the Industrial Revolution, the rich world's industries also have been responsible for producing four-fifths of greenhouse gases. The industrial nations' lifestyle, the focal point of economic development, was proven to be a fatal lure, and its promotion a promise as irresponsible as it is untenable. The use of fossil fuel energies in rich countries is particularly unsustainable, according to the partisans of the emergent ecologically focused political ideology.

At the 1992 Rio summit, the heating up of Earth's climate induced by the combustion of fossil fuel started to become the focus of all major ecological questions. Already identified as a potential risk by the Club of Rome report twenty years earlier, the greenhouse effect—caused by carbon dioxide released into the atmosphere, in ever increasing quantities, from the combustion of hydrocarbons and coal—appeared on the radar of the major media for the first time on June 24, 1988. On that day, scientist James Hansen was invited to testify before a US congressional committee. He testified that the temperatures in early 1988 were the highest recorded in 130 years, when in 1858 (one year before Colonel Drake drilled in Pennsylvania) meteorological readings were first systematically recorded. Hansen, director of the NASA Goddard Institute for Space Studies in New York, submitted that there was a 99 percent probability that the warming observed in recent years was the result of the combustion of carbon energy. "It is time to stop waffling so much and say that the evidence is pretty strong that the greenhouse effect is here," he asserted, before the elected representatives of the US Congress.[54] On December 6, 1988, in a resolution entitled "Protection of Global Climate for Present and Future Generations of Mankind," the UN General Assembly decided to establish an Intergovernmental Panel on Climate Change (IPCC) to provide "internationally co-ordinated scientific assessments of the magnitude, timing and potential environmental and socio-economic impact of climate change and realistic response strategies." A year and a half later, in May 1990, in a discreet countryside hotel west of London, the first hundred experts of the IPCC met to finalize their first report. At the back of the room sat seven scientists employed by the petroleum, coal, and chemical industries, including two from Exxon, one from Shell, and one from BP. Although they only had observer status, they were allowed to speak. One of these was Brian Flannery, a scientist in Exxon's employ and a representative of the International Petroleum Industry Environmental Conservation Association. While the experts focused on the crucial part of the report, its summary, Flannery seized the microphone on two occasions to say that the reduction from 60 percent to 80 percent of greenhouse gas emissions, presented as indispensable in the IPCC text, was based on an "uncertain" scientific diagnosis. "This should be stated as such in the executive summary," insisted the Exxon scientist.[55] For Exxon, heir of Standard Oil, as for all the other oil producers, a reduction of 60 percent to 80 percent of carbon dioxide emissions meant, at base, an equivalent sales reduction. From the very beginning of the fight against global warming, Big Oil challenged its scientific basis and legitimacy. Directly or indirectly, the oil industry's barrage continued. Soviet leader Mikhail Gorbachev had been the first head of state to show public concern about the warming of the atmosphere, prior to the creation of the IPCC.[56] But in May 1990, when the experts' first report, mandated by the United Nations, was submitted, it was Margaret Thatcher, whose concerns for the environment had thus far gone unvoiced, who spoke before the press. In an unexpected *aggiornamento*, the "Iron Lady" announced that the IPCC's climatologists had confirmed "that greenhouse gases are increasing substantially as a result of man's activities, that this will warm the Earth's surface with serious consequences for us all." About the diagnosis concerning future climate refugees (a diagnosis that would not substantially change), Thatcher warned: "There would surely be a great migration of population away from areas of the world liable to flooding, and from areas of declining rainfall and therefore of spreading desert. Those people will be crying out not for oil wells but for water."[57] The next day, the front page of the main pro-Thatcher tabloid, the *Daily Express*, ran the headline: "Race to Save Our World."[58]

In August, while Saddam Hussein's troops invaded Kuwait, the IPCC met again in the peaceful little Swedish city of Sundsvaal. The industrial front against climate change policy was represented by two powerful lobbies, the Global Climate Coalition and the Global Climate Council. The first included the American Petroleum Institute, the National Coal Association, most of the Western oil majors (Texaco, Shell, and BP, among others), and some of the major fossil fuel consuming sectors (Association of International Automobile Manufacturers and even the Edison Electric Institute). The second lobby, the Global Climate Council, was much more opaque but seemed to defend the interests of many of the large Arab petroleum countries, notably Kuwait and Saudi Arabia.[59] This lobby was headed by a former Reagan administration high official, Don Pearlman. The latter refused to disclose his clients; he was a partner of Patton, Boggs & Blown, a powerful firm of lawyers and lobbyists located in Washington and counting among its customers Exxon, Shell, BP, the giant of the chemical giant Dupont, and (until its dissolution) the BCCI.[60] However, in 1990, in Sundsvaal, the most influential fossil fuel industry lawyers were on the same side as the members of the US delegation sent by President George H. W. Bush's administration. The diplomat at the head of this US delegation, Frederick Bernthal, regretted that the IPCC had focused on the negative aspects of climate warming, without considering its positive aspects. He insisted upon the social and economic cost of emissions reductions, and concluded its first intervention in these terms: "All countries need to consider how they can adapt to climate change."[61] Exaggerated perils, exorbitant cost, the need to take action—all were founded on an uncertain scientific foundation: This was the major argument Big Oil used in the fight against the reduction of greenhouse gas emissions in the decades to come.

1990 proved to be the warmest year recorded to that point (many other record years followed). Before the industrial era, approximately 580 billion tons of carbon were present in the atmosphere, in the form of carbon dioxide gas. There were 750 billion in 1990, and humanity was emitting 6 billion additional tons each year, the IPCC reported. With an additional 220 billion tons, global warming risked reaching 2°C by the end of the twenty-first century, that is to say the limit beyond which catastrophic consequences are probable, well before 2100. Some 4 trillion tons of exploitable carbon remained beneath the Earth's surface, in the forms of oil, gas, and coal, three-quarters of which was coal. The sources of carbonaceous energies (and oil, especially) were still the matrix of our world.

Shortly before the 1992 Rio Earth Summit, the media quoted George H. W. Bush, president of the nation that was then emitting by far the largest quantities of greenhouse gases, as saying, "The American way of life is not negotiable."[62] The White House position relaxed significantly the following year when President Bill Clinton assumed power. Al Gore, his vice president, entered the public eye a few months before the Earth Summit by publishing a book on the ecological challenge, in which he pled for a departure from petroleum use within the next quarter-century.[63] Early in the presidential campaign, the man who was still only a Tennessee senator received a standing ovation from activist Democrats in Florida when he exclaimed: "We have the equivalent in our civilization of a ten-pack-a-day habit!"[64]

Opposing Bill Clinton and Al Gore, the lobbyists who journalists nicknamed the "Carbon Club" closed ranks; the Global Climate Coalition, for example, recruited Harlan Watson, an elegant scientist with a stentorian voice who had been one of George H. W. Bush's primary climate advisors in the White House.[65] The lobbyists were particularly effective on Washington's congressional hill. For a number of elected American Republicans as well as Democrats, the oil, gas, and coal industries weighed heavily, in terms of votes and campaign funds.

Eighty-three countries signed the Kyoto Protocol on climate change on December 11, 1997. It was the delicate fruit of an international compromise; there were many who felt that it was too flawed an instrument to allow an effective fight against the emission of greenhouse gases. The protocol exempted the transport sector from all the efforts and asked little of developing countries. And the United States remained on the sidelines. The Clinton administration, however, was among the signatories. This was purely symbolic. On July 25, 1997, the senators of the number-one worldwide industrial power—the oil country where the two largest companies were General Motors and Exxon—had voted for a resolution that prohibited any ratification of the protocol that was signed five months later in the ancient imperial capital of Japan. Participation in this protocol "would result in serious harm to the economy of the United States," the senators affirmed unanimously in the Byrd-Hagel resolution.* On the eve of the Kyoto summit, a *New York Times* survey indicated that 65 percent of Americans wanted their country to reduce its greenhouse gas emissions as quickly as possible, "regardless of what other countries do."[66] The problem of the hole in the ozone layer, which only touched minor aspects of the industrial system, was in the process of being regulated. As for the problem of climate change, it touched the very heart of the system of power.

1997, the year of the Kyoto Protocol, was also the year of the Asian financial crisis that caused brutal recessions, from the Philippines to South Korea, leading to a sharp fall in oil prices the following year.

---

* The amendment was named after Robert Byrd, a Democratic senator from West Virginia, and Chuck Hagel, a Republican senator from Nebraska who became Barack Obama's defense secretary from January 2013 to November 2014.

That year, average global temperatures set new records. Researchers were preparing to reveal that, according to data recorded by American nuclear submarines, 40 percent of the North Pole's ice pack had melted in forty years. American manufacturer Ford withdrew from the industry's Global Climate Coalition lobby, and the Rockefeller Foundation itself joined the fight against global warming. Barrel prices for oil fell to just above $10, rapidly stimulating a vigorous return to economic growth in Western industrial countries as well as for a new, heavyweight player in the game of worldwide capitalism: China. Oil consumption and intense economic activity were more entangled than ever. Paradoxically, this new low-water mark in crude oil prices was the departure point for the most powerful surge ever seen—the beginning of a new period in the history of oil and humankind, from which there was no turning back.

In 1859, "Colonel" Edwin Drake (*right, foreground*) drilled the Pennsylvania well that caused the first black gold rush. IMAGE COURTESY OF WIKIMEDIA COMMONS

Petroleum saves the whales: a cartoon from *Vanity Fair* magazine in 1861. IMAGE COURTESY OF WIKIMEDIA COMMONS

John D. Rockefeller, probably the richest man of all time. PHOTO COURTESY OF OSCAR WHITE, WIKIMEDIA COMMONS

Calouste Gulbenkian: After the First World War, the legendary Mr. Five Percent helped negotiate for allied powers Great Britain, France, and the United States, which competed for access to Iraq's black gold. PHOTO COURTESY OF WIKIMEDIA COMMONS

Los Angeles: This archetypal twentieth-century megalopolis grew from the riches of its oil fields during the Roaring Twenties, which came along at just the right time to postpone a decline of oil production in the United States. PHOTO COURTESY OF ORANGE COUNTY ARCHIVES

A US government poster promoting ridesharing in 1943. Image courtesy of National Archives and Records Administration

Mohammad Mossadegh, Iranian prime minister named man of the year by *Time* magazine in 1952 for his role in the democratic awakening in his country. In 1953 he was the victim of a coup conducted under the auspices of a CIA team directed by a cousin of Franklin D. Roosevelt. The American oil majors were then able to gain an upper hand on Iranian petroleum. PHOTO © AFP

A scene from *Easy Rider*, Dennis Hopper's 1969 cult film, depicts a different version of the American dream.

Abdullah Tariki, *above*, known as "the red sheik," and Juan Pérez Alfonzo, *below right*, the ecologist: These two founding fathers of OPEC found their dreams broken. ABOVE PHOTO FROM GETTY

Left, Marion King Hubbert, the Shell engineer who in 1956 correctly predicted that American crude production would reach a critical peak in 1970. No one heeded his warning at the time. PHOTO COURTESTY OF AMERICAN HERITAGE CENTER

George W. Bush, a pure product of the American black gold aristocracy, talks with workers in Texas oil fields while campaigning for Congress in 1978. Photo courtesy of George Bush Presidential Library and Museum

The Oval Office in 1975: President Gerald Ford (*center*), the only American president never directly elected to the office, flanked by his vice president, Nelson Rockefeller (*left*), and Nelson's protégé, Henry Kissinger, secretary of state. Photo courtesy of National Archives

American soldiers in front of the Iraqi Ministry of Oil after taking Bagdad in 2003. The oil ministry was one of the few public buildings protected by the US military during the pillage that followed the fall of Saddam Hussein. PHOTO © CRIS BOURONCLE/AFP

The crest for CENTCOM, or the US Central Command, depicts an American eagle spreading its wings over the Persian Gulf. IMAGE COURTESY OF US DEFENSE DEPARTMENT

The April 2010 explosion of the Deepwater Horizon oil platform caused the worst oil spill in US history. The platform, rented by BP, held the world record for drilling depth and became a symbol for the end of "easy" oil, whose decline would now have to be offset by unconventional and extreme hydrocarbon extraction. PHOTO COURTESY OF US COAST GUARD

A line of Chinese taxis waiting for gas in 2013. China has become the leading importer of petroleum and the leading emitter of greenhouse gasses on the planet. At what point will the "fuel tanks" of world growth be satisfied? And at what price? PHOTO REPRINTED BY PERMISSION FROM REUTERS

PART FOUR

AUTUMN

1998–20??

TWENTY-SEVEN

Oil's Future Decline Is Announced: The Persian Gulf Returns to the Center of the Chessboard

Despite the new set of realities, the same old system kicked back into gear. In 1998, following the Asian financial crisis, the old Western industrial powers, more greedy for oil than ever, reveled in the surplus crude oil on the market, while Venezuela and Saudi Arabia fought for Iraq's market share, which was plagued by an ongoing embargo. This battle helped lower the barrel price to only $12, a price no one had seen since the first oil crisis. Due as much to the meager oil price as to the "internet bubble" poised to explode, the United States, Great Britain, and France briefly enjoyed at the end of the twentieth-century record growth rates unheard of since the euphoria of the oil countershock in the Reagan years, or even since the 1960s. The hope of a new economic golden age resurfaced. In the OPEC countries, on the other hand, particularly in Saudi Arabia, the low crude oil values shook the wealthy welfare states, which struggled to buy social stability at much greater expense. Among the twenty-two Arab League nations, which had benefitted from the oil windfall in ways large or small, the suffering was immense: As the price of crude oil reached its nadir, the total GDP of the Arab League did not even equal the GDP of Spain, and in these oppressive outskirts, the ranks of voiceless millions were ready to unravel.* In the Far East, the low price of shipping fuel spurred a historic amplification of delocalization, a primary factor for globalization. China, which was preparing to make a resounding entry into an economic war that from this point on would be global, saw its carbon dioxide emissions begin to soar.

---

* Established in Cairo on March 22, 1945, the Arab League (or League of Arab States, formally) united the Arabic nations and has, as an organization, observer status in the United Nations.

For the oil majors, the perilous consequence of this brief oil glut resembled the one at the beginning of the century and that which occurred on the eve of the 1929 crisis: They had to join forces or risk perishing from lack of profits and therefore investment capacity. At just $10 or $13 per barrel, the Western companies had no other choice than to reduce their exploration budgets: Henceforth, "few investments outside the Middle East will any longer make sense," the *Economist* warned.[1] The London-based weekly feared that the barrel price could drop even further, down to $5; financial journalists did not perceive many limits to global oil-extraction capabilities. A more accurate vision of reality preoccupied the masters of black gold.

Megamergers of the Majors: Confronting Their Limit of Easily Extracted Reserves

On December 1, 1998, at its New York headquarters, JP Morgan held a press conference that could have been titled "Back to the Future." Exxon, Standard Oil's eldest daughter, announced that it was absorbing its "little" sister, Mobil. The deal, at $80 billion in shares, "re"created the largest private enterprise in the world. The marriage of Exxon and Mobil reconstituted, through its two main offspring, the bulk of John Rockefeller's empire. The Clinton administration voiced no complaints. ExxonMobil's megamerger succeeded—and was, in part, a response to the merger of BP and Amoco (formerly Standard Oil of Indiana), announced four months earlier. At the dawn of the new millennium, after many vicissitudes, the Western oil industry found itself once again dominated by three companies that were almost identical to the behemoths that prevailed on the eve of the First World War: ExxonMobil, BP-Amoco, and Shell, direct descendants of Standard Oil, of the Anglo-Persian Oil Company, and of the apparently unalterable Royal Dutch Shell. The Darwinian need for oil-company megamergers continued in the following years. In 1999, Total absorbed the Belgian Petrofina, and then the formerly public French company, Elf Aquitaine. Finally, after purchasing Gulf Oil in 1984, in October of 2000, Chevron consumed the other great Texan company, Texaco.

The Western oil-production sector was at the epicenter of a mergers movement that also involved several related sectors. In February 1998, the world's oil infrastructure leader, the Texan giant Halliburton, headed by a tireless bull, Dick Cheney, offered $7 billion to purchase one of its main competitors, also located in Dallas: Dresser Industries, the company that had allowed George H. W. Bush to enter the petroleum industry in the late 1940s. In September 2000,

Chase Manhattan Bank bought JP Morgan for $33 billion in shares, giving birth to JP Morgan Chase. This united two banks that for many years had served Big Oil, acting as both financiers and tributaries for the profits of the largest and richest of all industries. JP Morgan Chase competed successfully with Citigroup, another Wall Street super-heavyweight created two years earlier, also by the supernova of the old Rockefeller empire.

The leviathans of energy persevered, animated by their own unshakeable will. Exxon and Mobil, in particular, for over a century had remained firmly at the top of the largest planetary enterprises, perpetuating, generation after generation, a hegemonic culture of secrecy and a refusal to compromise, whether regarding the climate of the planet or their own survival. ExxonMobil never ceased to intensify its old policy of recruiting former high-level government workers, intelligence officers, and senior civil servants from the United States and elsewhere.

But its essential strategic preoccupation, the very reason for the merger between Exxon and Mobil, was access to global crude oil reserves. During the December 1998 conference where the merger was revealed, Lou Noto, Mobil's CEO, laid out a new truth: "We need to face the facts. The world has changed. The easy things are behind us. The easy oil, the easy costs savings, they're gone."[2] Few analysts or journalists made much of this statement, undoubtedly because the price of crude oil remained incredibly low throughout 1998. But on the Exxon side, the fundamental priority, the real obsession of Lee Raymond (nicknamed "Iron Ass"), the firm's CEO since 1987, was to reconstitute as quickly as possible the stocks of black gold, pumped and dispelled year after year.[3] Indeed, beginning in the 1990s, Exxon, like other Western oil companies deprived of almost all access to Persian Gulf's assets, found it tremendously difficult to replace, somewhere else around the planet, their reserves of conventional fuel: classic liquid petroleum.[4] The threat was clear; it was even written in black and white on the balance sheets. In 1996, the conventional reserves declared by Exxon fell, a fact that did not fail to worry the analysts of the largest Wall Street banks. Exxon's CEO, Raymond (who had long sat among the directors of JP Morgan), responded promptly: The eldest daughter of Standard Oil was at the forefront of investing hugely in Canada's tar sands, the future primary source of nonconventional oil.[5] Known and marginally exploited since the industry's beginnings, this "heavy" oil was harvested by huge excavators: a technique more similar to the mining industry than to that of the hydrocarbon industry.

However, it was mainly by swallowing its little sister, Mobil, that Exxon was able to regenerate its business in 1998. Mobil was four times smaller than

Exxon, but it presented the distinct advantage of possessing, from Kazakhstan to Qatar, reserves of petroleum (but especially natural gas) in all the new spheres of the planetary oil empire. If ExxonMobil had been a nation, its revenues would have ranked it the twenty-first richest in the world.[6] Its shareholders were delighted. It had become much easier to grow the company's reserves of crude oil and gas—the vital underpinning of an oil company's value—by buying up competitors than making its own new discoveries.

Offshore Drilling Cannot Compensate for Declines in Indonesia and the United States

It was rough going for the oilmen. After picking the most beautiful, ripe, low-hanging fruit, they had to seek new fields, less and less accessible, their yields more uncertain, their harvest more perilous and expensive. At sea, companies had to situate their platforms—ever more sophisticated, doomed to drill deeper and deeper—increasingly far from shore. The intensive exploitation of the North Sea, which began in the late 1970s, had already revealed itself to be perilous: In July 1988, the explosion of the Piper Alpha platform, fixed by huge stacks of steel to a concrete base that had been sunk to a depth of around 100 meters, caused 167 deaths. But shallow waters were no longer sufficient, and the North Sea production began to decline in 1999. It was necessary to drill even deeper. The fruits of very costly technical feats, the new floating platforms were anchored in large numbers to the west of the African coast and far off the coast of Texas. Exploitation in the Gulf of Mexico accelerated in 1996 when the Mars-Ursa field, located under more than 1,000 meters of water, entered production; then again in 2000 with Hoover-Diana, an ultradeep field, buried under 1,500 meters of water. The new floating platforms continued to be anchored in mind-boggling depths. Before the decade's end, they reached down beyond 2,500 meters (more than a mile and half deep), which added to the overall depth the drill had to travel. The distance between the crude oil reservoir and the platform commonly exceeded the height of Mount Everest.

During this time, some of the oldest oil countries—the historic suppliers of the most beautiful and low-hanging fruit—found themselves in serious difficulty, incapable of maintaining their production. The 1973 oil crisis had been preceded by record production growth, doubtless unsustainable. During the following decades, the progression of production slowed, despite the almost steady incentive of black gold prices that were radically elevated over those of the past century. After reaching record levels in 1977, Indonesia's extractions diminished a little more every year, in spite of the development of offshore fields. In 2002, the archipelago's oil production was surpassed by the oil consumption of its expanding population. Inevitably, Indonesia became an oil importer, and only retained its place in OPEC's cartel of exporter countries until 2008 to keep up appearance and, doubtless, to ensure its capacity to borrow. A highly experienced French oil reserve engineer, responsible for managing several important declining wells, stated: "For all companies, it was the same story: superiors constantly pressured us to maintain production, while we were happy when it did not drop too much from one month to the next."[7] The former Dutch colony, fertile cradle of Royal Dutch Shell, was undergoing the effects of a universal law: Production of any limited resource over time follows a bell curve, beginning with a phase of expansion, reaching its maximum when the resource matures and is roughly half exhausted, and then declining until its final extinction. In the absence of technical remedies, the so-called natural production decline of mature wells is generally on the order of 4 to 5 percent per year, but in many cases the declines prove to be much more rapid. Now that, in the words of Mobil's CEO, the era of easy oil was over, companies had no choice but to race on a treadmill that moved faster and faster in the opposite direction, wherever their wells became "mature."

Uncle Sam's oil treasury was subject to the same unalterable physical law. This was the principle cause of the decline in Exxon's conventional oil reserves, and of their pressing need to merge with Mobil. Beginning in 1985, by the time the Alaskan oil fields reached "maturity," American production had started to crumble, and the ever-deeper, ultracomplex platforms off the shores of the Gulf of Mexico weren't sufficient to change the situation. In 1993, when George H. W. Bush left the White House, the United States for the first time imported more oil than it produced: more than 7.6 Mb/d. Five years later, in 1998, US production continued to decline, supplying no more than 40 percent of American consumption. This consumption, strong from the reinvigorated economy, showed no sign of slowing its ascent in the years that immediately followed. On the contrary, this consumption reached the unprecedented level of 20 Mb/d in 2003, no less than a quarter of the world's oil production! From then on, the US economy has imported more or less half the crude oil it required. However, with its nuclear aircraft carriers continually crisscrossing the Persian Gulf, the United States was still the master of black gold. The only remaining global superpower, it had never been more dependent on oil pumped outside of its borders.

The Peak Oil Question Is Posed: Will the Halftime Whistle Blow before 2010?

Yet, in the oil industry, as elsewhere, faith in technological progress was greater than ever on the eve of the year 2000. The accuracy of seismographic devices, the effectiveness and power of computers modeling the structure of crude oil reservoirs, the power of injectors that pushed crude oil to the surface and accelerated the flow of the wells seemed unlimited. Although the decline of the North Sea reserves appeared inevitable to many engineers, some industry pundits cried out in the press that the improvement in pumping techniques could delay this decline by at least a decade. "Speeding up the pumping will only speed the decline!" retorted Jean Laherrère, a retired expert from Total who had participated in the discovery of Algerian oil almost half a century earlier. Laherrère's bright eyes, thick hair, and straight posture gave him the allure of an eternal science student. His uncompromising intellectual acuity, always at the alert, throughout his life had earned him great respect, as well as some enemies. Born in Pau, France, in 1931, Jean Laherrère was licensed in geology and a graduate of the École Polytechnique, the premier French engineering institution. His hikes as a young scout in the Pyrenees had bred in him a lifelong appreciation for wild lands and a love of magnificent natural beauty. Hired in 1955 by Compagnie Française des Pétroles, Total's ancestor, he had traversed the globe many times in search of new oil sources. His innovations in the field of seismic refraction played a decisive role in the discovery of Algeria's two "elephants": Hassi Messaoud and Hassi R'mel. From then on, from Labrador to the Australian desert to Patagonia, Laherrère had built his reputation as a first class international prospector, becoming the director of the Total group's technical exploration services, then deputy exploration manager. At the end of the 1980s, he became aware that new conventional oil discoveries were dwindling, that they no longer could replace the oil consumed. Therefore, he recommended to Total's directors that they make their initial investment in Canadian oil sands, around the same time that Exxon did so.

After his retirement in 1992, Laherrère expanded research that he had undertaken on the exact status of global oil reserves when consulting for one of the most renowned independent analysis companies, Petroconsultants. He worked with one of the most reputable French petroleum geologists, Alain Perrodon, former head of the exploration for Elf. The two men were soon joined by another retired geologist, Colin Campbell. This British oilman, born in Berlin in 1931, also very experienced, was notably the former boss of operations for the Belgian Fina company in the North Sea. Campbell wrote of having been fully aware, as early as the late 1960s, during the course of a study on the state of world petroleum resources he was conducting at Amoco's Chicago headquarters, that "the age of consumerism, nowhere more evident than in Chicago, could not continue indefinitely."[8] Feared just as much for his independent thinking as Laherrère was, Campbell chose to live, at the end of his professional life, in a small remote village in an ancient southwest port of Ireland constructed of stones and mossy hedges, Ballydehob. The prophetic tone this professor assumed was ridiculed when his theses became famous throughout the industry. In March 1998, Laherrère and Campbell published the synthesis of their common research in *Scientific American*. Their article, "The End of Cheap Oil," led to an unprecedented conclusion, which continues to echo throughout the industry: "Global production of conventional oil will begin to decline sooner than most people think, probably within 10 years."[9] This conventional (classic) oil then represented nearly nine-tenths of worldwide extractions: Humanity had less than a decade to begin weaning itself from petroleum! Campbell and Laherrère reminded readers that the annual global crude oil discoveries had "peaked in the early 1960s and has been falling steadily ever since." They emphasized, "About 80 percent of the oil produced today flows from fields that were found before 1973." They accurately predicted the North Sea's imminent decline and warned: "By 2002 or so the world will rely on Middle East nations, particularly five near the Persian Gulf (Iran, Iraq, Kuwait, Saudi Arabia, and the United Arab Emirates), to fill in the gap between dwindling supply and growing demand." The two geologists, who had spent their lives decrypting the industry's often opaque data, concluded: "The world is not running out of oil—at least not yet. What our society does face, and soon, is the end of the abundant and cheap oil on which all industrial nations depend." Campbell and Laherrère had such strong reputations that some of their results found their way into the pages of the International Energy Agency's 1998 annual report. A passage from this report, titled "The Pessimists' View," reviewed their conclusions.[10] Although substantially watered-down, this review infused a certain panic throughout the directorship of the oil majors, where some experts had, of course, already arrived at similar conclusions. This panic did not, however, reach the economic press.

In 2000, Colin Campbell invented the expression "peak oil": For a century and a half, humanity had climbed a mountain, reached the pinnacle, and would necessarily be hurtling down the slope, once it had extracted half of the recoverable oil. In May 2002, around a network of experts of the first order, independent or—like Pierre-René Bauquis, former strategic director of the Total group—recently retired, Campbell and Laherrère assembled the first symposium of the Association for the Study of Peak Oil (ASPO) at the University of Uppsala, Sweden, at the invitation of nuclear physics professor Kjell Aleklett. ASPO has been speaking out on "peak oil" ever since.

Dick Cheney, Halliburton's CEO, and the Persian Gulf Oil Fields' Return to Center Stage

While the United States never stopped digging itself deeper into the hole of dependence on foreign oil, the risk Campbell and Laherrère had pointed out, of the importance of the Persian Gulf's steadily mounting role in the worldwide petroleum game, was seen clearly by Washington. On January 26, 1998, two months before the *Scientific American* article's publication, a neoconservative think tank called Project for the New American Century (PNAC) sent President Bill Clinton an open letter calling for the "removal of Saddam Hussein's regime from power" in Iraq.[11] A rising force within the Republican Party, the neoconservatives championed a grandiose vision in which the United States would dominate the world for a century more. "The policy of 'containment' of Saddam Hussein has been steadily eroding over the past several months," wrote the letter's authors.[12] "It hardly needs to be added that if Saddam does acquire the capability to deliver weapons of mass destruction, as he is almost certain to do if we continue along the present course, the safety of American troops in the region, of our friends and allies like Israel and the moderate Arab states, and a significant portion of the world's supply of oil will all be put at hazard." Among the signatories were three "hawks" of earlier Republican administrations: Richard Perle, Paul Wolfowitz, and Donald Rumsfeld, president Gerald Ford's former cabinet chief and the man Ronald Reagan sent to Baghdad in 1983 to meet with Saddam Hussein and try to sell him a Bechtel pipeline—a meeting after which the White House long closed its eyes (and perhaps worse) on Saddam Hussein's use of chemical weapons.

As Halliburton's CEO since 1995, Dick Cheney favored relieving the sanctions against Iraq, sanctions that in his eyes penalized US oil companies—especially his. In addition, Cheney belonged to the group (consisting mainly of Pentagon alumni like himself) who had founded the PNAC in 1997. Probably more than any of the signatories of the 1998 letter addressed to President Clinton by the neoconservative think tank, Cheney seemed perfectly aware of the Persian Gulf's vital role for the future of the world's petroleum reserves. In 1992, shortly before the end of his mandate as secretary of defense, Dick Cheney had supervised Paul Wolfowitz's draft of a "defense planning guidance" policy statement, which described the Pentagon's prime objective as preventing "the re-emergence of a new rival." Sometimes called the "Wolfowitz doctrine," this guide intensified the Carter doctrine that was enacted twelve years earlier, explicitly specifying: "In the Middle East and Southwest Asia, our overall objective is to remain the predominant outside power in the region and preserve U.S. and Western access to the region's oil."[13]

In November 1999, in the banquet room of London's famous Savoy Hotel, one of the British Empire's favorite spots, before an audience of four hundred oilmen attended by a squad of security officers, Dick Cheney gave a long speech that amply demonstrated the shrewdness he had gained as head of Halliburton, a firm that had enabled him to acquire a fortune of several tens of millions of dollars.[14] His powerful baritone voice strangely muted by the lack of emotion capable of washing over his wide, smooth, expressionless face, the former Pentagon boss described the thorn in the side of the oil industry, the origin of its anxiety: "For over a hundred years we as an industry have had to deal with the pesky problem that once you find oil and pump it out of the ground you've got to turn around and find more or go out of business. Producing oil is obviously a self-depleting activity. Every year you've got to find and develop reserves equal to your output just to stand still, just to stay even."[15] The CEO of the world's largest petroleum service company cited the example of ExxonMobil, the large oil dragon rejuvenated nearly a year earlier: "ExxonMobil will have to secure over a billion and a half barrels of new oil equivalent reserves every year just to replace existing production." Dick Cheney's concern flowed out of his own experience: Halliburton's primary business was to design, sell, and install the equipment that was used to delay the decline of petroleum fields as they reached maturity, and then to curb this decline. At a time when world production approached 80 Mb/d, Cheney's speech indicated the fundamental issue of the coming decade: "By some estimates there will be an average of two percent annual growth in global oil demand over the years ahead, along with, conservatively, a three percent natural decline in production from existing reserves. That means by 2010 we will need on the order of an additional fifty million barrels a day. So where is the oil going to come from?"

It was then that, after two decades of respite and diversion offered by offshore oil, Dick Cheney confirmed the return to center stage of the Persian Gulf fields, to which the Western oil companies had been denied almost all access since the first oil crisis. He argued: "Governments and the national oil companies are obviously controlling about ninety percent of the assets. Oil remains fundamentally a government business. While many regions of the world offer great oil opportunities, the Middle East, with two thirds of the world's oil and the lowest cost, is still where the prize ultimately lies." He continued, "even though companies are anxious for greater access there, progress continues to be slow." Addressing an equally delicate subject, Cheney lamented that (according to him) up to that point, oil had been "the only large industry whose leverage has not been all that effective in the political arena." Whether or not it was true, this state of affairs was on the point of being changed, by Dick Cheney himself.

Big Oil Takes the Helm at the White House: The Sinking of Enron

Richard Cheney had, so to speak, grown up on an oil field. He was raised in Casper, a small Wyoming city at the foot of the Rocky Mountains, nicknamed "Oil City" when an oil field was discovered within its borders, a common occurrence in the United States at the end of the nineteenth century. Wyoming's crude oil production entered into decline in 1971, at the same time the total US production did. Two-thirds of Wyoming's oil flow and almost half of the United States' output had been burned up in chimneys and exhaust pipes when, on July 25, 2000, presidential candidate George W. Bush surprised the political world by announcing Dick Cheney, then fifty-nine years old, as his vice-presidential running mate.

Dick Cheney was not, of course, the only person on the Bush team closely connected to the American hydrocarbon industry. Far from it. George W. Bush, whose spin doctors transformed his vulgar expressions and incredible gaffes to manufacture a decent Mr. Everyman who was presentable in the eyes of the electorate, was the ultimate insider. No US government, not even Warren Harding's in the early 1920s, had been as close to Big Oil as the one Bush and Cheney established in the aftermath of their controversial victory during the presidential election of November 7, 2000.* Donald Evans, George W. Bush's old crony and CEO of a Midland, Texas, oil company, was appointed secretary of commerce. Bush appointed a brilliant forty-six-year-old African American woman, Condoleezza Rice, as head of the National Security Council. She was formerly a part of George H. W. Bush's administration and once a member of

* Illustrating the strength of the Bush clan's political networks, it was John Prescott Ellis, first cousin to both George W. Bush and his brother Florida governor John Ellis (Jeb) Bush, who, at 2 a.m. on the night the ballots were counted, was the first to "give" the state of Florida to the Republican candidate. Ellis was an election analyst employed by Fox News. In the moments that followed, the other major television networks—ABC, NBC and CBS, which, hitherto, had considered the race too close to call or had reported Democratic candidate Al Gore in the lead—followed the powerful conservative channel's lead after Fox announced prematurely that, thanks to Florida, George W. Bush was the new president of the United States. Three months later during a House of Representatives committee hearing, the Fox News CEO, Roger Ailes (former media advisor to Nixon, Reagan, and Bush Senior), said, "We gave our audience bad information. Our lengthy and critical self-examination shows that we let our viewers down. I apologize for making those bad projections that night; it will not happen again." Fox News is situated in the center of American capitalism—housed in a Manhattan building with other divisions of Rupert Murdoch's News Corp, such as the Dow Jones Company and the *Wall Street Journal*. This building is part of the Rockefeller Center extension built in the 1960s, which also includes the Exxon tower.

Chevron's board of directors. As a leader at Chevron, Bush Junior's new security and foreign policy advisor even had an oil tanker named after her: a 129,000 ton double-hull supertanker registered in the Bahamas, although Chevron quickly redubbed it *Altair Voyager* after the election.[16]

Also in the first ranks of the Bush administration was Thomas White, secretary to the army, veteran officer of Vietnam and elsewhere, and former senior officer of a company whose trajectory was remarkable: Enron. Highly envied, this Houston firm, initially in the business of trading natural gas, began to produce electricity, and subsequently invested all around the globe, from California to India, counting James Baker among its top emissaries in the Persian Gulf. Kenneth Lay, Enron's CEO, was extremely close to the Bush clan, and one of George W.'s most generous campaign donors. Surfing the waves of one of the largest stock market expansions of all time, with records that incessantly beat the Dow Jones index (the internet bubble was on the point of bursting), Enron dazzled Wall Street with imaginary profits. On January 17 and 18, 2001, two days before the president's inaugural address, electricity supply cutbacks affected hundreds of thousands of people in the western United States. It was the first significant episode of California's series of massive power outages, from San Francisco to San Diego, artificially created by the shenanigans of Enron energy traders in early 2001. The victim of this farce was the most populous state in the union, the sixth largest economy on the planet. Taking advantage of California's deregulated energy market to manipulate its prices, Enron went so far as to shut down central power plants and close pipelines to increase the price of electricity. But already the huge bubble called Enron, regarded for years as the most high-performing large American company, was on the verge of imploding. The stunning revelation of the firm's fraudulent accounting practices—which implicated the highest executives of Wall Street's finest jewels, beginning with JP Morgan Chase and Citigroup—led Enron stocks to collapse during that summer. The newest American energy giant had until that time been hailed as a shining example of the success of the deregulated energy market. Enron ended by declaring bankruptcy on December 2, 2001.* Facing up to forty-five years in prison, Kenneth Lay died of a heart attack in 2006, while his right-hand man,

* As Julius Caeser said, "Men believe what they want to believe." This wishful thinking defined Enron years. Enron's audacious leaders were a source of fascination: They loved to flex their macho muscles by taking dangerous motocross journeys through the desert. Lou Pai, one of Enron's "magicians," received more than $25 million by opportunely selling his shares a few weeks before the company's collapse began. His taste for strippers stirred the press's curiosity: In order to conceal their scent from his wife, he habitually stopped at service stations on the way home, to perfume himself with gasoline. (See Alex Gibney, *Enron: The Smartest Guys in the Room* (documentary), 2005.)

Jeffrey Skilling, remains in jail at the time of this writing. The scandal that followed, and lasted for several years, was a great source of embarrassment for the George W. Bush administration.

The connections between the Bush administration and the energy industry went further. Among the advisers of its energy transition team, the Bush administration included the CEO of Peabody Energy, the world's largest coal company. The ample financial contributions the coal and hydrocarbon lobby paid to the Republican Party were an effective investment, as scarcely two months after arriving in the White House, Bush reneged on his campaign promise to impose regulations on greenhouse gas emissions from power plants.

This timely backtracking on the climate change front relieved, among others, ExxonMobil's CEO, Lee Raymond. He was an old acquaintance of Dick Cheney: They had been neighbors in Dallas and hunted quail and pheasant together; their wives were friends.[17] On February 8, 2001, nineteen days after the inauguration, Halliburton's former CEO received Lee Raymond, one of the largest customers of his former firm, at the White House. "Look," Raymond recalled telling the vice president, "my view is this country is going to be an importer [of oil] for as far as the eye can see. If you believe that to be the case, what things can we do?"[18] It was precisely this that Cheney and his collaborators sought to determine in the weeks that followed.

In the first few days after his arrival in the White House, Dick Cheney established the Energy Task Force to seek advice from the masters of the American energy industry. Why was the vice president, barely installed in power, in such a hurry to launch this task force, populated by Big Oil representatives? The new state of affairs was no doubt worrisome. The end of "easy oil," announced in December 1998 by Mobil's CEO when it merged with Exxon, seemed to be approaching. Hurried along by the vigorous resumption of growth and the first wave of globalization (itself facilitated by the fall in crude oil prices in 1998), the oil market had bounced back, much faster than many had anticipated. The barrel price shot up, doubling in the space of two years! Set at $12 in 1998, barrel prices had risen to more than $20 in the year 2000, nearly flirting with $30. This ferocious spike in crude oil prices meant more for Big Oil than a return of fantastic profits. Just as clouds signal rain, the spike brought with it the promise of imminent tensions in the vital flow of the economy's energy.

After a long legal battle, certain documents that circulated within the Energy Task Force were made public. The nation's main environmentalist organizations, some of which had brought the lawsuit to release the documents, were kept out of these secret task force consultations. Among the documents dating back to March 2001, released by the American courts after the 2003 invasion of Iraq, were maps of the Middle East and in particular of Iraqi oil resources—fields, pipelines, refineries, terminals, areas of exploration—as well as a long table analyzing Saddam Hussein's relationships with a handful of oil companies, mainly European—particularly Total and ENI and (already) some in China.[19] The leaders of Exxon, Chevron, Shell, and the American branch of BP who met with Dick Cheney's White House staff for a long time refused to admit they had participated, going so far as to testify (although not under oath) that they had not appeared before a congressional committee in November 2005. A week later, the *Washington Post* revealed internal Bush administration documents showing that the opposite was true.[20]

A provisional, confidential version of the Energy Task Force report submitted to the White House on April 10, 2001, called not only for a reassessment of the sanctions against Iraq but also against Iran and Libya—sanctions that obliged American oil companies to stay away from "some of the most important existing and prospective petroleum producing countries in the world."[21] This plea for the relief of sanctions unfavorable to Big Oil corresponded to the policy Halliburton's former CEO had openly called for, before and after his arrival at the White House. A plea that made no mention of the terrorist threat.

TWENTY-EIGHT

September 11, 2001: A Rogue Pearl Harbor

At 8:46 a.m. on Tuesday, September 11, 2001, an American Airlines Boeing jet struck the north tower of New York's World Trade Center, killing all of the plane's passengers as well as hundreds of people inside the building. At 9:03, a United Airlines Boeing jet struck the World Trade Center's south tower. At 9:37 in Washington, an American Airlines Boeing jet crashed into the front of the Pentagon, deeply penetrating the building. Defense Secretary Donald Rumsfeld felt the whole building shake. At 9:59, the World Trade Center's south tower collapsed. At 10:07, a hijacked United Airlines plane crashed into a field in Pennsylvania, after its passengers attempted to regain control of the plane. At 10:28 in Manhattan, the north tower of the World Trade Center collapsed. Nearly three thousand people died in the September 11 attacks. Fifteen of the nineteen hijackers were Saudi nationals. It was the deadliest foreign attack ever perpetrated on American soil. Osama bin Laden's Islamist Al Qaeda terrorist movement, which quickly claimed responsibility, had chosen to target US financial and military centers.

A Moment of Truth

The shock of 9/11 landed in the middle of one of the most sensitive periods in the long "special relationship" between the United States and Saudi Arabia. At first, the election of George W. Bush seemed to imply a return to a perfect symbiosis between Washington and Riyadh. On January 19, 2001, on the eve of Bush's inauguration, Baker Botts, James Baker's legal firm, hosted a celebration in the imposing Ronald Reagan International Affairs Business Center, very close to the White House, with Bush Senior and Prince Bandar, nephew of King Fahd and ambassador of the Saudi Kingdom, as its honored guests. "Happy days are here again," announced an assistant to the man the Bush family had nicknamed "Bandar Bush."[1] Only, in Israel and in the occupied territories, violence was escalating. And the solidarity President Bush displayed with respect to Ariel Sharon, the conservative prime minister of the Jewish State, raised the ire of Saudi Arabia's Prince Abdallah. It was Prince Abdallah, aged seventy-six, who wielded the real power in Riyadh, not his Americanized elder brother King Fahd, who had become almost senile. In May, without precedent, Abdallah refused to honor an invitation to go to Washington. Opposed to the Bush Junior administration, the Saudi Arabian prince intended to influence American foreign policy; without a doubt, he believed he had the ability to ask the impossible. The following month, Bush Senior tried to pick up the pieces. In the presence of his son, the president, Bush Senior called Prince Abdallah and assured him that George W. Bush's heart was "in the right place."[2] On September 11, Bush Senior was still an adviser to the Carlyle Group, which counted the Saudi bin Laden family, supposed to have wholly cut ties with Osama, among its investors. The morning of the attacks, the firm held a meeting at Washington's Ritz-Carlton hotel, ten minutes from the Pentagon; participants included James Baker and the bin Laden family's representative, Shafiq bin Laden, Osama's half-brother. According to a witness, at the moment American Airlines flight 77 crashed into the Pentagon, Shafiq bin Laden ripped off his name tag and stole away in a car, as did James Baker.[3]

In the days following the attacks, Richard Clarke, the White House's director of counterterrorism, gave the FBI the green light to authorize the departure from American territory of at least 160 Saudi nationals, including many members of the royal family and several members of the bin Laden family. None were interrogated before flying out of Los Angeles, Orlando, or Las Vegas. *Vanity Fair* and the *New York Times* argued that some of the planes evacuating the Saudis left before September 14, that is to say at a time when the Federal Aviation Administration was still prohibiting all planes from taking off.[4] This allegation caused considerable turmoil in American public opinion, in particular when Michael Moore's 2004 documentary, *Fahrenheit 9/11*, was released. The FBI refuted this allegation, reinforcing their denial with a series of documents published in 2005 in the American courts. The *New York Times* insisted that these documents were heavily redacted before publication.[5] This turned out to be the case: Saudis who feared for their safety had requested and received FBI protection to quickly leave American soil.[6] An anecdote reveals the solidarity found within the close caste of oil potentates: One of the aircraft that transported Saudis impatient to leave Las Vegas was a DC8 equipped with two bedrooms that belonged to Gabon's president, Omar Bongo.[7]

Several leads seem to indicate that for years after Osama bin Laden was stripped of his Saudi nationality in 1994—and even after the terrorist organization claimed the October 12, 2000, attack on the American destroyer USS *Cole* in Yemen—links still remained between Al Qaeda and the powerful Saudis. One of these links even led to Prince Bandar's wife herself, who helped to finance (without her husband's knowledge) the ringleader of two of the September 11 attacks.* But the most intriguing lead vanished behind the walls of the US military prison in Guantanamo Bay. Believing that he was dealing with Saudi agents, an Al Qaeda leader named Abu Zubaydah, captured in March 2002 by Pakistani commandos and American special forces, provided the CIA, from memory, with the phone number of a distinguished member of the House of Saud. "He will tell you what to do," said Abu Zubaydah to the false Saudi investigators, according to anonymous sources cited by Gerald Posner, an American investigative reporter.[8]† Zubaydah then told the investigators that in the late 1990s, Osama bin Laden had made a pact with the House of Saud in order to obtain Riyadh's support, in exchange for the assurance that Al Qaeda would spare Saudi Arabia. The contact Zubaydah provided was that of Prince Ahmed bin Salman Al Saud, grandson of King Ibn Saud and son of the man who became King Abdullah's successor in January 2015, Salman Al Saud. Until his unexpected death less than a year after the September 11 attacks, Prince Ahmed was especially known for his passion for horse racing. His horse War Emblem won the May 7, 2002, Kentucky Derby. Ahmed bin Salman Al Saud's family vehemently contested Gerald Posner's allegation, championing the prince's sincere love for the United States.

Al Qaeda members' attempts to poison relationships were numerous, and their hatred of the House of Saud's relationship with Washington was well known. The fact remains, on July 2, 2002, in Riyadh, Prince Ahmed died suddenly in his sleep. He was forty-three years old. In the days that followed, two other Saudi princes named by Abu Zubaydah, according to Posner, also died: One was killed in a car accident while en route to Prince Ahmed bin Salman's funeral, while the other was found dead in the desert, dehydrated.[9] Abu Zubaydah is, as of this writing, still detained at Guantanamo Bay, without ever having been tried or charged, tortured by CIA officers on numerous occasions.[10]

---

* Princess Haifa was accused of having provided funds to a Saudi Arabian, Omar al-Bayoumi, who in San Diego helped two of the 9/11 suicide bombers, Khalid al-Mihdhar and Nawaf al-Hamzi; he himself was suspected of being in contact with the Saudi secret services. See in particular Philip Shenon, *The Commission: The Uncensored History of the 9/11 Investigation*, Twelve, New York, 2008, p. 50–55.

† Gerald Posner was previously known for having published debunkings of some of the conspiracy theories surrounding the assassinations of John F. Kennedy and Martin Luther King Jr.

The Leniency 9/11 Investigation; the Opening and Closing of Saudi Arabia's Gates

For several months, the White House attempted to restrict the 9/11 investigation to secret congressional committees with limited power, until, pressured by lawmakers and the victims' families, an ad hoc national commission was established on November 27, 2002, more than a year after the attacks. George W. Bush announced that Henry Kissinger would lead the commission. Two weeks later, in the cold of the New York winter, a group of the victims' widows, the "Jersey Girls," was received by the former American secretary of state in the overheated offices of his international relations company, on the twenty-sixth floor of a high-security Park Avenue building. The shrewd Kissinger, now seventy-nine years old, had always refused to make his client list public. However, it was well known that he counted several oil majors among his clientele. Sitting with a cup of coffee on one of the conference room sofas, facing a Kissinger who was extremely courteous and affirming his humility in the face of the historic task that awaited him, one of the Jersey Girls, Lorie Van Auken, asked the dreaded question: "With all due respect, Dr. Kissinger, I have to ask you: Do you have any Saudi clients? Do you have any clients named bin Laden?" The next morning, Henry Kissinger called the White House to announce his resignation.[11]

The chairmanship of the 9/11 Commission was finally assigned by President Bush, with the agreement of Congress, to a former Republican governor of New Jersey, Tom Kean: a friend of Bush's father, who, by his own admission, was ignorant of intelligence and national security issues.[12] To analyze the circumstances of the worst crime ever committed in US history, the commission had a budget of $3 million, thirteen times less than was allotted for the investigation of the space shuttle *Challenger* explosion, and half as much as had been spent on the Monica Lewinsky scandal. In January 2003, the key position of executive director of the commission was handed over to a professor of history and political science from the University of Virginia, Philip Zelikow. Contrary to what one might think, based on his "old European"-style tweed jackets, Zelikow was a Texan from the world oil capital, Houston. Author or editor of fourteen books and dozens of scientific articles (not without significant flaws*), he was also an old colleague of Condoleezza Rice: They had worked together on George H. W. Bush's National Security Council. In 1995, they had even co-authored a book on the reunification of Germany.[13]

* Significant errors have been identified in his transcripts of tapes (admittedly bad quality) of Kennedy during the Cuban missile crisis.

In January 2001, during the transfer of power from the Clinton administration to the Bush administration, Zelikow had sat beside Condoleezza Rice while Sandy Berger, the new head of the National Security Council, explained that Al Qaeda embodied the number one threat to the United States.[14] The conclusions of the *9/11 Commission Report*, published in July 2004, were very lenient toward Condoleezza Rice and the Bush administration in general, even though they had ignored warning signs as numerous as they were specific during the months preceding the attacks.[15]* Many who worked on and witnessed the efforts of the 9/11 Commission accused Zelikow of shielding the White House from blame.†

The most troubling potential conflict of interest between Philip Zelikow and his friend Condoleezza Rice did not see the light of day until March 2004, when it did not cause much of a stir. Zelikow was the primary author (anonymous) of the "preemptive war" doctrine, drafted at Rice's request following September 11 and signed by George W. Bush on September 17, 2002, two months before the 9/11 Commission was created.[16] A dramatic reversal of US military policy, this text served to legitimize the invasion of Iraq. The 9/11 Commission's first public hearings took place on March 31, 2003, a week after

---

* On August 6, 2001, when he was away on his cattle ranch in Crawford, Texas, President Bush received a CIA memorandum—the later publication of which Vice President Cheney attempted to thwart (see the *New York Times*, May 20, 2002)—titled: "Bin Laden Determined to Strike the United States." Other earlier and apparently more precise reports remained secret for a time but continued to surface in the press for a long time after the commission's report was published. (See Kurt Eichenwald, "The Deafness Before the Storm," *The New York Times*, September 10, 2012.)

† The White House refused to declassify twenty-eight censored pages of a confidential investigation of the American Congress. These pages—released after lawmakers pushed to make them public—referenced the Californian Al Qaeda cell with which Prince Bandar's wife was found to be connected. They evoked the possible complicity of a Saudi consultate officer in Los Angeles. When Zelikow learned that one of the investigators under his command had obtained—legally but in defiance of her instructions—a copy of the censored pages, he dismissed her on the spot (Philip Shenon, *The Commission*, op. cit., p. 109–112). Various potentially problematic aspects of the September 11 attacks were left out of the Commission's inquiry report. In particular, the report scarcely mentioned the existence of tower No. 7 in the World Trade Center complex, the third building that collapsed that day, at approximately the speed of free fall and in the direction of the strongest resistance. After a long delay, the final, official, technical report on the collapse of the World Trade Center towers, made public in September 2005, did not analyze the behavior of the towers' structures as they fell, "after the conditions for collapse initiation were reached." (See the *Final Report on the Collapse of the World Trade Center Towers*, NIST-NCSTAR 1 National Institute of Standards and Technology, September 2005, note 13, p. 82.) Yet it is this technical aspect (with the discovery of melted steel at Ground Zero, considered in the February 2, 2002, *New York Times* as "perhaps the deepest mystery uncovered in the investigation," since that the fire caused by the crash had not reached a temperature sufficient to melt steel; see the NIST final report, p. 29) that caused a group of American architects and civil engineers to vehemently call for a reopening of the investigation, and even support the most sophisticated, frenzied conspiracy theory: that of a controlled demolition of the towers. Without these blind spots in the official investigation, the "conspiracy theory movement" around the September 11 attacks, which persists today, would probably not be as intense.

the American army began their invasion of Iraq. Abraham Sofaer, the expert who was first called to testify, argued in favor of the doctrine of preemptive war in Iraq; he was even one of its most ardent advocates. Iraq was occupied for eleven months before Zelikow was revealed as the doctrine's creator.

Many accused the Bush administration of favoring the Saudis, before and after September 11, despite the high probability that millions of Saudi dollars had flowed to Al Qaeda, and also in spite of Riyadh's poor cooperation in the aftermath of the attacks.[17] On May 1, 2001, according to classified documents that were nonetheless leaked, the CIA had informed President Bush that "a group presently in the United States" was preparing a terrorist operation, and then, on June 22, that Al Qaeda's attack could be "imminent."[18] But during that same month, the US Embassy in Saudi Arabia indicated that its new Visa Express program permitted all nationals of the Kingdom of Saudi Arabia to obtain a visa without even having to apply to the American Consulate. Three of the 9/11 terrorists used this program to begin their journey to the United States without waiting in line or being questioned by American consular officers.[19] As unbelievable as it may seem, Visa Express was maintained after September 11: During the thirty days following the attacks, the American Consulate in Jeddah interviewed only 2 out of 104 visa applicants, and did not reject a single request.[20]

Why show so little concern? In addition to the large capital investments Saudis had made in the United States, part of the answer may reside in Big Oil's impatience to resume business in Saudi Arabia. On September 12, the day after the World Trade Center and the Pentagon were attacked, President Bush designated a man he trusted to be the new US ambassador in Saudi Arabia. He was none other than Robert Jordan, a petroleum industry lawyer, a partner in James Baker's firm, Baker Botts—and the man who in 1991 had defended George W. Bush when the latter was suspected of insider trading in the Harken affair. Also on September 12, Saudi Arabia published a declaration in which it ensured that Saudi Aramco's exports to the United States would remain stable, and committed to ensuring that any possible fall in international market exports would be offset by OPEC. "These are welcome words, indeed," Robert Jordan responded, shortly afterward.[21] It turns out that during the time of 9/11, a series of important oil contracts were in the midst of being finalized with Saudi Arabia. In particular, a deal worth $25 billion for the exploration of natural gas in Saudi Arabia, with Baker Botts as the brokers and Exxon at the helm of the Western oil companies' interests.* This was probably not the opportunity of the century—without a doubt it was "money tossed in the air"

* Among the other major oil companies were BP, Shell, and Total.

or even a "bitter potion" for American leaders, according to a source familiar with the deal—but what Western petroleum company would have refused to be ushered into Saudi Arabia through an open door, more than twenty years after Aramco's full nationalization wrested away the American oil companies' control of the planet's most important reserves? Less than two weeks after September 11, during the congressional hearing required for his confirmation to the post of ambassador to Saudi Arabia, Robert Jordan stated: "[I] have been really satisfied, Senator, to note the gas concession that has been granted to three consortiums, two of which are led by Exxon-Mobil, into development of the gas fields in Saudi Arabia. . . . I certainly will have this high on my agenda."[22] The importance of this opportunity to regain footing in Saudi Arabia could not be underestimated by a government like that of George W. Bush.

Another potential source of conflict of interest awakened the suspicions of the press after the October 2001 invasion of Afghanistan, where the Taliban regime had offered asylum to Osama bin Laden's terrorist organization. After taking power in 1996, radical Islamist Afghans had attempted to negotiate plans, with a Unocal executive, for a northern gas pipeline project to bring natural gas from the Caspian Sea to Turkmenistan via Pakistan's shores along the Indian Ocean, through unstable Afghan territory. This project would have opened a path, outside post-Soviet Russia's sphere of influence, to access some nearly untouched hydrocarbon resources in and around the Caspian, presented as fabulously promising by a motley crowd of promoters, unreliable and hungry for commissions. In addition to Unocal, Halliburton and Enron were among the firms interested in the trans-Afghanistan pipeline project. After the terrorist attacks in Dar es Salaam and Nairobi, in August 1998, and Bill Clinton's refusal to grant full diplomatic recognition to the Taliban regime, Unocal's directors resolved on December 5, 1998, to announce its withdrawal from the consortium that had planned to build the pipeline. But their connections with Kabul's reactionary Muslim masters—monstrous offspring, like the Al Qaeda terrorists, of the groups of Islamist fighters financed by Saudi oil money and equipped by Washington in 1980 to combat the Red Army—remained active. In March 2001, five months after Al Qaeda attacked the American destroyer USS *Cole* in the port of Aden, Yemen, an emissary of the Taliban government was received in Washington, while at the same time, the head of American counterterrorism, Richard Clarke, was advocating for an immediate attack on Al Qaeda's bases in Afghanistan. September 11 ended this waffling. Once the Taliban was driven out of Kabul in December 2001, the Bush administration supported the rise to power of future Afghan president Hamid Karzai, whom *Le Monde* described as a former Unocal consultant, something the Afghan politician denied just as vigorously as the American oil firm did.[23] Unocal recognized, however, that Zalmay Khalilza, an American strategist of Afghan origin sent by the White House as special envoy to assist Karzai, endorser of the neoconservative think tank Project for the New American Century, was employed as a consultant by Unocal when the trans-Afghanistan pipeline project began, in the mid-1990s.[24] The project briefly resurfaced in 2005, with President Karzai's blessing, but the seemingly perpetual war that ravaged Afghanistan ultimately blocked the path to the pipeline's construction.

The Iraq War: A War for Oil?

Was the invasion of Iraq, unleashed on March 19, 2003, by the American army and to a much lesser extent by the British army (and to an even lesser extent by the meager coalition forces of Poland, Italy, South Korea, and a handful of other nations), a war for oil? This question was constantly swept aside by the Bush administration, by the British government of Prime Minister Tony Blair's "New Labour" party, and by Big Oil.[25] And yet, each reason the Bush administration evoked to justify the Iraq War, before the American Congress as well the United Nations, proved to be groundless and, in the end, one of the most serious diplomatic scandals in history.

Even before the invasion began, the charge that Saddam Hussein had provided support to Al Qaeda appeared to be a smokescreen. An alternative explanation, the threat of "weapons of mass destruction" allegedly possessed by Iraq, was also an untruth. On February 5, 2003, Colin Powell, secretary of state, reluctantly presented this explanation at a plenary meeting of the UN Security Council; it proved false. The meticulous research nearly 1,400 international inspectors conducted throughout Iraq after the war failed to find any weapons of mass destruction, which in reality had been virtually put out of service on the day after the Gulf War in 1991, under the scrupulous gaze of Colin Powell, chairman of the joint chiefs of staff at that time.[26*]

---

* The *New York Times*, which in this case was itself misled, revealed in 2014 that the American soldiers and their allies in Iraq's regular army, post-Saddam Hussein, from 2004 to 2011 did discover stocks of chemical weapons in Iraq. These were not the famously sought "weapons of mass destruction" but were only artifacts of the Iran-Iraq War, out of use and often rust-covered (but nevertheless dangerous—several accidents took place): weapons designed at the time by the United States, manufactured in Europe, and assembled by Iraqis. The Pentagon and the Bush administration deemed it appropriate to keep this fact secret. (See C. J. Chivers, "The Secret Casualties of Iraq's Abandoned Chemical Weapons," *The New York Times*, October 14, 2014.)

Powell later blamed the CIA and the Pentagon for having misled him, and for having pushed him to mislead the international community with a fiction vital to winning the United Nation's support for the invasion (though in the end, he did not succeed).* A few months before September 11, however, Powell had stated: "Frankly, the sanctions [against Iraq] have worked. Saddam has not deployed any significant capability with respect to weapons of mass destruction. He is unable to project conventional power against his neighbors."[27] At the White House's request, Colin Powell resigned from his position of secretary of state on November 15, 2004. He subsequently qualified his spectacular statement before the UN Security Council, a month before the war began, as a "blot" on his career.[28] Finally, even George Bush Junior's reference, during a September 2002 speech in Houston, to a 1993 conspiracy in which Saddam Hussein supposedly ordered an assassination attempt on Bush Senior ("After all, this is the guy who tried to kill my dad"), appears unfounded, among other reasons because the master of Baghdad at that time looked favorably upon the White House.[29]

If the link with Al Qaeda and the threat of weapons of mass destruction adulterated the truth—or were even, at times, simply lies—what fundamental aim did the Bush administration have to invade Iraq? A dense cluster of clues supports the thesis that its aspiration was to take hold of the oil in Iraq, a country sitting on what constituted, by far, the most important reserves of known underexploited or unexploited crude oil reserves on the Earth.

In September 2000, in the middle of the presidential campaign, the Project for the New American Century (which two years previously had called upon then-President Bill Clinton to chase Saddam Hussein from power) published a report that, at the time, went unnoticed outside the highest circles of the military-industrial complex. PNAC, whose members were promised great influence after George W. Bush's victory, recalled then that "the United States has for decades sought to play a more permanent role in Gulf regional security." And continued: "While the unresolved conflict with Iraq provides the immediate justification, the need for a substantial American force presence in the Gulf transcends the issue of the regime of Saddam Hussein."[30] Regretting the reduction in military expenditure during the Clinton era, the authors of that PNAC report, titled "Rebuilding America's Defenses," continued to insist on the need for the US Army to refuse any "strategic pause," to go to the front in order to "create tomorrow's dominant force" in the context of what the

* Is it perhaps relevant to note in this context that Pentagon strategists found it sensible to engage many fewer troops to take control of Iraq in three weeks than they did to liberate the small emirate of Kuwait in 1991?

Pentagon then called the "revolution in military affairs": new technologies, cyberwar, war in space, and so on.[31] The US military's deployment of such a revolution, ambitious and costly, would of necessity be very long, lamented the PNAC, which counted among its supporters luminaries such as Vice President Dick Cheney, Secretary of Defense Donald Rumsfeld, and Florida's governor, Jeb Bush, the president's younger brother. A passage in the report, however, evoked a possible alternative: "The process of transformation, even if it brings revolutionary change, is likely to be a long one, absent some catastrophic and catalyzing event like a new Pearl Harbor."[32]

On the evening of September 11, President Bush noted in his diary: "The Pearl Harbor of the 21st century took place today."[33]

Had the September 11, 2001, attacks played the role of a kind of perverted, misguided Pearl Harbor? Should we consider as a mere anecdote the fact that the doctrine of preemptive war, having made possible the intervention in Iraq, was drafted by the executive director of the 9/11 Commission?

From the Bush administration's point of view (publicly, at any rate), it was Iraq, not Saudi Arabia, that posed a problem. Upon their arrival at the White House, George W. Bush and Dick Cheney assessed all possible options regarding Saddam Hussein. Discharged from his government position on December 31, 2002, Secretary of the Treasury Paul O'Neill later revealed that the eviction of the Iraqi president was among George W. Bush's first priorities, only ten days after his swearing-in and eight months before September 11.[34] However, until the attack on the World Trade Center and the Pentagon, the Bush administration's executive branch was torn between two factions: on one side, former petroleum industry leaders who defended the effectiveness of the sanctions against Iraq (such as Condoleezza Rice) or championed adjusting those sanctions in a way that was favorable to the industry (such as Dick Cheney); and on the other side, the leader of the hardliners, Secretary of Defense Donald Rumsfeld. Rumsfeld, who had served as President Ford's chief of staff three decades earlier, continued to advocate for a direct military approach to topple Saddam Hussein.* Both factions, however, considered Iraq a problem of prime importance. In April 2001, five months before September 11, a report from the Baker Institute, a prestigious political institute created by James Baker, underlined the risk represented by Saddam Hussein's stranglehold on Iraq's oil. "Iraq

* With a certain prescience, Rumsfeld dictated, on July 23, 2001, a memorandum in which he refused to be held responsible for a potential "future Pearl Harbor post mortem." Five days later, he dictated another memorandum intended for Condoleezza Rice, on the subject of Iraq: "Sanctions are being limited in a way that cannot weaken Saddam Hussein." He said that the United States should stop claiming they had a policy "that is keeping Saddam 'in the box,' when we know he has crawled a good distance out of the box." (See Errol Morris' documentary, *The UNKNOWN known*, 2013.)

remains a destabilizing influence to U.S. allies in the Middle East, as well as to regional and global order, and to the flow of oil to international markets from the Middle East," the institute's experts wrote.[35]

For the first time, echoing Dick Cheney's 1999 London speech but with a much more alarmist tone, the Baker Institute voiced its concern that the petroleum industry would be unable to cope with the surge in world demand. Remember that since 1999, barrel prices had climbed significantly: They had doubled since the short-lived $12 price in 1998, and were firmly established at well beyond $20, as at the time of the second oil crisis. "Perhaps the most significant difference between now and a decade ago is the extraordinarily rapid erosion of spare capacities in critical segments of energy chains," the PNAC report said, stating that the most "extraordinary" of these losses were in the in the oil arena. "Today, shortfalls appear to be endemic."[36]

Next, the Baker Institute warned: "And with spare capacity scarce and Middle East tensions high, chances are greater than at any point in two decades of an oil supply disruption that would even more severely test the United States' security and prosperity."[37] This warning came during the general time frame that Dick Cheney's Energy Task Force consulted privately with the leaders of Big Oil. Such a finding could not fail to gain the White House's attention, all the more so because the study was co-published by the Council on Foreign Relations, always a vital factor in US foreign policy. Like many players clustered around the Bush administration, the Baker Institute accused Baghdad of seeking to manipulate the oil market.[38] This threat was, without a doubt, overestimated because of the draconian restrictions that limited the Iraqi industry. It was true, however, that Saddam Hussein proclaimed, loud and clear, his intention to sell his oil henceforth in euros and not, like the rest of the world, in dollars (again, the threat appeared mainly in vain, taking into account the Food for Oil program).[39] According to the institute, "Tight [oil] markets have increased U.S. and global vulnerability to disruption and provided adversaries undue potential influence over the price of oil. Iraq has become a key 'swing' producer, posing a difficult situation for the U.S. government."[40] The Baker Institute report, reflecting the strategic thinking of a contingent at the heart of the American oil industry, remained ambiguous regarding the most effective way to address this "difficult problem." They settled for calling on the United States to "conduct an immediate policy review toward Iraq including military, energy, economic and political/diplomatic assessments."[41] In its final recommendations, the institute argued (in support of the entire American oil industry and Vice President Cheney) in favor of a prudent review of the sanctions against Iraqi oil, in order to facilitate its access to the American industry.[42]

The Baker Institute report is a public document. Almost a year after the invasion of Iraq, the *New Yorker* magazine affirmed it had knowledge of a top-secret document from the National Security Council (headed by Condoleezza Rice) dated February 3, 2001, fourteen days after President George W. Bush's inauguration. According to the *New Yorker*, this document bore the signature of a National Security Council high official and concerned Dick Cheney's newly formed Energy Task Force. It directed the NSC staff "to cooperate fully with the Energy Task Force as it considered the 'melding' of two seemingly unrelated areas of policy: 'the review of operational policies towards rogue states' such as Iraq, and 'actions regarding the capture of new and existing oil and gas fields.'"[43]

"Of Course It's about Oil, We Cannot Really Deny It" (General Abizaid, Commander of CENTCOM)

The attacks of September 11 appeared for a time to disrupt any strategic perspective. In Washington, ten months after the attacks, a Frenchman who for years had navigated the American neoconservative networks, Laurent Muraviec, made the mistake of pointing the finger at the state that could legitimately be regarded as that most directly responsible for this new Pearl Harbor: Saudi Arabia. At the same time, the White House was working to concoct a fictitious connection between Al Qaeda and Saddam Hussein. On July 10, 2002, this fifty-one year old French researcher employed as an analyst by the RAND Corporation (the temple of military research in the United States) committed the blunder of the century. In a briefing that he submitted to the Pentagon, Muraviec, who had built his reputation thanks to his reflections on the "Revolution in Military Affairs," dared to describe that faithful ally, Saudi Arabia, as "a kernel of evil, the prime mover, the most dangerous opponent" of the United States.[44] Referring to the many proven links between Al Qaeda and Saudia Arabia, Muraviec asserted: "The Saudis are active at every level of the terror chain." He suggested taking control of Saudi fields outright: "What the House of Saud holds dear can be targeted: Oil. The oil fields are defended by U.S. forces, and located in a mostly Shiite area."[45] This suggestion, which figured in the conclusions of his twenty-four page PowerPoint presentation, cost Laurent Muraviec his contract with RAND. His last PowerPoint slide presented Iraq as the "tactical pivot" of the Middle East, whereas Saudi Arabia was, according to him, the "strategic pivot." Doubtless Muraviec's analysis was in step with the subtle, cryptic protocols of those who aspired to control the source of power buried in the Persian Gulf. An aspiration for which President Bush Junior's administration provided the ultimate vehicle, but which had resided at the heart of American foreign policy at least since the Iran-Iraq War, and which had begun to shape certain vital Western geopolitical axes in 1918.

Various statements and documents bear witness to this aspiration expressed during the outbreak of the war on Iraq, by a coalition of governments tied to the United States, as well as by the oil companies that hid behind them. In early July 2003, a few weeks after the culmination of the invasion of Iraq, the minister of foreign affairs of Poland, part of the "coalition of the willing" laboriously constituted by the Bush administration, told the press: "We have never hidden our desire for Polish oil companies to finally have access to sources of commodities."[46] This minister, Wlodzimierz Cimoszewicz, specified that the oil fields constituted the "ultimate objective" of his government.[47] The head of Polish diplomacy expressed this at a time when several of his country's enterprises were in the process of signing partnership agreements with one of the giants in American oil services: Kellogg, Brown & Root.* On March 8, 2003, before the combat began, the US Army Corps of Engineers (without a call for bids) gave this Halliburton affiliate an umbrella contract for the rehabilitation of the oil industry in Iraq.[48] The contract's value was estimated at no less than $7 billion over two years and represented a potential profit of nearly half a billion dollars for the firm directed from 1995 to 2000 by the man who was currently the vice president of the United States, Dick Cheney.[49] Kellogg, Brown & Root was also responsible for feeding the American soldiers; Department of Defense auditors later accused the firm of having overbilled the Pentagon by millions of dollars.[50]

In May 2005 an investigation by Italy's public TV station, RAI, accused Silvio Berlusconi's government of associating himself the "coalition of the willing" in order to salvage a 1997 agreement between the Italian company ENI and Saddam Hussein, for the exploitation of abundant oil of excellent quality that lay buried in the vicinity of Nasiriya, a city bordering the Euphrates River in the south of Iraq. The RAI documentary highlighted an Italian report drafted shortly before the beginning of the conflict, advocating the securing of the Nasiriya area and, at the same time, the preservation of an agreement whose value was assessed at no less than $300 billion.[51]

In September 2007, at the conclusion of the penultimate chapter of his memoirs, a long chapter devoted to the present and future global energy

---

* Brown & Root, a Halliburton subsidiary since 1962, merged in 1998 with Kellog, another branch of the firm then directed by Dick Cheney. Dresser Industries had purchased Kellogg in 1987 and was itself absorbed by Halliburton in 1997.

situation, Alan Greenspan, chairman of the Federal Reserve from 1987 to 2006, wrote: "I am saddened that it is politically inconvenient to acknowledge what everyone knows: the Iraq war is largely about oil."[52] Greenspan, who since the time of President Ford had maintained close links with all the Republican administrations, continued: "Thus, projections of world oil supply and demand that do not note the highly precarious environment of the Middle East are avoiding the eight-hundred-pound gorilla that could bring the world economic growth to a halt."[53] A few pages earlier, the man who had succeeded Paul Volcker as the chairman of the Federal to serve as the dreaded oracle of American capitalism for two decades, demonstrated that, like others in Washington before him, he never lost sight of the nature of the most powerful raw material, the source of economic power, nor the inevitable contingencies that control of this resource imposed on the world's remaining superpower: "How many years will the oil last? Supplies will shrink well before the end of this century, most experts now say. . . . Crude-oil production peaked in the United States' lower forty-eight states in 1970. Ultimately, all reservoirs will peak, and few large new ones are likely to be found in the now thoroughly drill-pocked developed world."[54]

During the brief controversy born during the publication of Greenspan's memoirs, Republican senator Chuck Hagel, President Obama's future secretary of defense, came to the economist's rescue, declaring: "People say we're not fighting for oil. Of course we are."[55] General John Abizaid himself, the head of CENTCOM (the joint chiefs of staff of the United States for the Middle East) from 2003 to 2007, affirmed at a roundtable organized by Stanford University in 2008, just after he retired: "Of course it's about oil, we can't really deny that."[56]

In 2011, on Fox News, one of the Bush administration's most zealous hawks, John Bolton, former US ambassador to the United Nations, referred to the "critical oil and natural gas producing region we fought so many wars to try and protect our economy from the adverse impact of losing that supply or having it only available at very high prices."[57]

In April 2011, the notes from secret meetings organized by Tony Blair's government during the months preceding the invasion became public.[58] In October 2002, the director of the British Foreign Office in the Middle East had written: "Shell and BP could not afford not to have a stake in [Iraq] for the sake of their long-term future. . . . We were determined to get a fair slice of the action for UK companies in a post-Saddam Iraq."[59] On October 31, Baroness Elizabeth Symons, minister of British trade, met with representatives of BP, Shell, and British Gas. According to the meeting's report, she agreed that "it would be difficult to justify British companies losing out in Iraq in that way if the UK had itself been a conspicuous supporter of the US government throughout the crisis."[60] Tony Blair's minister agreed to "report back to the companies before Christmas," in order to inform them of the results of her efforts to obtain guarantees from the White House. On November 6, the British Foreign Office invited BP to discuss opportunities in Iraq "post regime change." A synthesis of these exchanges indicated: "Iraq is the Big Oil prospect. BP is desperate to get in there and anxious that political deals should not deny them the opportunity."[61] BP feared that the outcome of the Washington invasion would leave the French company Total free to assert the validity of its agreements with Saddam Hussein. This fear was clearly dispelled by Paris's refusal to participate in the war. On February 6, 2003, Tony Blair proclaimed that "the oil conspiracy theory is honestly one of the most absurd when you analyse it." On March 12, two weeks before the war began, Lord John Browne, BP's CEO, swore, "It is not in my or BP's opinion, a war about oil."

Before and after the invasion of Iraq, the Anglo-Saxon economic press rustled with signs of the Western petroleum industry's whetted appetites. In July 2002, a pragmatic *Times of London* headline proclaimed: "West Sees Glittering Prizes Ahead in Giant Oilfield."[62] The *Washington Post* noted in September that "a U.S.-led ouster of Iraqi President Saddam Hussein could open a bonanza for American oil companies."[63] In England, the *Guardian* reported on a conference titled "Invading Iraq: Dangers and Opportunities for the Energy Sector," organized in October 2002 at Chatham House, the grand English geopolitical institution. For a delegate, the debates could be summed up this way: "Who gets the oil?"[64] Although its production had declined since 1980 and the beginning of the Iran-Iraq War, Iraqi oil, in particular that of southern Iraq, was still one of the best sources in the world. It could cost less than a $1 a barrel to extract (in 2002, the crude oil price rose to more than $25 a barrel). "That's why people are interested," exclaimed an enthusiastic Irish oilman in a *New York Times* article published on April 10, 2003, the day Baghdad fell: "it's the cheapest oil in the world!"[65] Two weeks later, using the vocabulary of finance, the *Wall Street Journal* brazenly stated that "the Pentagon is embarking on one of the most audacious hostile takeovers ever: the seizure and rejuvenation of Iraq's huge but decrepit state-run petroleum industry."[66]

According to its proponents, the war would quickly establish democracy, peace, and prosperity in Iraq. Estimates of the violent deaths that occurred in the course of the US military's invasion and occupation of Iraq vary from 100,000 to 600,000 people. These were mainly Iraqi civilians, combatants or not. How many more died? The death count for the US Army and coalition soldiers was higher than 4,500. On both sides, the great majority were killed not during the brief invasion of Iraq, launched on March 20, 2003, and lasting scarcely a month, but during the period of insurgency and terrorist chaos that endured, in spite of rare lulls, after the withdrawal of American troops in 2011. On March 26, 2003, while the coalition troops progressed rapidly toward Baghdad, the Pentagon's Office for Reconstruction and Humanitarian Aid issued a list of sixteen sites that "merit securing as soon as possible."[67] The first of these was the central bank, which, however, was looted and ruined. The second was the National Museum of Iraq, the sanctuary of many archaeological treasures dating from the dawn of civilization, including writing and some of the oldest myths that enliven the human story today, doubtless generated by a symbiosis between primitive Semitic peoples and a people who perhaps came from elsewhere, the Sumerians. The museum also housed invaluable evidence dating back to the caliphs of Baghdad's School of Abbasid, the origin of the modern science responsible for the revision of Ptolemy's *Almagest*, of the first translation of Euclid, or even of the invention of algebra and chemistry. The American army did not even attempt to protect the National Museum of Iraq, which was also extensively ravaged by looters.

The sixteenth and last place on the list was one of only two, including the Ministry of the Interior, that was immediately secured by the American invaders. It was the Ministry of Oil.

TWENTY-NINE

Shocks and Ruptures: The Occupation of Iraq and Crisis of 2008

It was time. Because of the hellish mess that exploded almost as soon as the American occupation began, the exploitation of Iraqi oil resumed very slowly. But the return of Iraq's black gold nonetheless coincided with the timely resurgence of global extractions that marked the year 2003. Indeed, during the two previous years, crude oil production had dropped 1.2 Mb/d, equivalent to Libya's entire production at that time.[1] Such a thing had not been seen since the second oil crisis. The West was primarily responsible for this collapse. And this time, the exhaustion of "easy" oil reserves was clearly the cause. The apparently inexorable decrease in aging American production continued, and even more significant, the predicted decline of the British and Norwegian fields on the North Sea began. This was further aggravated by the collapse of some African wells feverishly operated by the oil majors. Thus, extractions in Gabon, dear to the Françafrique, began to drop in 1997.

The weakening of another country weighed heavy on the West's balance: President Hugo Chávez's Venezuela. Since 1998, Venezuelan production, controlled by the national company, PDVSA, had begun to decline in alarming fashion. Founded in 1976, PDVSA had long been criticized for corruption and inefficiency. Was Venezuela, historically a producer country and the number one world exporter during the Second World War, a victim of the depletion of its oil resources or only prey to bad management? In November 2001, President Chávez, a close ally of Fidel Castro, signed a law doubling the royalties that foreign companies were required to pay, while increasing national control. Five months later, on April 12, 2002, a failed coup, mounted by a network of entrepreneurs and industrialists who favored foreign investment, ousted President Chávez from power for two days. The CIA had been apprised of all details in advance.[2] The Bush administration immediately recognized the short-lived provisional government of the perpetrators. Exxon, former master of the region, tried to patch things up with Chávez. Venezuela, which provided 15 percent of US crude oil imports, saw its production continue to wither, due to the lack of sufficient investments to maintain the extractions from fields exploited for nearly a century.

The Decisive Issue of Liberation and of the Insurgency in Iraq

For some time after the United States seized Baghdad, Uncle Sam's long-term interests appeared to be well-satisfied by Iraq. The Pentagon quickly implemented a strategic change that was clearly essential and long overdue: On April 19, 2003, barely a week after the taking of Baghdad and a year and a half after September 11, the United States announced the withdrawal of all their armed forces stationed in Saudi Arabia (apart from five hundred military instructors).[3] According to renowned journalist Bob Woodward (of Watergate fame), Saudi Arabia's ambassador to the United States, Prince Bandar ("Bandar Bush"), promised that the House of Saud would help in the reelection of President George W. Bush in 2004 by lowering the price of crude oil before the elections.[4]

On May 1, off the coast of the American naval base in San Diego, George W. Bush celebrated his victory in Iraq on the flight deck of the nuclear-powered aircraft carrier USS *Abraham Lincoln*, adorned for the occasion with a banner proclaiming "mission accomplished." In the south of Iraq, at the foot of Etemennigur, the Great Ziggurat of the ancient city of Ur, the American military built its formidable and dusty Tallil Air Base, on the site of a former Iraqi base largely destroyed twelve years earlier during the 1991 Gulf War. Equipped with two 3-kilometer landing strips and abundant fuel reserves, loaded with ultrasophisticated electronic equipment deployed in the middle of the desert, this base was situated near the richest oil fields of Mesopotamia, including that of Rumaila, involved in the invasion of Kuwait in 1990, and a key location for the defeat of the Iraqi army in 2003. "Shock and awe": Such was the Pentagon's description of their powerful, crushing strategy. On the wall of the Etemennigur temple, a massive vestige of the Sumerian civilization more than forty centuries old, whose name means "house whose foundations inspire terror," a few US marines graffitied their motto: *Semper Fidelis* (Always Faithful). In October 2003, the US Air Force celebrated the opening of a second Burger King on the Tallil base.[5]

The majority of foreign firms who landed in Iraq came from the United States. Naturally, all agreements signed with the deposed regime were considered obsolete (in particular, of course, those of the Total group, Jacques Chirac's French government having refused to join the coalition formed by the United States). It was not the members of the Iraqi "government council" established by the coalition who awarded the first contracts related to the country's reconstruction but Washington, and without a call for bids. Additionally, even before fighting ended, Bechtel, which laid its first pipeline in Iraq in 1950, obtained a contract for the repair or construction of much of the vital infrastructure throughout the devastated country: ports, airports, sanitation systems, and agricultural structures, much of which had been destroyed in 1991 during the bombing of the Gulf War.[6] The American construction and oil infrastructure giant monopolized the bulk of the new market and quickly made it clear that they were hiring subcontractors, not searching for partners.[7] "It's just like in the old days under British mandate," satirized the *Los Angeles Times*.[8]

On September 20, 2003, George W. Bush's presidential envoy to Iraq, Paul Bremer, former assistant to Henry Kissinger, launched the forced privatization of an economy that until that point had been largely owned by the state of Iraq, with practically two-thirds of the active population unemployed. Two days later, during a meeting of the World Bank and the IMF held in Dubai, Iraq's finance minister, installed by the White House, announced that all sectors would be fully opened to foreign capital. Petroleum was the only exception: The Iraqis would continue to own the oil. Concessions of no fewer than forty years could, however, be granted to petroleum producers from elsewhere.[9] But this was only a draft, the skin of the bear was far from sold. In the meantime, business was already prolific for the American industries contracted to repair wells, pumps, and pipelines. The worksites accounted for tens of billions of dollars. The Citigroup bank announced that it was ready to open lines of credit guaranteed by future crude oil deliveries.[10] Halliburton shares doubled between October 2002—the date on which the US Congress authorized the war—and January 2004. On December 11, 2003, the Pentagon revealed that it was opening up a new investigation of the tens of millions of dollars that Kellogg, Brown & Root, one of Halliburton's main subsidiaries, had overcharged the military for a contract to supply Iraqi fuel valued at greater than $1 billion (the American army eventually suspended Halliburton's contract).[11] The US vice president personally encouraged investors to take a chance on Iraq. "We ought to be able to get their production back up in order of 2.5 (to) 3 million bbl. a day within, hopefully, by the end of the year," Dick Cheney assured investors during the spring of 2003.[12]

The imposition of chaos in the realms of politics and domestic security meant that Iraqi production took not one but eight years to reach 2.5 Mb/d.[13] Those who were counting on it little by little renounced the American takeover of Iraq's black gold. The shock and awe were not enough, and the politics of peacemaking were fruitless. The Iraqi insurgency, joined by veterans of Saddam Hussein's army as well as by religious radicals, quickly grew to a nightmarish magnitude. The insurgents had recourse to all the guerrilla tactics, mortar or rocket attacks, snipers, remote-controlled bombs, and suicide bombings via car bombs (thanks to the rudimentary and formidable mixture of chemical fertilizer and oil popular with terrorists since the 1970s: nitrate-fuel).[14] Not to mention the sabotage of petroleum equipment. Beginning in spring 2004, the insurgency swelled almost out of control. In that year alone, forty attacks against oil pipelines and other facilities took place in the Kurdish north and the Shiite south. These acts of sabotage were aimed at destabilizing American authority. But most importantly, the control of oil became the key issue of political divisions and of fighting between the armed factions of different communities who wanted to form a free nation in Iraq. From all directions, Iraq attracted those who wished to be martyrs. On the outskirts of Baghdad, fighting raged around the miserable neighborhood of Sadr-City, named after the Shiite leader Muqtada al-Sadr, leader of the Mahdi Army.* The insurgency also saw the emergence of Abu Musab al-Zarqawi's branch of Al Qaeda in Iraq, made up of Sunni Islamists recruited across the Arab world and largely funded through Persian Gulf oil. After sovereignty was transferred to the Iraqi interim government on June 28, 2004, the insurgency morphed into an intense civil guerilla war between Sunni and Shiite factions, which peaked in 2006 and 2007.

"If Zarqawi and [Osama] bin Laden gain control of Iraq, they would create a new training ground for future terrorist attacks," President George W. Bush said. "They'd seize oil fields to fund their ambitions. They could recruit more terrorists by claiming a historic victory over the United States and our coalition."[15]

War for a Paycheck

Donald Rumsfeld resigned from his position as secretary of defense on December 18, 2006. Iraq had not, as promised, morphed into a peaceful democracy. The true resumption of crude oil production was as eagerly awaited as the vote for a new law establishing the sharing of oil between foreign investors

* From a family believed to have descended directly from the Prophet Muhammad, Muqtada Al-Sadr is the son-in-law of Mohammed Bakir Al-Sadr, martyr of the Baathist Repression in 1980.

and the Iraqi state, but also and especially between rival warring Iraqi communities. For the United States, the occupation of Iraq proved to be much more expensive than the invasion. The total bill for the war was estimated at over $700 billion, more than for the Vietnam War.[16] Bitter critics complained that, at such a price, hundreds of millions of poor households around the world could have been equipped with solar panels. In the Iraqi campaign and all around the world, the US Army's oil consumption reached approximately four hundred thousand barrels per day, more than the needs of a country like Sweden or Greece: The Pentagon, out of all organizations, had the greatest thirst for oil.[17] Nothing was more greedy for energy than US efforts to sustain leadership in the Persian Gulf. For many, especially in Washington, the United States' strategic rationale in the region, created in the aftermath of the Second World War and continually supported since, was quickly collapsing in a headlong rush when George W. Bush's second presidential term ended in 2008.

The kidnapping of foreigners was common in Iraq, and beheadings became almost commonplace. Many Western businessmen took refuge in the Green Zone, an ultrasecure international compound in the heart of Baghdad, much more extensive than the colonial concessions established in Shanghai by the Europeans in the nineteenth century. For their protection, oil companies, American diplomats, and even the US Army Corp of Engineers made wide use of mercenaries.[18] The "mercs," driving heavily armored vehicles, escorted their clients through lethal roads, where they shot their assault rifles in response to the slightest gesture, often without a reason, while undergoing daily attacks sometimes organized with the complicity of Iraqi insurgents who had infiltrated their number. The most dangerous of the routes these paramilitary convoys traveled—sometimes, if necessary, driving in the wrong direction—led from the airport in Baghdad to the Green Zone. The official American military main supply route name for this road was Route Irish, but the American soldiers and security officers nicknamed it "IED Alley," after the improvised explosive devices, or simply "Death Street."

For a long time, oil constituted the most lucrative source of business for private security companies, in particular across the turbulent African continent. Nothing, however, compared with the manna that was offered in Iraq. Powerful firms thrived. Some, such as Vinnell, enjoyed a historic reputation around the Persian Gulf. Others, such as the American Blackwater or the British Aegis, had been born almost ex nihilo. The companies were so numerous (there were nearly two hundred of them) that they created an "Iraqi bubble" in the industry.[19] Starting in 2008, they employed 190,000 people. This was 30,000 more than the number of American soldiers present in Iraq.[20]

Among the war professionals recruited by mercenary groups were many American veterans, British, Australians, Nepalese Gurkhas, and even, some claimed, former members of the Shining Path from Peru.[21] Small makeshift American companies recruited decommissioned and alcoholic police officers or former delinquents, ex-GIs, or trigger-happy amphetamine addicts, risk-takers full of bravado. Among the managers of one of these companies, under contract with Halliburton, one might find a former ranger whose neck displayed the words "Christian Crusader" tattooed in Arabic.[22]

The neoliberalism advocated by the Bush administration lent Iraq an aspect of jungle law. Just before leaving his position as head of the Coalition Provisional Authority in June 2004, Paul Bremer signed Order 17, by which he granted mercenaries total immunity with respect to Iraqi justice. This fact did not greatly interest the world until, after many other fatal incidents, scandal eventually burst: In Baghdad, on September 16, 2007, stuck in a traffic jam on the roundabout of Nisour Square while escorting American diplomats, Blackwater mercs fired without reason and did not stop. Seventeen civilians lost their lives.

Professional warriors sometimes left their trades for the oil industry, potentially much more lucrative but not less risky (financially in any case). At the head of a small business that he baptized Excalibur Ventures, a former American special forces soldier eventually lost his case in a trial after he accused Gulf Keystone Petroleum, a firm originating in Texas and registered in Bermuda, of having stolen his company's rights to important petroleum fields located in Iraqi Kurdistan, acquired thanks to contacts the soldier had forged during the war.[23] A young American student on break, a former parachutist employed by a small, rather seedy security company, Jon Coté was kidnapped in November 2006 along with four of his colleagues. All were savagely murdered by their captors, who demanded the withdrawal of American troops and the liberation of Iraqi prisoners. Sometime earlier, Jon had confided to a journalist who followed him: "This place is a money-making *machine*. . . . There is just so much of it. It really amazes me: a war, how it creates money, generates it, how a war can be profited off. All you have to do is look around: the amount of food and fuel and oil and shit that we use over here. There's so much money that it's fucking ridiculous."[24]

China Wins the Race for Iraq's Black Gold

In June 2008, at the beginning of the final six months of George W. Bush's second term in the White House, the press revealed that Washington had tried to impose no fewer than fifty military bases on the new government of liberated Iraq.[25] The cost of the war in Iraq that year reached a record amount for the United States: $142 billion. During the transition period that followed Barack Obama's victory in the November presidential race, the Bush administration had conceded to Iraq a plan for the withdrawal of American forces.

Not happy with the results of Bush's policy in Iraq, Vice President Dick Cheney searched for ways to destabilize the other great enemy still in place in the Persian Gulf: the regime of the ayatollahs in Iran. Cold and impassioned (a strange combination that also characterized John Rockefeller), in 2005 Dick Cheney had repeatedly insisted about the Iranian nuclear program that: "They're already sitting on an awful lot of oil and gas. Nobody can figure why they need nuclear as well to generate energy."[26] (In the 1970s, Washington attempted to "understand" why Iran, for example, wished to master the atom.) The CIA was under significant pressure to say that Tehran was preparing to develop a nuclear bomb.[27] But this time, the leaders of American intelligence did not yield to pressure.* According to a *New Yorker* investigation, in 2007 Congress approved the financing of secret operations to destabilize Iran. In early 2008 in Vice President Cheney's office, a meeting was held to study various means to produce a false *casus belli* between the United States and Iran.[28] The journalist who headed the *New Yorker* investigation, Seymour Hersh, reported that the meeting, for example, assessed the possibility of US Navy commandos disguising American torpedo boats as Iranian, which the US Navy would then be approved to attack in the Straits of Ormuz.[29] No plan was adopted, in the end. The inner workings of Dick Cheney and his old crony Donald Rumsfeld's psyches remain unclear, as if obliterated by the large machinery for which these men served as top minions. Born in 1941, the year of Pearl Harbor, and raised in the Midwestern state of Nebraska, perhaps Dick Cheney viewed himself as a brilliant captain promoted to high general during the conquering of the West, in the pioneers' day: a delirium born of the omnipotent myth of the frontier?

On June 29, 2009, at the heart of the Green Zone, in the Al Rashid Hotel ballroom, which had been entirely redecorated in green, the color of Islam, and glowing in the light of the national television program that broadcast the scene live, the new Iraqi government placed a glass urn on a dais. Outside in the sky, helicopters patrolled. The competition to obtain the rights of Iraqi oil field exploitation had begun. The representatives of the oil majors of the West arrived in armored 4x4 vehicles, but also giant national companies controlled by

* In December 2007, a report whose publication was long delayed, released by the US intelligence agencies, indicated that the military nuclear program in Iran had been shut down in 2003.

Russia and China arrived, one after the other, and each presented an envelope. On the inside of the envelope, two numbers only: the desired production target for the oil field, and the royalty amounts they wished to claim for each barrel extracted, expressed in US dollars.[30] Crippled by a wave of hate that had been stirred up once again, Baghdad's new regime made a point of guaranteeing perfect transparency, as they prepared to put in play a large share of the second or third most important crude oil reserves in the world, after those of Saudi Arabia and Iran. A crude oil of the best quality, spared during a quarter century of war and embargo. In the eyes of the oilmen (and in particular, according to an American diplomat, in Exxon's view), Iraq was the last source of "easy oil."[31]

Except that the conditions proposed by Baghdad were far from generous. In the face of the potential risks their employees would be taking, most of the Western oil majors bridled at the meager royalties: barely $2 per barrel, a profitability disaster. The French company, Total, had to resort to staying out of the deal, to the great chagrin of its director general, Christophe de Margerie, Total's former manager for the Middle East.[32] Exxon and Shell shared the exploitation of one of the best fields of Iraq, West Qurna-1. But the agreements signed were only "service" contracts. This meant that Exxon and Shell did not achieve their most vital goal: to include on their balance sheets fresh reserves of crude oil, and increase their own productions.

Even more than Russia, China was the great winner after months of bidding. Driven into the most formidable economic boom of all time, facing enormous energy needs, China provided itself the means to impose itself as the number one oil operator in Iraq. BP's boss, Englishman Tony Hayward, played a decisive auxiliary role in this victory, raising the ire of the leaders of other old major players. In order to be able to associate itself with the Chinese state-owned company NCPC in the exploitation of the famous Rumaila field, BP was the first private firm to accept a pay rate of only $2 a barrel, thus setting a dangerous precedent. His counterpart from a rival company called Hayward to ask: "Tony, have you gone mad?"[33] From the Chinese point of view, the investments' profitability was secondary. What mattered was taking control of the flow of crude oil, wherever possible. For the new Titan of the world economy, it was imperative to ensure security of supply in the long term. At great expense (doubtless much less, however, than the cost of the war for the United States), China gained control of almost a quarter of the petroleum projects in Iraq. China was soon the recipient of nearly half of all the oil extracted in that country. "It is enough to say that if America went to war in the hope of securing cheap oil, we failed miserably," American economist Joseph Stiglitz recapped ironically.[34]

Amid Fears of Peak Oil the Bush Administration Pushes Back

While the Iraqi disaster took root, fear spread through the economic world. Recurring signs, each time more pressing, persuaded numerous experts from all sides that a decline in global crude oil production was likely to happen soon. In 2005 and 2007, Fatih Birol, scientific director of the International Energy Agency, the great oil oracle in the eyes of the media, warned that world crude oil production outside OPEC would decline "just after 2010" and the global oil market would hit a "wall" without a rapid reintroduction of Iraqi extractions.[35]

For five years, the international press had rippled with alarming analyses. They repeated the same phrase to the public: peak oil. Twenty-six "super-giant" fields had been discovered between 1950 and 1980, and only four since that time. The last of these was located in the Caspian Sea. This new "elephant," Kashagan, perhaps one of the last that remained intact on Earth, was particularly difficult to exploit, due to the sensitive geology and extreme climate of the territorial waters of the new Republic of Kazakhstan. Around the great, landlocked sea, the high-stakes game for the control of the former Soviet Empire's hydrocarbon reserves got tougher. This intensification was evident in BP's mighty entry into Azerbaijan (nicknamed "BP Land" three-quarters of a century after the Western companies had departed), discernible again in the backdrop of the horrendous counterinsurgency operations Russia's Vladimir Putin conducted in Chechnya.[36] But despite all fatally sharpened appetites, American diplomacy confirmed in November 2002 that the Caspian Sea, the last area to attract prospectors and emissaries of major oil groups following the fall of the USSR, was much poorer in black gold than what had been promised by promoters.[37] Hardly anyone in the industry could ignore that nine-tenths of the sedimentary oil sands basins around the globe had already been exploited, intensely, for many decades.

And it was at this point that the economic pages of the *New York Times*, after two years of very worrisome decline in crude oil production, reported the unreserved declaration made by Exxon's CEO himself, Lee "Iron Ass" Raymond, just weeks before the beginning of the Iraq war, while at an energy conference in February 2003: "When we consider that as demand increases, our existing base production declines, we come squarely to the magnitude of the task before us. About half the oil and gas volume needed to meet demand 10 years from now is not in production today."[38] Knowing that it takes approximately ten years to launch the production of any new field, the alert broadcast by the head of the world's number one major oil company did not go unnoticed in Houston or in London, and was loud and clear in Washington. About six weeks after the invasion of Iraq began, a US naval intelligence officer made an impromptu visit to geologist Colin Campbell—the Cassandra who had invented the expression "peak oil," and the leader of those who announced the imminent end of cheap oil—in the small south Ireland village where Campbell had retired. "You know that the admiral is very interested in what you guys are doing," said the American military spy who wore, as one might expect, a trench coat, black glasses, and a fedora that hid part of his face.[39]

On January 9, 2004, the entire petroleum planet shuddered when a scandal broke and confirmed the well-founded fears that until then had been fleeting. On that day, Shell suddenly announced that they had overstated their reserves of exploitable hydrocarbons. Immediately, Shell's stock plummeted. The company recognized that a fifth of its "proven" reserves had vanished. The production manager had been warning his directors of this fact for quite some time. He was quickly shown the door and left Shell with an iron-clad confidentiality agreement and a golden parachute: a check four times higher than his CEO, Philip Watts, received when he, too, was shown the exit door. At the root of the deception was an attempt to hide the oil peak reached by Oman, a second-tier producer since the early 1970s—a sophisticated, ancient nation located at the end of the Arabian Peninsula, southeast of Saudi Arabia.* In 1997, the wellhead pressure of Oman's main oil field, Yibal, operated by Shell, began a sharp decline. The British-Dutch firm deployed the newest techniques in its arsenal, including horizontal drilling, and sent more pressure underground to try to forestall degeneration. In vain. Like the North Sea, Oman had begun a geological decline. Technological progress could then accomplish nothing.

Confronted with such bad news, Bush administration analysts naturally looked to their old ally, Saudi Arabia, who held one-quarter of the planet's reserves. In April 2004, the US Department of Energy published a report in which it presented a table that showed a mind-boggling progression of Saudi Arabian extractions, purported to be capable of achieving 22.5 Mb/d, no less than double the Kingdom's production capacity. The Saudis' informal response was scathing, even humiliating. Referring to Washington's report, geologist Sadad al-Husseini, who had just moved on from his post as Saudi Aramco's executive director of extraction operations, declared, "The whole industry laughs at it."[40] There was "more chance" that he would be "living on

---

* Having previously nearly ignored Oman's existence, in 1974 Washington was advised of the urgent need to appoint an ambassador to Muscat, Oman's capital, after David Rockefeller, returning from Oman, sent word to Henry Kissinger. (See David Rockefeller, *Memoirs*, op. cit., p. 298.)

the moon" by then, insisted this perfect gentleman from Syria, a graduate of Brown University, and considered by some to be the voice of the engineers from Saudi Aramco (now the most powerful oil company in the world, by far).[41] Sadad al-Husseini added that the decline in the production of crude oil fields that already existed around the world would accelerate, inevitably. "Should we be worried? Yes," al-Husseini responded, sitting across from the minister of petroleum in an upscale neighborhood of Dhahran, Saudi's oil capital.[42] A few months later, the Bush administration scaled back its projection of Saudi future production to 16.3 Mb/d. Once again, Sadad al-Husseini replied that even the largest tree in the world, one that grows in the desert of Arabia, cannot rise up to the sky. Through the press, he indicated that Washington's new estimate was still "huge," and specified: "The Kingdom today lacks the capacity to mount to 15 Mb/d. It would take years and years to get there. . . . If we succeed, say, in 2015, this would be tenable for ten to fifteen years before decades of declining output begin."[43] This resounding false note exposed the great gap that separated the actual gross production capacities and the potential needs, ever increasing, of the ensemble of industrial nations. On April 25, 2005, Crown Prince Abdullah, who was preparing to succeed King Fahd, spoke at length with Vice President Dick Cheney and President George W. Bush at Bush's ranch in Crawford, Texas.* Paramount was the question of future developments in Saudi production in the face of escalating barrel prices.[44] On this occasion, before photographers, Bush ostensibly held hands with Saudi Arabia's future king.

The following month, it was a Texan who completely sent the derrick up in flames. An independent Houston banker specializing in oil, Matt Simmons, published *Twilight in the Desert*, a firebrand book in which he purported that Saudi production itself was on the brink of decline![45] Simmons enjoyed a solid reputation in the heart of Big Oil. He was among the members of the Bush administration's Energy Task Force, and had been one of the experts consulted by the Baker Institute for the 2001 report that presented the threat Saddam Hussein's Iraq posed for the global oil market. Simmons's prognosis of an imminent cataclysm was not realized. His book was nonetheless sufficient to transform, for a time, the confidential question of peak oil into a minor but intense media phenomenon. To anyone who listened, Simmons repeated: "Since the nationalization of Aramco, no foreign geologist has been able to verify any of the data" in Saudi Arabia.[46] Cornered, Saudi Aramco lifted part of the veil. There was bad news: Ghawar, the largest field on the planet, had exhausted 48 percent of its reserves. To compensate for the decline of a number of its old

* Victim of a stroke ten years earlier, King Fahd died in Riyadh on August 1, 2005.

wells, Saudi Aramco had to drill enough each year to provide nearly 1 Mb/d of new capacity. Officially, Aramco preached optimism and, more precisely, a proactive approach. From May 9 to 18, 2005, at the same time Simmons's book landed in US bookstores, the bulk of Aramco's board of directors traveled from Wall Street to Houston in order to raise at least $623 billion of foreign investment for the next few years, primarily destined for the development of Saudi hydrocarbon resources.[47] Coincidentally, on May 19, in Lisbon, the Association for the Study of the Peak Oil (ASPO) held a symposium quietly attended by observers from various major industrial groups. At the end of May, while Iraqi production was still stagnant and US production began a new severe downturn, barrel prices for the first time rose substantially above the $50 mark. This price had been considered totally implausible as recently as a year or two earlier, by the most prominent centers for analysis (apart from ASPO).

At the White House, far from the cameras, the peak oil hypothesis had become a throbbing concern. In February 2005, American physicist Robert Hirsch, the former head of the US nuclear fusion program and a former Exxon executive, delivered to the US Department of Energy the Bush administration's first and last report explicitly considering the question of the decline of world crude oil production.[48] According to Hirsch and his co-authors, peak petroleum would very likely become a reality by 2025, perhaps much earlier. Its cost would run to thousands of billions of dollars for the United States alone. To prepare for it would take at least ten years, and would require the establishment of a draconian "crash program." As peak petroleum would approach, the price of liquid fuels, as well as their instability, would increase dramatically, and "without timely mitigation, the economic, social, and political costs would be unprecedented," wrote Robert Hirsch, who concluded: "Previous energy transitions (wood to coal and coal to oil) were gradual and evolutionary; oil peaking will be abrupt and revolutionary."[49] Doubtless to the chagrin of the White House, the message contained in Hirsch's report rippled across the internet and gained a wide following. Very quickly, Robert Hirsch recalled: "After the work we did on the 2005 study and the follow-up of 2006, the Department of Energy headquarters completely cut off all support for oil peaking and decline analysis. The people that I was working with at the National Energy Technology Laboratory were good people, they saw the problem, they saw how difficult the consequences would be, you know, the potential for huge damage, yet they were told: 'No more work, no more discussion.'"[50] Thus, what Hirsch termed a "conspiracy of silence" took effect.

With or without the Hirsch report, the signs were clear enough to make peak oil a primary recurring theme. On June 23, 2005, in a Washington hotel conference room, two former CIA directors, Robert Gates and James Woolsey, participated with other former high-functioning American civil servants in a role-playing game called "Oil Shockwave." It was designed to foresee the consequences of supply shocks on the petroleum market. The "scenarios portrayed were absolutely not alarmist; they're realistic," said Robert Gates, who was appointed Donald Rumsfeld's successor as Pentagon chief the following year (a position that was renewed in 2009 by President Barack Obama).[51]

Until his second term ended, introducing new oil resources to the market was ever more pressing for Bush's presidency. During the eight years that George W. Bush occupied the White House, he and the Republican Party had, in vain, led an almost uninterrupted congressional fight to lift drilling restrictions in the Arctic National Wildlife Refuge, located in the extreme north of Alaska, near the declining wells of Prudhoe Bay.

In July 2005, when the barrel price had just crossed the $50 mark, Congress adopted a new energy law pushed by Dick Cheney. This act opened the door for two radical solutions. First, it planned a sweeping development of agrofuels, much faster than Europe's and far more massive than Brazil's; five years later, 40 percent of the United States' gigantic (genetically modified) corn crops were destined to produce ethanol, a "renewable" fuel then presented as making no impact on the climate.[52] Second, the July 2005 law exempted from the control of almost all environmental law a whole new, promising technique developed within the US industry: the hydraulic fracturing, or fracking, of the oil bedrocks associated with horizontal drilling. This exemption was nicknamed the "Halliburton loophole."[53] It catalyzed a new phenomenon that bore fruit after Barack Obama's 2009 arrival in the White House: the advent of shale oil and gas.

The 2008 Crisis: The Third Oil Shock?

The price of crude oil shot up as never before from 2005 to 2008, rising from nearly $50 to $100 per barrel, on average. This skyrocketing was amplified by the rush on new financial products that allowed investors to speculate on the apparently limitless increase in the price of raw materials, and of oil in particular.

In 2005, Shell figured among the pioneers of this juicy speculative market by creating a partnership in London with one of the sector's leaders, ETF Securities. For Shell, it amounted to making a kind of oil loan to investors, who in exchange loaned money to the old oil majors that lacked the capital necessary to interrupt the decline in their reserves.[54] The largest business banks followed suit, diverting an unprecedented flow of capital to the petroleum market and contributing to inflating the bubble that soon burst. As they had throughout history, the giants of finance and oil often shared directors. BP and Goldman Sachs International shared the same honorary president until 2009: Irishman Peter Sutherland, who was former European commissioner for competition and former director general of the World Trade Organization. This businessman, approaching his sixties in 2005, also sat on the board of directors of the illustrious and sprawling Royal Bank of Scotland, which proclaimed itself the "Oil & Gas Bank" until it teetered on the brink of bankruptcy in 2008, just before it was saved, in extremis, by the British government. Sutherland was also a member of the steering committee of the Bilderberg Group, along with nearly one-hundred-year-old David Rockefeller, Italian economist and politician Mario Monti, and neoconservative American Richard Perle.[55]

Below the bustling speculative market lay the rampant growth of the "emerging" economies, with China in the lead, which pushed the petroleum market to rise ever further, pushing it nearly to a breaking point. On the face of it, world crude oil production had stagnated and even seemed trapped in a downward spiral: It had been pulled toward the bottom by the United States and the North Sea, and decreased almost 2 Mb/d between May 2005 and September 2007, a drop close to 2.6 percent.[56*] This drop was even more disturbing than that of 2001–2002. Yet the valves remained wide open everywhere, except where impeded by political restrictions in Iraq or by security concerns in Iraq and Nigeria.[†] Clearly, the industry's operating margins were diminishing.

Because of its gravity, the 2008 economic crisis was often compared to the crisis of 1929. Perhaps it distinguished itself by a fundamental difference. While in several respects the stock market crash of 1929 was a crisis of new abundance, it is possible that this crisis revealed a collision with a certain physical, ecological limit to economic growth.

In 2008, barrel prices began by transcending the $100 mark, while in the United States the subprime mortgage crisis punctured the enormous bubble of the real estate market, dragging all financial institutions in its wake. In January, and then in May, President Bush implored Saudi Arabia's King Abdullah to loosen the vise by increasing Saudi Aramco's production. In vain. The airlines,

* This decline went unnoticed in the press, doubtless because it was masked by the takeoff of agrofuels. Indeed, during this period, the total production of oil and its substitutes stagnated but did not decrease.

† See in particular the explanations provided by Christophe de Margerie, president of Total, before the French Finance Committee of the National Assembly on June 4, 2008.

faced with a rise in jet-fuel prices, were among the first to suffer, including the prestigious American Airlines, Quantas, and SwissAir. The situation reached a breaking point that summer. After briefly negotiating the barrel price to $147 on July 11—an absolute record—the value of the barrel collapsed, precipitated by the recession and the fiasco of speculative capital. In December, prices fell to around $45. A phantom fleet of empty tankers—counting more ships than the US Navy—anchored off the coast of Singapore, pending a price escalation.[57] It didn't take long. During this crazy year, 2008, hunger riots erupted in many poor countries (Egypt, Tunisia, the Sudan, Somalia, Yemen, Cameroon, Mozambique, Haiti, and India, among others); some economists didn't hesitate to contemplate the quasi-parallelism between evolving oil prices and the evolution of the food price index published by the FAO.

Did there exist a cause-and-effect link between the continuing rise in oil prices, interrupted by the crisis of 2008 (in ten years, the value of the barrel had been multiplied by seven), and the outbreak of the so-called subprime crisis? It is quite possible. The first viable explanation is that the costly rise in the price of filling gas tanks and heating homes broke the budgets of many modest, indebted American households, some of which may have been forced to miss their mortgage payments. Plausible, this link however has barely been substantiated by solidly researched studies or mentioned in the press. A second explanation lies in the link between the oil prices and interest rates. The evolution of crude oil prices has an impact, direct or indirect, on the price of countless commodities. From 2004 to 2006, the US Federal Reserve (the Fed) continued to raise, bit by bit, its principal interest rate, the rate on which it bases the credit it grants to commercial banks. The aim: to fight inflation. This prime interest rate (the federal funds target rate) bears directly on the lending policies of the banks, and at the same time on the ability of borrowers to repay their credit at a variable subprime rate. In less than two years, the prime rate went from 1 percent to more than 5 percent, until it exploded the US housing bubble. Now, almost every time the members of the Fed decided to raise the prime rate, they explicitly referenced the increase in energy prices (in other words, the barrel price) as an essential factor of the inflation they purported to be fighting.[58]

The increase in barrel prices strongly contributed to inflation, which then triggered an escalation in interest rates, an increase that caused the explosion of the subprime bubble: The possibility of such a chain reaction has captured the attention of only a small number of economists. It appears, however, as logical as it is well documented. If validated, this would make the 2008 crisis the third oil crisis; the burst subprime bubble and ensuing financial crash would be merely its results, its symptoms.* This would not have been a sudden oil shock triggered by obvious political circumstances but a crisis that came from far away; not a steep groundswell but a long and powerful swell caused, however, as in 1973, by colliding with a new historic threshold, not limited to the United States this time but throughout the world, allowing access to still-intact, expensive reserves of oil, unconventional and extreme.

Implausible? The 1973 oil crisis, we have seen, had very serious consequences on inflation and the credit market, which just missed bankrupting Italy and did splinter the nascent and already very speculative mortgage credit market, nearly causing the failure of New York City and the Chase Manhattan Bank, bailed out in extremis by American taxpayers. It is notable that the increase in barrel prices had in the past an almost systematic, deleterious impact on economic growth. For example, since 1945, of eleven US economic crises of varying severity, ten were associated with a strong increase in oil prices, while eleven out of twelve major fuel price increases were accompanied by economic crises.[59] More specifically, nine of these crises were *preceded* by soaring pump prices.

The International Energy Agency: In a Cold Sweat

With the petroleum market white-hot and Wall Street groping in the dark at the start of the fatal summer of 2008, the International Energy Agency's (IEA's) experts struggled to keep their cool behind the large windows of a concrete building on the banks of the Seine, in Paris. They were in the process of finalizing their annual report, doubtless the most anticipated report in the history of the IEA, which had been created by the rich countries in the wake of the first oil crisis, at Kissinger's behest, and was primarily financed by the United States. In the agency's spacious offices above the Australian Embassy, near the Eiffel Tower, the atmosphere was anything but optimistic. The tension was distinctly palpable. The IEA already had become much more disillusioned over the past years, and—just like the US Department of Energy—had never stopped revising its downward forecasts of future crude oil production, while revising its price predictions upward. The succession of events in 2008 generated further concern. If the demand for oil had a tendency to slow down in the rich countries, its growth was furious in China, India, Brazil, and elsewhere. The price

* One can see the stabilization of world crude oil demand and its exploding prices in the course of the years that followed (between $100 and $120 a barrel) as a massive adjustment of the world economy in order to restore the balance between crude oil supply and demand.

of crude oil had never been as high, and investments were unprecedented, yet extractions were stagnating and even declining. Bizarre. Was the "end of cheap oil" that Colin Campbell and Jean Laherrère predicted in 1998 about to occur?

In order to strengthen the report scheduled to be released in November 2008, the man responsible for the IEA studies, Fatih Birol, a Turkish economist and former OPEC employee, considered it necessary to publish an extremely thorough analysis of the state of world oil production. This was a first. The initiative aroused curiosity. Insiders whispered that the drafting of IEA reports was under the scrupulous, confidential control of a handful of discreet bureaucrats who conducted their careers between the oil industry and international diplomacy. The IEA was responsible for taking the pulse of the global energy infrastructure for the investors. The methodology of their diagnosis was seldom clear. "It is never good to say too much to the patient," joked one expert. Until that time, no report released by the IEA had contained even a hint of an alarmist message. If such a message existed, it was more or less cryptic and buried among hundreds of pages: strictly reserved for insiders.

But this time was different; the risk was too urgent. At only fifty years old, with a round face and a warm, soothing voice, a fan of the Turkish Galatasaray soccer club, Fatih Birol could not have been more aware of the problem of peak oil, which he finally acknowledged, after eluding it for a long time. He had participated in the preparation of the 1998 IEA report, in which Campbell and Laherrère's analysis had blazed a trail. Birol embarked on the analysis of no fewer than 798 of the world's petroleum fields, relying on a database reinforced for the occasion. A whole chapter was assigned to this analysis, with an unprecedented level of detail. The balance sheet: More than half of the fields studied (precisely 479) were already in decline, at a dizzying average rate of 5.8 percent per year.[60]

In Washington, Fatih Birol's plan to publish this information was ill received. Several times during the summer, unofficial emissaries traveled to Paris in order to firmly encourage the IEA to water down its message. The pressure was hard to withstand.* The Bush administration appointed a new deputy executive director to the IEA. Richard Jones, a highly experienced diplomat, arrived before the summer's end, several weeks before the agreed-upon date. Jones, a former oil analyst at the American Embassy in Saudi Arabia during the delicate time of the oil countershock, and former ambassador to

* According to stories that provide consistent evidence from two sources who wished to remain anonymous, notably for security reasons, one in Paris, the other in Washington (interviews with the author in 2011 and 2014). The existence of the Bush administration's pressures on the IEA was revealed by *The Guardian*. (See Terry Macalister, "Key Oil Figures Were Distorted by US Pressure, Says Whistleblower," *The Guardian*, November 9, 2009.)

Kuwait and Kazakhstan, among other countries, had assisted Paul Bremer in Baghdad in 2003 and 2004, before taking over the coordination of the Bush administration's policy in Iraq in 2005. Upon his arrival in Paris, he dove into projects that were already underway.

After many adjustments, the summary of the report finally provided to the press on November 6, 2008, was a skillful mix of alarm signals with unheard of magnitude ("Current global trends in energy supply and consumption are patently unsustainable") and mollifying counter arguments ("The world's total endowment of oil is large enough to support the projected rise in production beyond 2030)."[61] While the price of oil dove beneath $50, the IEA still promised, correctly, that there would be a return to three-digit barrel prices. To describe the fight against the existing production decline, the agency referred—with a question mark—to the need to run "faster to stand still."[62] The real stark news had to wait two years before becoming public.

Drill, Baby, Drill: The 2008 Presidential Campaign in the Land of Black Gold

On November 4, 2008, two days before the IEA report was published, Barack Obama was elected president of the United States. He succeeded oilman George W. Bush, who two years earlier, during ongoing unrest in Iraq and just after launching the most ambitious of agrofuel development programs, had pronounced the obvious in his State of the Union address, declaring: "America is addicted to oil." The future president Obama was no less forthright when, on a frosty February day in 2007, in the middle of an impassioned, wholesome speech during which he officially declared himself a candidate for the White House, he proclaimed: "Let's be the generation that finally frees America from the tyranny of oil!"

At the beginning of the election campaign, Democrat Barack Obama's Republican opponent, John McCain, passed for a candidate favorable to the environmental cause, breaking ranks with President Bush. Like Obama, McCain supported the creation of a cap-and-trade system for greenhouse gas emissions. In the land of black gold, people had been deeply shocked by news of Hurricane Katrina, which devastated New Orleans in August 2005. Along the Louisiana coast, where hundreds of small, abandoned oil platforms had been built on oil fields that had run dry, many perceived the ravages of Hurricane Katrina as clear harbingers of climate change. Yet McCain proposed in April to suspend the tax on fuels, and then he followed up with a genuine about-face on June 16, 2008, when the barrel price exceeded $120. In a speech in Houston, the Republican candidate said he was in favor of abandoning a twenty-five-year moratorium on offshore drilling throughout a large portion of the American coast. In the weeks to follow, his campaign coffers were filled like never before, thanks in particular to donations from Texas.[63]

Out of the blue, an unexpected vice-presidential candidate appeared by McCain's side. A former beauty queen who had become governor of Alaska, Sarah Palin at the time appealed, paradoxically, to both grassroots white conservatives and powerful industrialists; after their defeat, she became an emblem of the pro-business, populist Tea Party and was financed, among others, heavily and silently, by two American fossil fuel billionaires, the Koch brothers.[64] Alaska, Sarah Palin's home state, presented another paradox. Its economy and population were largely financed and subsidized by petroleum royalties from the declining Prudhoe Bay, and it was also the area of the world most unquestionably affected by the warming climate. In the north, Inuit fishing villages were being eroded by the rising Arctic Ocean. To the south, the endless spruce forests were dying, gnawed away by an influx of bark beetles. Everywhere, melting permafrost resulted in tens of millions of dollars in infrastructure damage each year, in particular for the spectacular pipeline that crossed the state from end to end. As governor, Sarah Palin supported opening the Arctic National Wildlife Refuge to oil prospecting. Her own husband was a BP oil operator in Prudhoe Bay. Prudently, Palin appeared agnostic regarding the changing climate (this changed after the election). During a televised debate, in the heat of the campaign, she definitively reached celebrity status by taking up the risqué slogan that for some time had circulated through the Republican camp: the famous "Drill, baby, drill."

Among the victims left to bleed dry by the 2008 crisis, Detroit suffered more than most. Nicknamed "Motor Town," and having enjoyed the highest standard of living in the United States during the 1960s, the city saw its seminal company, General Motors, file for bankruptcy on June 1, 2009, following another local automobile giant, Chrysler, which had filed on April 30. Washington's decision to bail out the automobile giants, the incarnation of powerful, industrialized capitalist America, was one of the Obama presidency's first salient acts. Four years later, the municipality of Detroit itself suffered bankruptcy, as New York City had nearly done in 1975. The heart of the world's automotive industry virtually ceased to beat. In New York, the heart of finance, thanks to sensational monetary trickery, had narrowly escaped failure. Among the factories lying in ruins in Detroit, some optimistic environmentalists hoped to see the seeds of an agrarian, postindustrial urban society.

Another spectacular decision was made by Barack Obama at the beginning of his presidency: On a military air base, before one of the fighter aircraft equipped to operate partly on ethanol, Obama announced that he would lift the moratorium on drilling off the American coast—the wish of his opponent, John McCain. Scarcely three weeks later, on April 20, 2010, the semisubmersible Deepwater Horizon platform, holding a record for drill depth, exploded in the Gulf of Mexico off New Orleans, killing eleven people.* Several executives from BP, who leased the platform, were aboard to celebrate a safety bonus. After burning for two days, the Deepwater Horizon sank, perhaps because the Coast Guard had made the mistake of allowing a dozen ships to try to extinguish the fire with seawater and had fatally overfilled the ballasts.[65] The well shaft broke. Thus began the worst black tide in United States history.

* In August 2009, the Transocean company, owner of the platform that at that time operated at a depth of 1,260 meters in the Gulf of Mexico, drilled a well more than 10,680 meters deep for BP. (See "*Deepwater Horizon* Drills World's Deepest Oil and Gas Well," *Beacon: Transocean in the Spotlight*, Fall 2009, no. 3, searchable at beaconmag.com/deepwaterhorizon.html.)

THIRTY

Winter, Tomorrow?

And the truth of the morning will be the error of the evening.

—Carl Gustav Jung[1]

Our solution is our problem.

—Richard Heinberg[2]

It took BP eighty-seven days of colossal effort to seal the Deepwater Horizon well head, which had been leaking at exceedingly high pressure 1.6 kilometers below the surface. The world watched as a series of failed attempts, some shockingly ridiculous, exposed how the quest for new crude oil sources had become extreme.

The great majority of extraction projects were located further and further offshore, in ever deeper waters. The most recently developed area was situated almost 200 miles from São Paulo, Brazil's economic capital. Drill bits descended through nearly 2 kilometers of water, then drilled through 3 kilometers of rock and another 2 kilometers of salt. Ultratechnical and financially risky prospecting sites became the norm, necessary to compensate for an ever increasing decline of oil production.

From the time of the Gulf of Mexico oil spill until autumn 2014, the barrel price climbed, reaching and staying at an unprecedented level: A large number of new crude oil sources were demanding a barrel price around $100, at minimum. But the 2008 crisis clearly demonstrated that, beyond the $120 barrel price, on the demand side there were also problems. And it is this supply-and-demand stranglehold that may constrain the next chapter of humankind's history with oil. There are still a lot of hydrocarbons in the ground. However, as Fatih Birol, executive director of the International Energy Agency aptly summarized, the question "is not the size of the glass, but the diameter of the straw."

The Peak of Conventional Oil Is Confirmed

Ever more constrained, barrel prices henceforth will strictly reflect the limits of industrial capacity (technically, physically, ecologically) to procure enough energy, at a sustainable price, to fulfill the demands of this supposedly indispensible growth. If the oil price drops too far, many new extraction projects will be halted and production will decrease due to the so-called natural decline of existing wells, and therefore, sooner or later, the price will rebound. If the price rises too much, the economy will quickly stall, along with new extraction projects and so forth. According to many analysts, world oil production is expected to be an "undulating plateau," extending until it reaches its long, inevitable final descent.

The collapse in crude oil prices that began in June 2014 was the consequence of, on the one hand, a global growth slowdown and, on the other, the surprising boom of a new form of nonconventional shale oil. This price collapse promised to renew growth, saving each motorist a few hundred dollars per year, which economists predicted would lead to the purchase of other consumer goods (and, eventually, a new car).

Additionally, the fall of the barrel price posed a great threat to the sustainability of oil-dependent industries, since it undermined the oil executives' capacity to invest in new projects and thus compensate for the so-called natural decline of existing production. In its November 2014 annual report, the International Energy Agency warned: "Given the long lead times for upstream projects, the consequences of a shortfall in investment may not be apparent for some time. But clouds are starting to form on the long-term horizon for oil supply, holding out the possibility of stormy conditions ahead."[3]

An endemic disease should not be confused with its temperamental, cyclical symptoms. A low oil price may not suffice to revive sustainable global economic activity. What economic growth requires is more energy. In this case, our daily medicine is a rectangular column of black gold, 100 meters by 1.4 kilometers: 90 Mb/d, the approximate worldwide consumption of crude oil and its substitutes. Consider the year 2010: In Washington, Obama administration experts were perfectly aware of the growing difficulty of finding abundant new oil sources. The Department of Energy's most prominent oil analyst, Glen Sweetnam, recognized a month before the Gulf of Mexico oil spill that "a chance exists that we may experience a decline" of world liquid fuel production between 2011 and 2015 "if the investment is not there."[4]* A study Sweetnam had produced a few months earlier

* A few days after the publication of this interview, to the surprise of his colleagues in the Department of Energy, Glen Sweetnam was given a strategic confidential job: senior director for energy at the National Security Council.

recognized a strong drop in world production starting as early as 2012, unless "unidentified" projects capable of filling the shortfall were put into production.[5] Before Obama appointed him secretary of energy, Steven Chu (a Nobel Prize–winning physicist), reported the disturbing predictions of the Association for the Study of Peak Oil (ASPO).[6] In 2008, and again in March 2010, two future-oriented reports of the US joint chiefs of staff concluded that the market could suffer a shortfall equal to the amount of Saudi yearly production by 2015.[7] Pentagon analysts missed on a grand scale with this report; but their error was instructive. The caveat included in the 2010 report was valid: "A severe energy crunch is inevitable without a massive expansion in production and refining capacity. [I]t surely would reduce the prospects for growth in both the developing and developed worlds. Such an economic slowdown would exacerbate other unresolved tensions, push fragile and failing nations further down the path toward collapse, and perhaps have serious economic impact on both China and India. At best, it would lead to periods of harsh economic adjustment." They concluded, "One should not forget that the Great Depression spawned a number of totalitarian regimes that sought economic prosperity for their nations by ruthless conquest."[8]

Several other reports of this ilk followed in 2010, while after months of air escaping the subprime crisis bubble, oil prices rose sharply. One of these alarmist reports was co-published by Chatham House (the old London publisher of the British geostrategy) and Lloyd's (one of the oldest, most prestigious insurance companies in the world).[9] Another emanated from the German army's chiefs of staff.[10] An HSBC Bank study concluded that the last drop of oil could be exhausted as early as 2060.[11]

The International Energy Agency rounded out the picture with its annual report, published in November 2010. This agency of the OECD, located in the fifteenth arrondissement of Paris, produced the clearest warning, thoroughly examining the situation that had been incompletely articulated two years earlier at the dusk of the Bush administration. The peak of "conventional" or "classic" crude oil production, which constitutes four-fifths of the world's production, had now been reached. This production would never increase again, insisted the IEA: Any future increase would have to come from new, extreme, and "unconventional" sources. The founders of ASPO, Colin Campbell and Jean Laherrère, were right in their 1998 article, "The End of Cheap Oil." The story of petroleum had indeed been shaken somewhere between 2006 and 2008, and was ushering in a new, fateful era.* On October 8, 2010, at the annual

* In its 2010 report, the IEA placed the peak of conventional oil in 2006, at 70 Mb/d. It corrected this in its 2012 report, naming the date as 2008, also at 70 Mb/d, and explicitly excluding from its accounting the shale oil and other forms of "tight oil" that soon followed; chemically, these are like conventional

convention of ASPO-USA in Washington, James Schlesinger, the former head of the CIA and the Pentagon for whom President Jimmy Carter had created the US secretary of energy office, stated that the debate on oil peak was over, and "the peakists have won."[12]

Transition, Climate, Subsidies: Big Oil 3 / Barack Obama 0

Just as the summit of an airplane's arc gives the illusion of weightlessness, a suspension point appeared to have been reached at the hour of these calculations, accurate or not. On June 15, 2010 at the White House, while the black tide that followed the *Deepwater Horizon*'s explosion in the Gulf of Mexico flowed on, Barack Obama delivered an exceptional televised speech. "After all," noted the president, "oil is a finite resource." And then he laid it out: "We consume more than 20 percent of the world's oil, but have less than 2 percent of the world's oil reserves. And that's part of the reason oil companies are drilling a mile beneath the surface of the ocean—because we're running out of places to drill on land and in shallow water. For decades, we have known the days of cheap and easily accessible oil were numbered. For decades, we've talked and talked about the need to end America's century-long addiction to fossil fuels. And for decades, we have failed to act with the sense of urgency that this challenge requires. Time and again, the path forward has been blocked—not only by oil industry lobbyists, but also by a lack of political courage and candor. And today, as we look to the Gulf, we see an entire way of life being threatened by a menacing cloud of black oil. We cannot consign our children to this future." As emphatic as it was, this speech poorly masked the persistence of a sincere ambivalence. "I believe that we're going to need to increase domestic oil production," President Obama had admitted during an interview the previous week, as he prepared public opinion for the inevitable resumption of offshore drilling.[13]

On July 22, US senators killed, once and for all, a proposal Obama had defended to establish market quotas for greenhouse gas emissions. The naive planetary hope born in December 2009 during the International Conference on Climate Change in Copenhagen was also killed: The world's number one petroleum consumer refused to establish a serious policy to combat the anthropogenic heating of the atmosphere. Obama was left in the lurch by his own political party—due to the crisis. Faced with the consequences of

---

petroleum, but the IEA considers them nonconventional petroleum due to their specific and sophisticated mode of production. (See IEA, *World Energy Outlook 2010*, "Executive Summary," p. 6, and *Word Energy Outlook 2012*, p. 81.)

the worst recession since the time of the 1970s oil crises, the majority of US lawmakers could not support such constraint on American industry in general, and on fossil fuels in particular. Isolated within his party, John Boccieri, a newly elected Democrat who favored market quotas, bitterly complained that he was tired of fighting "petrol dictators and Big Oil." The price of remaining in power is measured in many ways: In October 2010, Washington announced a new program of arms sales to Saudi Arabia, a record amount of $60 billion.

On January 25, 2011, midway through his first term, Obama launched what already appeared to be his last stand on the energy and climate change front. That evening, opposite Congress, in his annual State of the Union speech, the president attempted to invoke what he called a "Sputnik moment," encouraging Congress to "eliminate the billions of dollars" that American taxpayers paid to oil companies every year.* "I don't know if you've noticed, but they're doing just fine on their own. So instead of subsidizing yesterday's energy, let's invest in tomorrow's." The CNN news team mocked, "Good luck. Obama has been down this road before and run out of fuel." In fact, it failed again. The American subsidies intended for the energy companies continued to progress, and exceeded $21 billion in 2013, according to one report.[14] Those in the United States who designed an alternative to "business as usual," however, got a boost in November 2011 when Obama suspended the Keystone XL Pipeline project, intended to carry crude oil from the Canadian oil sands to refineries in Houston, a distance of 3,400 kilometers. This was the only tangible (though short-lived) policy victory of the Occupy Wall Street protest movement that had launched that month: Dependence on fossil energy then became the number one target of anticapitalism.

Nonconventional Petroleum to the Rescue?

It was a crossroads. Just as the IEA had announced, the barrel price again rose above $100 in early 2011, allowing the highest investment levels on record. In ten years, from 2002 to 2012, the amount spent each year to seek and especially to extract oil and gas rose from $200 billion to $700 billion, double the central government revenue in France, seventy times the price of the Channel Tunnel or a nuclear reactor, or twenty-six times the budget of China's Three Gorges Dam. "Nonconventional" and extreme petroleum finally knew the long-awaited, thundering development Houston had promised.

* A Sputnik moment is a reference to the beneficial impact for the American space program, following the Soviet Union's sending the first artificial satellite into orbit in 1957.

In 2010, the Canadian oil sands were the number one source of petroleum imports for the United States. On December 12, 2011, the conservative Canadian government of Stephen Harper announced that the great northern neighbor of the United States had formally renounced its commitments to the Kyoto Protocol regarding the reduction of greenhouse gas emissions. Canada not only had originally ratified this protocol but had actively participated in the international community's 1997 effort to adopt a single comprehensive framework in the fight against global warming, a warming of the type that the boreal regions would be the first to face. The reserves of crude oil that Alberta's tar sands contained, tremendously exploitable thanks to the new price hikes, were estimated to be more important than those of Iran and Iraq. Those reserves were considered the third most important in the world, behind the conventional oil of Saudi Arabia and the "extra-heavy" oil of the Orinoco River belt in Venezuela, another gigantic nonconventional source that, for many reasons, remained delicate to exploit. Meanwhile, from Alaska to Colorado and even beyond, on both sides of the Rocky Mountain border, bark-eating beetles swarmed in response to the warming atmosphere and killed millions of acres of forest (where the already dead wood was then burned during summer fires of frightening magnitude). East of the Rockies, in the plains that surrounded Canada's Athabasca River, large swathes of forests were razed to make room for the titanic excavating machines and 400-ton trucks of Exxon, Chevron, Shell, BP, Total, Canada's Suncor and Syncrude, and China's Sinopec and Petrochina.[15] Upstream of the Athabasca River, whose water was pumped to generate steam necessary for oil sands extractions, the Athabasca Glacier retracted at a "surprising" pace.[16] Further north, the melting permafrost at times created spongy, steel-blue pools as far as the eye could see.

Another unconventional source of oil production, in discrete development for several years throughout Texas, Louisiana, North Dakota, and the wooded hills of Pennsylvania (where the first forest of oil derricks had grown a century and a half earlier), suddenly ramped up significantly in 2011. The hydrocarbon boom from "shale" or "tight" oil had begun. The story began a decade earlier.[17] In Texas, during the winter of 2002, a large independent company, Devon Energy, started to deploy a new method of extraction, on a grand scale. This method had been developed for years within the corporation of a brilliant, independent oil man, George P. Mitchell. Devon purchased the corporation for more than $3.5 billion in 2001. Invented by an engineer named Nick Steinsberger, the method combined techniques that had long been used in the oil fields: horizontal drilling and the hydraulic fracturing of rock. Resorting to the careful mixture of highly pressurized water, sand, a resin from a red bean cultivated in India for livestock (guar), and a cocktail of various chemicals, a multitude of independent petroleum companies followed Devon Energy's example. They fractured layers of bedrock that had not yet released all their hydrocarbons into the permeable rocks that the conventional industry targeted. Thanks to sophisticated trials, engineers found favorable zones, the "sweet spots" inside or in the immediate periphery of these tight bedrocks—in order to create, a few meters around each drilled drain, microcracks the thickness of a human hair, through which the hydrocarbons could rise to the surface from a depth of approximately 2.5 kilometers.[18]

What became the shale gas and oil revolution caught much of the world off-guard, beginning with the directors of the major oil companies. In 2003, Exxon's CEO, Lee Raymond, had said that natural gas had reached its production peak in the United States.[19] In effect, after it started to fall in 1973 (three years after oil), the production of conventional gas in America after 2001 entered a second phase of decline that was of great concern to the Bush administration. It was precisely then that the first wells of shale gas took hold, first in the Barnett field (around and soon beneath the town of Fort Worth, Texas), and then in the fields of Haynesville, in Louisiana, and in Marcellus, Pennsylvania. Thanks to hydraulic fracturing, US natural gas extractions began to increase starting in 2005. Even more stunning: US oil halted in 2009 its four decades of slow agony, and in 2011 rose rapidly like reborn phoenix, launched into flight by a barrel price of $100. Here were the "unidentified projects" that Glen Sweetnam, the Obama administration's number one oil expert, had not seen coming in 2009.

The new manna rapidly grew a tangle of drilling equipment, the scale of which the United States had not seen since the 1930s. A region as isolated as the state of North Dakota metamorphosed into a place of fabulous wealth. On its road, one might cross paths with a petroleum worker traveling up from the Gulf of Mexico with his family, whose car had run out of fuel and who, naturally, begged for a few dollars to refill his tank and resume his journey. Because of the lack of sufficient pipelines, shale oil producers had to resort to trains to transport crude oil, as in Rockefeller's day. Accidents were not unusual. On a hot night in July 2013, a convoy of seventy-two cars of light crude oil (crude oil from bedrock is generally very light) caught fire in the quiet community of Lac-Mégantic, in Canada, killing forty-seven people.

The economic viability of fracturing, or fracking, tight hydrocarbons, like its ecological impact, is the source of much debate. By its nature, fracking often calls for much more drilling than conventional oil or gas to obtain an equivalent production in the long run. It only releases hydrocarbons through restrained passageways. Tight oil and gas production usually collapses after a few months (or a couple of years in the best cases) and the well becomes a "stripper," a marginal well that isn't very productive. To provide a high extraction level, it is therefore necessary to continuously drill new wells. Many experts critical of fracking see this as a fatal economic vulnerability. But the production of gas and oil from tight resources has not stopped escalating, advancing in great strides toward, and exceeding, peaks that have been considered irreversible since the 1970s. Perhaps the oil industry is simply in the process of losing its exceptional status, of becoming a normal industry, clamoring, like the coal mines, for a constant investment flow in order to maintain production, whereas drilling conventional oil guaranteed a fortune for many years.

The Western oil majors were largely on the side of the shale oil and gas revolution. They were also confronted with historic inflation of the investment needed to maintain their conventional oil extractions. Although just as cold-blooded as his predecessors, the rugged Rex Tillerson, CEO of Exxon for eight years before becoming President Donald Trump's secretary of state, exclaimed in June 2012 before the Council on Foreign Relations in New York, while speaking of the shale gas boom heating up the industry: "We are all losing our shirts today. We're making no money. It's all in the red."[20] Important independent companies in the boom's outskirts—Devon Energy and Chesapeake, among others—got into deep debt and were deemed unsustainable by many observers. The boom continued, nonetheless.

Where was the oil planet when this book went to press, in the spring of 2018? Is the oil-made man rushing forward toward an incredibly perilous forced weaning from oil, likely to start well before the politics of the energy transition can have serious impact? Although global, this issue is quite simple. It is about is rudimentary arithmetic. Shell's former CEO, Peter Voser, summarized it in terms corroborated by all of the major sources as well as statistical data: The production of existing fields declines by 5 percent per year as exploited reserves are produced and thoroughly depleted, so much so that the world would have to add perhaps as much as four Saudi Arabias or ten North Seas every ten years just to maintain the supply at its current level, before even considering a way to meet any increase in demand.[21]* Up to half of world crude oil production would need to be renewed with each new decade.

According to the IEA, any eventual growth of extractions, a growth necessary to feed the current machine of worldwide economic growth, has to

* The industry's reference source, IHS, renowned for its quality as much as for its optimism, identified the average rate of "natural" decline in existing production as between 5.6 percent and 3.7 percent per year, approximately the equivalent of just over the North Sea's yearly production. (See Peter Jackson and Leta K. Smith, "Exploring the Undulating Plateau: The Future of Global Oil Supply," *Philosophical Transactions of the Royal Society A*, vol. 372, no. 2006, January 13, 2014.)

come from unconventional and extreme sources, which not only are much more complex, difficult, and costly to exploit, but also require much greater energy than conventional crude oil sources for the extraction itself. The two main unconventional and extreme sources currently available? Tight oil and oil from tar sands and mother rock. There are, in addition, deep and ultradeep offshore wells, but their exorbitant cost and slow production pace limit their contributions. The strong ramp-up of the ultra-deep offshore production in Brazil, despite cost overruns and severe financial scandals, constitutes the only other significant good news in recent years for the worldwide oil industry, apart from the shale oil boom. Yet new offshore projects remain rare and tricky to fund. In 2017, only four countries account for almost 90 percent of deepwater and ultra-deepwater production: Brazil, the United States, Angola, and Nigeria.

At present, the other alternatives that can provide hydrocarbon fuel have little chance of increasing at a much greater rate than they have. The production of coal-based synthetic oils begun by Nazi Germany has hardly been developed further since the Second World War. The development that did occur was in the 1980s by South Africa, which, due to apartheid, was subject to an oil embargo imposed by the United Nations in November 1987.[22] China would have established large programs in this area but has not launched any on a significant scale, because the return on energy invested for the manufacture of coal-based or natural gas-based synthetic fuels is at best mediocre (on the order of 2 joules retrieved for 1 joule invested). The gigantism of necessary infrastructure provides another powerful brake on the industry. As for agrofuels: Given their competition with the food supply for source material, their need for abundant cropland, and their nonneutral impact on the climate, the production of ethanol and vegetable oils developed intensively during the last decade are likely to see relatively limited progress. Prowess in organic chemistry and genetic manipulation may still transform algae and micro-algae into gasoline or oil. Again, though, the requisite infrastructure and poor energy efficiency make it unlikely that this will be accomplished on a significant worldwide scale in the next two decades.

As for ultraheavy petroleum like that found in the Canadian tar sands, or even the extra-heavy petroleum of Venezuela and elsewhere: Expansion will probably remain constrained by the problem of profitability posed by the high energy cost of extraction and by the environmental impact of exploiting these resources on a massive scale. To justify the abandonment of an $11 billion project in Alberta in May 2014, the regional manager of Total said: "We are still in the cycle within this industry where cost inflation in general is going much faster than price adjustments."[23] The barrel prices had then lifted well above the $100 mark.

The Uncertain Future of Extreme Petroleum: Shale Oil and the Arctic Ocean

The future of the shale oil boom remains a great unknown. Though they came around slowly, the major oil companies believe in it strongly as of 2018. Total's scientific director, Jean-François Minster, has argued that "the theme of peak oil production is obsolete because of the shale oil boom in the USA and the prospects of this kind in many countries."[24] Never before have experts of this type been so divided. Richard Miller, author of an analysis published in January 2014 by the Royal Society in London, declared to the *Guardian*: "We are probably in peak oil today, or at least in the foot-hills." About petroleum from shale and tar sands, Miller, now retired from his position overseeing BP's prospecting, quipped ironically: "We're like a cage of lab rats that have eaten all the cornflakes and discovered that you can eat the cardboard packets too."[25]

Following a universal tendency, prospectors always attack the most easily accessible resources first. It will be more and more vital to compensate for the natural decline of the already fracked wells with wells that are necessarily more delicate. All this is happening very quickly. The gap between the output of stripper, or marginal, wells and the opening of new wells to maintain production levels widens as production increases. The resulting situation is similar to running on a treadmill that's going ever faster. The levels of US oil production anticipated in the scenarios published by the US Energy Information Agency remain extremely uncertain. The most pessimistic scenarios, under low-price and low-exploitable-resources assumptions, continue to show a resumption of the decline of production in the early 2020s, after reaching a peak at a level of production close that of 1970, due to the absence of enough easily exploitable sweet spots.[26*] But these same experts, who initially hadn't seen any of this coming, also put forth "high resources and technology" scenarios in which the extractions would grow without end, well beyond the peak of 1970. What's certain is that the future of tight-oil production remains very unknown. On the side of the glass half empty, it is worth noting that in 2014 Washington revised its initial estimate of recoverable shale oil resources in California: down by 96 percent. Or that extractions from the first two fields of shale gas developed in the United States—those of Barnett, Texas, and Haynesville, Louisiana—were in early 2018 far below the peaks each of them reached in late 2011.

"If there are ten or twenty cases similar to North Dakota on the planet, this will not raise the peak more than around 5 Mb/d and wouldn't delay the

* Estimates in early 2018 indicate that 1970 levels will be reached or surpassed within the year, albeit with nonconventional resources.

decline by more than four to five years," said Pierre-René Bauquis, former strategic director of Total and one of the pillars of the French ASPO.[27] In France, the Paris Basin might have the potential to produce 100,000 barrels of shale oil each day, but the French do not want that (as with the GMO cereals from abroad that feed their livestock, they prefer that others are responsible for production). Not that this could, in any case, offset the decline of several of France's major foreign historical supply sources. Initially, Poland seemed like a good candidate for another "miracle." But the larger companies, including Exxon, quickly threw in the towel. The resources were buried too deeply: They weren't profitable enough. China has been studying exploitable sites closely, but they are widely scattered, in areas that are difficult to access.[28] Geology imposes its conditions; it always has the last word.

Although isolated and distant, Argentina's desert plateaus, in the Vaca Muerta region, seem to offer better prospects and have already attracted several important companies, including Shell, Chevron, and Total. Last but not least, there is Russia. Below or close to fields of conventional oil and gas often pronounced in decline, the bedrock of Bazhenov, the most extensive on the planet, might have the most potential. Russian companies more or less directly under the Kremlin's thumb have been exploring possibilities with the Western oil majors, seeking expertise and capital: Rosneft with Exxon and Statoil, Gazprom with Shell, Lukoil with Total, and so on. The supposed flight of foreign capital driven by the crisis that broke out at the end of 2013 between Russia and the Ukraine may have affected the Russian petroleum industry less than it seems. For instance, in March 2014, while Berlin was displaying its firmness with regard to Moscow, the director general of the German group Siemens made the effort to shake Vladimir Putin's hand in order to assure him that his engineering firm would continue to collaborate with Gazprom. Yet the turbulence on the steps of the new Russian empire could not have reassured the bankers and insurers in the Western world. During this time in the United States (in North Dakota, Texas, and elsewhere) drilling and extraction techniques ceaselessly advanced and productivity improved rapidly, simultaneous with profit levels. However, in spite of its rampant development, which reached a production rate of about 5 Mb/d in 2017, American shale oil still provided only a small fraction of the crude oil that the world consumed that year: 96.5 Mb/d.

The ultimate oil horizon, once again, was located in Russian territory as well as to the north of Canada: the Arctic Ocean, accessible "thanks" to the melting ice caused by the warming climate (itself induced by the consumption of carbon-based fossil fuels). In a historic reconciliation both ironic and staggering, on August 30, 2011, Exxon's CEO, Rex Tillerson, under the watchful gaze of the former KGB officer Vladimir Putin, signed an agreement with the Russian state petroleum company Rosneft in the seaside resort of Sochi, beside the Black Sea. A day earlier, Putin had paraded in front of the Russian warships, riding a Harley-Davidson. The day after the signing, BP's offices in Moscow (which, incidentally, had been in the process of being pushed out by the Kremlin) were searched by police commandos armed with assault rifles. The event (according to the *New York Times*) doubtless reminded Exxon's leaders that they were "no longer in Texas."[29] An exploratory drilling campaign was to be carried out in the Kara Sea. St Petersburg's shipyards built floating barges equipped with nuclear submarine reactors in order to generate electrical power in an area devoid of infrastructure. In the event of a sufficient discovery of oil, the agreement provided for an investment plan that could go as high as $500 billion. But all of these projects were interrupted in 2014 when the Obama administration imposed strict sanctions on Putin's Russia for its invasion of Crimea. Wherever it lies, the ultimate polar horizon might in the future turn out to be a mirage. The technical (and therefore geological) risks are enormous, with what remains merely theoretical results. Further south, off the Newfoundland coast in an area known as "Iceberg Alley," where Total pioneered in the 1980s, oil tankers are forced to use tugboats to deflect the path of icebergs that frequently threaten to strike the platforms. The IEA does not foresee any significant production at the North Pole before at least 2035.[30] Warned off, perhaps, by the disastrous financial impact the oil slick in the Gulf of Mexico had for BP, Total has already officially abandoned the idea of drilling for oil in the Arctic Ocean. Shell froze its exploration projects in Alaska after one of its platforms, enhanced to cope with the ice, was destroyed, drifting to the coast in a winter storm on the last day of 2012.

Conventional Oil: A Global Race on the Treadmill of Natural Decline

Shortly after his 1999 arrival in power, Vladimir Putin began seizing the Russian hydrocarbons industry, halting the sale of assets to foreign companies, and appointing lieutenants and elders of the FSB (the former KGB) to all key positions. For Putin, oil was a personal matter, if we are to believe an old rumor, denied but nevertheless spread through the heart of American diplomacy (and then via Wikileaks), that the Russian president took a cut from each barrel exported from Russia, through the Swiss trading company Gunvor, which was once headed by a former KGB colleague. Putin's Russia unhesitatingly has advanced its pawns to gain hydrocarbon access (whether in Chechnya or Georgia), or to utilize these hydrocarbons, of which Russia is once again one of the primary producers, as unstoppable geostrategic weapons to intimidate Ukraine and the European Union, which both have a vital need for Russian oil and gas. At the end of 2005, Gazprom, the Russian natural gas giant, succeeded in recruiting former German chancellor Gerhard Schröder shortly after he left office, entrusting him with the development of Nord Stream, a major gas pipeline laid in the bottom of the Baltic Sea and leading into a German port. The same year, Putin made a clear example of Mikhail Khodorkovsky, the CEO of the giant Russian oil company Yukos, who was sentenced to eight years in prison for embezzlement, and locked behind a half-dozen armored doors in a Siberian penal colony near the largest uranium mine in Russia. The young, ambitious tycoon had without doubt committed an inexcusable error: When trying to sell Yukos shares to the American companies Exxon and Chevron, he had demonstrated a desire for raw political power.* In 2011, BP was shooed away from the Russian consortium TNK-BP: Rosneft's purchase of its shares had deprived the old English company of at least 40 percent of its production!

In such a context, why, since it was a question of developing arctic Russian petroleum and shale oil (both still hypothetical), did the Kremlin consent to invite several Western oil majors to dance at its ball, jealous of their capital and especially their expertise? Doubtless, because, like a number of other old producing countries, Russia, at the time the world's second largest producer behind Saudi Arabia, had seen its production of conventional oil—the classic liquid oil that constituted the majority of its production—threatened by decline. Total's scientific director, who believed so strongly in shale oil, Jean-François Minster, reported that in Russia, it was not unusual to find fields declining by more than 6 percent per year.[31] For several years, the IEA has been promising that Russia is bound for a slow but continuous decrease in extractions in the coming years due to a lack of sufficient readily exploitable resources.[32]

Conversely, what did the Western oil majors see in Russia, after the painful, extremely costly misadventures of BP (and others)? Doubtless they had no choice: The cumulative crude oil production of the five largest private Western oil companies—ExxonMobil, Chevron, Shell, BP, and Total—had reached a peak in 2004. From then until 2016, their production had been reduced by 23.4 percent.[33] Not surprisingly, it was the most powerful of the

* In March 2006, the fall of Yukos was precipitated by a consortium of Western banks led by Société Générale, very active in Russia, which called for Yukos to file for bankruptcy ("whores ," said the former president of Yukos, Victor Guerachtchenko). (See Lorraine Millot, "Yukos démantelé, la fin du hold-up de Poutine," *Libération*, May 11, 2007.)

companies, ExxonMobil, that performed most strongly. Shell had reached its peak in 2003, precisely the year of the scandal around its "proven reserves." The French Total had experienced a 25 percent plunge since its record level of 2004. And yet, since the mid-2000s, capital expenditures ("capex") of the largest listed oil companies had almost tripled! Plant more and more trees, add more and more fertilizer, but the crops wither year after year: Might there be a problem underground? Many shareholders are impatient with the mediocrity of these oil output results.[34] Beginning in 2014, most of the oil majors—Total and Exxon, in particular—indicated that they were preparing to reduce their profitable investments. This did not help anything. And this was months before oil prices started to plummet. When a company decides to invest less, it is often because it lacks investment opportunities.

Despite the spectacular takeoff of unconventional petroleum after barrel prices shot up to around $100 in early 2011, conventional oil continued to provide approximately 80 percent of the world's liquid fuel supply. And there are many countries whose conventional production has been declining for a long time, or has recently started declining in a worrying way, or threathens to decline more or less imminently. As a matter of fact, the vast majority of today's major producers, worldwide, fall into one of these three categories.

Latin America, the North Sea, Africa, and Central Asia: Searching Everywhere for "the Devil's Excrement"

In 2013, Mexico, the first country to be free of Anglo-Saxon oil companies, abandoned the strongest symbol of its independence, modifying no less than its constitution to allow foreign capital in. The extractions from Mexico's national petroleum company, Pemex, the financial pillar of the "institutional revolution" regime and the inexhaustible source (so to speak) of corruption, had peaked in 2004 and subsequently declined by 36 percent. Venezuela, whose regime has discouraged foreign investors since Hugo Chávez had established the seminationalization of its oil, has also seen its production decline sharply since a peak in 2006; 2016 production was down 28 percent from 2006. Mexico and Venezuela are losing the treadmill race.

Across the Atlantic, the North Sea represents a textbook case of a decline unanimously deemed irreversible, of a geological oil zone that was prolific not so long ago. Off the coast of Aberdeen in Scotland, a truism that could be applied to any finite resource was illustrated better than anywhere else: The more we drink, the faster the glass is emptied. Offshore oil called for massive investments that creditors were often anxious to recover, and its facilities, severely tested, had a limited life span: All this encouraged companies to press on greedily to acquire the expensively harvested fruit. Great Britain's offshore production had collapsed 65 percent since its peak in 1999. More conservative, the Norwegians limited their losses: They were down "only" 42 percent in 2016 since their peak production of 2001. The high oil prices of the early 2010s brought a surge of investments into the declining oil fields of the North Sea. That surge has prompted a slight increase of production since 2014, giving a so-called "camel back" shape to the North Sea production profile, a classic shape for large oil-producing zones, where investors seldom surrender to geological declines without a fight. Yet, in 2016 the overall North Sea production remained 50 percent below its record level, reached at the turn of the century. Preserved from the political hassles that other regions experienced, the North Sea is also the area that had weighed most heavily in the decline of the oil majors' crude oil production throughout the previous decade. It is, however, far from being the only one.

Let's consider Africa. Algeria was a vital producer country for Europe. But the former French colony is experiencing a worrisome waterslide concerning its oil extractions (which passed a peaked in 2007.) The former vice president of the national company Sonatrach, Tewfik Hasni, said that "all serious experts know that our reserves, including shale gas, do not guarantee more than twenty years of consumption at their current pace of exploitation."[35] The notable Algerian daily *El Watan* summarized: "The reserves are stagnating and exploration efforts only bring even smaller fruits. This is portending nothing good and is likely to induce a rapid production decline."[36]

In Libya much has occurred since the time Great Britain and France were linked to the Gaddafi regime, deposed in 2011 at the end of the war that these European countries, extremely concerned with Libyan petroleum, had started. Libyan petroleum was of the highest quality, and Libya was one of Europe's most accessible sources. In 2010, BP was accused of successfully pressuring British prime minister Tony Blair to obtain the release of the former Libyan chief of air security, who was condemned as responsible for the 1988 attack in Lockerbee, Scotland. The French judicial system also investigated charges that Gaddafi had funded Nicolas Sarkozy's 2007 presidential campaign.[37] With the 2011 military intervention scarcely over, French minister of foreign affairs Alain Juppé stated in a public radio broadcast: "The NTC [Libyan National Transition Council] said in a very official capacity that in the reconstruction, it would apply a preferential manner to those who have supported him, which seems to me quite logical and fair." Pragmatic and without false modesty,

Juppé added, "We were told that this operation in Libya is expensive, but it is also an investment for the future."[38] All this already seems long ago and far away, at a time when Libya was transformed into a kind of small Iraq in a state of civil war that is rarely dormant. The interruption of exports repeats itself. But the problem might be more profound: In 2010, a year before the war, production already had experienced a significant downturn compared to its peak in 2007–2008. Many experts fear that this was the marker of an underlying structural decline.

Jumping over to the Sahara: Nigeria, although an OPEC member, had always managed to accommodate foreign oilmen. Oil was at the heart of massive corruption and many horrors, yet at the same time served as an escape valve from misery for those who pierced pipelines or siphoned oil from reserves, often risking their lives, and those who practiced clandestine refining in the most archaic fashion. Repeated oil spills, representing the equivalent of an Exxon-Valdez black tide each year—all occurring within a half-century—did not inhibit investment in the African continent's biggest producer country.[39] However, in early 2014, Shell, the dominant Western oil company in Nigeria, had stated that it needed its investments to increase even more. At issue was the problem that in some fields, the rate of decline in so-called natural production could reach up to 15 percent to 20 percent per year.[40]

Angola was no more squeamish about foreign companies who, even before the decades of civil war had ended, rushed to pump the crude oil of the former Portuguese colony's rich territorial waters. A French diver testified that he had led a flotilla of rubber boats responsible for dispersing oil slicks at night. Angola's capital, Luanda, is one of the most expensive cities in the world for expatriates. In the "bad slum" on a gentle Boavista slope, journalists reported, "the more fortunate live in the barracks of the summit, just below the ledge of Miramar, because each night the embassies settled up there empty their garbage onto the heads of the people below."[41] The petroleum age will come to an end there, too. Angola has to compensate for an estimated "natural" decline of 200,000 barrels per day, emphasized a 2014 study produced by JP Morgan bank analysts, on a production that remains, with difficulty, a little below its record level of 2008.[42]

Take a detour to the Françafrique. Once again, the news is not good for producers of black gold. In the Gabon of Ali Bongo, Omar's son, with great effort Total and other companies hardly manage to stabilize offshore production to more than a third below the 1997 peak level. In Congo-Brazzaville since 2011, the regime of the steely Denis Sassou Nguesso has entered an unexpected and sharp oil decline.[43]

Before turning to look at the center of the chessboard, detour to Central Asia. In the ultracomplex exploitation of one of the last major discoveries, the dangerous offshore field of Kashagan, in the Caspian Sea—where until recently one fiasco followed another—has become a prime example of the end of the "easy oil." After a delay that took years and cost tens of billions of dollars, the production finally nears in 2018 a target set when the project was launched—in 2002—however not without technical issues. The cause of the latest big delay: corrosion problems involving leakage that was very difficult to stop. Kashagan took a heavy toll on its three main operators: Total, Shell, and the Italian company ENI; it has been nicknamed "cash all gone."[44] No doubt tired of the enormous, mounting delay, in March 2014 the "kleptocracy" of former Soviet apparatchik Nursultan Nazarbayev, Kazakhstan's first and only president since 1990, claimed $737 million in reparations from the Kashagan consortium for "ecological damage." On the other side of the Caspian Sea lies Azerbaijan, the land with the earliest discovered petroleum source, with which we began this book. In 2012, in a rare instance, the local potentate, Ilham Aliyev, publicly expressed his anger toward BP, the industry's major leader in that country: "It is absolutely unacceptable. . . . Investors who cannot stick to their obligations and contract terms must learn lessons. Serious measures must and will be taken."[45] The Azeri president (successor to his father, Heydar Aliyev) was furious about BP's inability to maintain production in a new highly technical field in the Caspian. The complex Azeri-Chirag-Guneshli field, approximately 50 miles off the Baku coast, had helped restore life to an industry that had been left for dead after the USSR fell, in the absence of easily exploitable oil after over a century of pumping. But this complex offshore site had peaked in 2010, before beginning a brutal decline. Since then, BP had done everything possible to stop the decline. Without great success: In 2016, the Azeri oil production was 19 percent below its 2010 record level.

In the Center of the Chessboard: Saudi Arabia, the Cautious Queen; and the Key Player—The Bishop, Iraq

Finally, on to the Persian Gulf. At the center of the chessboard: two rooks (at least one of which seemed fragile), the queen, and the bishop. Kuwait and especially Iran were threatened by a structural decline due to lack of investments. The IEA referred in 2012 to the state of Kuwait's "uncertain" prospects, due to a lack of sufficient efforts to develop heavy oils that were necessary to maintain the emirate's production capacity.[46] The IEA points at the "uncertainties" regarding Iran, where international sanctions have held back "the pace of anticipated investment in new projects," a policy that the IEA has warned may lead to a reduced capacity of production, for a country where oil is extracted, for more than a century."[47]

In Saudi Arabia, in 2010, King Abdullah had indicated that he had "ordered a halt to all oil explorations so part of this wealth is left for our sons and successors, God willing."[48] This decision was mainly symbolic, considering that the queen of the oil-producing nations held the territory that was likely already the most systematically prospected in the world (along with that of the United States). Nevertheless, Riyadh has started to produce heavy, lower quality petroleum. To this end, Total and Saudi Aramco have collaborated to build the Jubail megarefinery, which opened in 2013: $12 billion for 20,000 kilometers of tangled lines, on 500 hectares. Various forms of the following expression, perhaps first uttered by a wise neo-Bedouin, were often applied to certain Saudi personages: "My father climbed on a camel, I ride in a Mercedes, my son flies in a jet, but his son will climb on a camel." By far the largest holder of the world's excellent conventional oil reserves, the Saudis had repeatedly shown their desire not to push their production beyond its current capabilities. The last barrel of conventional crude oil pumped on Earth will very likely be a barrel from Saudi Arabia. But we should not rely on Riyadh to fill the decline of other petroleum producers that were less well-endowed. For Riyadh (like most of the oil giants) its own energy opulence is a double-edged sword: The unrestrained growth of Saudi Arabia's domestic consumption of crude oil—to run air conditioners or desalinization plants—encouraged by substantial governmental grants, more and more seriously impede the kingdom's ability to export its oil.

Iraq's resources, preserved during a quarter century of war and embargo, make Iraq the only major global conventional oil producer whose extractions are likely to significantly increase in the years to come. Tomorrow even more than today, Iraq should be the cornerstone of the oil planet, insisted the IEA's Fatih Birol in November 2014.[49] Its crumbling cornerstone. Despite its perpetual chaos, the country's production is at its all-time high. Yet it has constantly revised its published short- and long-term production objectives downward. And is it still a country? Not a nation, in any case. The 2014 offensive conducted in northwest Iraq by the fundamentalist Sunni army of Daesh, the Islamic State of Iraq, and the Levant (ISIL), and their involvement in the quagmire of Syria's civil war, was doubtless doubly financed by oil: on one hand, through rich Sunni faithful of the Persian Gulf, Wahhabism's usual funding source; on the other, according to Western intelligence, thanks to wells in eastern Syria that were conquered by Daesh, whose production, according to Western intelligence, was notably resold to Bashar al-Assad, the tyrant enemy from Damascus, in exchange for covert compensation to their forces, a plausible *modus vivendi*.[50] Some people interpreted the situation in Iraq as a proxy war between the Sunni petroleum masters of Saudi Arabia and the Shiite masters of Iran's petroleum. A war that has not yet finished taking new, apparently absurd turns. Riyadh has undertaken the construction of a kind of Great Wall at the Iraqi border: One wonders if its Wahhabite financers are more worried about the Shiites in Iraq than by the Sunni jihadists they once funded. Saudi Arabia and the emirate of Qatar claim they oppose Daesh. Confidential French secret service files note that funding from sources within these two countries that benefit jihadists networks from Iraq to Mali do not prevent successive French governments from making unbridled efforts to attract the petrodollars of Riyadh and Doha.[51]*

As for the United States, ultimate masters of the Persian Gulf for a half-century, it worries about advancing too far into the region. The world the Obama administration had to navigate was still the foreign policy world Dick Cheney and George W. Bush left behind them. The tactical paths carved out by Washington appear infinitely tortuous and perilous. In March 2011, for example, during the Arab Spring, the Obama administration looked the other way when Saudi Arabia lent a forceful hand, using American equipment, to the Sunni princes of their small island neighbor, Bahrain, to squelch the rebellion of the majority Shiite population by torturing them (according to Doctors Without Borders) even in hospitals. Launched in 2009, the nuclear aircraft carrier USS *George H. W. Bush*, the last of the Nimitz class, crossed the Persian Gulf in June 2014 to carry out its first air strikes against the areas controlled by Daesh. Today, the Trump administration supports, without reservation, the House of Saud, including in the very questionable war it wages in Yemen—even while the proxy war between Iran and Saudi Arabia continues to degenerate, assuming at times the character of a latent World War Three.

In Baghdad, in the face of the self-proclaimed Daesh "Islamic Caliphate," Prime Minister Haider al-Abadi's central government, deeply divided but dominated by the Shiite majority, cannot handle a large undertaking alone, among other reasons because its military wings had been meticulously clipped

* The propensity of the West's powerful old guard to choose their customers on the basis of petroleum wealth—former French prime minister Dominique de Villepin with the sovereign funds of QIA in Qatar, former British prime minister Tony Blair with the sovereign funds of Mubadala in the United Arab Emirates, former German chancellor Gerhard Schröder in Russia, or former US secretary of state Henry Kissinger in Saudi Arabia—highlights once again where they currently tend to apply certain major modalities of economic and political power in the world.

by the United States. Iran and its Syrian ally, Bashar al-Assad, once offered to help the United States come to Baghdad's rescue, and it is notably thanks to Shiite militias funded and armed by Iran that Mossul was retaken in 2017 by the Iraqi regular army and by the US-led international coalition against Daesh. In another incredible irony, the Kurds of Iraq have had no other choice, if they hope to win the full independence of which they dream, than to side with the old Turkish ogre to sell their crude oil. Iraq seems doomed for a time to be the ultimate cesspool that one of OPEC's two founding fathers referred to as the "devil's excrement." And it is also in Iraq that the beginning of the end will come for those who partake, in one manner or another, in what remains of the planet's most important oil reserves.

China will, of course, be a decisive player in this endgame. Over the past decade, while its consumption of crude oil more than doubled, Beijing advanced its pawns everywhere, from Canada to Siberia, passing into Venezuela, woeful North and South Sudan, and above all in the Persian Gulf.* Chinese oil companies, among the most powerful, have also prospected on the doorsteps of the Middle Kingdom, for instance in the disputed territorial waters of Vietnam. China is a major producer of crude oil, but its economy consumes more than it produces, even more so than with coal. And its domestic oil production may have already started a decline, which the IEA considers very likely in the next decade. China now tops the United States as the world's number one importer, according to Washington. However, despite the resurgence of American production by the fracking of shale, the United States is more dependent on Persian Gulf petroleum today than in 1973 or 1979, and only slightly less than it was in the early 2000s. The competition between these two great gluttons has probably only just begun. The European Union, the third great glutton, is in a very unfortunate situation. Its nearest three major petroleum sources are in bad shape (the North Sea, Algeria, and Libya). Meanwhile, Europe's principal source, the territory of the former USSR, is at the mercy of the Russian Federation, whose oil production may also be on the verge of decline.

The Logical Slope of Decline in Humanity's Total Energy Production

When will the world's petroleum peak, the beginning of the end of oil, take place? Today, only Iraq, the United States, Brazil, and Canada seem able to significantly

* Chinese companies are among Saudi Aramco's main clients. Combined, they are also the largest oil operator in Iraqi.

increase their production. The prospects of most of the major producer countries are worrisome. The two to four Saudi Arabias that would be required over the next ten years to compensate for the declines, equivalent to up to half of current global production, have not been found and made ready for production.

After arguing for a long time that the trees would continue to climb to the sky indefinitely, even the "optimistic" analyses emanating from the oil industry are currently willing to discuss an "undulating plateau" of total oil production that would start in 2020 and last for several decades, thanks in particular to unconventional petroleum.[52] Beyond the 2020s, however, lies the "question mark" acknowledged at a 2012 Conference in Qatar by Christophe de Margerie, former CEO of Total (the French company whose crude oil production collapsed by a quarter between 2004 and 2016).[53] Total's retired strategic director and a pillar of ASPO in France, Pierre-René Bauquis, has referred to a 2020 peak, followed by a decline that may be slowed down if nonconventional petroleum sources are quickly developed.

The industry's "optimistic" institutional sources have almost systematically lowered their predictions for almost twenty years.* Despite the vigor of demand, particularly in China, crude oil production has progressed much more slowly in the past ten years than during previous periods. Outside of the United States' shale oil and the Canadian tar sands, the remaining crude oil production—essentially conventional oil production—remains essentially flat, stuck on an "undulating plateau" since 2005. As the IEA has indicated since 2010, the peak in classic liquid oil seems to have been crossed.

We will know for sure the exact date of peak oil only when the total oil production starts to irremediably decline. Most independent experts, at least those who publish and those who gravitate around ASPO from near or far, are for the most part pessimistic. For example, according to Olivier Rech, who was responsible for oil forecasts during the drafting of the IEA's 2008 report, world production will probably enter decline before 2020. That is to say: soon. In such a case, the consequences, inherently chaotic, would certainly be at their most severe.

There will still be oil for a long time. But how, for whom, and at what price? In any case, the limits of accessible reserves are likely to naturally reduce global extraction capabilities to around 25 Mb/d in 2100, or the production level of 1960, approximately four times less than today's.† By then, the Earth's

---

* Why "optimistic"? At least because it is necessary to win over the banks—and the insurance companies. The oil industry cannot cry wolf; if it does, the wolf will appear.

† According to a plausible estimate advanced by Pierre-René Bauquis, former strategic director of Total.

population may be nine billion people, three times greater than in 1960. The amount of oil available per person in 2100 would thus be reduced to the level available in 1900.* This is an average, of course: There is likely to be vastly greater disparity in access to the mother of all raw materials than there was at the beginning of the twentieth century.

Building on the assumption that we will reach a peak in petroleum production in 2020, Bernard Rogeaux and Yves Bamberger, respectively advisor and director of the French EDF Group's research and development department, concluded in 2007 that technical humanity is inevitably moving, by the next generation, toward a decline of its total energy output (not just oil).[54] They have shown that, even by pushing projections to their extreme limit, a world with five times as much coal and civilian nuclear energy production, plus twice as many dams and a twenty-five-fold increase in renewable energies, would still not compensate for the post–peak oil decline. World production of all energy forms would flatten out by somewhere between 2035 and 2040—and much earlier, surely by the next decade, if humanity as a whole were to abandon the greatest offender, coal, while strictly reducing its carbon dioxide gas emissions. Without abundant energy, there will be no more deep mines, nor distant whims. Whether we like it or not, the laws of nature appear to be on the brink of imposing a radical change in the course of the brief history of human technology.†

Basically, talk of energy "production" is nonsense: Energy, by definition, is preserved. We can only transfer energy or change its form (solar, chemical, electrical, kinetic, or otherwise) to finally dispel it. Nature (human societies included) tend to follow the path of least resistance. Animals (humans included) have always tended to begin by picking the first very ripe fruit at hand, before resorting to other means, or somehow decreasing their consumption. Unconventional oil occupies a central position among the technical expedients at our disposal. No alternative source, whether it be nuclear or renewable energy, offers the cocktail of convenience, flexibility, and power that is provided by the energy contained in the electromagnetic links of hydrocarbon molecules—natural or synthesized. Because of their

* In 1900, production was on the order of 400,000 barrels per day, for a world population estimated at 1.5 billion people.

† If it exists, the path of breeder reactors or of nuclear fusion won't change much. It would solve the problem of limited uranium reserves but not the problem of power production. Climate engineering, or geoengineering—a process of exploring the possibilities of artificially cooling the atmosphere, developed in the late 1990s following reflections by, for example, Edward Teller (see chapter 14), father of the hydrogen bomb and inspirer of cinematographer Stanley Kubrick's *Dr. Strangelove*—has hardly produced convincing, and certainly not ethically defensible, results; although several governments (the United States, China, and India, among others) encourage experiments.

inherent physical qualities, these hydrocarbons remain the foundational axle of progress, a progress we have designed with pragmatism and opportunism: by using our intelligence to increase vital energy, taking the slope of least resistance—as dictated by *physis*, or nature. In the face of limits and of the high cost of oil, coal is the energy source that has gained the most ground over the past two decades, and by far—certainly not renewable energies. China's primary recourse for development, "old king coal" (also fed by Western capital disguised as sustainable-development funding) remains close to oil's supreme position as the primary fuel of the thermo-industrial machine.

"Dissipative Structures" and the Red Queen Effect

An empirical analysis conducted beginning in 2013 by French economist Gaël Giraud, director of research at the National Center for Scientific Research, led (as others had) to the conclusion that, in utter contrast to what had been written in economics textbooks for the past two centuries, energy—rather than capital or work, which without energy would remain inert—seemed to be the essential, decisive factor in economic activity.[55] This factor fundamentally eludes humankind. If Gaël Giraud's conclusion is correct, then our efforts to interpret economic phenomena have been in vain, much like the efforts of astronomers who sought to understand planetary motion before Copernicus. Classic economic theories—Marxist and capitalist alike—would in this case be no more than anthropocentrist opinions. The only things left for us to do would be to erase the dubious artificial border that separates economy from ecology and often renders them incompatible and, in our endeavor to understand the ways and means of human enterprises, to rely first on physical laws—foremost on the doors that are opened and closed before us by the laws of thermodynamics.

Another potential consequence, more direct than the first, of Gaël Giraud's empirical conclusions: Without radical new gains in the field of energy efficiency, technical advancements that are far from assured today, it would be illusory to imagine a return to the rhythms of growth experienced in the second half of the twentieth century, yearned for by political and economic leaders, without a highly significant increase in energy production—a massive supplementary infusion of honey on the anthill.

Although they do not, in general, connect it to the energy problem, many economists evoke "secular stagnation" to describe the weak trends in the rhythm of world growth since the 2008 crisis, a weakening that actually took root some time ago—as early as the oil crisis of 1973 and the end of the Thirty Glorious Years.[56] In a 2012 prospective report from the CIA, one can read that if recoverable reserves below American soil are lower than forecasted and, as a result, the oil industry cannot prolong the shale oil boom, the United States will be unable to maintain their power in the decades to come.[57] In contrast to a great number of economists, the authors of this analysis, doubtless pragmatic people, persisted in considering economic (and political) power as being, first and foremost, a direct function of energy power, both its possession and the knowledge of how to use it.

Climate, access to intact energy sources, contingencies imposed by the imperative to grow the economy, wars: Diverging from the slope of least resistance opened for us by fossil fuels could be the ultimate challenge of the technological world, of which we are the agents. In 1922, during the Roaring Twenties, American physicist Alfred Lotka, famous for his work in population dynamics and particularly for his systems model of the interactions between predators and prey, advanced the hypothesis that natural selection favors the organism that dissipates, or captures and uses, energy most efficiently.[58] In 1973, the year of the first oil crisis, American biologist Leigh Van Valen noted that when the environment changes more quickly than a species is able to adapt, that species is doomed to disappear. He calls this the Red Queen effect, referring to the enigmatic warning pronounced by the Red Queen in *Through the Looking Glass*, the second volume of Lewis Carroll's *Adventures of Alice in Wonderland*: "Now, here, you see, it takes all the running you can do, to keep in the same place."[59] Ilya Prigogine, a Russian-born Belgian scientist, for decades explored his empirical concept of "dissipative structures," which he had come up with in 1969 and which won him the Nobel Prize in Chemistry in 1977. Prigogine's concept refers to the structures that, in one form or another, seem to appear spontaneously within open energy systems that benefit from a continuous energy flow.[60]

The stars are dissipative energy structures; they transform gravitational energy through nuclear reactions and dissipate in the form of radiation. A cyclone is another form of dissipative energy structure, unfolding thanks to the temperature differences between the equator and the poles. Evidently, living beings, and even more so human societies, are dissipative energy structures. In 2001, American astrophysicist Eric Chaisson noted that if a human being had comparable mass, he or she would dissipate ten thousand times more energy than a star like the sun.[61] "Dissipative structures self-organize in a way that maximizes the flow of energy traveling through them," wrote French astrophysicist François Roddier in 2012. Roddier added that these structures also maximize the speed at which the energy dissipates as it passes through them. This empirical observation, sometimes called the "law of maximum entropy production," has been studied for a generation by physicists, chemists, biologists, and cyberneticists.* Some researchers are no longer reluctant to consider it an authentic third law of thermodynamics: The energy is preserved (1), it dissipates (2), and, finally, it allows the emergence of dissipative structures which, in order to continue to exist, minimize their own entropy by maximizing the energy dissipation around them (3). Roddier formulated the assumption that what is called "evolution" was in fact a physical process during which dissipative structures acquired the ability to increasingly maximize the entropy around them; the instrument that allows humans to achieve this is information, which those structures would organize in order to store it in ever growing quantities.

Here we find the Red Queen effect, as well as the treadmill race of the petroleum producers: The more efficiently a structure dissipates energy, the more quickly it alters its environment, and the more quickly it must acquire information from this environment in order to adapt and evolve accordingly. Inevitably. Roddier drew the following conclusion: "Nourished up to this point by fossil energy, a kind of mother's milk provided by the Earth that generated it, humanity has been able to develop. It will soon undergo the test of weaning. As an adult, humanity will have to learn to feed itself. Humanity will come to realize that only solar energy will assure its long-term survival."[62]

The dissipative energy structure called the sun—the primary energy source of any dissipative structure on Earth, and of life in particular—has reached approximately the midpoint of its existence. There remain about five billion years before the sun engulfs the Earth. Appearing on Earth a mere two hundred thousand years ago, Homo sapiens can hope it is merely at the beginning of its own history. In this time frame, the fossil hydrocarbon molecules will soon be gone. With them will disappear, probably, the expensive and inconsistent feats of humanity. Who knows? Elsewhere, other compelling life forms already may have fallen into this same juvenile trap.

* American Rod Swenson, for example, believes that the law of maximum entropy production, which he calls the "maximum power principle," explains why, instead of living in a world where the appearance of order is highly unlikely, we live in and are the products of a world that we can in fact expect to produce "as much order as possible." (See Rod Swenson, "Autocatakinesis, Evolution, and the Law of Maximum Entropy Production: A Principled Foundation Towards the Study of Human Ecology," *Advances in Human Ecology*, vol. 6, 1997, p. 1–47, searchable at spontaneousorder.net.) The act of maximum production of entropy is also a physical response to the immortal philosophical question: Why is there something rather than nothing?

Denial or About-Face

Could peak oil touch us before the first massively undeniable consequences of climate change? It is quite possible. In addition to their cause, the two phenomena have in common their unending consequences. On the climate front, the latest decade has been full of missed opportunities, false pretenses, and bad excuses. The November 12, 2014, agreement between Barack Obama and Chinese president Xi Jinping (the United States committed to cut its greenhouse gas emissions 26 to 28 percent below its 2005 levels by 2025, while China promised to stabilize its own emissions "toward 2030") was rightly described as historic, at least in the sense that it endorsed the near certainty that the climate will be irretrievably altered. The hypocrisy is stunning. Preserving the climate means leaving the majority of remaining fossil fuel reserves underground. On a worldwide scale, the only serious weapon that currently exists to fight climate change is Europe's emission-trading scheme. But that scheme was rendered powerless by industrial lobbyists and governmental complacency even before the 2008 crisis, when the ramp up of crude oil prices and its consequences on energy consumption in the rich countries provided the pretext for inaction in those countries, while allowing them to fault the developing countries, to which our financial institutions outsourced the industry so that they could profit from social dumping thanks to the low cost of supertanker fuel. Among other rich countries, France congratulates itself on reducing its greenhouse gas emissions. But if (to be honest) we take into account the emissions caused by the manufacture of all the things they import, the French population's emissions have in reality increased by at least 25 percent since 1990.[63] Shame on the idiots who merely blame the "Made in China" label.

Failing to see reality is common. Consider this taunt from a local leader in the American Tea Party movement—intoxicated by some climate-change-denialist back room financed by Big Oil (no longer Exxon but others, less visible): "They're trying to use global warming against the people. It takes away our liberty." And those who, in the name of their Christian religion, say, "I cannot help but believe the Lord placed a lot of minerals in our country and it's not there to destroy us."[64] Are we who laugh at this any less naive? In France, on a daily basis, eight million poor people are struggling to afford their basic fuel needs.[65] Referring to the "unavoidable" event of a liter of gasoline rising to 2 euros, former Total CEO Christophe de Margerie noted in 2011: "It is to be hoped that this does not happen too quickly, otherwise the consequences will be dramatic."[66] Although from a very different ideological background, Bolivian president Evo Morales in 2011 had to abandon his administration's strategy of drastically reducing the subsidies on gasoline prices and was forced to continue the subsidies in response to the angry protests of his people, including his own supporters (a story repeated elsewhere, in Venezuela or in Iran). Ecuador's president, Raphael Correa, who since 2010 had asked the rich, oil-consuming countries to financially assist his country so he would not have to authorize oil exploration in Yasuni National Park—one of the many improbable places where oilmen are reduced to hunting black gold—had to abandon this sustainable development goal in 2013, in the absence of sufficient support.

Such blind spots lead to cascading problems. Syria passed its oil peak in 2002. The erosion of financial and energy resources, caused by the hydrocarbon sales decline, have contributed to the country's economic slump, which led to the frustration of a segment of the population and then to civil war. (In May 2008, the Bashar al-Assad regime had to considerably reduce substantial gasoline subsidies, which accounted for no less than 15 percent of GDP; the fuel price tripled from one day to the next, resulting in a high inflation of agricultural prices.)[67] In the course of twelve years, Iran's largest lake, Lake Urmia, almost dried up; experts explain that this resulted from the changing climate, excessive irrigation, and an overly ambitious dam construction program to provide energy to a large, often destitute population, excessively dependent on gas and fuel oil. In Algeria, with oil money, a government agency called ANSEJ buys social peace at a high price for an immense youth population who would otherwise be pushed over the edge by unemployment and idleness. Since at least the 1990s, in France, Germany, Italy, Ireland, Portugal, Spain, and Greece—countries that are completely dependent on imported oil—economic growth, sovereign debt, and energy bills are progressing at startlingly similar orders of magnitude.[68]

The possibility of a collapse of technological society, as predicted by the Club of Rome, has all but vanished.[69] It is probably already too late to prevent heating the atmosphere by more than 2°C by the century's end, warned the IEA on numerous occasions, without taking into account the strong plausibility of catastrophic "positive feedback loops," such as a runaway melting of permafrost. Despite the extremely ambitious objectives outlined in Europe, the direction of our technological society remains unchanged. Even a radical recycling effort can be no more than a short-term measure if we do not renounce the growing consumption of nonrenewable raw materials, a top executive of the French waste treatment company Veolia stressed in 2010.[70] Wind turbines are made of Kevlar (a petroleum derivative) and contain considerable quantities of imported ores from the other side of the planet. A hybrid car has two engines. It is possible that what is designated under the heading of "sustainable development" will lead to an increase, rather than to a decline, in material needs. The so-called dematerialization of the economy seems no less illusory than the systemic decrease in oil consumption that some claim is occurring in rich countries. In the United States, this consumption appears to be rising simultaneously with the shale oil boom. Ants, whenever they find more to eat, pick up their already frenzied pace. Do we not have more common sense?

The search for homeostasis promises to be long, painful, and subtle, requiring great ingenuity. What other quest could be more urgent, more essential? By its nature and by excelling, humankind may have evolved to maximize energy consumption, at the price of a critical expansion of chaos in the world. If humankind's intelligence led us here, there is no reason it cannot now find a different, more positive path.[71] I worry about the signs from the deafening, universal din of a society that continually itches to consume. Oil has stimulated the emergence of our technological society, so beautifully capable of reversing much of the suffering that nature inflicts. Within the destiny that the thirst for oil imposes, I believe, lies the greatest threat to this generation and those to follow. At the very least, this threat requires us to turn our backs on greed, slow down, and simplify—all, in practice, the same thing.

Afterword

Oil, Power, and War tells the story of humanity's unprecedented expansion of power over the past century and a half, through the lens of oil—or, more precisely, of access to oil, the energy source indispensable to the expression of modern economic, political, and military power. It is an exploration of past events that tries to provide a context for understanding current times and future challenges.

History always repeats itself twice, wrote Karl Marx: first in the form of a tragedy, then in the form of a farce. Our twenty-first century opened with the 2003 invasion of Iraq—a war, perhaps for oil, that has not finished metastasizing in new, monstrous conflicts. If that invasion was incontestably a total tragedy, what does Donald Trump's rise to power represent?

In his campaign speeches, Trump spoke not of achieving American energy independence but rather of imposing American energy "dominance" on the world. Before becoming president of the United States, he repeatedly declared that had he been in power when the United States pulled out of Iraq, he would have seized Iraq's oil. "You know, it used to be to the victor belong the spoils," he said during NBC's September 2016 Commander-in-Chief Forum. "There was no victor there, believe me. There was no victory. But I always said, 'Take the oil.'" And, voilà, once elected, Trump appointed Rex Tillerson, the CEO of Exxon, to head American diplomacy.

As I explain in the last chapter, Tillerson made a deal with Putin and Rosneft in 2011 to develop the possible oil reserves of the Russian Arctic, as well as other potential oil resources. The joint venture involved several hundred billion dollars of potential investments—way beyond any usual scale, even for Big Oil. It appears that Exxon and the Kremlin both saw each other as a lifeline, because both are having huge trouble replenishing their oil reserves. Exxon has seen its crude oil production slowly but surely decline since a peak in 2006, from 2.68 Mb/d that year down to 2.36 Mb/d in 2016, a situation it had never experienced in its long history, and this took place despite a huge ramp up of capital expenditures. Russia is an old oil producer, too—of a different kind and, of course, much larger. But it is confronted with a similar issue. Although its production is not declining, the IEA has been warning for several years now, in many reports, that its production is bound to decrease sometime in the coming years because of the so-called "natural" decline of a large number of old, "mature" conventional oil fields, mainly in Western Siberia.

The bottom line: Exxon badly needed Russia's potential new oil plays to replenish its declining production, and Moscow badly needed Exxon's access to capital and expertise to exploit new, complex, and expensive sources of oil. Such a deal would have kept Russia from seeing its main source of cash weaken in the not-too-distant future. This grand plan met a roadblock, though, when the Obama administration placed economic sanctions on Russia for seizing Crimea in 2014. Most likely, both Tillerson and Trump would have been pleased to see these sanctions lifted—a possibility that seemed to vanish amid suspicion of collusion between the Trump campaign and Russia to influence the 2016 presidential election. Less than two weeks after the Exxon joint venture with Rosneft was quietly abandoned, Tillerson's White House exit was underway. It's not just because of Trump's persona that this little story has the taste of a farce.

Built on the might bestowed by oil and dollars, US "dominance" is neither invincible nor unchallenged. The struggle for world hegemony could easily unsettle the physical foundation of today's economic order. Washington's present will to prolong or extend US economic sanctions against Russia and Iran increases the probability of long-feared structural declines in the oil-production capacity of both these old, major producers—much like the way the crisis and economic isolation of Venezuela is terribly accelerating the oil-production decline there. Washington's policy toward Russia, Iran (and to a lesser extend Qatar) also opens up a door for China's oil diplomacy, as the giant oil importer increasingly incites its suppliers to sell their crude in yuan, or else lose market share.

Donald Trump chose in April 2018 to name John Bolton as his national security advisor. Bolton is a person who, when interviewed on Fox News in 2011 about US plans to pull troops from Iraq, proudly referred to the Persian Gulf as the "critical oil- and natural gas–producing region that we fought so many wars to try and protect our economy from the adverse impact of losing that supply or having it available only at very high prices." Bolton embodies in the most candid and obscene way the absolute will of the most powerful country on Earth to impose its energy "dominance" over the world. His appointment raises, just a bit more, the very obvious odds of arriving at some sort of Third World War triggered by the desire to control the vital fossil fuel resources of the Middle East. Whether farce or tragedy, the story remains the same: The sweeping power of human greed always needs to secure access to the queen of energy sources to exert its will.

This book presents the idea that the oil shock of 1973, a crucial event in the twentieth century, was a direct consequence of the decline of US oil production that began in 1970—a geological problem (ecological, one might say) that has triggered an avalanche of upheavals in human history. It is also a problem likely to revisit us: Sooner or later, oil supplies will be exhausted everywhere, if we don't turn away from them beforehand. Oil's production seems to be already on the slope of decline in many major countries where—as in Venezuela, Mexico, or Algeria—civil peace maintains a precarious foothold. We continually refuse to deal concretely with the inexorable peril of peak oil, just as we refuse to deal concretely with global warming. We have preferred, up to this point, to charge ahead with the status quo. How can this approach be considered anything other than pure madness?

Thanks to the boom in shale oil, the United States is on track to beat the crude oil production level reached in 1970, long considered to be unbeatable. Nobody knows where this boom will lead us. Nobody knows to what extent it is reproducible elsewhere. One thing is clear: Shale oil appears to be the fatal last "fix" relaunching in extremis the addiction of a society drugged up on oil. Any reasonable person can see that it's time to detoxify.

Numerous studies have shown that a global oil detox would cost no more than a few percent of each nation's wealth. And a cure is possible, as long as the process changes a good number of habits—habits of production, habits of consumption, and, especially, habits of thinking.

In a finite world, everything that is taken here and now is lost for elsewhere and later. As we approach the physical limits to growth—in a vast swamp of indifference, unconsciousness, and cognitive dissonance—we arrive at conflicting responses to our predicament. While a few logical and powerless souls might respond with sobriety and a move toward simplicity, many others will adopt a different attitude: "everyone for himself," or "after us, the deluge."

The IEA, in its 2016 annual report, gave its sharpest warning to date. Here are a few of the more significant alerts:

- More than 50 percent of producing oil fields in the world have crossed their peak of production and will decline in the future.
- Annual discoveries of new oil fields are at the lowest level in seventy years.
- By 2025 there could be an estimated gap of approximately 16 Mb/d—equal to the production of Saudi Arabia and Iran combined—between

the level of expected production and the decline of existing production (which was 94.5 Mb/d in 2015) or production under development.

- This gap could be filled by new resources, provided that investments quickly rise to more than $700 billion, their record level before the oil price drop that started in 2014.
- The longer investments remain low, the more likely a constraint on future supplies becomes.
- The future potential from American shale oil, which is dependent on the evolution of the price of crude, remains very uncertain, and 90 percent of American shale operators had a negative cash flow even when crude prices were highest.

More alarming yet, the HSBC Bank stressed the following facts in a September 2016 report titled "Will Mature Field Declines Drive the Next Supply Crunch?":

- At least 64 percent of world oil production is in decline.
- By 2040, there will be a need to develop more than 40 Mb/d of new resources (nearly half of the world's production, or the equivalent of four Saudi Arabias) to maintain current production levels.
- Small petroleum fields begin to decline generally two times faster than large fields, while world production of crude depends more and more on small fields.
- The significant improvements in production and drilling efficiency undertaken in response to price drops have masked the underlying rate of decline, but the degree to which these improvements can continue is limited.

Both the IEA and HSBC Bank warn that barrel prices must rise, and quickly, in order to revive investment, otherwise peak oil will take hold soon.

IEA executive director Fatih Birol has been repeating (for three years now) that his main concern lies in the fact that investments have collpased since the oil price crash of 2014, "which could have major implications for the security of the supply in the years to come."

When the oil price was dwelling around $50 in 2016, this warning was loudly echoed by prominent industry actors, such as Saudi minister of energy Khalid al-Falih ("If we under-plan, the world would pay an enormous price in terms of oil supply shortages that would lead to certain price spikes"), Total CEO Patrick Pouyanné ("We are not investing enough. . . . In 2020, the supply will be insufficient"), or the esteemed French trader Pierre Andurand ("production is headed toward a 'structural deficit'").

But can barrel prices rise quickly enough?

Despite production reductions by OPEC members and several other producing countries, including Russia, at the end of 2017, no major source of analysis anticipates a return by 2018 to $100 a barrel, the level that prevailed when investments were at their peak.

On the demand side, the world economy is not evolving toward favorable conditions for a sharp increase in the barrel price. Yet the Chinese economy, by far the prime engine of global growth but heavily dependent on the rest of the world's economic situation, could easily find itself in a fragile position. The reason: concerns about the rise of Chinese debt, and an international environment made more uncertain by the election of US president Donald Trump. According to Standard & Poor's, the debt level of Chinese companies reached 171 percent of Chinese GDP in 2015, two times higher than that of the United States or Europe. Before the crisis of 2008, China needed $1 of debt to generate $1 of GDP, but the ratio is $6 of debt for $1 of GDP noted Morgan Stanley in 2016.

The Trump presidency risks locking in the effects of the 2008 financial crisis. That crisis, which we are still experiencing today, was prima facie a crisis of debt resolved through more debts. It could prove to be the grand crisis marking the point at which we reached physical limits to growth. In other words, the global economy's potential for growth could be too weak to maintain world oil production, our primary source of energy. This possibility seems to arise, on the one hand, from the stagnation of middle-class income in rich countries, particularly in the United States, and on the other, from an overly rapid acceleration of diverse costs induced by the growing complexity of our technical societies in general, and the extraction of abundant sources of energy in particular.[1]

The crisis could be, in the end, enlightening—revealing that finite resources of fossil energy have been the fundamental driver of world economic growth for the past 150 years.

As the saying goes, "May you live in interesting times." Indeed, the times ahead promise to be instructive. Facing indications of turbulence and the dissolution of growth, our developed nations have before them the necessity and the moral obligation to invent the post-fossil-fuel economy.

To have a shot at meeting this unavoidable challenge, human intelligence must urgently confront the physical conditions that make its dreams and its projects sometimes reasonable, sometimes futile. For example, are quests for the widespread use of electric and driverless cars harmful pipe dreams, vain attempts to avoid the question at hand? More generally, to what extent does betting on a technology that requires enormous amounts of raw materials and energy to escape the limitations of our energy resources make sense? The cheating by some car manufacturers revealed by Dieselgate constitutes the only logical response to the impossible problem posed by contradictory demands of consumers. No, a car that is more powerful and burdened by ever more gadgetry and sophisticated equipment cannot consume and pollute less. Of course not.

Our magical thinking is the most perverse and naive trap of all. Germany has been forced to slow down its costly effort to develop wind and solar, while its emissions of greenhouse gases, by far the highest in Western Europe, have not declined further since 2010. Unless it is proven otherwise, it appears that if you want a lot of economic growth, you need a lot more cheap and abundant energy. Which means you need more fossil energy.

We fought for oil and other finite sources of power when they were abundant and easy to find. What will happen when, sooner or later, these resources become rare? Several recent studies show that without the very favorable fiscal status it enjoys, the American oil industry would collapse, undoubtedly taking the economy and power of the United States with it.[2]

Some implications? No need to complicate matters—or, as we say in French, to look for noon at two o'clock—with carbon pricing: Simply removing the crutch of subsidies (and telling the oil producers to "clean up their mess") would lead to a brutal decarbonization of the American economy, which would likely lead to a collapse of world crude production that has been propped up by American shale oil since 2005. Above all, these subsidies guarantee the danger (climate and otherwise) toward which we run if the irrational policies of Trump go forward. Alas, this lunacy follows its own logic, though it may eventually lead to the establishment of war economies: Oil is the historical foundation of US power, and the agents of this power may have demonstrated in Iraq in 2003 what they can do when they feel threatened.

If the crisis of 2008 was indeed—as seems plausible to me—the first great crisis stemming from the physical limits to growth, what does that say of current times? There has been no gulf between the world of 2008 and the world of today. In particular, quantitative easing created a bubble of expensive oil, in the form of shale oil, that partially popped at the end of quantitative easing in 2014.[3] A bubble that Donald Trump, or what he embodies, seems ready to do anything to keep afloat—whether that's condemning the Paris Agreement, eviscerating the Environmental Protection Agency, rolling back numerous regulations that were "constraining" the oil industry, opening refuges and vast tracts of coastline to drilling, or eventually venturing deep into the territory of negative interest rates. It's as if the regime perpetuating thermo-industrial power has all kinds of spontaneous immune systems, and it's a gigantic and perilous conundrum.

Let's imagine that, in the absence of sufficient extractable reserves, US oil production or Chinese coal production begin to decline (which is quite possible).[4] Voluntarily or forcibly, we would need to confront the drivers that perpetuate power: the growth that demands energy; the energy that demands growth; and the complexity born of, and also feeding, both. It's a cycle that grows more turbulent, more vicious as it evolves. Breaking free of it would require a fundamental shift. But to devise a more sober society is to devise a more robust society.

MATTHIEU AUZANNEAU
Paris, April 19, 2018

Acknowledgments

This project would not have seen the light of day without the "oilmen" who patiently enlightened my understanding of the principles, specificities, and limits of their industry, especially Jean Laherrère. For his valuable proofreading of the French edition, a certain person both brilliant and caring, an expert on the nuances of the petroleum industry, knows my gratitude. I am also grateful to the engineers, economists, and physicists (energy experts, geologists, ecologists) who helped me understand the implications of the intransigence of the physical laws that underlie technical progress: I want to acknowledge Jean-Marc Jancovici, Jean-Charles Hourcade, Bernard Durand, Yves Bamberger, Thierry Solomon, François Grosse, Mickael Kumhof, Robert Hirsch, Robert Ayres, Benjamin Warr, Lester Brown, Mark Lewis, Fatih Birol, André-Jean Guérin, Philippe Bihouix, François Roddier, and Gaël Giraud. A special acknowledgment to the late and much missed Bernard Rogeaux, for his talent but even more so for the depth of his humanity. A big thank you to Dominique Bourg, to Jean-François Mouhot for their encouragement, as well as to Jamal Najim, Michel Lepetit, and Jean-Marc Manach for their helpful comments. I would also like to thank *Le Monde* newspaper and its attentive and faithful readers of my blog: Without their confidence, tireless throughout the years, I would not have been able to complete this long project. For their support, I salute friends and fellow journalists from all sides who are trying to explore the inextricable and important links that (we believe) unite economy and ecology. I would also like to thank the small band of friends from the cafés of Belleville. For Greg, Toussaint, Alex, Bastien, Annie, Alain, Zoé, and Camille, at long last: Thank you for the encouragement, patience, and love.

For the English edition, a very warm thank you to translator John Reynolds and editor Joni Praded for their dedication and tireless efforts to make this an outstanding translation, and to Margo Baldwin and Chelsea Green Publishing for having believed in this ambitious project, and to the entire Chelsea Green team. Gratitude also goes to the Post Carbon Institute and Roger C. Baker Jr. for making the English edition possible and to J. David Hughes for a technical review.

Appendix

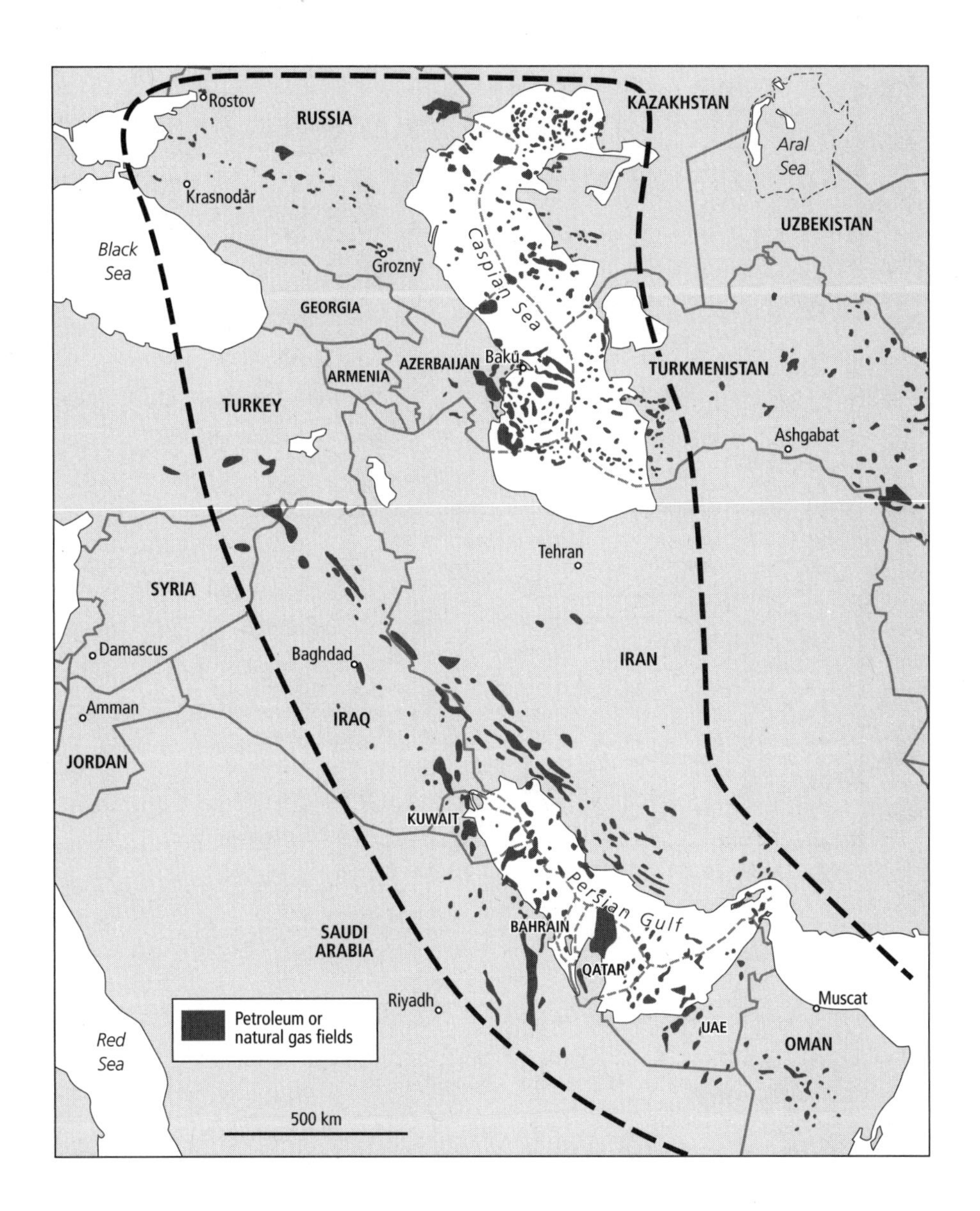

Figure A.1. The Oil Corridor

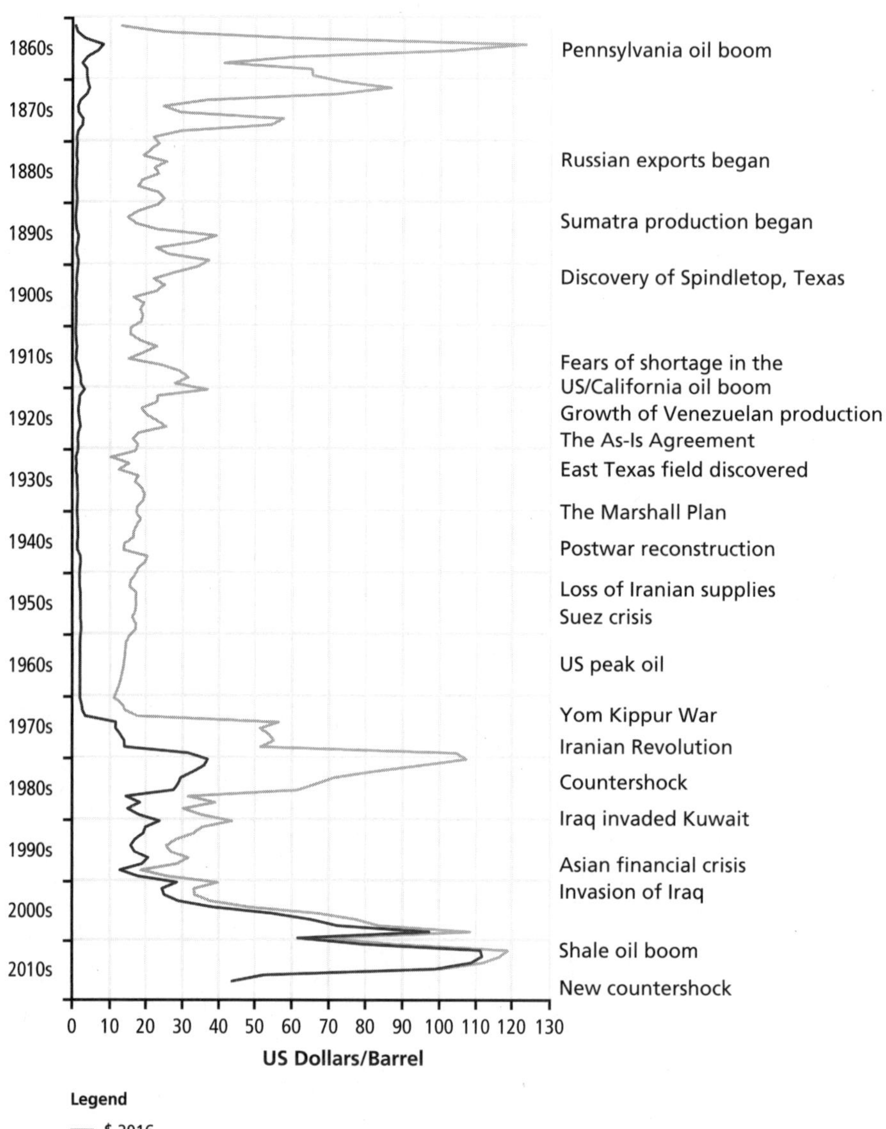

Legend

$ 2016
(deflated using the consumer price index for the United States)
Money of the day

Prices based on

1861–1944 US average
1945–1983 *Arabian Light* crude posted at Ras Tanura
1984–2016 Dated *Brent* (today's most widely used pricing benchmark)

Figure A.2. The Price of Crude Oil: 1861–December 2016

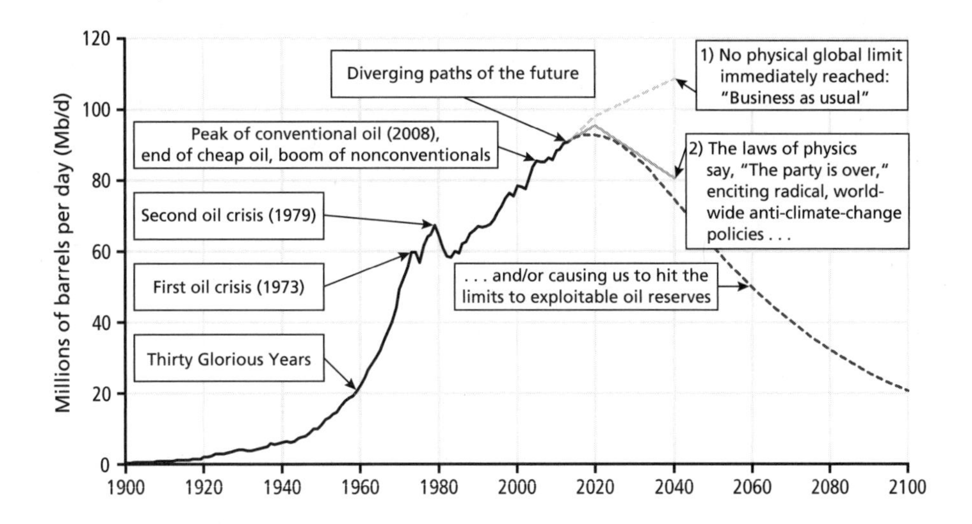

Legend

— Historical production (1900–2013)

--- International Energy Agency, *World Energy Outlook 2014.*
New policies scenario to 2040: extension of current trends (if possible).

— International Energy Agency, *World Energy Outlook 2014.* 450 ppm scenario to 2040, compatible with keeping warming to under 2°C by 2100:
Humanity must drastically restrict fuel consumption to around 2020 levels.

--- Jean Laherrère, president of ASPO France, 2014. Scenario based on an assumption of recoverable oil resources of 3 trillion barrels. World production of all types of fuels would decline from 2019/20 (global peak oil) in the absence of sufficient exploitable resources of crude.
(In 1998, Jean Laherrère correctly predicted the peak of conventional oil as 2008, see chapter 27.)

Figure A.3. World Production of Petroleum and Substitutes: Historical and Projected

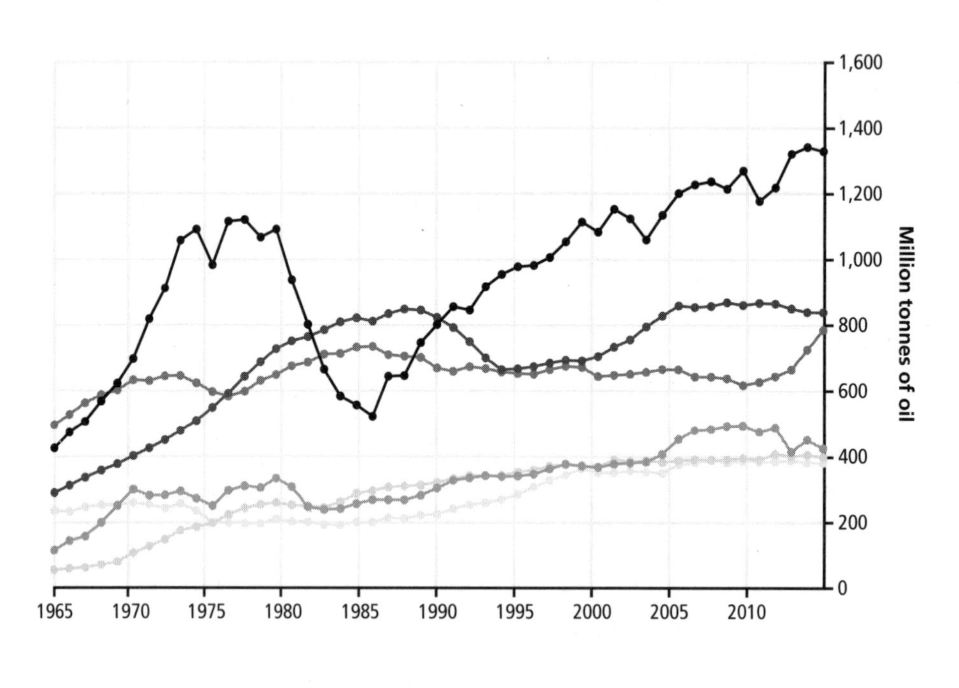

Figure A.4. Production by Region, 1965–2013

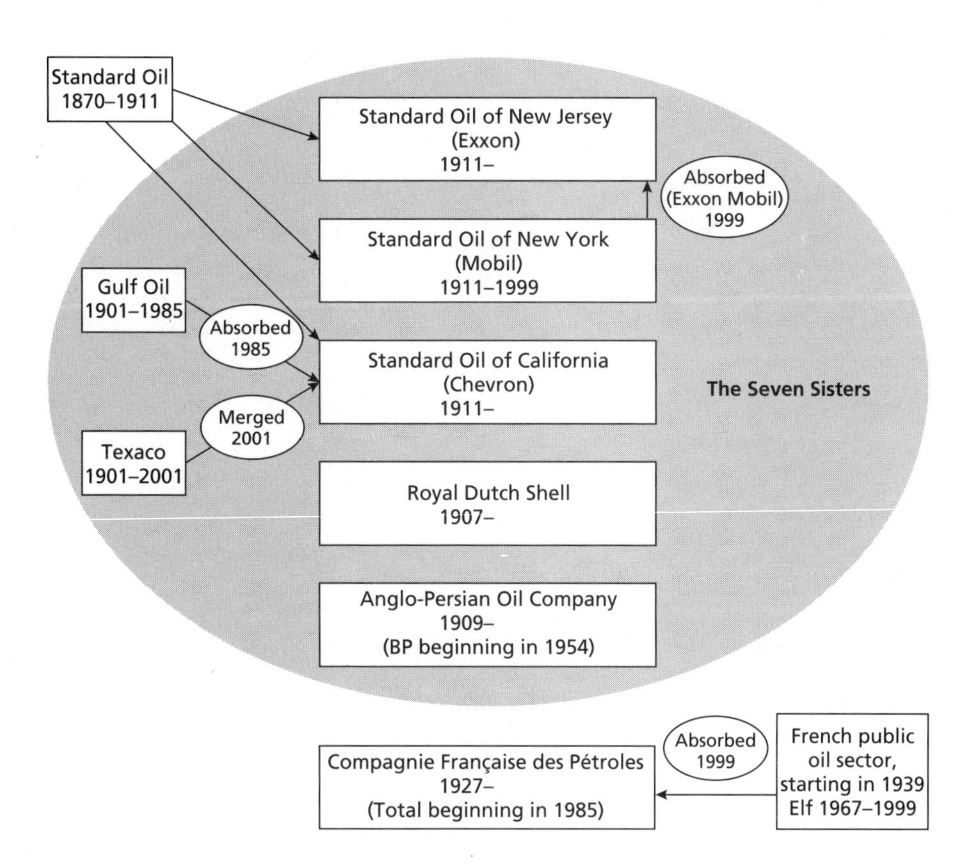

Figure A.5. The Oil Majors

IRAQ
Iraq Petroleum Company 1928
Anglo-Persian 23.75%
Shell 23.75%
CFP 23.75%
American companies 23.75%
Gulbenkian 5%

Nationalization 1972
Iraq National Oil Company
9% of worldwide reserves in 2014

KUWAIT
Kuwait Oil Company 1934
Anglo-Persian 50%
Gulf 50%

Nationalization 1975
Kuwait Petroleum Corporation
6% of worldwide reserves in 2014

IRAN
Anglo-Persian Oil Company 1909
Iranian Oil Participants 1954
BP 40%
Jersey Standard 7% – Mobil 7%
SoCal 7% – Gulf 7% – Texaco 7%
Independent American companies 5%
[American companies 40%]
Shell 14% – CFP 6%

Nationalization 1979
National Iranian Oil Company
9% of worldwide reserves in 2014

SAUDI ARABIA
Aramco 1948
Jersey Standard 30%
SoCal 30%
Texaco 30%
Mobil 10%

Nationalized from 1973 to 1980
Saudi Aramco
16% of worldwide reserves in 2014

Figure A.6. The Persian Gulf Giants

Notes

Introduction

1 Georges Bataille, "L'Économie à la Mesure de l'Univers—Notes Brèves, Préliminaires à la Rédaction d'un Essai d'Économie Générale," under the title: *La Part Maudite, Œuvres Complètes*, vol. VII, Gallimard, Paris, 1976, p. 9.

2 Frank Herbert, "La Catastrophe de Dune," *Les Enfants de Dune* [*The Children of Dune*] (1976), Pocket, Paris, 2010, p. 473.

3 The dissipation of energy, with the concept of "entropy," by the German physicist Rudolf Clausius (1822–1888), which followed the work of French physicist Sadi Carnot (1796–1832); the depletion of fossil resources by the English economist Stanley Jevons (1835–1882); and the increase in the greenhouse effect due to the emission of carbon dioxide by the Swedish winner of the Nobel Prize for Chemistry, Svante Arrhenius (1859–1927).

Chapter 1: A Seed Is Planted

1 Concerning the link between fire, the idea of soul, spirit, or God, and the intuitive idea of energy conservation, see in particular Carl Gustav Jung, "L'Inconscient Individuel et l'Inconscient Collectif ou Supra-individuel," (1942), *La Réalité de l'Âme. Structure et Dynamique de l'Inconscient*, vol. 1, Le Livre de Poche, Paris, 2008, p. 61.

2 Steve LeVine, *The Oil and the Glory: The Pursuit of Empire and Fortune on the Caspian Sea*, Random House, New York, 2007, p. 4.

3 Ciar Byrne, "Man-Made Wonders of the World under Threat from War, Want and Tourism," *The Independent*, February 14, 2005.

4 Azeri production rose to a peak of 1.08 million barrels per day (Mb/d) in May 2010, falling to 0.82 Mb/d in 2016; see in particular: John Roberts, "At the Wellhead: BP's Dudley Has Explaining to Do in Baku," Platts Agency, December 3, 2012.

5 For his valuable information, a big thank-you to Bernard Durand, former director of the geology-geochemistry division of the French Petroleum Institute and former director of the l'École Nationale Supérieure de Géologie.

6 *Epic of Gilgamesh*, Librio, Paris, 2008, p. 54.

7 Genesis, III, 11.3 and 11.4.

8 Herodotus, *Histories*, I, p. 179–184.

9 Exodus, II, 2.3.

10 Romain Pigeaud, "Du Bitume dans les Momies?," *La Recherche*, March 2002.

11 Edward Burtynsky, *Oil*, Steidl, London, 2010, p. 1.

12 Marco Polo, *The Travels*, Penguin UK; Reprint edition (February 26, 2015).

13 Hans Ulrich Vogel, "Bitumen in Pre-modern China," *Encyclopedia of the History of Science, Technology, and Medicine in Non-Western Cultures*, Kluwer Academic Publishers, Dordrecht, 1997, p. 161–162.

14 Ibid.
15 Zayn Bilkadi, "The Oil Weapons," *Saudi Aramco World*, January/February 1995, p. 20–27.
16 Hans Ulrich Vogel, "Bitumen in Pre-modern China," op. cit.
17 Zayn Bilkadi, "The Oil Weapons," op. cit.
18 Ibid.
19 Véronique Grandpierre, "En Irak, le Califat Détruit le Patrimoine de la Mésopotamie au Nom d'un Monde Nouveau," *Le Monde*, August 17, 2014.
20 Hans Ulrich Vogel, "Bitumen in Pre-modern China," op. cit.
21 Constantin Porphyrogénète, *De Administrando Imperio*, cited in Anselmo Banduri, *Imperium Orientale*, vol. I, Paris, 1711, p. 64.
22 Jose Federico Fino, "Le Feu et ses Usages Militaires," GLADIUS, IX, 1970, p. 15–30.
23 Egyptian historian al-Maqrizi (1364–1442), cited in Zayn Bilkadi, "The Oil Weapons," op. cit.
24 Cited in Pierre Juhel, *Histoire du Petrole*, Vuibert, Paris, 2011, p. 9.
25 Hans Ulrich Vogel, "Bitumen in Pre-modern China," op. cit.
26 Pierre Fontaine, *L'Aventure du Pétrole Français*, Les Sept Couleurs, Paris, 1967, p. 14.
27 John A. Harper, "Yo-Ho-Ho and a Bottle of Unrefined Complex Liquid Hydrocarbons," *Pennsylvania Geology*, vol. 26, no. 1, 1995, searchable at www.oil150.com /essays/article?article_id=68.
28 P. H. Giddens, *Pennsylvania Petroleum, 1750–1872: A Documentary History*, Drake Well Memorial Park, Pennsylvania Historical and Museum Commission, Titusville, 1947 cited in John A. Harper, "Yo-Ho-Ho and a Bottle of Unrefined Complex Liquid Hydrocarbons," op. cit.
29 D. A. Katpehko, Черное золото, Научно-популярная библиотека, vol. 52, Moscow, State Publishing House, CRP, 1953, p. 8; Miryusif Mirbabayev, *Concise History of Azerbaijani Oil*, Bakou, 2010, p. 13–14, searchable at socar-aqs.com.
30 Mike Wysatta, "66-Year-Old Oil Well Largely Unknown to West until 2002," *Reservoir Solutions*, vol. 15, no. 3, Ryder Scott Co., Houston, September–November 2012, p. 4–7, searchable at ryderscott.com.
31 Elizabeth Kolbert, "Unconventional Crude: Canada's Synthetic Fuel Boom," *The New Yorker*, November 12, 2007.
32 Jeff Lund, "The First Oil and Gas Well," *Geologic News*, Subsurface Consultants Associates, LLC, 3rd quarter 2011, p. 7.
33 Ron Chernow, *Titan: The Life of John D. Rockefeller, Sr.*, Vintage, New York, 2004, p. 33.
34 John Huston, *Moby Dick*, United Artists, 1956.
35 John Bockstoce, *Whales, Ice, & Men: The History of Whaling in the Western Arctic*, University of Washington Press, Washington, 1986, p. 94; and Walter Sheldon Tower, *A History of the American Whale Fishery*, University of Philadelphia, IHP, 1907, p. 126–127.
36 *Vanity Fair*, April 20, 1861.
37 Aristide and Stanislas Frézard, "Chronique Forestière, Revue des Eaux et Forêts, Annales Forestières," 1868, vol. 7, p. 175.
38 Ibid.
39 John H. A. Bone, *Petroleum and Petroleum Wells*, Philadelphia, 1865, cited in Anthony Sampson, *The Seven Sisters: The Great Oil Companies and the World They Made*, Viking Press, New York, 1975, p. 20.
40 Ron Chernow, *Titan*, op. cit., p. 100.
41 Daniel Yergin, *The Prize: The Epic Quest for Oil, Money and Power*, Simon and Schuster, New York, 1991, p. 13 (the pages cited here correspond to the 2008 rerelease).
42 Ron Chernow, *Titan*, op. cit., p. 101.
43 Daniel Yergin, *The Prize*, op. cit., p. 14.

44 Hildegarde Dolson, *The Great Oildorado: The Gaudy and Turbulent Years of the First Oil Rush, Pennsylvania, 1859–1880*, Random House, New York, 1959, p. 99.
45 Daniel Yergin, *The Prize*, op. cit., p. 15.

Chapter 2: John D. Rockefeller, the Power of Petroleum, and the Spiral of Expansion

1 Daniel Yergin, *The Prize*, op. cit., p. 15.
2 Ron Chernow, *Titan*, op. cit., p. 73.
3 Ibid.
4 Ibid., p. 64.
5 Ibid., p. 63.
6 Ibid., p. 77.
7 Samuel Andrews, *The Enyclopaedia of Cleveland History*, searchable at case.edu/ech.
8 Ron Chernow, *Titan*, op. cit., p. 77.
9 Ibid., p. 78.
10 Ibid., p. 76.
11 Ibid., p. 111.
12 Ibid., p. 102.
13 John Cassidy, "Rich Man, Richer Man," *The New Yorker*, May 11, 1998, p. 101.
14 Ron Chernow, *Titan*, op. cit., p. 105.
15 Ibid., p. 113.
16 Ibid., p. 132.
17 Ibid., p. 130.
18 Ibid.
19 Ibid.
20 Ibid., p. 134.
21 Ibid., p. 137.
22 Ibid.
23 Ibid., p. 143.
24 Ibid., p. 159.
25 Ibid.
26 Ibid., p. 160.
27 Ibid., p. 180.
28 Ibid., p. 182.
29 Ibid., p. 139.
30 William Inglis, "Genesis of Standard Oil," notes 4–9, the Rockefeller Archive Center Rockefeller, Sleepy Hollow.
31 Ida Tarbell, *The History of the Standard Oil Company*, McClure, Philipps & Co., New York, 1904, p. 5, searchable at archive.org/stream/historyofstandar01tarbuoft?ref=ol#page/n9/mode/2up.
32 Karl Marx, *Capital: A Critique of Political Economy*, Progress Publishers, Moscow, vol. 1, p. 435, searchable on marxists.org.
33 Ron Chernow, *Titan*, op. cit., p. 151.
34 Joseph A. Schumpeter, *Capitalism, Socialism, and Democracy*, Payot, Paris, 1990.
35 Robert Tombs, *The War Against Paris, 1871*, Cambridge University Press, Cambridge, 1981; and Gay Gullickson, *Unruly Women of Paris: Images of the Commune*, Cornell University Press, Ithaca, 1996, cited in the article "Pétroleuses," searchable at en.wikipedia.org.
36 Paul Lafargue, *Le Droit à la Paresse*, searchable at marxists.org.
37 Samuel Andrews, *The Enyclopaedia of Cleveland History*, op. cit.

38 Harold F. Williamson, Ralph L. Andreano, and Carmen Menezes, "The American Petroleum Industry," *Output, Employment, and Productivity in the United States after 1800*, National Bureau of Economic Research, 1966, p. 374, searchable at nber.org/chapters/c1572.
39 Ron Chernow, *Titan*, op. cit., p. 180.
40 Yakov Rabkin, "La Chimie et le Pétrole: Les Débuts d'une Liaison," *Revue d'Histoire des Sciences*, vol. 30, no. 4, 1977, p. 303–336.
41 Ron Chernow, *Titan*, op. cit., p. 284.
42 Charles Frederick Chandler, "Dangerous Kerosene," *American Chemist*, vol. 1, 1871, p. 123–129, cited in Yakov Rabkin, "La Chimie et le Pétrole," op. cit.
43 Yakov Rabkin, "La Chimie et le Pétrole," op. cit.
44 Ron Chernow, *Titan*, op. cit., p. 153.
45 Robert Ayres, Leslie Ayres, and Benjamin Warr, "Energy, Power and Work in the US Economy. 1900–1998," Center for the Management of Environmental Resources, Insead, Fontainebleau, 2002, fig. 6.
46 Ron Chernow, *Titan*, op. cit., p. 153, 181.
47 Daniel Yergin, *The Prize*, op. cit., p. 158–159.
48 Ron Chernow, *Titan*, op. cit., p. 181.
49 Ibid., p. 79.
50 Peter Mathias, *The First Industrial Nation: Year Economic History of Britain, 1700–1914*, University paperbacks, London, 1983 (second edition), p. 126–127.
51 Clifford D. Conner, *Histoire Populaire des Sciences*, éditions L'Échappée, Paris, 2011, p. 397.
52 Joseph A. Tainter and Tadeusz W. Patzek, *Drilling Down: The Gulf Oil Debacle and Our Energy Dilemma*, Springer-Verlag, New York, 2011, p. 105.
53 Peter Mathias, *The First Industrial Nation*, op. cit., p. 10–11.
54 Ibid., p. 122.
55 Eric Hobsbawm, *L'Ère des Révolutions: 1789–1848*, Fayard/Pluriel, Paris, 2011 (1996).
56 Peter Mathias, *The First Industrial Nation*, op. cit., p. 111.
57 Eric Hobsbawm, *L'Ère du Capital: 1848–1875*, Fayard/Pluriel, Paris, 2010 (1997), table 1.
58 Daniel Yergin, *The Prize*, op. cit., p. 10.
59 Eric Hobsbawm, *L'Ère du Capital*, op. cit., table 1.
60 Harold F. Williamson, Ralph L. Andreano, and Carmen Menezes, "The American Petroleum Industry," op. cit., table A-3., p. 400.
61 Kevin Butcher and Matthew Ponting, "The Roman Denarius under the Julio-Claudian Emperors: Mints, Metallurgy and Technology," *Oxford Journal of Archeology*, vol. 24, no. 2, May 2005, p. 163–197.
62 Barry Yeoman, "The Mines that Built Empires," *Archeology*, Archaeological Institute of America, September/October 2010, searchable at archive.archaeology.org/1009/abstracts/rio_tinto.html.
63 See for example Claude Levi-Strauss, *Tristes Tropiques* (1955), pocket, Paris, 2009, p. 99.
64 William Giles Nash, *The Rio Tinto Mine: Its History and Romance*, Simpfin, Marshall, Hamilton, Kent Co., London, 1904, p. 208, searchable at archive.org/details/riotintomineitsh00nashrich.
65 History searchable at riotinto.com.
66 Benjamin Warr, searchable at sites.google.com/site/benjaminwarr.
67 Joseph A. Tainter and Tadeusz W. Patzek, *Drilling Down*, op. cit., p. 65–134.
68 Joseph Tainter, "Collapse of Complex Societies," conference, 2010, searchable at youtube.com.
69 Stanley Jevons, *The Coal Question: An Inquiry Concerning the Progress of the Nation, and the Probable Exhaustion of Our Coal-Mines* (1865), The Macmillan Company, London, 1906, p. 200.

Chapter 3: Sharing the World Market

1 Steve LeVine, *The Oil and the Glory*, op. cit., p. 7.
2 Ibid., p. 15.
3 Robert W. Tolf, *The Russian Rockefellers: The Saga of the Nobel Family and the Russian Oil Industry*, Hoover Press, Stanford, 1976, p. 69.
4 Steve LeVine, *The Oil and the Glory*, op. cit., p. 17.
5 Daniel Yergin, *The Prize*, op. cit., p. 44.
6 Steve LeVine, *The Oil and the Glory*, op. cit., p. 19.
7 Ibid.
8 Cited in Alexei Muratov, "Baku: City and Its Time," *Stadt Bauwelt*, no. 183, September 25, 2009, p. 42.
9 Steve LeVine, *The Oil and the Glory*, op. cit., p. 11–13.
10 Ibid., p. 21–22.
11 Ralph Hewins, *Mr. Five Percent: The Story of Calouste Gulbenkian*, Rinehart and Co., New York, 1958, p. 21, cited in Steve Levine, *The Oil and the Glory*, op. cit., p. 23.
12 Ralph W. Hidy and Muriel E. Hidy, *History of Standard Oil Company New Jersey: Pioneering in Big Business, 1882–1911*, Harper, New York, 1955, p. 135.
13 Steve LeVine, *The Oil and the Glory*, op. cit., p. 23–24; and Daniel Yergin, *The Prize*, op. cit., p. 46.
14 James D. Henry, *Baku: An Eventful History*, A. Constable & Co., London, 1905, p. 104–105 searchable at archive.org/details/cu31924028739088.
15 Ron Chernow, *Titan*, op. cit., p. 284.
16 Daniel Yergin, *The Prize*, op. cit., p. 46.
17 See notably, Ida M. Tarbell, *The History of the Standard Oil Company*, op. cit.
18 Ibid., p. 215.
19 Ibid., p. 223.
20 Ron Chernow, *Titan*, op. cit., p. 207–208.
21 Ibid., p. 208.
22 Ibid., p. 208–209. During this period, Mark Twain, close to several senior directors of Standard Oil, exclaimed during a banquet: "There is a Congressman—I mean to say a son of a bitch—but why do I repeat myself?" cited in Chernow, ibid., p. 209.
23 Daniel Yergin, *The Prize*, op. cit., p. 46.
24 Ron Chernow, *Titan*, op. cit., p. 247.
25 Pierre Fontaine, *L'Aventure du Pétrole Français*, op. cit., p. 27–29.
26 Ron Chernow, *Titan*, op. cit., p. 283.
27 Thanks to the emergence of geophysics, the progress in drilling technology, and especially the introduction of hydraulics, the production of Pennsylvania and New York would be restarted during the 1920s, but would never reach 1890s production levels, and would decline definitively at the end of the 1930s. Michael Caplinger, *A Contextual Overview of Crude Oil Production in Pennsylvania*, 1997, quoted in James Hamilton, *Peak Oil in Pennsylvania*, 2010, searchable at econbrowser.com.
28 Ron Chernow, *Titan*, op. cit., p. 284.
29 Ibid., p. 283.
30 Ibid., p. 284.
31 Ibid., p. 287.
32 Ralph W. Hidy and Muriel E. Hidy, *History of Standard Oil Company New Jersey*, op. cit., p. 122–124.
33 Robert Blake, *Disraeli*, St. Martin's Press, New York, 1966, p. 581–587, cited in "Benjamin Disraeli," searchable on wikipedia.org.
34 Anthony Sampson, *The Seven Sisters*, op. cit., p. 45.

35 Robert Henriques, *Bearsted: A Biography of Marcus Samuel, First Viscount Bearsted and Founder of "Shell" Transport and Trading Company*, Barrie & Rockliff, 1960, p. 109, cited in Anthony Sampson, *The Seven Sisters*, op. cit., p. 46.
36 Ron Chernow, *Titan*, op. cit., p. 248.
37 Daniel Yergin, *The Prize*, op. cit., p. 56.
38 Ron Chernow, *Titan*, op. cit., p. 248.
39 Daniel Yergin, *The Prize*, op. cit., p. 98.
40 Robert Henriques, *Bearsted*, op. cit., p. 183; and Daniel Yergin, *The Prize*, op. cit., p. 102.
41 Robert Wooster, "Anthony Francis Lucas," Texas State Historical Association.
42 Daniel Yergin, *The Prize*, op. cit., p. 70.
43 Kenneth E. Hendrickson, *The Chief of Executives of Texas: From Stephen F. Austin to John B. Connally, Jr.*, Texas A&M University Press, College Station, 1995, p. 127.
44 Anthony Sampson, *The Seven Sisters*, op. cit., p. 48.
45 Frederik Carel Gerretson, *History of the Royal Dutch*, Brill Archive, Leiden, 1958, vol. 4, p. 8.
46 Anthony Sampson, *The Seven Sisters*, op. cit., p. 39.
47 "Elgood C. Lufkin, Oil Leader, Dead," *The New York Times*, October 10, 1935.
48 "Chauncey S. Lufkin, Well Finder of Standard, Dead," *The Boston Globe*, February 23, 1918.
49 *American Petroleum Industry, Facts and Figures*, Centennial edition, 1959.
50 "The Baku Oilfields. Great Fires Will Cripple Russia's Oil Trade," *The New York Times*, September 10, 1905.
51 Ibid.
52 Jean-Jacques Marie, *Stalin, 1878–1953: Mensonges et Mirages*, Autrement, Paris, 2013, p. 19–20.
53 Ibid., p. 21.
54 Steve LeVine, *The Oil and the Glory*, op. cit., p. 30.
55 Jean-Jacques Marie, *Stalin*, op. cit., p. 22.
56 Ibid.
57 Svante Cornell, *Small Nations and Great Powers: A Study of Ethnopolitical Conflict in the Caucasus*, Routledge Curzon, Oxford, 2000, p. 55.
58 James D. Henry, *Baku*, op. cit., p. 156.
59 Ibid., p. 157–158.
60 Ibid., p. 175.
61 Ibid., p. 181.
62 Ibid., p. 179.
63 Ibid., p. 183–184.
64 "Revolution and Fire Are Devastating Baku," *The New York Times*, September 7, 1905.
65 "German Trust to Oppose Standard Oil," *The New York Times*, November 11, 1906.
66 Maurice Pearton, *Oil and the Romanian State: 1895–1948*, Oxford University Press, Oxford, 1971, p. 1–45, cited in Daniel Yergin, *The Prize*, op. cit., p. 116.
67 Brita Åsbrink, "Branobel Celebrates Its 30th Anniversary—But the Baku Oil Is Running Out," searchable at branobelhistory.com/themes/the-branobel-company/Branobel-celebrates-its-30th-anniversary.
68 Pierre Jaloustre, "The Oil Interests of the Rothschild of Paris. Between World Competition and the Russian Government (1883–1912)," European Business History Association.

Chapter 4: The Automobile

1 Pierre Fontaine, "L'Aventure du Pétrole Français," op. cit., p. 20.
2 Michel Freyssenet, *Production Automobile des Principaux Pays Constructeurs, 1898–2009*, searchable at freyssenet.com.

3 "Hart Parr #3 Tractor," searchable at americanhistory.si.edu.
4 Ron Chernow, *Titan*, op. cit., p. 373.
5 Michel Freyssenet, *Production Automobile des Principaux Pays Constructeurs, 1898–2009*, op. cit.
6 Harold F. Williamson, Ralph L. Andreano, and Carmen Menezes, *The American Petroleum Industry*, op. cit., Table A-3., p. 400.
7 Vladimir Grigorievich Shukhov, searchable at shukhov.org.
8 No. HS-41. "Transportation Indicators for Motor Vehicles and Airlines: 1900 to 2001," US Census Bureau, Washington, 2003.
9 Cited in *Histoire du Pétrole*, op. cit., p. 70.
10 Ibid.
11 Léon Bloy, *Quatre Ans de Captivité à Cochons-sur-Marne* (1905), Robert Laffont, Paris, 1999, p. 388.
12 Viva Posada, *The Insomniac*, Montreuil, 2006, p. 53.
13 A concept developed notably in France by the engineer Jean-Marc Jancovici: See in particular Jean-François Mouhot, *Des Esclaves Énergétiques: Réflexions sur le Changement Climatique*, Champ Vallon, Seyssel, 2011.
14 Adam Hochschild, *Bury the Chains: Prophets and Rebels in the Fight to Free an Empire's Slaves*, Houghton Mifflin, Boston, 2006, cited in ibid., p. 16.
15 Michel Foucault, *Les Mots et les Choses: Une Archéologie des Sciences Humaines* (1966), Gallimard, Paris, 1990, p. 328–338.
16 Jean-François Mouhot, *Des Esclaves Énergétiques*, op. cit., p. 14.
17 Max Weber, *The Protestant Ethic and the Spirit of Capitalism,* Scribners, 1953, p. 181. See in particular in Ida M. Tarbell, *The History of the Standard Oil Company*, op. cit., p. 239.
18 See Ida M. Tarbell, *The History of the Standard Oil Company*, op. cit., p. 239.
19 Ron Chernow, *Titan*, op. cit., p. 212–213.
20 John T. Flynn, *God's Gold: The Story of Rockefeller and His Times*, Greenwood Press, Westport, 1932, p. 259, cited in Ron Chernow, *Titan*, op. cit., p. 254.
21 Ron Chernow, *Titan*, op. cit., p. 255–256.
22 Ibid., p. 259.
23 Ibid., p. 261.
24 Ibid.
25 Ibid., p. 295.
26 Ibid., p. 298.
27 Ibid., p. 336–337.
28 John T. Flynn, *God's Gold*, op. cit. and Anna Robeson Burr, *The Portrait of a Banker: James Stillman, 1850–1918*, Duffield, New York, 1927, p. 117, cited in Ron Chernow, *Titan*, op. cit., p. 338.
29 Ida M. Tarbell, *The History of the Standard Oil Company*, op. cit., p. 36.
30 Ibid., p. 37.
31 Ibid., p. 541.
32 Ibid., p. 543.
33 Ibid.
34 John T. Flynn, *God's Gold*, op. cit., p. 429, cited in Ron Chernow, *Titan*, op. cit., p. 543.
35 Bruce Bringhurst, *Antitrust and the Oil Monopoly: The Standard Oil Cases, 1890–1911*, Greenwood Press, Westport, 1979, p. 136, cited in Ron Chernow, *Titan*, op. cit., p. 545.
36 Ron Chernow, *Titan*, op. cit., p. 540–542.
37 Ibid., p. 549–550.
38 David Bryn-Jones, *Frank B. Kellogg: A Biography*, Putnam, New York, 1937, p. 66, cited in Daniel Yergin, *The Prize*, op. cit., p. 93.

39 Michael A. Whitehouse, "Paul Warburg's Crusade to Establish a Central Bank in the United States," Federal Reserve Bank of Minneapolis, 1989, searchable at minneapolisfed.org /publications /the-region/paul-warburgs-crusade-to-establish-a-central-bank-in-the-united-states.
40 Ron Chernow, *Titan*, op. cit., p. 377.
41 David Rockefeller, *Memoirs*, Random House, New York, 2002, p. 124–125.
42 Ron Chernow, *Titan*, op. cit., p. 377.
43 "Millions Paid to Spain," *The New York Times*, May 2, 1899.
44 Michael A. Whitehouse, "Paul Warburg's Crusade," op. cit.
45 Frank Arthur Vanderlip and Boyden Sparkes, *From Farm Boy to Financier*, D. Appleton-Century Co., New York, 1935, p. 218.
46 Bruce Bringhurst, *Antitrust and the Oil Monopoly*, op. cit., p. 173, cited in Ron Chernow, *Titan*, op. cit., p. 555.
47 Ron Chernow, *Titan*, op. cit., p. 556–557.
48 Gabriel Kolko, *The Triumph of Conservatism: A Reinterpretation of American History, 1900–1916*, Quadrangle Paperbacks, Chicago, 1967, p. 194, cited in Ron Chernow, *Titan*, op. cit., p. 555.
49 Bertrand Russell, *Freedom and Organization: 1814–1914*, Routledge, Oxford, 2013, p. 357.
50 Ron Chernow, *Titan*, op. cit., p. 558.

Chapter 5: The Tank

1 Yakov Rabkin, "La Chimie et le Pétrole: Les Débuts d'une Liaison," op. cit., p. 314.
2 Pierre Juhel, *Histoire du Pétrole*, op. cit., p. 42.
3 Anthony Sampson, *The Seven Sisters*, op. cit., p. 49.
4 Pierre Juhel, *Histoire du Pétrole*, op. cit., p. 44.
5 Anthony Sampson, *The Seven Sisters*, op. cit., p. 49.
6 Tony Gibbons, *The Complete Encyclopedia of Battleships and Battlecruisers*, Salamander Books, London, 1983, p. 137.
7 Anthony Sampson, *The Seven Sisters*, op. cit., p. 50.
8 Nubar Gulbenkian, *Pantaraxia*, Hutchinson, London, 1965, p. 81–82.
9 Joseph A. Tainter and Tadeusz W. Patzek, *Drilling Down*, op. cit., p. 23 sq.
10 Winston Churchill, *The World Crisis*, vol. I, Scribner, New York, 1923, p. 130–136, cited in Daniel Yergin, *The Prize*, op. cit., p. 140.
11 E. M. Earle, *Turkey, the Great Powers, and the Bagdad Railway: A Study in Imperialism* (1923), Russel & Russel, New York, 1966, p. 14.
12 André Giraud and Xavier Boy de la Tour, *Géopolitique du Pétrole et du Gaz,* Éditions Technip, Paris, 1987, p. 198.
13 Nubar Gulbenkian, *Pantaraxia*, op. cit., p. 166.
14 Ibid., p. 83.
15 Ibid., p. 84.
16 Arlette Estienne Mondet, *Le Général J. B. E. Estienne "Père des Chars,"* L'Harmattan, Paris, 2010, p. 65.
17 Henri Ortholan, *La Guerre des Chars: 1916–1918*, Bernard Giovanangelis Éditeur, Paris, 2007, p. 25, cited in the article "Jean- Baptiste Eugène Estienne," searchable at fr.wikipedia.org.
18 Henri Ortholan, *La Guerre des Chars*, op. cit., p. 29.
19 *American Petroleum Industry, Facts and Figures*, Centennial Edition, op. cit.
20 Roberto Nayberg, "Qu'Est-ce Qu'un Produit Stratégique? L'Exemple du Pétrole en France, 1914–1918," 2005, searchable at institut-strategie.fr.
21 "Standard Oil Takes Million of the Loan," *The New York Times*, May 5, 1917.
22 Pierre Fontaine, *L'Aventure du Pétrole Français*, op. cit., p. 40–41.

23 Archives du Ministère des Affaires Etrangères Français, Correspondance Diplomatique, Série "Guerre," 1CPCOM/1331.
24 Roberto Nayberg, " Qu'Est-ce Qu'un Produit Stratégique?," op. cit.
25 Ibid.
26 Ibid.
27 Hector C. Bywater, *A Searchlight on the Navy*, Constable & Co. Ltd., London, 1934, p. 113.
28 "Chambre des Députés, Séance du 7 Mars 1895," cited in Jean Jaurès, *Rallumer Tous les Soleils*, Omnibus, Paris, 2006, p. 243 sq.

Chapter 6: The Roaring 1920s

1 B. Traven, "Contraste," in *Dans l'État le Plus Libre du Monde, L'Insomniaque*, Montreuil, 2011, p. 73.
2 See in particular François Furet, *Le Passé d'Une Illusion. Essai sur l'Idée Communiste au XXe Siècle*, Robert Laffont, Paris, 1995.
3 Gustav Landauer, Aufruf zum Sozialismus, Cologne, 1923, p. 47–48, cited in Allgemein, "En Avant à Toute Vapeur," *Espace contre Ciment*, raumgegenze-ment.blogsport.de/index.php?s=En+Avant+à+Toute+Vapeur.
4 Karl Marx and Friedrich Engels, *Manifeste du Parti Communiste*, 1848, p. 12, searchable at classiques.uqac.ca/classiques/labriola_antonio/essais_materialisme_historique/Essai_3_Manifeste_PC/Le_manifeste_PC.html.
5 Fritz Lang, *Interviews*, University Press of Mississippi, Jackson, 2003, p. 69.
6 Friedrich Nietzsche, "Loisir et Oisiveté," *Le Gai Savoir*, GF Flammarion, Paris, 1997, p. 265.
7 Walter Benjamin, *Sur le Concept d'Histoire*, no. XI, Payot, Paris, 2013, p. 68–69.
8 B. Traven, *Le Vaisseau des Morts*, La Découverte, Paris, 2010, p. 178.
9 B. Traven, *Rosa Blanca*, La Découverte, Paris, 2010.
10 Steve LeVine, *The Oil and the Glory*, op. cit., p. 33.
11 Antony C. Sutton, *Wall Street and the Bolshevik Revolution*, Buccaneer Books, Cutchogue, 1999, chapter 8.
12 Nubar Gulbenkian, *Pantaraxia*, op. cit., p. 103.
13 Steve LeVine, *The Oil and the Glory*, op. cit., p. 42–45.
14 Antony C. Sutton, *Western Technology and Soviet Economic Development, 1917 to 1930*, Hoover Institution Press, Stanford, 1968, p. 16–40.
15 "Sinclair Works Big Oil Grant in Africa," *The New York Times*, June 29, 1923.
16 Steve LeVine, *The Oil and the Glory*, op. cit., p. 47.
17 Ibid., p. 47–48.
18 Frank C. Hanighen, *The Secret War*, John Day & Co., New York, 1934, p. 109–119, cited in Steve LeVine, *The Oil and the Glory*, op. cit., p. 48.
19 Alexander Igolkin, "Oil of Russia," Learning from American Experience, no. 1, 2006, searchable at oilru.com.
20 Fritz Lang, *Interviews*, op. cit., p. 69.
21 Ron Cowen, "George Gershwin: He Got Rhythm," *The Washington Post*, November, 1998.
22 Ibid.; and Louis-Ferdinand Céline, *Voyage au Bout de la Nuit*, Le Livre de Poche, Paris, 1966, p. 225.
23 Bennett H. Wall and George S. Gibb, *Teagle of Jersey Standard*, Tulane University, New Orleans, 1974.
24 Ludwell Denny, *We Fight for Oil*, A. Knopf, New York, 1928, p. 25, searchable at cdn.preterhuman.net/texts/religion.occult.new_age/occult.conspiracy.and.related/Denny%20-%20We%20Fight%20for%20Oil%20(1928)%20+%20America%20Conquers%20Britain%20(1930).pdf.
25 Harold F. Williamson, Ralph L. Andreano, and Carmen Menezes, *The American Petroleum Industry*, op. cit., table 16, p. 382.

26 Energy Information Administration, searchable at eia.gov.
27 "Future Oil Supply Needed by Country," *The New York Times*, January 5, 1920.
28 David White, "The Petroleum Resources of the World," *The Annals of the American Academy of Political and Social Science*, vol. 89, May 1920, p. 121, cited in US Senate, *The International Petroleum Cartel*, Staff Report to the Federal Trade Commission, US Government Printing Office, Washington, 1952, p. 37–46, searchable at mtholyoke.edu.
29 Ibid.
30 Margaret L. Davis, *Dark Side of Fortune: Triumph and Scandal in the Life of Oil Tycoon Edward L. Doheny*, University of California Press, Berkeley, 1998, p. 8.
31 Daniel Yergin, *The Prize*, op. cit., p. 198.
32 Margaret L. Davis, *Dark Side of Fortune*, op. cit., p. 257–258.
33 "Oil in California," Paleontological Research Institution, Ithaca, searchable at priweb.org /index.php/ed-woo-home/582-ed-woo-s6-california.
34 Robert W. Cherny, *Graft and Oil: How Teapot Dome Became the Greatest Political Scandal of Its Time*, Gilder Lehrman Institute of American History, searchable at new.gilderlehrman.org /history-by-era/roaring-twenties/essays/graft-and-oil-how-teapot-dome-became-greatest -political-scand.
35 Greg Lange, "US President Warren G. Harding Makes His Last Speech in Seattle on July 27, 1923," February 10, 1999, searchable at historylink.org.
36 Francis Russell, *The Shadow of Blooming Grove: Warren G. Harding in His Times*, Easton Press, Norwalk, 1968, p. 588–589, cited in the article "Warren G. Harding," searchable at en.wikipedia.org.
37 Harvey M. Beigel, *Battleship Country: The Battle Fleet at San Pedro-Long Beach, California, 1919–1940*, Pictorial Histories Publishing Co., Missoula, 1983, p. 1–3.
38 Louis Adamic, "Los Angeles. There She Blows," *Outlook and Independent*, August 13, 1930.
39 Ibid.
40 Mike Davis, *City of Quartz: Los Angeles, Capitale du Futur*, La Découverte, Paris, 2006, p. 105.
41 See Heading Out, "Tech Talk. The Oil under Los Angeles," June 12, 2011, searchable at theoildrum.com.
42 Marc Wanamaker, *Early Beverly Hills*, Arcadia Publishing, Mount Pleasant, 2006, p. 17–18.
43 Mike Davis, *City of Quartz*, op. cit., p. 57.
44 Ibid.
45 See particularly "Fear McAdoo Candidacy Doomed," *The New York Times*, February 2, 1924.
46 Mike Davis, *City of Quartz*, op. cit., p. 103; and Jules Tygiel, "What a Money-Gusher," *The Los Angeles Times*, December 3, 2006.
47 Jules Tygiel, "What a Money-Gusher," op. cit.
48 Louis Adamic, "Los Angeles. There She blows," op. cit.
49 Jules Tygiel, "What a Money-Gusher," op. cit.
50 Mike Davis, *City of Quartz*, op. cit., p. 106.
51 Upton Sinclair, *Oil!* (1927), Penguin Books, London, 2008. This book was the main inspiration for the movie by Paul Thomas Anderson, *There Will Be Blood*, released in 2008.
52 Mike Davis, *City of Quartz*, op. cit., p. 104.
53 John Steinbeck, *Les Raisins de la Colère* [*The Grapes of Wrath*] (1939), Gallimard, Paris, 1964, chapter 12.
54 Report of the Federal Oil Conservation Board, Part 1, September 1926, p. 1, cited in Ludwell Denny, *We Fight for Oil*, op. cit., p. 160, searchable at cdn.preterhuman.net.
55 Calvin Coolidge, December 19, 1924, letter to the US secretaries of war, navy, interior, and commerce, as it appears in Report of the Federal Oil Conservation Board to the President

of the United States, Part I, September 1926, Interior Department Building, Office of the Board, Department of the Interior, Washington, Government Printing Office, 1926.

Chapter 7: Birth of a Petrol-Nation

1 James Barr, *A Line in the Sand: Britain, France and the Struggle that Shaped the Middle East*, Simon & Schuster, New York, 2012, p. 31.
2 Ibid. p. 33.
3 Ibid., p. 66.
4 Ibid.
5 Ibid.
6 Ibid., p. 72.
7 Ibid., p. 82.
8 Ibid., p. 84.
9 Ibid., p. 105.
10 Edgar Sperling's journal, September 1919, cited in US Senate, "The International Petroleum Cartel," op. cit., p. 37–46.
11 Archives of the French Minister of Foreign Affairs series. "Irak," 48CPCOM/32.
12 Nubar Gulbenkian, *Pantaraxia*, op. cit., p. 67.
13 Anthony Sampson, *The Seven Sisters*, op. cit., p. 66.
14 Daniel Yergin, *The Prize*, op. cit., p. 181.
15 Subcommittee on Multinational Corporations, Multinational Oil Corporations and US Foreign Policy, report of the Senate of the United States, US Government Printing Office, Washington, 2 January 1975, searchable at mtholyoke.edu/acad/intrel/oil1.htm.
16 Anthony Sampson, *The Seven Sisters*, op. cit., p. 66.
17 "Churchill to Trenchard," August 29, 1920, Churchill Archive Center, cited in James Barr, *A Line in the Sand*, op. cit., p. 113.
18 "War Office Minute," May 12, 1919, cited in Martin Gilbert and Randolph S. Churchill, *Winston S. Churchill: The Official Biography*, Heinemann, London, 1976, companion vol. 4, part 1.
19 Niall Ferguson, *The War of the World: History's Age of Hatred*, Allen Lane, London, 2006, p. 412.
20 Thomas Edward Lawrence, *Letters*, David Garnett, New York, 1939, p. 311.
21 James Barr, *A Line in the Sand*, op. cit., p. 113.
22 Ibid., p. 108–109.
23 "Mr Lloyd George in the Garden of Eden," *The Times*, June 24, 1920, cited in ibid., p. 109.
24 James Barr, *A Line in the Sand*, op. cit., p. 121.
25 Ibid., 123.
26 Daniel Yergin, *The Prize*, op. cit., p. 175, 177.
27 Peter Mathias, *The First Industrial Nation*, op. cit., p. 399–400.
28 Nubar Gulbenkian, *Pantaraxia*, op. cit., p. 89.
29 Ibid.
30 James Barr, *A Line in the Sand*, op. cit., p. 155.
31 Daniel Yergin, *The Prize*, op. cit., p. 185.
32 Senate Hearings on Petroleum Resources, Washington, June 27 and 28, 1945, cited in Anthony Sampson, *The Seven Sisters*, op. cit., p. 66.
33 Pierre Juhel, *Histoire du Pétrole*, op. cit., p. 113.
34 Nubar Gulbenkian, *Pantaraxia*, op. cit., p. 127.
35 Subcommittee on Multinational Corporations, *Multinational Oil Corporations and U.S. Foreign Policy*, op. cit.
36 Nubar Gulbenkian, *Pantaraxia*, op. cit., p. 96.

37 James Barr, *A Line in the Sand*, op. cit., p. 157.
38 Ibid.
39 E. P. Fitzgerald, "Business Diplomacy. Walter Teagle, Jersey Standard, and the Anglo-French Pipeline Conflict in the Middle East, 1930–1931," *Business History Review*, vol. 67, no. 2, summer 1993, p. 222.
40 Ibid., p. 223.
41 James Barr, *A Line in the Sand*, op. cit., p. 163.
42 "Chronologie," searchable at total.com.
43 "Trouble in Paradise," *Time*, April 21, 1941, cited in James Barr, *A Line in the Sand*, op. cit., p. 163.

Chapter 8: The Majors Band Together

1 Anthony Sampson, *The Seven Sisters*, op. cit., p. 70; and US Senate, "The International Petroleum Cartel," op. cit., p. 198.
2 Anthony Sampson, *The Seven Sisters*, op. cit., p. 72.
3 Bennett H. Wall and George S. Gibb, *Teagle of Jersey Standard*, op. cit., p. 261.
4 Anthony Sampson, *The Seven Sisters*, op. cit., p. 72.
5 "Draft Achnacarry Agreement," August 18, 1928, cited in J. H. Bamberg, *The History of the British Petroleum Company, vol. 2, The Anglo-Iranian Years, 1928–1954*, Cambridge University Press, Cambridge, 1994, p. 528–534, searchable at mtholyoke.edu.
6 US Senate, "The International Petroleum Cartel," op. cit., p. 200.
7 John M. Blair, *The Control of Oil*, Vintage Books, New York, 1978, p. 113–114; and Anthony Sampson, *The Seven Sisters*, op. cit., p. 73–74.
8 John M. Blair, *The Control of Oil*, op. cit., p. 57.
9 US Senate, "The International Petroleum Cartel," op. cit., p. 244.
10 John M. Blair, *The Control of Oil*, op. cit., p. 62.
11 US Senate, "The International Petroleum Cartel," op. cit., p. 245.
12 Geoffrey Jones, *The State and the Emergence of the British Oil Industry*, Macmillan Business History Unit, University of London, London, 1981, p. 22.
13 "Draft Achnacarry Agreement," op. cit.
14 Angus Maddison, "Historical Statistics for the World Economy: 1-2006 AD," op. cit.
15 Bureau of Economics Analysis, Federal Reserve.
16 Jacques Brasseul, "La Crise de 1929," searchable at brasseul.free.fr/HFEt3.htm.
17 Bennett H. Wall and George S. Gibb, *Teagle of Jersey Standard*, op. cit., p. 98.
18 Herbert Hoover, *The Memoirs of Herbert Hoover: The Great Depression, 1929–1941*, MacMillan Company, New York, 1952, p. 30.
19 Executive Order 5327 by the president of the United States.
20 "Drilling Sideways. A Review of Horizontal Well Technology and Its Domestic Application," US Department of Energy, April 1993, p. 13, searchable at eia.gov.
21 Julia Cauble Smith, "East Texas Oilfield," Texas State Historical Association, 1999, searchable at tshaonline.org.
22 Ibid.
23 Julia Cauble Smith, "East Texas Oilfield," op. cit.
24 William S. Osborn, "Curtains for Jim Crow: Law, Race, and the Texas Railroads," *Southwestern Historical Quarterly*, vol. 105, no. 3, January 2002, p. 392–427, searchable at texassantafehistory.com/q%20osborn.pdf.
25 Daniel Yergin, *The Prize*, op. cit., p. 233.
26 Julia Cauble Smith, "East Texas Oilfield," op. cit.

27 British Petroleum, *Statistical Review of World Energy Workbook*, 2013.
28 "J. A. Moffett Quits Standard Oil Post," *The New York Times*, July 29, 1933.
29 British Petroleum, *Statistical Review of World Energy Workbook*, 2013.
30 "James A. Moffett Jr. Succeeds F. D. Asche," *The New York Times*, 1924.
31 Anthony Sampson, *The Seven Sisters*, op. cit., p. 72.
32 "J. A. Moffett Quits Standard Oil Post," op. cit.
33 "Supper Dance Held at Palm Beach," *The New York Times*, December 24, 1933; and "James A. Moffett Host on His Yacht," *The New York Times*, December 29, 1933.
34 Ron Chernow, *Titan*, op. cit., p. 195–196.
35 Anthony Sampson, *The Seven Sisters*, op. cit., p. 87.
36 Michael Moore, *Fahrenheit 9/11* (documentary), 2004.
37 Harold Ickes, "After the Oil Deluge, What Price Gasoline," *The Saturday Evening Post*, February 16, 1935.
38 Energy Information Administration, searchable at eia.gov.
39 Daniel Yergin, *The Prize*, op. cit., p. 279.
40 Nubar Gulbenkian, *Pantaraxia*, op. cit., p. 107.
41 Rafael Archondo, "La Guerra del Chaco: ¿Hubo Algún Titiritero?," *Poblaciaon y Desarrollo*, no. 34, Universidad Nacional de Asuncion, p. 29–39, searchable at revistascientificas.una.py/index.php/RE/article/view/736; and Herbert Klein, *Parties and Political Change in Bolivia: 1880–1952*, Cambridge University Press, Cambridge, chapter 5.
42 "Paraguay to Accuse Oil Concern in Chaco," *The New York Times*, July 23, 1934; and "Russians Berate Helsinki Leaders," *The New York Times*, November 19, 1939.
43 Robert Vitalis, *America's Kingdom: Mythmaking on the Saudi Oil Frontier*, Verso, London, 2009, p. 22.
44 Ludwell Denny, *We Fight for Oil*, op. cit., p. 28.
45 Ibid., p. 28–29; and H. C. F. Bell, *Woodrow Wilson and the People*, Kessinger Publishing, Whitefish, 2005, p. 151–152.
46 Andrew Garfield, "Dragging Lazards into the 21st Century," *The Independent*, June 25, 1999.
47 Ibid., p. 105.
48 Nubar Gulbenkian, *Pantaraxia*, op. cit., p. 65.
49 Ibid., p. 65–66.
50 "Mexican Expropriation of Foreign Oil, 1938," Office of the Historian, US Department of State, searchable at history.state.gov/milestones/1937-1945/mexican-oil.
51 Ibid.
52 Robert Huesca, "The Mexican Oil Expropriation and the Ensuing Propaganda War," Latin American Studies Institute, University of Texas, Austin, 1988, p. 3, searchable at lanic.utexas.edu/project/etext/llilas/tpla/8804.pdf.
53 "Reich Increasing Fuel Oil Imports," *The New York Times*, February 16, 1939.

Chapter 9: The Persistent Alliance of Big Oil with Nazi Germany

1 Daniel J. Kevles, *Au Nom de l'Eugénisme: Génétique et Politique dans le Monde Anglo-Saxon*, Presses Universitaires de France, Paris, 1995.
2 Edwin Black, "Eugenics and the Nazis—the California Connection," *The San Francisco Chronicle*, November 9, 2003.
3 Michael Dobbs, "Ford and GM Scrutinized for Alleged Nazi Collaboration," *The Washington Post*, November 30, 1998.
4 "Reich Oil Monopoly Sought by Deterding," *The New York Times*, October 26, 1934.

5 "Sir H. Deterding Provides Money," *The Manchester Guardian*, December 29, 1936; and "Henri Deterding Dies in St. Moritz," *The New York Times*, February 5, 1939.
6 "Deterding Honored by Nazis at Funeral," *The New York Times*, February 11, 1939.
7 Herbert Feis, *The Spanish Story*, New York, 1948, p. 269–271; Luis Bolin, *Spain: The Vital Years*, London, 1967, p. 224–225; and Hugh Thomas, *The Spanish Civil War*, London, 1975, cited in Anthony Sampson, *The Seven Sisters*, op. cit., p. 81–82.
8 "U.S. Oil Executive Recent Visitor to Reich, Calls on President, Raising Talk of Peace," *The New York Times*, January 30, 1949.
9 "Ethiopia Reported Giving Huge Grant to Anglo-U.S. Group," *The New York Times*, August 31, 1935.
10 "Ethiopia Rich in Oil, Rickett Declares," *The New York Times*, September 2, 1935.
11 "Backers Are Revealed," *The New York Times*, September 4, 1935.
12 Joseph M. Levy, "Rickett Confident on Oil Concession," *The New York Times*, September 9, 1935.
13 "Teagle Disclaims Share in Oil Deal," *The New York Times*, September 10, 1935.
14 "War over Oil," *The New York Times*, December 1, 1935.
15 "Complexity Delays Oil Embargo Moves," *The New York Times*, December 8, 1935.
16 "Ickes Now Denies He Urged Oil Ban," *The New York Times*, December 4, 1935.
17 "Standard Oil Denies Deal with Italians," *The New York Times*, December 4, 1935.
18 "Jersey Standard Denies War Trade," *The New York Times*, December 6, 1935.
19 "Oil Sales to Italy Jump," *The New York Times*, December 7, 1934; and "Oil 'War' Exports to Italy Increase," *The New York Times*, December 22, 1934.
20 "Rickett Visits Rome on Mystery Errand," *The New York Times*, January 8, 1935.
21 G. L. Steer, "Ethiopia Makes New Denial," *The New York Times*, March 19, 1936.
22 Robert Goralski and Russell W. Freeburg, *Oil and War: How the Deadly Struggle for Fuel in WWII Meant Victory or Defeat*, William Morrow & Co., New York, 1987, p. 23–24.
23 Raymond Guglielmo, *La Petrochimie dans le Monde*, Presses Universitaires de France, Paris, 1958, p. 10.
24 Ibid., p.7.
25 Daniel Yergin, *The Prize*, op. cit., p. 313.
26 David Edgerton, *The Shock of the Old: Technology and Global History since 1900*, Profile Books, London, 2008, p. 120.
27 *Nuernberg Military Tribunal*, vol. VII, The Mazal Holocaust Library, p. 15.
28 Ibid., p. 19.
29 Ibid., p. 25.
30 "Oil as a Factor in the German War Effort, 1933–1945," Chiefs of Staff Committee, Washington, 1946, p. ii–18.
31 Ibid.
32 Mira Wilkins, *The History of Foreign Investment in the United States, 1914–1945*, Harvard University Press, Cambridge, 2004, p. 241.
33 B. J. Woolf, "Teagle of the Dynasty of Oil Kings," *The New York Times*, June 15, 1930; and Mira Wilkins, *The History of Foreign Investment in the United States, 1914–1945*, op. cit., p. 333.
34 Ibid., p. 242.
35 Anthony Sampson, *The Seven Sisters*, op. cit., p. 79.
36 Frank L. Kluckhohn, "Arnold Says Standard Oil Gave Nazis Rubber Process," *The New York Times*, March 27, 1942.
37 J. H. Carmical, "Synthetic Rubber to Be Made Here," *The New York Times*, February 11, 1940.
38 Frank L. Kluckhohn, "Arnold Says Standard Oil Gave Nazis Rubber Process," op. cit.

39 "Farish Says German Deals Speeded Our War Industry," *The New York Times*, April 1, 1942.
40 Ibid.
41 J. H. Carmical, "Along Wall Street," *The New York Times*, January 28, 1940; and "Rumania Decrees Oil Line's Seizure," *The New York Times*, December 5, 1940.
42 "Deception Is Laid to Standard Oil," *The New York Times*, June 1, 1942.
43 Bennett H. Wall and George S. Gibb, *Teagle of Jersey Standard*, op. cit., p. 297–319.
44 "Wallace Charges Standard Stalled," *The New York Times*, September 14, 1943.
45 US Government Printing Office, Washington, 1953, VII, p. 1309, NI-10, 551, cited in Joseph Borkin, *The Crime and Punishment of IG Farben*, Free Press, New York, 1978, chapter 4, searchable at bibliotecapleyades.net/sociopolitica/sociopol_igfarben02.htm. See also "U.S. Firms Fueled Germany for War," *The New York Times*, October 19, 1945.
46 Joseph Borkin, *The Crime and Punishment of IG Farben*, op. cit., chapter 4.
47 US Senate, "Scientific and Technical Mobilization: Hearing before a Subcommittee of the Committee on Military Affairs," US Government Printing Office, Washington, 1943, p. 939. See also Joseph Borkin, *The Crime and Punishment of IG Farben*, op. cit., chapter 4.

Chapter 10: The Enablers of the Second World War

1 Daniel Yergin, *The Prize*, op. cit., p. 283.
2 "Japanese Seeking Mexican Oil Firm," *The New York Times*, April 8, 1938.
3 Jean Lopez, "Les Alliés, Maîtres du Jeu Pétrolier," *Guerres and Histoire*, no. 9, October/November 2012, p. 40.
4 *Twentieth Century Petroleum Statistics*, DeGolyer & MacNaughton, cited in Robert Goralski and Russell W. Freeburg, *Oil and War*, op. cit., appendix 1.
5 Karl-Heinz Frieser, *Blitzkrieg-Legende. Der Westfeldzug 1940*, Oldenbourg Industrieverlag GmbH, Munich, 1996 (2nd edition), p. 5.
6 Robert Goralski and Russell W. Freeburg, *Oil and War*, op. cit., p. 56.
7 Jean Lopez, "Les Alliés, Maîtres du Jeu Pétrolier," op. cit., p. 42.
8 Robert Goralski and Russell W. Freeburg, *Oil and War*, op. cit., p. 42.
9 Ibid., p. 47–48.
10 "Oil as a Factor in the German War Effort, 1933–1945," op. cit., ii–7.
11 Robert Goralski and Russell W. Freeburg, *Oil and War*, op. cit., p. 275–277.
12 International Military Tribunal, vol. XXVI, 873-PS, cited in ibid., p. 47.
13 Robert Goralski and Russell W. Freeburg, *Oil and War*, op. cit., p. 92.
14 Ibid., p.102 and 348, appendix 14.
15 "Roosevelt's Talks Center on Crises," *The New York Times*, September 13, 1937.
16 Robert Goralski and Russell W. Freeburg, *Oil and War*, op. cit., p. 93–96.
17 Nubar Gulbenkian, *Pantaraxia*, op. cit., p. 119–120.
18 Ibid., p. 98.
19 Saburō Hayashi and Alvin D. Coox, *Kōgun: The Japanese Army in the Pacific War*, Greenwood Press, Westport, 1959, p. 23–24.
20 Robert Goralski and Russell W. Freeburg, *Oil and War*, op. cit., p. 87.
21 Daniel Yergin, *The Prize*, op. cit., p. 306.
22 Ibid., p. 299.
23 Robert Goralski and Russell W. Freeburg, *Oil and War*, op. cit., p. 54–55.
24 "Oil as a Factor in the German War Effort, 1933–1945," op. cit., vii–18.
25 Robert Goralski and Russell W. Freeburg, *Oil and War*, op. cit., p. 62.
26 Ibid., p. 278.
27 Jean Lopez, "Les Alliés, Maîtres du Jeu Pétrolier," op. cit., p. 38.

28 Robert Goralski and Russell W. Freeburg, *Oil and War*, op. cit., p. 82.
29 Ibid., p. 82–83.
30 Ibid., p. 178–179.
31 Ibid., p. 183.
32 Ibid., p. 174.
33 Jean Lopez, "Et Si l'Allemagne Avait Pris Maïkop et Grozny?," *Guerres and Histoire*, no. 9, October/November 2012.
34 Benoist Bihan, "L'Armée Rouge a Changé l'Art Militaire pour l'Emporter," *Guerres & Histoire*, no. 11, February/March 2013.
35 Jacques Benoist-Méchin, "Rapport à l'Amiral Darlan," January 9, 1942, translated by Benoist-Méchin, p. 344, cited in Jacques Benoist-Méchin, *Ibn Séoud ou la Naissance d'un Royaume*, Albin Michel, Paris, 1990, p. 334.
36 Robert Goralski and Russell W. Freeburg, *Oil and War*, op. cit., p. 127.
37 Edward R. Stettinius, *Lend-Lease, Weapon for Victory*, Macmillan, New York, 1944, p. 252–253, cited in Jacques Benoist-Méchin, *Ibn Séoud ou la Naissance d'un Royaume: Le Loup at le Léopard*, op. cit., p. 339–341.
38 Robert Goralski and Russell W. Freeburg, *Oil and War*, op. cit., p. 209.
39 Ibid., p. 211.
40 Ibid., p. 200.
41 B. H. Liddell-Hart, *The Rommel Papers*, Harcourt, Brace & Co., New York, 1953, p. 298.
42 Robert Goralski and Russell W. Freeburg, *Oil and War*, op. cit., p. 216–217.
43 B. H. Liddell-Hart, *The Rommel Papers*, op. cit., p. 261.
44 Robert Goralski and Russell W. Freeburg, *Oil and War*, op. cit., p. 170, 173.
45 "The Big Inch and Little Big Inch Pipelines: The Most Amazing Government-Industry Cooperation Ever Achieved," Texas Eastern Transmission Corporation, 2000.
46 American Petroleum Industry, *Facts and Figures Centennial Edition*, op. cit.
47 Albert Speer, *Au Cœur du Troisième Reich*, Fayard, Paris, 1971, p. 489.
48 Ibid., p. 491.
49 Robert Goralski and Russell W. Freeburg, *Oil and War*, op. cit., p. 279.
50 Ibid., p. 291.
51 Ibid., p. 303.
52 Ibid., p. 277.
53 Ibid., p. 282.
54 Primo Levi, *Survival in Auschwitz and the Reawakening: Two Memoirs*, Summit Books, New York, 1985, cited in Daniel Yergin, *The Prize*, op. cit., p. 328–329.
55 Robert Goralski and Russell W. Freeburg, *Oil and War*, op. cit., p. 279.
56 Ibid., p. 254.
57 Robert W. Coakley and Richard M. Leighton, *Global Logistics and Strategy: 1943–1945*, US Government Printing Office, Washington, 1969, appendix A, cited in ibid., p. 260.
58 Robert Goralski and Russell W. Freeburg, *Oil and War*, op. cit., p. 300.
59 Ibid., p. 297–298.
60 Ibid., p. 284–289.
61 Ibid., p. 310.
62 Ibid., p. 309.
63 Daniel Yergin, *The Prize*, op. cit., p. 340.
64 Robert Goralski and Russell W. Freeburg, *Oil and War*, op. cit., p. 318.
65 Daniel Yergin, *The Prize*, op. cit., p. 345–346.
66 Warren Moscow, "City's Heart Gone," *The New York Times*, March 9, 1945.

67 Mark R. Wilson, "Making 'Goop' out of Lemons: The Permanent Metals Corporation, Incendiary Bombs, and the Costs of Industrial Overexpansion during World War II," *Enterprise & Society*, vol. 12, March 2011, p. 10.
68 Mike Davis, *City of Quartz*, op. cit., p. 305.
69 Mark R. Wilson, "Making 'Goop' out of Lemons," op. cit.
70 Robert Goralski and Russell W. Freeburg, *Oil and War*, op. cit., p. 298.
71 Ibid., p. 304.
72 Ibid.
73 Errol Morris, *The Fog of War: Eleven Lessons from the Life of Robert S. McNamara*, documentary film, Sony Pictures, 2003.

Chapter 11: After Yalta

1 Robert Vitalis, *America's Kingdom*, op. cit., p. 64.
2 Bascom N. Timmons, *Jesse H. Jones: The Man and the Statesman*, Holt, New York, 1956, p. 236.
3 "Politics Has a Part in International Oil," *Life*, March 28, 1949.
4 Harold L. Ickes, *The Secret Diary of Harold L. Ickes, 1933–1936*, Simon and Schuster, New York, 1953, p. 646.
5 Daniel Yergin, *The Prize*, op. cit., p. 375.
6 Leonard M. Fanning, *Foreign Oil and the Free World*, McGraw-Hill, New York, 1954, p. 359.
7 Cited in Daniel Yergin, *The Prize*, op. cit., p. 375.
8 Everette Lee DeGolyer, "Preliminary Report of the Technical Oil Mission to the Middle East," *Bulletin of the American Association of Petroleum Geologists*, vol. 28, no. 7, July 1944, p. 919–923, cited in Anthony Sampson, *The Seven Sisters*, op. cit., p.97.
9 Daniel Yergin, *The Prize*, op. cit., p. 377.
10 Letter from Harold Ickes to Franklin D. Roosevelt, August 18, 1943 (FDR Library), cited in Rachel Bronson, *Thicker Than Oil: America's Uneasy Partnership with Saudi Arabia*, Oxford University Press, Oxford, 2006, p. 39.
11 Harold L. Ickes, "We're Running Out of Oil," *American Magazine*, January 1944.
12 Cited in Anthony Sampson, *The Seven Sisters, op. cit.*, p. 95.
13 Ibid.
14 James W. Knight, "Arabian Oil Deal Opposed. Plan to Seek Reserves Abroad Held Move Toward Imperialism," (tribune), *The New York Times*, March 10, 1944.
15 Herbert Feis, *Seen from E. A.: Three International Episodes*, Alfred A. Knopf, New York, 1947, p. 129.
16 "Oil and the Near East," *The New York Times*, February 25, 1944.
17 David Painter, *Oil and the American Century: The Political Economy of the US Foreign Oil Policy, 1941–1954*, Johns Hopkins University Press, Baltimore, 1986, p. 55.
18 "Anglo-U.S. Oil Pact Called a *Cartel*," *The New York Times*, October 26, 1944.
19 Anthony Sampson, *The Seven Sisters*, op. cit., p. 98.
20 Rachel Bronson, *Thicker Than Oil*, op. cit., p. 38.
21 Anthony Cave Brown, *Oil, God, and Gold: The Story of Aramco and the Saudi Kings*, Houghton Mifflin, Boston, 1999, p. 117.
22 Rachel Bronson, *Thicker Than Oil*, op. cit., p. 38.
23 See Pascal Ménoret, *L'Énigme Saoudienne: Les Sasudiens et le Monde, 1744–2003*, La Découverte, Paris, 2003.
24 Daniel Yergin, *The Prize*, op. cit., p. 271.
25 Jacques Benoist-Méchin, *Ibn Séoud ou la Naissance d'un Royaume*, op. cit., p. 125.

26 Nubar Gulbenkian, *Pantaraxia*, op. cit., p. 161.
27 Anthony Sampson, *The Seven Sisters*, op. cit., p. 90.
28 Harry Philby, *Arabian Oil Ventures*, Middle East Institute, Washington, 1964, p. 125.
29 Searchable at saudiembassy.net.
30 Daniel Yergin, *The Prize*, op. cit., p. 386.
31 Ibid.
32 Ibid.
33 "Arabian Oil," editorial, *The New York Times*, February 23, 1944.
34 Elliot M. Senn, "Taking Roosevelt to Yalta: A Personal Memoir," *Washington Star*, January 30, 1966.
35 Letter from King Abd al-Aziz Al Saoud to Roosevelt, May 1943, cited in Rachel Bronson, *Thicker Than Oil*, op. cit., p. 41.
36 Letter from Colonel William A. Eddy to the State Department, March 3, 1945, documents of the diplomatic branch 890F.001 Abdul Aziz/3-345, National Archives, Washington, DC.
37 C. L. Sulzberger, "Turkey Believed Preparing War," *The New York Times*, February 22, 1945.
38 Daniel Yergin, *The Prize*, op. cit., p. 387.
39 Ibid., p. 387.
40 Jacques Benoist-Méchin, *Ibn Séoud ou la Naissance d'un Royaume*, op. cit., p. 400–401.
41 Ibid., p. 388–391.
42 Karl Twitchell, *Saudi Arabia: With an Account of the Development of Its Natural Resources*, Princeton University Press, Princeton, 1958, p. 237.
43 Robert P. Grathwol and Donita M. Moorhus, *Bricks, Sand, and Marble: U.S. Army Corps of Engineers Construction in the Mediterranean and Middle East, 1947–1991*, Center of Military History, Washington, 2009, p. 29.
44 Anthony Sampson, *The Seven Sisters*, op. cit., p. 97.
45 Ibid.
46 Daniel Yergin, *The Prize*, op. cit., p. 394.
47 Subcommittee on Multinational Corporations, *Multinational Oil Corporations and U.S. Foreign Policy*, op. cit.
48 Anthony Sampson, *The Seven Sisters*, op. cit., p. 102.
49 Daniel Yergin, *The Prize*, op. cit., p. 397.
50 Nubar Gulbenkian, *Pantaraxia*, op. cit., p. 225.
51 Ibid., p. 227.
52 Ibid.
53 Subcommittee on Multinational Corporations, *Multinational Oil Corporations and U.S. Foreign Policy*, op. cit.
54 This was General de Gaulle's only statement about Algeria, 1955 to 1958. See Guy Pervillé, "General de Gaulle and the Independence of Algeria, 1943–1962," May 1976.
55 Leonard M. Fanning, *Foreign Oil and the Free World*, McGraw-Hill, New York, 1954, p. 359; and "Venezuela: International Partnership," *Time*, October 29, 1951.

Chapter 12: Washington Gives Absolute Power to American Petroleum

1 Cordell Hull, *The Memoirs of Cordell Hull*, MacMillan, New York, 1948, cited in Rachel Bronson, *Thicker Than Oil*, op. cit., p. 20.
2 Walter J. Levy, "Oil and the Marshall Plan," article presented to the American Economic Association, December 28, 1988.
3 *The Secret Story of the Oil Companies in the Middle East*, vol. 1, Documentary Publications, Salisbury, 1980, cited in Peter Tertzakian, *A Thousand Barrels a Second: The Coming Oil*

Break Point and the Challenges Facing an Energy Dependent World, McGraw-Hill, New York, 2007, p. 52.

4 Aaron David Miller, *Search for Security: Saudi Arabian Oil and American Foreign Policy, 1939–1949*, University of North Carolina Press, Chapel Hill, 1980, p. xiii.

5 George F. Kennan, "Policy Planning Study 23 (PPS23)," Foreign Relations of the United States (FRUS), vol. 1, part. 2, 1948, p. 510–529, searchable at swans.com/library/art11/ga192.html.

6 Rómulo Betancourt, *Venezuela: Oil and Politics*, Houghton Mifflin, Boston, 1979.

7 Adam Bernstein, "George C. McGhee Dies; Oilman, Diplomat," *Washington Post*, July 6, 2005.

8 Daniel Yergin, *The Prize*, op. cit., p. 429.

9 Ibid., p. 429.

10 *Foreign Relations of the United States*, vol. 5, US Government Printing Office, Washington, 1950, p. 107.

11 "Malgré ses Profits Records, Total Ne Paie Pas d'Impôts en France," *La Tribune*, December 20, 2010.

12 Anthony Sampson, *The Seven Sisters, op. cit.* p. 117.

13 Manucher Farmanfarmaian, *Blood and Oil: Inside the Shah's Iran*, Modern Library, 1999, p. 184–185, cited in Stephen Kinzer, *All the Shah's Men: An American Coup and the Roots of Middle East Terror*, John Wiley & Sons, Hoboken, 2003, p. 67.

14 William Engdahl, *Pétrole, une Guerre d'un Siècle: L'Ordre Mondial Anglo-Américain*, Jean-Cyrille Godefroy, Paris, 2007, p. 113.

15 Daniel Yergin, *The Prize*, op. cit., p. 433.

16 Ibid., p. 437.

17 Mark J. Gasiorowski, "The 1953 Coup d'État in Iran," *Journal of Middle East Study*, no. 19, 1987, Cambridge University Press, p. 261–286.

18 Henry Longhurst, *Adventure in Oil: The Story of British Petroleum*, Sidgwick and Jackson, London, 1959, p. 144, cited in Anthony Sampson, *The Seven Sisters*, op. cit., p. 119.

19 "Mohammed Mossadegh, Person of the Year 1951," *Time*, January 7, 1952.

20 Stephen Kinzer, *All the Shah's Men*, op. cit., p. 135.

21 Ibid., p. 136.

22 "The CIA in Iran," *The New York Times*, searchable at nytimes.com.

23 Daniel Yergin, *The Prize*, op. cit., p. 450.

24 Stephen Kinzer, *All the Shah's Men*, op. cit., p. 5.

25 Ibid., p. 163.

26 Ibid., p. 7–8; and ibid., p. 102.

27 William Roger Louis, *End of British Imperialism: The Scramble for Empire, Suez, and Decolonization*, I. B. Taurus, London, 2007, p. 776.

28 Stephen Kinzer, *All the Shah's Men*, op. cit., p. 9–10.

29 Ibid., p. 10.

30 Ibid., p. 166.

31 Ibid., p. 168.

32 Ibid., p. 169.

33 Ibid., p. 171–172.

34 "Reversal in Iran," editorial, *The New York Times*, August 23, 1953.

35 Stephen Kinzer, *All the Shah's Men*, op. cit., p. 180.

36 Kim Roosevelt, *Countercoup: The Struggle for the Control of Iran*, McGraw-Hill, New York, 1979.

37 Ibid.

38 "Obama Admits US Involvement in 1953 Iran Coup," AFP, June 4, 2009.

39 Cole C. Kingseed, *Eisenhower and the Suez Crisis of 1956*, Louisiana State University Press, Baton Rouge, 1995, p. 94.
40 Roger Morris, "A Tyrant 40 Years in the Making," *The New York Times*, March 14, 2003.
41 Ibid.
42 Ibid.
43 Ibid.
44 Richard Sale, "Saddam Key in Early CIA Plot," UPI, April 10, 2003.
45 Roger Morris, "A Tyrant 40 Years in the Making," op. cit.
46 Abdulhay Yahya Zalloum, *Oil Crusades: America through Arab Eyes*, Pluto Press, London, 2007, p. 62–63.
47 Kenneth J. Conboy and James Morrison, *Feet to the Fire: CIA Covert Operations in Indonesia, 1957–1958*, US Naval Institute Press, Annapolis, 2000.
48 David Brichoux and Deborah J. Gerner, *The United States and the 1958 Rebellion in Indonesia*, Institute for the Study of Diplomacy, Georgetown University, Washington, 2002, p. 8.
49 Katy Kadane, "Ex-agents Say CIA Compiled Death Lists for Indonesians," *San Francisco Examiner*, May 19, 1990.
50 Howard Zinn, *Le XXe Siècle Américain: Une Histoire Populaire de 1890 à Nos Jours*, Agone, Paris, 2003, p. 211.
51 Philip Shabecoff, "Critics of US Oil Policies in Vietnam Shift Focus of their Attacks," *The New York Times*, April 2, 1971.
52 "Saigon Opens Coasts for Oil Exploration," *The New York Times*, June 10, 1971.
53 Philip Shabecoff, "Critics of US Oil Policies in Vietnam Shift Focus of their Attacks," op. cit.
54 "Vietnam: Exploration / Development History," Coordinating Committee for Geoscience Programs in East and Southeast Asia, searchable at ccop.or.th.

Chapter 13: Big Oil's Planetary Empire and the Rockefellers' Hegemonic Ambitions

1 Henry Luce, "The American Century," *Time*, February 17, 1941.
2 Martin Heidegger, "La Question de la Technique," essay and conference, translated by André Préau, Gallimard, Paris, 2004, p. 26, 36.
3 Anthony Sampson, *The Seven Sisters*, op. cit., p. 106.
4 "Moffett Charges Oil Overpayment," *The New York Times*, May 22, 1948.
5 John M. Blair, *The Control of Oil*, op. cit., p. 74.
6 Searchable at petrole.blog. lemonde.fr.
7 John M. Blair, *The Control of Oil*, op. cit., p. 72.
8 Anthony Sampson, *The Seven Sisters*, op. cit., p. 123–124.
9 Ibid., p. 124.
10 Ibid., p. 125.
11 Daniel Yergin, *The Prize*, op. cit., p. 455.
12 Subcommittee on Multinational Corporations, "Report of the Attorney General to the National Security Council Relative to the Grand Jury Investigation of the International Oil Cartel. January 1953," *The International Petroleum Cartel, the Iranian Consortium and US National Security*, US Senate, US Government Printing Office, Washington, DC, 1974, p. 29–33, searchable at mtholyoke.edu / acad / intrel / Petroleum / ag.htm.
13 Daniel Yergin, *The Prize*, op. cit., p. 456.
14 Subcommittee on Multinational Corporations, *Hearings on Multinational Petroleum Corporations and Foreign Policy, US Senate*, US Government Printing Office, Washington, DC, 1974, part 7, p. 107.
15 Ibid., p. 109.

16 Anthony Sampson, *The Seven Sisters*, op. cit., p. 129–130.
17 John M. Blair, *The Control of Oil*, op. cit., p. 73–74.
18 "Eisenhower's Farewell Address to the Nation, January 17, 1961," searchable at mu.edu.
19 Subcommittee on Multinational Corporations, *Hearings on Multinational Petroleum Corporations and Foreign Policy*, op. cit., part 7, p. 304.
20 Ibid.
21 Ibid.
22 Ibid., part 7, p. 254.
23 John M. Blair, *The Control of Oil*, op. cit., p. 104 sq.
24 Ibid., p. 109.
25 Subcommittee on Multinational Corporations, *Hearings on Multinational Petroleum Corporations and Foreign Policy*, op. cit., part 7, p. 309.
26 Ibid.
27 John M. Blair, *The Control of Oil*, op. cit., p. 100–103.
28 Daniel Yergin, *The Prize*, op. cit., p. 412.
29 John M. Blair, *The Control of Oil*, op. cit., p. 40.
30 Ibid., p. 49.
31 Ibid.
32 Subcommittee on Multinational Corporations, *Hearings on Multinational Petroleum Corporations and Foreign Policy*, op. cit., part 4, p. 68.
33 "Preliminary Federal Trade Commission Staff Report on Its Investigation of the Petroleum Industry," 1973, p. 13–22, cited in John M. Blair, *The Control of Oil*, op. cit., p. 129.
34 John M. Blair, *The Control of Oil*, op. cit., p. 139.
35 BP, *Statistical Review of World Energy*, 2010.
36 John M. Blair, *The Control of Oil*, op. cit., p. 139.
37 Ibid., p. 147 sq.
38 Ibid., p. 142 sq.
39 David Rockefeller, *Memoirs*, op. cit., p. 315.
40 Ibid., p. 264.
41 Subcommittee on Multinational Corporations, *Hearings on Multinational Petroleum Corporations and Foreign Policy*, op. cit., part 5, p. 114, cited in John M. Blair, *The Control of Oil*, op. cit., p. 63.
42 Ibid., part 5, p. 134.
43 Gareth Williams, *Paralysed with Fear: The Story of Polio*, Palgrave, Basingstoke, 2013, cited in Wendy Moore, "Paralysed with Fear: The Story of Polio by Gareth Williams—Review," *The Guardian*, July 17, 2013.
44 Milton Friedman, *Newsweek*, June 26, 1967, cited in Éric Laurent, *La Face Cachée du Pétrole*, Plon, Paris, 2006.
45 Subcommittee on Multinational Corporations, *Hearings on Multinational Petroleum Corporations and Foreign Policy*, op. cit., part 4, p. 104.
46 David Rockefeller, *Memoirs*, op. cit., p. 88.
47 Ibid., p. 433.
48 Hannah Arendt, *Les Origines du Totalitarisme: L'Impérialisme*, Points, Paris, 2010, p. 391.
49 Annual Ambassadors' Dinner, September 14, 1994, C-SPAN, searchable at youtube.com.
50 John Ensor Harr and Peter J. Johnson, *The Rockefeller Century: Three Generations of America's Greatest Family*, Charles Scribner's Sons, New York, 1988, p. 432–433.
51 Peter Grose, "Continuing the Inquiry: War and Peace," CFR History, searchable at cfr.org.
52 Ibid.

53 David Rockefeller, *Memoirs*, op. cit., p. 149.
54 James Srodes, *Allen Dulles. Master of Spies*, Regnery Publishing, Washington, 1999, p. 207.
55 David Rockefeller, *Memoirs*, op. cit., p. 410; and Michael Gama, *Rencontres au Sommet: Quand les Hommes de Pouvoir se Réunissent*, L'Altiplano, Paris, 2007.
56 David Rockefeller, *Memoirs*, op. cit., p. 411.
57 Douglas Martin, "Prince Bernhard, Father of Dutch Queen, Dies at 93," *The New York Times*, December 2, 2004.

Chapter 14: Big Oil Asserts Itself

1 David Rockefeller, *Memoirs*, op. cit., p. 154.
2 Walter Isaacson and Evan Thomas, *The Wise Men: Six Friends and the World They Made*, Simon & Schuster, New York, 2012, p. 124.
3 "In Die Speichen des Kriegsrades Gegriffen," *Der Spiegel*, no. 6, 1980, searchable at spiegel.de /spiegel/print/d-14322484.html.
4 John Simkin, "John McCloy," searchable at spartacus.schoolnet.co.uk.
5 John McCloy, "Approach to Clemency Decisions," *Information Bulletin*, May 1951, searchable at images.library.wisc.edu.
6 Kai Bird, *The Chairman John J. McCloy: The Making of the American Establishment*, Simon & Schuster, New York, 1992, p. 368.
7 James H. Critchfield, *Partners at the Creation. The Men Behind Postwar Germany's Defense and Intelligence Establishments*, US Naval Institute Press, Annapolis, 2003, p. 124.
8 Kai Bird, *The Chairman John J. McCloy*, op. cit., p. 343; Robert Wolfe, *Analysis of the Investigative Records Repository (IRR) File of Klaus Barbie*, The Nazi War Crimes and Japanese Imperial Government Records Interagency Working Group, 2001, searchable at archives.gov/iwg /research-papers/barbie-irr-file.html.
9 Subcommittee on Multinational Corporations, *Hearings on Multinational Petroleum Corporations and Foreign Policy*, op. cit., part. 6, p. 290; and Anthony Sampson, *The Seven Sisters*, op. cit. p. 166.
10 Peter Drucker, *Témoin du 20e Siècle: De Vienne 1920 à la Californie 1980*, Village Mondial, Paris, 2001, p. 163–180.
11 David Rockefeller, *Memoirs*, op. cit., p. 407; and Special Studies Project, searchable at rockarch.org.
12 Cary Reich, *The Life of Nelson A. Rockefeller: Worlds to Conquer, 1908–1958*, Doubleday, New York, 1996, p. 617 and 835, note 558.
13 Philip Benjamin, "Arms Rise Urged Lest Reds Seize Lead in 2 Years," *The New York Times*, January 6, 1958.
14 Walter Isaacson, *Kissinger: A Biography*, Simon & Schuster, New York, 2005, p. 90–93.
15 Robert Bryce, "The Candidate from Brown & Root: Bush Doesn't Know Dick," *The Austin Chronicle*, August 28, 2000; "George R. Brown, Industrialist, Dies," *The New York Times*, January 24, 1983; and "George R. Brown, 84, Lyndon Johnson Mentor," *The Miami Herald*, January 23, 1983.
16 John Bainbridge, *The Super Americans: Picture of Life in Texas, Land of Millionaires*, Victori Gollancz, 1962, cited in Anthony Sampson, *The Seven Sisters*, op. cit., p. 133.
17 Anthony Sampson, *The Seven Sisters*, op. cit., p. 144; and Robert Bryce, *Cronies: Oil, the Bushes, and the Rise of Texas, America's Superstate*, Public Affairs, New York, 2004, p. 93.
18 Robert Metz, "Tax Bill's chance," *The New York Times*, November 27, 1963.
19 Russ Baker, *Family of Secrets: The Bush Dynasty, the Powerful Forces That Put It in the White House, and What Their Influence Means for America*, Bloomsbury Press, London, 2009, p. 166.
20 Ibid., p. 162–165.

21 Anthony Sampson, *The Seven Sisters*, op. cit., p. 203–204; and Russ Baker, *Family of Secrets*, op. cit., p. 191.
22 Kevin Phillips, *American Dynasty: Aristocracy, Fortune, and the Politics of Deceit in the House of Bush*, Penguin Books, New York, 2004, p. 21.
23 Russ Baker, *Family of Secrets*, op. cit., p. 71.
24 Duncan Campbell and Ben Aris, "How Bush's Grandfather Helped Hitler's Rise to Power in Washington," *The Guardian*, September 25, 2004.
25 Joseph J. Trento, *Prelude to Terror: Edwin P. Wilson and the Legacy of America's Private Intelligence Network*, Carroll & Graf, New York, 2005, p. 3.
26 Ibid.
27 Peter Gross, *Gentleman Spy: The Life of Allen Dulles*, Houghton Mifflin, Boston, 1994; and Joseph J. Trento, *The Secret History of CIA*, MJF Books, New York, 2007, p. 29.
28 Kevin Phillips, *American Dynasty*, op. cit., p. 194–198; and Joseph J. Trento, *Prelude to Terror*, op. cit., p. 7–12.
29 Joseph J. Trento, *Prelude to Terror*, op. cit., p. 7.
30 Ibid., p. 10–11.
31 Russ Baker, *Family of Secrets*, op. cit., p. 162–165.
32 Ibid., p. 24.
33 Ibid., p. 25.
34 Joseph J. Trento, *Prelude to Terror*, op. cit., p. 13–14.
35 Russ Baker, *Family of Secrets*, op. cit., p. 24.
36 Ibid., p. 165.
37 Jean Laherrère, "US Wells. Average Depth from EIA," September 2012, searchable at aspofrance.viabloga.com.
38 George H. W. Bush and Victor Gold, *Looking Forward: An Autobiography*, Doubleday, New York, 1987, p. 49–50.
39 Russ Baker, *Family of Secrets*, op. cit., p. 34.
40 Monica Perin, "Adios, Zapata! Colorful Company Founded by Bush Relocates to N.Y.," *Houston Business Journal*, April 25, 1999; and Michael Hasty, "Secret Admirers. The Bushes and the Washington Post," *Online Journal*, February 5, 2004.
41 "Memorandum: To: Deputy Director of Operations; Subject: George Bush and Thomas J.," CIA, November 29, 1975, registration no. NARA: 104-10310-1027, searchable at maryferrell .org, and cited in Russ Baker, *Family of Secrets*, op. cit., p. 12–16.
42 "Ex-Howard Hughes Spokesman, CIA Operative Dies," Associated Press, August 6, 2008, searchable at usatoday.com.
43 Russ Baker, *Family of Secrets*, op. cit., p. 34–35.
44 Joseph J. Trento, *Prelude to Terror*, op. cit., p. 16.
45 Ibid.
46 Russell Bowen, *The Immaculate Deception: The Bush Crime Family Exposed*, Bridger House Publishers, 1991, p. 31.
47 Joseph J. Trento, *Prelude to Terror*, op. cit., p. 13–16; and William R. Corson, *The Betrayal*, Ace, New York, 1968.
48 Russ Baker, *Family of Secrets*, op. cit., p. 36.
49 R. Hart Phillips, "Cuba Limits Search for Oil: Nationalization Step Seen," *The New York Times*, November 22, 1959.
50 Russ Baker, *Family of Secrets*, op. cit., p. 83.
51 Fabián Escalante, *The Cuba Project: CIA Covert Operations 1959–62*, Ocean Press, New York, 2004, p. 43–44.

52 Russ Baker, *Family of Secrets*, op. cit., p. 507, note 28.
53 Stephen Birmingham, *"Our Crowd": The Great Jewish Families of New York*, Harper & Row, New York, 1967, p. 409–410, cited in Russ Baker, *Family of Secrets*, op. cit., p. 80.
54 Russ Baker, *Family of Secrets*, op. cit., p. 80.
55 "Oil Drilling Deal Set," *The New York Times*, November 30, 1956.
56 "Rise in Domestic Oil Flow Bolsters Cuba: Exploratory Capital Pouring into Island," *The New York Times*, January 5, 1956.
57 R. Hart Phillips, "Cuba Limits Search for Oil: Nationalization Step Seen," op. cit.
58 Joe Simnacher, "John Alston "Jack" Crichton: Oilman, Military Officer in WWII," *Dallas Morning News*, December 15, 2007.
59 Kevin Phillips, "Bush's Worst Political Nightmare," *The Los Angeles Times*, January 13, 1991; and Anthony Kimery, "A Well of a Deal," *Common Cause*, Spring 1991.
60 Russ Baker, *Family of Secrets*, op. cit., p. 38.
61 Joseph J. Trento, *Prelude to Terror*, op. cit., p. 17.
62 Jonathan Kwitny, "The Mexican Connection: A Look at an Old George Bush Business Venture," Barron's, September 19, 1988.
63 Joseph J. Trento, *Prelude to Terror*, op. cit., p. 18.
64 Jonathan Kwitny, "The Mexican Connection," op. cit.
65 Russ Baker, *Family of Secrets*, op. cit., p. 70.
66 Steve LeVine, *The Oil and the Glory*, op. cit., p. 34.
67 Russ Baker, *Family of Secrets*, op. cit., p. 75.
68 Ibid., p. 72–73.
69 Ibid., p. 81.
70 Ibid., p. 84.
71 Ibid., p. 59.
72 Ibid., p. 105–106 and 510, note 45.
73 Ibid., p. 161.
74 "Connally Is Linked to Oil Aid for Nixon," Associated Press, *The New York Times*, January 19, 1972.
75 Kitty Kelley, *The Family: The Real Story of the Bush Dynasty*, Doubleday, New York, 2004, p. 245.
76 Russ Baker, *Family of Secrets*, op. cit., p. 161.
77 Ibid., p. 170.
78 "Unneeded Import Quotas," editorial, *The New York Times*, August 16, 1969.

Chapter 15: Saudi Arabia and Gabon

1 Walt W. Rostow, memo addressed to President Lyndon Johnson, cited in Robert Vitalis, *America's Kingdom: Mythmaking on the Saudi Oil Frontier*, Verso, London, 2009 (2nd edition), p. 228.
2 Roy Lebkicher, *Handbooks for American Employees*, Moore, New York, 1952.
3 Anthony Cave Brown, *Oil, God, and Gold*, op. cit., p. 143.
4 Robert P. Grathwol and Donita M. Moorhus, *Bricks, Sand, and Marble*, op. cit., p. 37.
5 Robert Vitalis, *America's Kingdom*, op. cit., p. 101 and 274.
6 Ibid., p. 59.
7 Ibid., p. xxix.
8 Ibid., p. 171–172.
9 Ibid., p. 184.
10 Ibid., p. 263.

11 Ibid., p. 93.
12 Ibid., p. 98.
13 Ibid., p. 160.
14 Anthony Sampson, *The Seven Sisters*, op. cit., p. 95.
15 David Rockefeller, *Memoirs*, op. cit., p. 265 and 275.
16 Robert Vitalis, *America's Kingdom*, op. cit., p. 131.
17 Ibid., p. 141.
18 Letter dated August 28, 1986 found in the personal archives of William Mulligan, government relations officer for Aramco, cited in Robert Vitalis, *America's Kingdom*, op. cit., p. 120.
19 Robert Vitalis, *America's Kingdom*, op. cit., p. 69.
20 Ibid., p. 143–145.
21 Ibid., p. 34.
22 Richard Sanger, *Arabian Peninsula*, Ayer Company Publications, North Stratford, 1954.
23 Thomas W. Lippman, *Arabian Knight: Colonel Bill Eddy USMC and the Rise of American Power in the Middle East*, Selwa Press, Vista, 2008, p. 203.
24 Robert Vitalis, *America's Kingdom*, op. cit., p. 79.
25 Ibid., p. 144–145.
26 Thomas W. Lippman, *Arabian Knight*, op. cit.; and Robert Vitalis, *America's Kingdom*, op. cit., p. 144.
27 Ibid., p. 137.
28 Ibid., p. 138.
29 Ibid., p. 137.
30 Jane Mayer, "The Contractors," *The New Yorker*, May 5, 2003.
31 Robert Vitalis, *America's Kingdom*, op. cit., p. 75.
32 Jean-Pierre Penez and Maurice Jarnoux, *Paris Match*, March 6, 1954.
33 Jacques Benoist-Méchin, *Ibn Séoud ou la Naissance d'un Royaume*, op. cit., p. 366.
34 Robert Vitalis, *America's Kingdom*, op. cit., p. 65.
35 Ibid., p. 81.
36 Ibid., p. 79–86.
37 Adam Robinson, *Bin Laden: The Inside Story of the Rise and Fall of the Most Notorious Terrorist in History*, Arcade Publishing, New York, 2011, p. 27.
38 Laurie Kerr, "The Mosque to Commerce: Bin Laden's Special Complaint with the World Trade Center," December 28, 2001, searchable at slate.com.
39 Vivian M. Baulch, "Minoru Yamasaki, World-Class Architect," *The Detroit News*, August 14, 1998; and Robert Vitalis, *America's Kingdom*, op. cit., p. 143.
40 Ibid.
41 Robert P. Grathwol and Donita M. Moorhus, *Bricks, Sand, and Marble*, op. cit., p. 252.
42 Ibid., p. 251.
43 Ibid.
44 Irvine H. Anderson, *Aramco, The United States, and Saudi Arabia: A Study of the Dynamics of Foreign Oil Policy*, 1933–1950, Princeton University Press, Princeton, 1981; and Daniel Yergin, *The Prize*, documentary, 1992.
45 Hannah Arendt, *Les Origines du Totalitarisme*, op. cit., p. 384.
46 The Papers of Dwight David Eisenhower, vol. XVIII, part 1: "A New Beginning, Old Problems. January 1957 to May 1957," chap. 1: "The Mideast and the Eisenhower Doctrine," searchable at eisenhowermemorial.org.
47 "More US Arms Go to Saudi Arabia," *The New York Times*, May 17, 1956; "US Arms Sent to Saudi Arabia," Associated Press, May 17, 1956; and David B. Ottaway, *The King's Messenger:*

Prince Bandar Bin Sultan and America's Tangled Relationship with Saudi Arabia, Walker & Co, New York, 2008, p. 15.

48 Laurence Garey, "The Buraimi and Jebel Akhdar Crises, 1952–1959," Emirates Natural History Group, searchable at enhg.org.

49 "Sheikh Zayed bin Sultan Al-Nahyan," *The Daily Telegraph*, November 4, 2004.

50 Rachel Bronson, *Thicker Than Oil*, op. cit., p. 64–68.

51 J. B. Kelly, *Arabia, the Gulf and the West*, Basic Books, New York, 1980, p. 61.

52 "A Saudi Revolution," *The Economist*, April 30, 1960, cited in Robert Vitalis, *America's Kingdom*, op. cit., p. 214.

53 Robert Vitalis, *America's Kingdom*, op. cit., p. 217.

54 Diplomatic cable sent to Washington from the Djeddah embassy, June 16, 1960, cited in Robert Vitalis, *America's Kingdom*, op. cit., p. 217.

55 Robert Vitalis, *America's Kingdom*, op. cit., p. 217.

56 American classified documents revealed by Robert Vitalis, ibid., p. 217.

57 David Rockefeller, *Memoirs*, op. cit., p. 274.

58 Parker T. Hart, *Saudi Arabia and The United States: Birth of a Security Partnership*, Indiana University Press, Indianapolis, 1998, p. xvii.

59 Pierre Péan, *Affaires Africaines*, Fayard, Paris, 1983, p. 41.

60 Ibid., p. 45.

61 Charles Darlington and Alice B. Darlington, *African Betrayal*, David McKay Co., Philadelphia, 1968, p. 4–5; and Pierre-Michel Durand, *L'Afrique et les Relations Franco-Américaines des Années Soixante: Aux Origines de l'obsession Américaine*, L'Harmattan, Paris, 2007, p. 231.

62 Pierre Péan, *Affaires Africaines*, op. cit., p. 48.

63 Patrick Pesnot, *Les Dessous de la Françafrique: Les Dossiers Secrets de Monsieur X*, Nouveau Monde, Paris, 2008, p. 84.

64 Roger Faligot and Jean Guisnel (dir.), *Histoire Secrète de la Ve République*, La Découverte, "Poche," Paris, 2006, p. 162.

65 Pierre Péan, *Affaires Africaines*, op. cit., p. 52–53.

66 Ibid.

67 Ibid.

68 Ibid. p. 59.

69 Maurice Delauney, *De la Casquette à la Jaquette, ou de l'Administration Coloniale à la Diplomatie Africaine*, Éditions Pensée universelle, Paris, 1982, cited in Pierre Péan, *Affaires Africaines*, op. cit., p. 61.

70 Ibid., p. 63.

71 Pierre Péan, *Affaires Africaines*, op. cit., p. 64–65.

72 Ibid., p. 66–67.

73 Cited in ibid., p. 69.

Chapter 16: Cartel Against Cartel

1 Exxon, presented by ASPO (the Association for the Study of Peak Oil).

2 Kjell Aleklett, *Peeking at Peak Oil*, Springer Verlag, New York, 2012, p. 208.

3 Televised interview, March 23, 1961.

4 Andrew Gumbel, "Autopsy May Solve Deadly Mystery of the Mattei Affair," *The Independent*, August 29, 1997.

5 Maurice Ezran, *Histoire du Pétrole*, L'Harmattan, Paris, 2010, p. 180.

6 Frédéric Tonelli, *Le Secret des Sept Sœurs*, documentary, 2011.

7 Buscetta, "Cosa Nostra Uccise Enrico Mattei," La Repubblica, May 23, 1994, searchable at ricerca.repubblica.it.
8 Donato Firrao and Graziano Ubertalli, "Was There a Bomb on Mattei's Aircraft?," Convegno Nazionale IGF, 2009, searchable at gruppofrattura.it.
9 Edith Penrose, *The International Petroleum Industry*, 1968, p. 195, cited in Anthony Sampson, *The Seven Sisters*, op. cit., p. 15.
10 Wanda Jablonski, *Petroleum Weekly*, July 22, 1960.
11 Cited in Daniel Yergin, *The Prize*, op. cit., p. 504.
12 Pierre Terzian, *L'Étonnante Histoire de l'Opep*, Jaguar/Jeune Afrique, Paris, 1983.
13 Cited in Vijay Prashad, *The Darker Nations: A Biography of the Short-Lived Third World*, LeftWord Book, New Delhi, 2007, p. 197.
14 Robert Vitalis, *America's Kingdom*, op. cit., p. 137.
15 Ibid., p. 137.
16 *Time*, April 27, 1959, cited in Robert Vitalis, *America's Kingdom*, op. cit., p. 137.
17 Robert Vitalis, *America's Kingdom*, op. cit., p. 219.
18 J. E. Hartshorn, *Politics and World Oil Economics: An Account of the International Oil Industry in Its Political Environment*, (2nd ed.), Praeger, New York, 1967, p. 312.
19 Terry Lynn Karl, *The Paradox of Plenty: Oil Booms and Petro-States*, University of California Press, Berkeley, 1997.
20 Pierre Terzian, *L'Étonnante Histoire de l'Opep*, op. cit.
21 Ibid.
22 Daniel Yergin, *The Prize*, op. cit., p. 493.
23 Anthony Sampson, *The Seven Sisters*, op. cit., p. 163.
24 Subcommittee on Multinational Corporations, *Hearings on Multinational Petroleum Corporations and Foreign Policy*, op. cit., part 5, p. 255.
25 Daniel Yergin, *The Prize*, op. cit., p. 515.
26 Anthony Sampson, *The Seven Sisters*, op. cit., p. 168.
27 Subcommittee on Multinational Corporations, *Hearings on Multinational Petroleum Corporations and Foreign Policy*, op. cit., part 7, p. 309.
28 Ibid., p. 168.
29 Anthony Sampson, *The Seven Sisters*, op. cit., p. 172.
30 Subcommittee on Multinational Corporations, *Hearings on Multinational Petroleum Corporations and Foreign Policy*, op. cit., part 7, p. 720.
31 Anthony Sampson, *The Seven Sisters*, op. cit., p. 173.
32 Ibid.
33 Ibid.
34 Tad Szulc, "Industrialists Reported to Warn Nixon on Loss of Influence with Arabs," *The New York Times*, December 21, 1969.

Chapter 17: The Leapfrog Effect

1 Pierre Juhel, *Histoire du Pétrole*, op. cit., p. 120.
2 Ibid., p. 120.
3 Guy Pervillé, "Le Général de Gaulle et l'Indépendance de l'Algérie, 1943–1962," op. cit.
4 Hocine Malti, *Histoire Secrète du Pétrole Algérien*, La Découverte, Paris, 2010, p. 19.
5 Loi n° 57-27 du 10 Janvier 1957 Créant une Organisation Commune des Régions Sahariennes, *Journal Officiel de la République Française*, January 11, 1957, p. 578.
6 "Le Conseil Constitutionnel Est Formé," *Le Monde*, February 23, 1959.

7 Jean-Pierre Corcelette and Frédéric Abadie, *Georges Pompidou: Le Désir et le Destin,* Nouveau Monde, Paris, 2007; and Alain Frerejean, *C'Était Georges Pompidou*, Fayard, Paris, 2007.
8 Jean-Pierre Corcelette and Frédéric Abadie, *Georges Pompidou*, op. cit.
9 Roger Faligot and Jean Guisnel (dir.), *Histoire Secrète de la Ve République*, op. cit., p. 138.
10 Maurice Faivre, *Les Archives Inédites de la Politique Algérienne, 1958–1962,* L'Harmattan, Paris, 2000, p. 104.
11 Jean-Pierre Corcelette and Frédéric Abadie, *Georges Pompidou*, op. cit.
12 Maurice Faivre, *Les Archives Inédites de la Politique Algérienne, 1958–1962,* op. cit, p. 104.
13 "Attentat Devant la Banque Rothschild Dont le Directeur Général Est M. Pompidou," *Le Monde*, March 23, 1961.
14 "Un Groupement Clandestin, le RDL, Revendique la Responsabilité des Attentats de Mardi et de l'Incendie du Palais-Bourbon," *Le Monde*, April 6, 1961.
15 Maurice Faivre, *Les Archives Inédites de la Politique Algérienne, 1958–1962*, op. cit., p. 111.
16 Hocine Malti, *Histoire Secrète du Pétrole Algérien*, op. cit., p. 19.
17 Ibid.
18 Lounis Aggoun and Jean-Baptiste Rivoire, *Françalgérie, Crimes et Mensonges d'États: Histoire Secrète, de la Guerre d'Indépendance à la "Troisième Guerre" d'Algérie,* La Découverte, Paris, 2005, p. 44; and Vincent Jauvert, "Quand la France Testait des Armes Chimiques en Algérie," *Le Nouvel Observateur*, October 23, 1997.
19 Ibid., p. 25.
20 Hocine Malti, *Histoire Secrète du Pétrole Algérien*, op. cit., p. 36.
21 In the previous edition of this book, there was mention of the presence of Claude Allègre being in Algeria during this period. It was actually Maurice Allègre, director of the salaried workers of the organization of the mines, who was in Algiers from 1962 to 1964.
22 Hocine Malti, *Histoire Secrète du Pétrole Algérien*, op. cit., p. 44.
23 Ibid., p. 51.
24 Ibid., p. 54.
25 Ibid., p. 72–81.
26 Ibid., p. 82.
27 Ibid., p. 99.
28 Ibid., p, 112–113.
29 Hanafi Taguemout, *L'Affaire Zeghar, Déliquescence d'un État: L'Algérie Sous Chadli*, Publisud, Paris, 1994, cited in Lounis Aggoun and Jean-Baptiste Rivoire, *Françalgérie, Crimes et Mensonges d'États*, op. cit., p. 67.
30 Hocine Malti, *Histoire Secrète du Pétrole Algérien*, op. cit., p. 135.
31 Patrick Benquet, *La Françafrique*, Compagnie des Phares & Balises, France, 2010.
32 Jean Guisnel, "Derrière la Guerre du Biafra, la France," in Roger Faligot and Jean Guisnel (dir.), *Histoire Secrète de la Ve République*, op. cit., p. 149.
33 Total and Nigerian National Petroleum Corporation, "Total Upstream in Nigeria," 2004.
34 Jacques Foccart, "Foccart Parle: Interviews with Philippe Gaillard, tome I," *Fayard/Jeune Afrique*, Paris, 1995, p. 342.
35 Jean Guisnel, "Derrière la Guerre du Biafra, la France," op. cit., p. 148–151.
36 "Il y a Quarante Ans, le Biafra Renonçait à l'Indépendance," January 12, 2010, searchable at lemonde.fr/afrique/portfolio/2010/01/12/il-y-a-quarante-ans-le-biafra-renoncait-a-l-independance_1290483_3212.html.
37 Jean Guisnel, "Derrière la Guerre du Biafra, la France," op. cit., p. 150.
38 Jacques Foccart, "Foccart Parle," op. cit., p. 346.
39 Maurice Robert, *Ministre de l'Afrique: Entretiens avec André Renault*, Seuil, Paris, 2004, p. 180.

40 Jean Guisnel, "Derrière la Guerre du Biafra, la France," op. cit., p. 154.
41 Total and Nigerian National Petroleum Corporation, "Total Upstream in Nigeria," op. cit.
42 Subcommittee on Multinational Corporations, *Multinational Oil Corporations and US Foreign Policy*, op. cit.
43 Ibid.
44 John M. Blair, *The Control of Oil*, op. cit., p. 212.
45 Anthony Sampson, *The Seven Sisters*, op. cit., p. 211.
46 Joseph Trento, *Prelude to Terror*, op. cit., p. 71.
47 John M. Blair, *The Control of Oil*, op. cit., p. 221.
48 "Lybia in Accord with Occidental," *The New York Times*, September 5, 1970.
49 Anthony Sampson, *The Seven Sisters*, op. cit., p. 212.
50 "Lybia in Accord with Occidental," op. cit.
51 Subcommittee on Multinational Corporations, *Hearings on Multinational Petroleum Corporations and Foreign Policy*, op. cit., letter to Senator Church.
52 Cited in Anthony Sampson, *The Seven Sisters*, op. cit., p. 215.
53 Ibid.
54 Abdul Amir Kubbah, "OPEC Past and Present," Opec Research Center, Vienne, 1974, p. 54, cited in Anthony Sampson, *The Seven Sisters*, op. cit., p. 215.
55 Subcommittee on Multinational Corporations, "The Leapfrogging or Whipsaw Effect Came into Full Play," *Hearings on Multinational Petroleum Corporations and Foreign Policy*, op. cit., part 5, p. 248.
56 André Giraud and Xavier Boy de la Tour, *Géopolitique du Pétrole et du Gaz*, op. cit., p. 242.
57 See albartlett.org.
58 "La Libido au Sens d''Énergie Vitale,'" see Carl Gustav Jung, "L'Énergétique Psychique" (1928), *La Réalité de l'Âme. Structure et Dynamique de l'Inconscient, tome I,* Le Livre de Poche, Paris, 2008, p. 294–309 ("Application du Point de Vue Energétiste") and p. 324 sq ("La Formation des Symboles").

Chapter 18: The Golden Childhood of the Oil-Made Man

1 Raymond Guglielmo, *La Pétrochimie dans le Monde*, op. cit., p. 12 sq.
2 Robert U. Ayres, "Resources, Scarcity, Growth and the Environment," Center for the Management of Environmental Resources, Insead, Fontainebleau, 2000.
3 Following the idea of "hydrocarbon man," a term coined by Daniel Yergin (Daniel Yergin, *The Prize*, op. cit., p. 523).
4 Angus Maddison, "Historical Statistics for the World Economy," op. cit.; J. Bradford DeLong, "Estimating world GDP," U.C Berkeley, searchable at delong.typepad.com; Donella Meadows, Dennis Meadows, Jorgen Randers, and William Behrens, *The Limits to Growth*, Universe Books, New York, 1972, p. 38.
5 BP, *Statistical Review of World Energy Workbooks*, 2014.
6 Robert Ayres, Leslie Ayres, and Benjamin Warr, "Energy, Power and Work in the US Economy, 1900–1998," op. cit., fig. 6.
7 Olivier Rech, cited in Jean Laherrère, "Saving Energy: Reliability of National Energy Flows," Aspo, searchable at aspofrance.viabloga.com.
8 BP, *Statistical Review of World Energy* 2009.
9 Robert Ayres, Leslie Ayres, and Benjamin Warr, "Energy, Power and Work in the US Economy, 1900–1998," op. cit., fig. 11.
10 Energy Watch Group, "Coal: Resources and Future Production," 2007, searchable at energywatchgroup.org; Bernard Durand, "Les Combustibles Fossiles, Grands Oubliés

du Débat National sur la Transition Energétique," Conference at the Collège de France, December 12, 2013; and Jean-Marc Jancovici, "Le Pic de Production a-t-il Déjà Eté Observé dans le Charbon?", searchable at manicore.com.
11 Bernard Durand, "Les Combustibles Fossiles, Grands Oubliés du Débat National sur la Transition Energétique," op. cit.
12 Olivier Rech, cited in Jean Laherrère, "Saving Energy," op. cit.
13 "Energy Resources and Our Future," remarks by Admiral Hyman Rickover delivered in 1957, searchable at resilience.org/stories/2009-07-29/energy-resources-and-our-future-speech-rear-admiral-hyman-g-rickover.
14 Cited in Daniel Yergin, *The Prize*, op. cit., p. 528.
15 Henri Bergson, *L'Évolution Créatrice (1907)*, p. 147, searchable at classiques.uqac.ca/classiques/bergson_henri/evolution_creatrice/evolution_creatrice.html.
16 Jean-Marie Domenach, "Khrouchtchev en Amérique," *Esprit*, October 1959, p. 396.
17 François Roddier, *Thermodynamique de l'Évolution*, Parole éditions, Artignosc-sur-Verdon, 2012, voir notamment p. 124 sq.
18 Mike Davis, *City of Quartz*, op. cit., p. 305.
19 Vannevar Bush, *Science, the Endless Frontier*, 1945, cited in Joseph A. Tainter, "Why Societies Collapse," conference, December 2010.
20 James Truslow Adams, *The Epic of America*, Little, Brown and Co., New York, 1934 (2nd edition), p. 404–405.
21 Aaron Severson, "Radical Fighter Kick-Started Cadillac Design," *Autoweek*, April 18, 2013.
22 Errol Morris, *The Fog of War*, op. cit.
23 Dwight D. Eisenhower, *Mandate for Change, 1953–1956: White House Years*, Doubleday, Garden City, 1963, p. 547, cited in Daniel Yergin, *The Prize*, op. cit., p. 535.
24 Ibid.
25 Daniel Yergin, *The Prize*, op. cit., p. 532.
26 "The Ordeal of Engine Charlie: How One of Our Most Enlightened Business Leaders Became the Symbol of Corporate Ruthlessness," *American Heritage Magazine*, February/March 1995.
27 "United States Court of Appeals for the Seventh Circuit: United States versus National City Lines," January 3, 1951, searchable at web.archive.org.
28 Ibid.
29 Ibid., p. 235.
30 Ibid., p. 253.
31 Ibid., p. 247.
32 Ibid., p. 251.
33 Roland Barthes, *Mythologies*, Seuil, Paris, 1957.
34 Daniel Yergin, *The Prize*, op. cit., p. 534.
35 Food and Agriculture Organization (FAO).
36 Lester Brown, *Plan B 3.0: Mobilizing to Save Civilization*, W. W. Norton & Company, New York, 2008, p. 36.
37 Ibid.
38 International Fertilizer Association (IFA), searchable at fertilizer.org.
39 Vaclav Smil, *Enriching the Earth: Fritz Haber, Carl Bosch, and the Transformation of World Food Production*, The MIT Press, Cambridge, 2004.
40 Robert Ayres, Leslie Ayres, and Benjamin Warr, "Energy, Power and Work in the US Economy, 1900–1998," op. cit., fig. 4.

41 Mark Dowie, *American Foundations: An Investigative History*, MIT Press, Cambridge, Massachusetts, 2001, p. 105–140.
42 Ibid., p. 109.
43 Eric Ross, *The Malthus Factor: Poverty, Politics and Population in Capitalist Development,* Zed Books, London, 1998, searchable at thecornerhouse.org.uk.
44 R. A. Fischer, Derek Byerlee, and G.O. Edmeades, "Can Technology Deliver on the Yield Challenge to 2050?," FAO, 2009.
45 Mike Davis, *Le Pire des Mondes Possibles: De l'Explosion Urbaine au Bidonville Global*, La Découverte, Paris, 2006, p. 27.
46 Mike Davis, *Planet of Slums,* Verso, London, 2005.
47 Claude Lévi-Strauss, *Tristes Tropiques*, op. cit., p. 145.
48 Ibid., p. 171.
49 Louis Hyman, "American Debt, Global Capital. The Policy Origins of Securization"; and Niall Ferguson, *The Shock of the Global: The 1970s in Perspective,* Harvard University Press, Cambridge, 2011, p. 128–142.
50 US Government, "2009 Historical Budget Tables," p. 47–55, searchable at whitehouse.gov.
51 Erich Heinemann, "Moderate Voice Is Heard in Monetary Debate," *The New York Times*, February 25, 1968; and Eugene Birnbaum, *Changing the United States Commitment to Gold,* Department of Economics, Princeton University, Princeton, 1967, cited in Timothy Mitchell, *Carbon Democracy: Political Power in the Age of Oil*, La Découverte, Paris, 2013, p. 204.
52 "Washington: For the Record," *The New York Times*, September 22, 1971.
53 "Jack F. Bennett: Death Notice," *The Washington Post*, April 28, 2010.
54 "Funds for Mohole Urged by Johnson," *The New York Times*, May 19, 1966.
55 Rem Koolhaas, Stefano Boeri, Sanford Kwinter, Nadia Tazi, and Hans Ulrich Obrist, *Mutations,* Actar, Barcelona, Arc en Rêve, Bordeaux, 2000, p. 557.
56 Ibid., p. 547–548.
57 Ibid., p. 547 and 550.
58 Quinton Scott and Tinku Bhattacharyya, *Nigeria 70. The Definitive Story of 1970's Funky Lagos*, Strut Records, London, 2011.
59 Ibid.
60 Jean Fourastié, *Les Trente Glorieuses, ou la Révolution Invisible de 1946 à 1975*, Fayard, Paris, 1979.

Chapter 19: OPEC

1 BP, *Statistical Review of World Energy Workbook 2013*.
2 David Room and Steve Tanner, "Shell Execs Briefed on Peak Oil in 1956," searchable at thecuttingedgenews.com/index.php?article=476.
3 Marion King Hubbert, *Nuclear Energy and the Fossil Fuels*, Shell Development Company, Houston, 1956.
4 US Energy Information Administration, searchable at eia.gov.
5 Robert D. McFadden, "Rationing Urged for Oil and Coal," *The New York Times*, September 1, 1970.
6 Robert Hargreaves, *Superpower: A Portrait of America in the 70s*, St. Martin's Press, New York, 1973, p. 176.
7 The American Presidency Project, searchable at presidency.ucsb.edu.
8 Railroad Commission of Texas.
9 US Energy Information Administration.
10 Anthony Sampson, *The Seven Sisters*, op. cit., p. 216–219.

11 Subcommittee on Multinational Corporations, *Hearings on Multinational Petroleum Corporations and Foreign Policy*, op. cit., part 6, p. 71.
12 Anthony Sampson, *The Seven Sisters*, op. cit., p. 237.
13 Interview published in March 1969 in the *Financial Times*, cited in ibid., p. 232.
14 Paul H. Giddens, "Historical Origins of the Adoption of the Exxon Name and Trademark," *The Business History Review*, Harvard, vol. 47, no. 3, autumn 1973.
15 Edward Cowen, "Connally Stern on Expropriation; Urges More US Support for Corporations Abroad," *The New York Times*, April 19, 1972.
16 Anthony Cave Brown, *Oil, God, and Gold*, op. cit., p. 291.
17 Jim Akins, "The Oil Crisis: This Time the Wolf Is There," *Foreign Affairs*, April 1973.
18 Subcommittee on Multinational Corporations, *Hearings on Multinational Oil Corporations and US Foreign Policy*, op. cit., p. 136.
19 Ibid., p. 139–140.
20 Ian Seymour, "Arabs Increasing Pressure on Oil; The Supply-Demand Squeeze Oil Pressure Mounting," *The New York Times*, October 7, 1973.
21 William B. Quandt, *Peace Process: American Diplomacy and the Arab-Israeli Conflict Since 1967*, Brookings Institution, Washington, 1993, p. 164.
22 Alistair Horne, *Kissinger: 1973, the Crucial Year*, Simon & Schuster, New York, 2009, p. 264.
23 Rachel Bronson, *Thicker Than Oil*, op. cit., p. 118.
24 Clyde H. Farnsworth, "Shah and Faisal Were Prime Movers in Strategy on Oil," *The New York Times*, December 29, 1973.
25 Rachel Bronson, *Thicker Than Oil*, op. cit., p. 121 and 119.
26 Richard Eder, "4 More Arab Governments Bar Oil Supplies for US," *The New York Times*, October 21, 1973.
27 Anthony Cave Brown, *Oil, God, and Gold*, op. cit., p. 294.
28 Rachel Bronson, *Thicker Than Oil*, op. cit., p. 119.
29 Anthony Cave Brown, *Oil, God, and Gold*, op. cit., p. 296.
30 Rachel Bronson, *Thicker Than Oil*, op. cit., p. 120.
31 Ibid., p. 286, note 66.
32 Ibid., p. 118.
33 Jim Akins, "The Oil Crisis," op. cit.
34 Jim Hoagland, "Fayçal Warns US on Israel," *The Washington Post*, July 6, 1973; and Steven Emerson, *The American House of Saud: The Secret Petrodollar Connection*, Littlehampton Book Services, Worthing, 1985, p. 27.
35 Subcommittee on Multinational Corporations, *Hearings on Multinational Petroleum Corporations and Foreign Policy*, op. cit., part 7, p. 509.
36 Jim Hoagland, "Saudis Ponder Whether to Produce the Oil US Needs," *Washington Post*, July 11, 1973; and Nicholas C. Proffitt, "Fayçal's Threat," *Newsweek*, September 10, 1973.
37 Alistair Horne, *Kissinger*, op. cit., p. 269.
38 Abdul Amir Kubbah, *OPEC Past and Present*, Petro-Economic Research Center, Vienna, 1974, p. 78, cited in Anthony Sampson, *The Seven Sisters*, op. cit., p. 232.
39 David Hammes and Douglas Wills, "Black Gold: The End of Bretton Woods and the Oil-Price Shocks of the 1970s," *The Independent Review*, Washington, 2005,vol. IX, no. 4, p. 501–511.
40 Ibid.
41 David Bird, "Fuel Shortage May Force Rationing, Swidler Warns," *The New York Times*, September 19, 1970.
42 Dana Adams Schmidt, "US Authorizes Oil Import Rise," *The New York Times*, December 22, 1971.

43 "Nixon Suspends Oil Import Quota, Tariff," United Press International, April 18, 1973.
44 "Timely Quote," *The Victoria Advocate*, November 22, 1972.
45 Dillard Spriggs, in Subcommittee on Multinational Corporations, *Hearings on Multinational Petroleum Corporations and Foreign Policy*, op. cit., part 4, p. 61.
46 Christopher T. Rand, "Oil, Oil Everywhere," *The New York Times*, January 3, 1974.
47 "History of Crude Oil Prices," Illinois Oil and Gas Association searchable at ioga.com/index.php?option=com_content&view=article&id=41:history-of-crude-oil-prices&catid=20:2015-ioga-fall-meeting&Itemid=121.
48 According to the narrative provided a quarter of a century later by Jim Akins himself, in the Patrick Barbéris documentary *La Face Cachée du Pétrole* (Arte Editions, 2010), adapted from the book of the same name by Éric Laurent, *La Face Cachée du Pétrole*, op. cit.
49 André Giraud and Xaxier Boy de la Tour, *Géopolitique du Pétrole et du Gaz*, op. cit., *p. 243.*
50 Éric Laurent, *La Face Cachée du Pétrole*, op. cit., p. 175–177.
51 "History of Crude Oil Prices," op. cit.
52 Jim Akins, "The Oil Crisis," op. cit.
53 Bernard Weinraub, "Persian Gulf Oil Ministers Meet in Iran Amid Predictions of Major Price Rise," *The New York Times*, December 23, 1973.
54 Anthony Sampson, *The Seven Sisters*, op. cit., p. 258.
55 "United States-OPEC Relations: Selected Materials," US Government Printing Office, Washington, 1976, p. 287.
56 Bernard Weinraub, "Oil Price Doubled by Big Producers on Persian Gulf," *The New York Times*, December 24, 1973.
57 Ibid.
58 Gene Smith, "US Oilmen Find Prices Expectable, but 'Shocking'," *The New York Times*, December 24, 1973.
59 "Speed-up Is Studied for North Sea's Oil," Associated Press, London, December 24, 1973.
60 James P. Sterba, "Oil Shale: Leasing to Begin on Vast Reserves in West," *The New York Times*, January 6, 1974.
61 Gerd Wilcke, "Exxon Announces $6 Trillion Budget," *The New York Times*, December 21, 1973; and William Robbins, "Oil Profits Up 46% on 6% Volume Rise," the *New York Times*, January 23, 1973.
62 Jean-Marie Chevalier, *Le Nouvel Enjeu Pétrolier*, Calmann-Lévy, Paris, 1973 (English edition: *The New Oil Stakes*, Penguin Books, Middlesex, 1975, p. 49 and 130, cited in John M. Blair, *The Control of Oil*, op. cit., p. 295).
63 Oliver Morgan and Fayçal Islam, "Saudi Dove in the Oil Slick," *The Observer*, January 14, 2001.
64 Ibid.; and Patrick Barbéris, documentary, *La Face Cachée du Pétrole*.
65 Oliver Morgan and Fayçal Islam, "Saudi Dove in the Oil Slick," op. cit.
66 Henry Kissinger, *Years of Upheaval*, Little Brown, Boston, 1982, chapter 19.
67 David Rockefeller, *Memoirs*, op. cit., p. 283.

Chapter 20: Oil Money

1 International Energy Agency, *Key World Energy Statistics*, 2013.
2 David Frum, *How We Got Here: The 70s: The Decade That Brought You Modern Life—For Better or Worse*, Basic Books, New York, 2008, p. 320.
3 Ibid.
4 Gary B. Clark and Burt Kline, "Impact of Oil Shortage on Plastic Medical Supplies," *Public Health Reports*, Washington, vol. 96, no. 2, March/April 1981, p. 111–115.
5 Daniel Yergin, *The Prize*, op. cit., p. 611.

6 "Oil Crisis Déjà Vu," *Yomiuri Shimbun*, December 3, 2000.
7 Daniel Yergin, *The Prize*, op. cit., p. 598.
8 World Steel Association, searchable at worldsteel.org.
9 David Rockefeller, *Memoirs*, op. cit., p. 286.
10 William P. Avery, "The Origins of Debt Accumulation among LDCs in the World Political Economy," *The Journal of Developing Areas*, Nashville, vol. 24, no. 4, July 1990, p. 507.
11 Wolfgang Saxon, "James S. Rockefeller, 102, Dies; Was a Banker and a '24 Olympian," *The New York Times*, August 11, 2004.
12 David Rockefeller, *Memoirs,* op. cit., p. 311–317.
13 John Darnton, "Beame Excluded as Board Meets on Fiscal Crisis," *The New York Times*, August 21, 1975.
14 David Rockefeller, *Memoirs*, op. cit., p. 393; and Ron Chernow, *The House of Morgan: An American Banking Dynasty and the Rise of Modern Finance*, Grove Press, New York, 2009, p. 620–621.
15 David Rockefeller, *Memoirs*, op. cit., p. 395–396.
16 Ibid.
17 International Energy Agency, searchable at iea.org.
18 BP, *Statistical Review of World Energy Workbook 2013*.
19 International Energy Agency, searchable at iea.org.
20 "Iran Profile—Nuclear Chronology 1957–1985," Nuclear Threat Initiative, searchable at nti.org.
21 Dafna Linzer, "Past Arguments Don't Square with Current Iran Policy," *Washington Post*, March 27, 2005.
22 Jean-Paul Dufour, "Les Accords Nucléaires Franco-Irakiens de 1975," *Le Monde*, January 25, 1991.
23 Jean Guisnel, "La France, Premier Proliférateur Nucléaire," in Roger Faligot and Jean Guisnel (dir.), *Histoire Secrète de la Ve République*, op. cit., p. 246–247.
24 Georges Amsel, "Osirak, la Bombe et les Inspections," *Le Monde*, October 16, 2002.
25 Amir Taheri, "The Chirac Doctrine," *The National Review*, New York, November 4, 2002.
26 Georges-Henri Soutou, "La France et la Non-Prolifération Nucléaire," *Revue Historique des Armées,* no. 262, 2011, p. 35–45.
27 Bruce Riedel, "Saudi Arabia: Nervously Watching Pakistan," The Brookings Institution, Washington, January 28, 2008, searchable at brookings.edu/opinions/saudi-arabia-nervously-watching-pakistan.
28 Joseph J. Trento, *Prelude to Terror*, op. cit., p. 104.
29 Anthony Sampson, *The Seven Sisters*, op. cit., p. 268.
30 Reported by Olivier Appert, president of The French Institute of Petroleum.
31 "Chase Sees Funding in Oil Falling Short," *The New York Times,* March 30, 1973.
32 Mihajlo Mesarovic and Eduard Pestel, *Mankind at the Turning Point: The Second Report to the Club of Rome,* E. P. Dutton, New York, 1974, cited in James E. Akins,"World Energy Supply: Cooperation with OPEC or a New War of Resources," Énergie, Coopération Internationale ou Crise, *Les Presses de l'Université de Laval*, Québec, 1979, p. 228.
33 "Implications of Worldwide Population Growth for US Security and Overseas Interests," National Security Study Memorandum, Washington, December 1974, p. 41, searchable at pdf.usaid.gov/pdf_docs/Pcaab500.pdf.
34 "Oil Price Floor . . ." editorial, *The New York Times*, February 6, 1975.
35 "Cost of Oil Pipeline Worries Alaska," *The New York Times*, March 5, 1972.
36 "Alaska Oil-Carrying Cost Is Put at Over $6 a Barrel," Associated Press, June 2, 1977.
37 "The Dutch Disease," *The Economist*, November 26, 1977.

38 Mihajlo Mesarovic and Eduard Pestel, *Stratégie pour Demain: Deuxième Rapport au Club de Rome*, Club of Rome, 2nd report, Seuil, Paris, 1974, p. 178.
39 Martin Sandbu, "The Iraqi That Saved Norway from Oil," *The Financial Times*, August 29, 2009.
40 Bernard Gwertzman, "U.S. Oil Companies Are Negotiating with Vietnamese," *The New York Times*, April 24, 1976.
41 Richard Cooper, *Saturday Review*, January 25, 1975, cited in Antoine Ayoub, "Energy, International Cooperation or Crisis," 1979, p. 229.
42 "Jack F. Bennett: Death Notice," *The Washington Post*, op. cit.
43 Bernard Gwertzman, "Milestone Accord Is Signed by the US and Saudi Arabia," *The New York Times*, June 9, 1974.
44 Ibid.
45 Ibid.
46 See newdelhi.usembassy.gov.
47 Rachel Bronson, *Thicker Than Oil*, op. cit., p. 124.
48 David Rockefeller, *Memoirs*, op. cit., p. 285.
49 Ibid., p. 128.
50 Ibid., p.126.
51 Kenneth Labich, "Saudi Power," *Newsweek*, March 6, 1978, cited in Rachel Bronson, *Thicker Than Oil*, op. cit., p. 126.
52 Subcommittee on Multinational Corporations, *Hearings on Multinational Petroleum Corporations and Foreign Policy*, op. cit., part 7, p. 539, cited in John M. Blair, *The Control of Oil*, op. cit., p. 18.
53 Jack Anderson, "Reports Hit Aramco Haste on Oil" and "Details of Aramco Papers Disclosed," *The Washington Post*, January 11 and 28, 1974.
54 Subcommittee on Multinational Corporations, *Hearings on Multinational Petroleum Corporations and Foreign Policy*, op. cit., cited in Steven Emerson, *The American House of Saud*, op. cit., p. 131.
55 Steven Emerson, *The American House of Saud*, op. cit., p. 133.
56 Ibid., p. 134.
57 Ibid., p. 141.
58 Seymour Hersh, "US Experts Fear Saudi Troubles in the Oil Fields May Limit Output, *The New York Times*, December 25, 1977.
59 Central Intelligence Agency, "The Impending Soviet Oil Crisis," Intelligence Memorandum, ER 77-10147, March 1977.
60 *Oil Fields as Military Objectives: A Feasibility Study*, Committee on International Relations, Special Subcommittee on Investigations, US Government Printing Office, Washington, August 21, 1975; Franck Davis, "Oil War, Strategy & Tactics," September/October 1975.
61 *Business Week*, December 23, 1974.
62 Peter Goldman, "Ford Shakes Up His Cabinet," *Time*, November 17, 1975; and Sidney Blumenthal, "The Long March of Dick Cheney," *Salon*, November 24, 2005.
63 Richard Ben Cramer, *What It Takes: The Way to The White House*, Vintage, New York, 1992, p. 419.
64 Nicholas M. Horrock, "President Says He Will Issue 'Charters' to Secret Agencies," *The New York Times*, February 18, 1976.
65 Sam Tanenhaus, "The Hard Liner," *The Boston Globe*, February 11, 2003.
66 Russ Baker, *Family of Secrets*, op. cit., p. 280; and Craig Unger, *House of Bush, House of Saud: The Secret Relationship between the World's Two Most Powerful Dynasties*, Scribner, New York, 2004, p.19–20.
67 Craig Unger, *House of Bush, House of Saud*, op. cit., p. 33.

68 Russ Baker, *Family of Secrets*, op. cit., p. 145.
69 Craig Unger, *House of Bush, House of Saud*, op. cit., p. 53.
70 Russ Baker, *Family of Secrets*, op. cit., p. 297–298.
71 Craig Unger, *House of Bush, House of Saud*, op. cit., p. 34.
72 Rachel Bronson, *Thicker Than Oil*, op. cit., p. 130.
73 Ibid., p. 130.
74 Ibid., p.130.
75 Jim Hoagland and J. P. Smith, "Practicing Checkbook Diplomacy," *The Washington Post*, December 21, 1977.
76 Henry Kissinger, *Years of Upheaval*, op. cit., p. 663.
77 Alexandre De Marenches and David A. Andelman, *The Fourth World War: Diplomacy and Espionage in the Age of Terrorism*, William Morrow & Co, New York, 1992, p. 248–249.
78 Rachel Bronson, *Thicker Than Oil*, op. cit., p. 135.
79 Ibid., p. 131.
80 Leslie H. Gelb, "Will the Flag Follow US Arms Sales," *The New York Times*, August 8, 1976.
81 Bernard Weinraub, "US Said to Bar Sail of 250 Ordered by Iran," *The New York Times*, June 8, 1977.
82 John P. Miglietta, *American Alliance Policy in the Middle East, 1945–1992: Iran, Israel, and Saudi Arabia*, Lexington Books, Lanham, 2002, p. 212.
83 Ibid., p. 211.
84 Robert P. Grathwol and Donita M. Moorhus, *Bricks, Sand, and Marble*, op. cit., p. 399–444.
85 "US Company Will Train Saudi Troops to Guard Oil," Associated Press, February 9, 1975.
86 John Perkins, *Confessions of an Economic Hit Man*, Berrett-Kohler, San Francisco, 2004, p. 12.
87 Peter Osnos and David B. Ottaway, "Yamani Links F15s to Oil, Dollar Help," *The Washington Post*, May 2, 1978.
88 William B. Quandt, *Camp David: Peacemaking and Politics*, Brookings Institution Press, Washington, 1986, p. 296, cited in Rachel Bronson, *Thicker Than Oil*, op. cit., p. 143.
89 "Saudis Now Biggest US Arms Customer," *Armed Forces Journal*, October 1979.
90 Noam Chomsky, "The Carter Administration: Myth and Reality," Radical Priorities, 1981, searchable at chomsky.info/priorities01.
91 Michel Crozier, Samuel Huntington, and Joji Watanuki, *The Crisis of Democracy: Report on the Governability of Democracies to the Trilateral Commission*, New York University Press, New York, 1975.
92 John E. Peterson, *Saudi Arabia and the Illusion of Security*, Oxford University Press, New York, 2002, p. 46.

Chapter 21: The Second Oil Crisis

1 Robert Fisk, *La Grande Guerre pour la Civilisation: L'Occident à la Conquête du Moyen-Orient (1979–2005)*, La Découverte, Paris, 2005, p. 129.
2 Flora Lewis, "Itinerary Is Shifted," *The New York Times*, January 1, 1978; and Robert Fisk, *La Grande Guerre pour la Civilisation*, op. cit., p. 147.
3 Ibid., p. 131.
4 Daniel L. Byman, "The Rise of Low-Tech Terrorism," May 6, 2007, searchable at brookings.edu/articles/the-rise-of-low-tech-terrorism.
5 Daniel Yergin, *The Prize*, op. cit., p. 662.
6 David Rockefeller, *Memoirs*, op. cit., p. 363–375.
7 Sepehr Zabih, *Iran since the Revolution*, Johns Hopkins Press, Baltimore, 1982, p. 2.
8 Susan Straight, "That '70s Energy Crisis," The International Herald Tribune, May 6, 2011.
9 Judith Miller, "House Staff Asserts US Hid Oil Data," *The New York Times*, January 4, 1980.

10 Mark Bowden, *Guests of the Ayatollah: The Iran Hostage Crisis: The First Battle in America's War with Militant Islam*, Grove Press, New York, 2006, p. 158.
11 Gary Sick, *October Surprise: America's Hostages in Iran and the Election of Ronald Reagan*, Three Rivers, New York, 1991; and Gary Sick, "The Election Story of the Decade," *The New York Times*, April 15, 1991.
12 Investigating the October Surprise, PBS Frontline, 1990; and Robert Parry, "Key October Surprise Evidence Hidden," searchable at consortiumnews.com/2010/050610.html, May 6, 2010.
13 Codetec and Cerchar, *Injections de Charbon aux Tuyères des Hauts Fourneaux*, Éditions Technip, Paris, 1988, p. 113, cited in "Haut Fourneau," fr.wikipedia.org.
14 David Rockefeller, *Memoirs*, op. cit., p. 369.
15 Lawrence Wright, *La Guerre Cachée: Al-Qaïda et les Origines du Terrorisme Islamiste,* Robert Laffont, Paris, 2007, p. 92.
16 Ibid., p. 93.
17 Ibid., p. 95.
18 Ibid., p. 94.
19 Ibid., p. 97.
20 Craig Unger, *House of Bush, House of Saud*, op. cit., p. 95.
21 Lawrence Wright, *La Guerre Cachée*, op. cit., p. 98.
22 Ibid., p. 99–100.
23 Ibid., p. 100–101.
24 Vincent Jauvert, "Oui, la CIA Est Entrée en Afghanistan Avant les Russes . . .," interview with Zbigniew Brzezinski, *Le Nouvel Observateur*, January 15, 1998.
25 Ibid.
26 Daniel Yergin, *The Prize*, op. cit., p. 684.
27 Rachel Bronson, *Thicker Than Oil*, op. cit., p. 147.
28 Craig Unger, *House of Bush, House of Saud*, op. cit., p. 323, note 11.
29 Rory O'Connor, "The Arming of Saudi Arabia," *PBS Frontline* #1112, February 16, 1993, cited in Craig Unger, *House of Bush, House of Saud*, op. cit., p. 60–61.
30 Caspar W. Weinberger, cited in Rachel Bronson, *Thicker Than Oil*, op. cit., p. 151; and Halford John Mackinder, "The Geographical Pivot of History," *The Geographical Journal*, The Royal Geographical Society, London, 1904.

Chapter 22: The Long Iran-Iraq War

1 Howard Teicher and Gayle Radley Teicher, *Twin Pillars to Desert Storm: America's Flawed Vision in the Middle East from Nixon to Bush*, William Morrow, New York, 1993, p. 62.
2 Henry Kissinger, "Defining a US Role in the Arab Spring," *The International Herald Tribune*, April 2, 2012.
3 Cited in Saïd K. Aburish, *Saddam Hussein: The Politics of Revenge*, Bloomsbury, New York, 2000, p. 73–74.
4 Henry Kissinger, *Years of Upheaval*, op. cit., p. 669.
5 Ibid., p. 674.
6 Seymour Hersh, *The Price of Power: Kissinger in the Nixon White House*, Summit Books, New York, 1983, p. 542, cited in Larry Everest, *Oil, Power and Empire: Iraq and the US Global Agenda*, Common Courage Press, Monroe, 2003, p. 80–81; Henry Kissinger, *Years of Upheaval*, op. cit., p. 674–675.
7 Jim Hoagland, "Kurds Ready to Resume Fight as Pact with Iraq Crumbles," *The International Herald Tribune*, June 22, 1973.
8 *The Kurds and Kurdistan*, Interlink Publishing Group, Northampton, 1997, p. 169–170.

9 Ibid., p. 171.
10 Jean Gueyras, "La Proie pour l'Ombre," *Le Monde*, April 3, 1975; 300,000 according to the UPK, cited in Robert Fisk, *La Grande Guerre pour la Civilisation*, op. cit., p. 198.
11 Otis Pike, *CIA: The Pike Report*, Spokesman Books, Nottingham, 1977, p. 198 and 217, cited in Larry Everest, *Oil, Power and Empire*, op. cit., p. 83.
12 Robert Fisk, *La Grande Guerre pour la Civilisation*, op. cit., p. 199.
13 Ibid., p. 191.
14 Testimony of nuclear physicist Hussein Shahristani, in ibid., p. 185–189.
15 Anthony Shadid, *Night Draws Near: Iraq's People in the Shadow of America's War*, Holt, New York, 2005, p. 164.
16 Efraim Karsh, *The Iran-Iraq War 1980–1988*, Osprey, London, 2002, p. 13; and Charles Tripp, *A History of Iraq*, Cambridge University Press, Cambridge, 2000, p. 220–221, cited in Larry Everest, *Oil, Power and Empire*, op. cit., p. 94–95.
17 Robert Fisk, *La Grande Guerre pour la Civilisation*, op. cit., p. 211.
18 Craig Unger, *House of Bush, House of Saud*, op. cit., p. 65.
19 Ibid., p. 67.
20 Robert Fisk, *La Grande Guerre pour la Civilisation*, op. cit., p. 228.
21 Howard Teicher and Gayle Radley Teicher, *Twin Pillars to Desert Storm*, op. cit., p. 103.
22 Robert Fisk, *La Grande Guerre pour la Civilisation*, op. cit., p. 209.
23 Ronen Bergman, *The Secret War with Iran: The 30-Year Clandestine Struggle against the World's Most Dangerous Terrorist Power*, Free Press, New York, 2008, p. 40–48; and Trita Parsi, *Treacherous Alliance: The Secret Dealings of Israel, Iran, and the United States*, Yale University Press, Yale, 2007, p. 107.
24 Bruce Jentleson, *With Friends Like These: Reagan, Bush, and Saddam, 1982–1990*, W. W. Norton, New York, 1994, p. 35; Mark Phythian, *Arming Iraq: How the US and Britain Secretly Built Saddam's War Machine*, Northeastern Publishing, Holliston, 1996 p. 20; and Trita Parsi, *Treacherous Alliance*, op. cit., p. 106–107.
25 Murray S. Waas and Craig Unger, "In the Loop: Bush's Secret Mission," Annals of Government, *The New Yorker*, November 2, 1992.
26 "The Teicher Affidavit: Iraq-Gate," searchable at informationclearinghouse.info.
27 Robert Fisk, *La Grande Guerre pour la Civilisation*, op. cit., p. 238.
28 "Subject: Rumsfeld One-to-One Meeting with Deputy Prime Minister and Foreign Minister Tarek Aziz," State Department diplomatic cable, December 21, 1983, declassified, George Washington University, searchable at gwu.edu.
29 Ibid.
30 Jim Vallette, Steve Kretzmann, and Daphne Wysham, "Crude Vision: How Oil Interests Obscured US Government Focus on Chemical Weapons Used by Saddam Hussein," Institute for Policy Studies, Washington, March 2003, p. 11.
31 Dana Priest, "Rumsfeld Visited Baghdad in 1984 to Reassure Iraqis, Documents Show: Trip Followed Criticism of Chemical Arms' Use," *The Washington Post*, December 19, 2003; and Jim Vallette, Steve Kretzmann, and Daphne Wysham, "Crude Vision," op. cit., p. 11.
32 Craig Unger, *House of Bush, House of Saud*, op. cit., p. 67.
33 Richard Holloran, "2 Iranian Fighters Reported Downed by Saudi Air Force," *The New York Times*, June 6, 1984.
34 Nathaniel Hurd and Glen Rangwala, "US Diplomatic and Commercial Relationships with Iraq, 1980–August 2, 1990," cited in Larry Everest, *Oil, Power and Empire*, op. cit., p. 103.
35 Michael Dobbs, "US Had Key Role in Iraq Build-up: Trade in Chemical Arms Allowed Despite Their Use on Iranians Kurds," *The Washington Post*, December 30, 2002.

36 Nathaniel Hurd and Glen Rangwala, "US Diplomatic and Commercial Relationships with Iraq, 1980–August 2, 1990," op. cit.; "Top Secret Iraq Weapons Report Says the US Government & Corporations Helped to Illegally Arm Iraq," *Die Tageszeitung*, December 17, 2002; Tony Paterson, "Leaked Report Says German and US Firms Supplied Arms to Saddam," *The Independent*, December 18, 2002; and Philip Shenon, "Declaration Lists Companies That Sold Chemicals to Iraq," *The New York Times*, December 21, 2002.
37 "What Iraq Admitted about Its Chemical Weapons Program," *The New York Times*, April 13, 2003; and Jim Vallette, Steve Kretzmann, and Daphne Wysham, "Crude Vision," op. cit.
38 Robert Fisk, *La Grande Guerre pour la Civilisation*, op. cit., p. 241.
39 Chalmers Johnson, *Nemesis: The Last Days of the American Republic*, Metropolitan Books, New York, 2008, p. 165.
40 Karl Siegfried Guthke, *B. Traven: The Life behind the Legends*, Lawrence Hill & Co., Chicago, 1991, p. 184.
41 Philip Taubman, "US Officials Say C.I.A Helped Nicaraguan Rebels Plan Attacks," *The New York Times*, October 16, 1983; and "Exxon Stops Supplying Nicaraguans with Oil," *The New York Times*, October 16, 1983.
42 Howard Zinn, *A People's History of the United States*, op. cit., p. 332.
43 Ibid., p. 334.
44 Maura Reynolds, "Poindexter Expected to Resign in Futures Market Controversy," *Los Angeles Times*, August 1, 2003.
45 "Oliver North Reports from Iraq," Fox News, December 9, 2005, searchable at foxnews.com /transcript/2005/12/09/oliver-north-reports-iraq0.html.
46 Rachel Bronson, *Thicker Than Oil*, op. cit., p. 184.
47 Craig Unger, *House of Bush, House of Saud*, op. cit., p. 64.
48 Russ Baker, *Family of Secrets*, op. cit., p. 313–314.
49 John Bulloch and Harvey Morris, *No Friends but the Mountains: The Tragic History of the Kurds*, Oxford University Press, Oxford, 1993, p. 142, cited in Larry Everest, *Oil, Power and Empire*, op. cit., p. 112.
50 Charles Tripp, *A History of Iraq*, op. cit., p. 243, cited in Larry Everest, *Oil, Power and Empire*, op. cit., p. 112.
51 Craig Unger, *House of Bush, House of Saud*, op. cit., p. 79.
52 Robert Pear, "US Says It Monitored Iraqi Messages on Gas," *The New York Times*, September 15, 1988; and "Rewind: Al Gore Blasts G. H. W. Bush for Ignoring Iraq Terror Ties," CSPAN, September 29, 1992.
53 Craig Unger, *House of Bush, House of Saud*, op. cit., p. 79–80.
54 Joost R. Hiltermann, "Halabja: America Didn't Seem to Mind Poison Gas," *The New York Times*, January 17, 2003.
55 Robert Fisk, *La Grande Guerre pour la Civilisation*, op. cit., p. 244.
56 "National Security Directive 26," searchable at fas.org/irp/offdocs/nsd/nsd26.pdf.
57 Joseph Trento, *Prelude to Terror*, op. cit., p. 299.
58 Murray Waas and Craig Unger, "Annals of Government. How the US Armed Iraq," *The New Yorker*, November 2, 1992, cited in Larry Everest, *Oil, Power and Empire*, op. cit., p. 107.
59 Stephen Engelberg, "Iran and Iraq Got 'Doctored' Data, U.S. Officials Say," *The New York Times*, January 12, 1987.
60 "Khomeini's Poison," *The Boston Globe*, July 24, 1988.

Chapter 23: The Oil Countershock

1 Rachel Bronson, *Thicker Than Oil*, op. cit., p. 186.

2 FMI, *Oil Scarcity, Growth, and Global Imbalances*, April 2011.

3 Douglas Martin, "The Singular Power of a Giant Called Exxon," *The New York Times*, May 9, 1982 ; "Pullout by Exxon Jolts a Boomtown," Associated Press, October 10, 1982.

4 Paul Krugman, "An Economic Legend," *The New York Times*, June 11, 2004.

5 Joseph Stiglitz, "Nobel Prize Winner—Oil Wealth Was 'Squandered' by UK," August 25, 2010, searchable at newsnetscotland.com.

6 Congressional Budget Office, searchable at cbo.gov.

7 Peter Schweizer, *Victory: The Reagan Administration's Secret Strategy That Hastened the Collapse of the Soviet Union*, Atlantic Monthly Press, New York, 1994, p. 111.

8 Steven Rattner, "Britain Defying U.S. Restriction in Soviet Project," *The New York Times*, August 3, 1982.

9 Peter Schweizer, *Victory*, op. cit., p. 216.

10 Ibid., p. 126.

11 Interview with Roger Robinson, Patrick Barbéris, *La Face Cachée du Pétrole*, documentary.

12 Ibid.

13 Warren E. Norquist, "How the United States Used Competition to Win the Cold War," *American Society for Competitiveness*, 2002.

14 Peter Schweizer, *Victory*, op. cit., p. 205.

15 Peter Schweizer, *Reagan's War: The Epic Story of His Forty-Year Struggle and Final Triumph Over Communism*, Knopf Group E-Books, 2003.

16 Ibid.

17 Peter Schweizer, *Victory*, op. cit., p. 204.

18 Rachel Bronson, *Thicker than Oil*, op. cit., p. 186.

19 "Saudi Oil Output Surges," *Chicago Tribune*, October 17, 1985.

20 Robert Mabro, "Netback Pricing and the Oil Price Collapse of 1986," Oxford Institute for Energy Studies, WPM10, January 1987.

21 Interview with Mikhaïl Gorbatchev, Patrick Barbéris, *La Face Cachée du Pétrole*, documentary.

22 Daniel Yergin, *The Prize*, op. cit., p. 733.

23 Craig Unger, *House of Bush, House of Saud*, op. cit., p. 39.

24 Energy Information Administration, searchable at eia.gov.

25 Timothy J. McNulty, "Bush, Saudi Oil Talks Fail: No Break Seen in Pricing Policy," *Chicago Tribune*, April 8, 1986, cited in Daniel Yergin, *The Prize*, op. cit., p. 737.

26 Timothy J. McNulty, "U.S. Oil Plan Called Impossible," *Chicago Tribune*, April 4, 1986, cited in Daniel Yergin, *The Prize*, op. cit. p. 737.

27 Daniel Yergin, *The Prize*, op. cit., p. 737.

28 *The Washington Post*, April 10, 1986, cited in ibid., p. 738.

29 Terence Hunt, "Bush, Fahd Discuss Oil Situation; King Reportedly Complains of 'Bum Rap,'" Associated Press, April 7, 2006.

30 *The Washington Post*, April 10, 1986, cited in Daniel Yergin, *The Prize*, op. cit., p. 738.

31 *The Washington Post*, April 8, 1986, cited in Daniel Yergin, *The Prize*, op. cit., p. 738.

32 Daniel Yergin, *The Prize*, op. cit., p. 739.

33 Wanda Jablonski, "The Inside Story of Why Yamani Was Actually Dismissed," *Petroleum Intelligence Weekly*, November 24, 1986.

34 Geoff Simons, *Saudi Arabia: The Shape of a Client Feudalism*, St. Martin's Press, New York, 1998, p. 28.

35 Marie Colvin, "The Squandering Sheikhs," *The Sunday Times*, August 29, 1993.

36 David B. Ottaway, "Been There, Done That; Prince Bandar, One of the Great Cold Warriors, Faces the Yawn of an Era," *The Washington Post*, July 21, 1996, cited in Craig Unger, *House of Bush, House of Saud*, op. cit., p. 130.

37 Elsa Walsh, "The Prince: How the Saudi Ambassador Became Washington's Indispensable Operator," *The New Yorker*, March 23, 2003.
38 Patrick Tyler, "Officers Say U.S. Aided Iraq in War Despite Use of Gas," *The New York Times*, August 18, 2002.
39 Craig Unger, *House of Bush, House of Saud*, op. cit., p. 130.
40 Rachel Bronson, *Thicker Than Oil*, op. cit., p. 177–182.
41 Ibid., p. 173.
42 Ibid., p. 173–174.
43 Craig Unger, *House of Bush, House of Saud*, op. cit., p. 147.
44 James Rupert, "Dreams of Martyrdom Draw Islamic Arabs to Join Afghan Rebels," *The Washington Post*, July 21, 1986, cited in Rachel Bronson, *Thicker Than Oil*, op. cit., p. 174.
45 Lawrence Wright, *La Guerre Cachée*, op. cit., p. 108.
46 James Rupert, "Arab Fundamentalists Active in Afghan War; Use of Funds and Volunteers from Mideast Watched for Disruptive Influence," *The Washington Post*, March 2, 1989, cited in Rachel Bronson, *Thicker Than Oil*, op. cit., p. 174.
47 David Leigh, "Bush, the Saudi Billionaire and the Islamists: The Story a British Firm Is Afraid to Publish," *The Guardian*, March 31, 2004.
48 Ibid.
49 Craig Unger, *House of Bush, House of Saud*, op. cit., p. 112.
50 John Cooley, *Unholy Wars: Afghanistan, America, and International Terrorism*, Pluto Press, London, 2002, p. 93–94.
51 John Kerry and Hank Brown, "The BCCI Affair: A Report to the Committee on Foreign Relations," US Senate Congress, Senate Print 102-140, Washington, December 1992, searchable at fas.org/irp/congress/1992_rpt/bcci.
52 Ibid.
53 Richard Lacayo, "Iran-Contra: The Cover-Up Begins to Crack," *Time*, June 24, 2001.
54 Nathan Vardi, "Sins of the Fathers?," *Forbes*, March 18, 2002; and Craig Unger, *House of Bush, House of Saud*, op. cit., p. 78.
55 Steve Lohr, "World-Class Fraud: How BCCI Pulled It Off—A Special Report; At the End of a Twisted Trail, Piggy Bank for a Favored Few," *The New York Times*, August 12, 1991.
56 John Kerry and Hank Brown, "The BCCI Affair," op. cit.
57 Ibid.
58 Joseph J. Trento, *Prelude to Terror*, op. cit., p. 305.
59 Jonathan Beaty and S. C. Gwynne, "The Dirtiest Bank of All," *Time*, July 29, 1991.
60 Alfred McCoy, "Drug Fallout: The CIA's Forty-Year Complicity in the Narcotics Trade," *The Progressive Magazine*, August 1, 1997.
61 Alfred McCoy, *The Politics of Heroin: CIA Complicity in the Global Drug Trade: Afghanistan, Southeast Asia, Central America, Colombia*, Lawrence Hill Books, Chicago, 2003, p. 486.
62 Richard Lacayo, "Iran-Contra," op. cit.
63 Craig Unger, *House of Bush, House of Saud*, op. cit., p. 78.
64 Jonathan Beaty and S. C. Gwynne, "Not Just a Bank; You Can Get Anything You Want through BCCI," *Time*, September 2, 1991.
65 Nathan Vardi, "Sins of the Fathers?," op. cit.; and Craig Unger, *House of Bush, House of Saud*, op. cit., p. 127.
66 "Meet the Press," NBC News, October 14, 2001, cited in Craig Unger, *House of Bush, House of Saud*, op. cit., p. 111.
67 Ibid.

68 John M. Goshko and Don Oberdorfer, "Chinese Sell Saudis Missiles Capable of Covering the Middle East," *The Washington Post*, March 18, 1988; John Kerry and Hank Brown, "The BCCI Affair," op. cit.; and Adrian Levy and Catherine Scott-Clark, *Deception: Pakistan, the United States, and the Secret Trade in Nuclear Weapons*, Walker & Company, New York, 2007, p. 126–128.
69 Craig Unger, *House of Bush, House of Saud*, op. cit., p. 118.
70 Richard Behar, "The Wackiest Rig in Texas," *Time*, October 28, 1991; Russ Baker, *Family of Secrets*, op. cit., p. 338 and 342.
71 Ibid., p. 341.
72 Thomas Petzinger Jr., Peter Truell, and Jill Abramson, "How Oil Firm Linked to a Son of Bush Won Bahrain Drilling Pact," *The Wall Street Journal*, December 6, 1991.
73 Paul Krugman, "Succeeding in Business," *The New York Times*, July 7, 2002.
74 Thomas Petzinger Jr., Peter Truell, and Jill Abramson, "How Oil Firm Linked to a Son of Bush Won Bahrain Drilling Pact," op. cit.
75 Charles Strain, cited in Toni Mack, "Fuel for Fantasy," *Forbes*, September 3, 1990.
76 Michael Kranish and Beth Healy, "Board Was Told of Risks before Bush Stock Sale," *The Boston Globe*, October 3, 2002.
77 Craig Unger, *House of Bush, House of Saud*, op. cit., p. 123–124.
78 Ibid., p. 124.
79 "Jack F. Bennett: Death Notice," *The Washington Post*, op. cit.
80 Ron Chernow, *The House of Morgan*, op. cit., p. 588–589.
81 Jim Jelter, "Exxon Valdez and the Birth of Credit Default Swaps," MarketWatch, marketwatch .com/story/exxon-valdez-and-the-birth-of-credit-default-swaps-2010-05-03, May 3, 2010.
82 Ibid., p. 699.
83 Ron Chernow, *The House of Morgan*, op. cit., p. 693.
84 Colin J. Campbell and Jean H. Laherrère, "The End of Cheap Oil," *Scientific American*, March 1998, p. 78–83.
85 Harold L. Ickes, "We're Running Out of Oil," op. cit.
86 Matthew R. Simmons, *Twilight in the Desert: The Coming Saudi Oil Shock and the World Economy*, Wiley, Hoboken, 2006.

Chapter 24: Dear Saddam

1 Thomas C. Hayes, "Confrontation in the Gulf: The Oil Field Lying below the Iraq-Kuwait Dispute," *The New York Times*, September 3, 1990.
2 Larry Everest, *Oil, Power and Empire*, op. cit., p. 119.
3 "National Security Directive 26," op. cit., and see supra, chapter 22.
4 Craig Unger, *House of Bush, House of Saud*, op. cit., p. 131.
5 Victor Marshall, "The Lies We Are Told about Iraq," *The Los Angeles Times*, January 5, 2003.
6 Kevin Phillips, "Bush's Worst Political Nightmare," op. cit.; Anthony Kimery, "A Well of a Deal," op. cit., see supra, chapter 14.
7 Michael T. Klare, *Resource Wars: The New Landscape of Global Conflict*, Holt Paperbacks, New York, 2002, p. 61–62.
8 Bob Woodward, *The Commanders*, Simon & Schuster, New York, 2002, p. 214.
9 Ibid., p. 252.
10 James Baker, *The Politics of Diplomacy*, Putnam, New York, 1995, p. 289.
11 Robert Fisk, *La Grande Guerre pour la Civilisation*, op. cit., p. 552.
12 Ibid., p. 552–556; and Sohbet Karbuz, "The U.S. Army Oil Consumption," searchable at energybulletin.net.
13 Larry Everest, *Oil, Power and Empire*, op. cit., p. 136.

14 James Baker and Steve Fiffer, *Work Hard, Study . . . and Keep Out of Politics!: Adventures and Lessons from an Unexpected Public Life,* Northwestern University Press, Evanston, 2008, p. 295.
15 Thalif Deen, "U.N. Credibility at Stake over Iraq, War Diplomats," *Inter Press Service*, October 1, 2002.
16 Craig Unger, *House of Bush, House of Saud*, op. cit., p. 145.
17 Ibid., p. 139.
18 Ibid.
19 Ibid., p. 140.
20 Bob Woodward, *The Commanders*, op. cit., p. 226.
21 Joseph Caroll, "Americans on Iraq: Military Action or Diplomacy?," searchable at gallup.com.
22 John MacArthur, "Remember Nayirah, Witness for Kuwait?," *Seattle Post-Intelligencer*, January 12, 1992.
23 Joseph Caroll, "Americans on Iraq: Military Action or Diplomacy?," op. cit.
24 Tim Weiner, "Smart Weapons Were Overrated, Study Concludes," *The New York Times*, July 9, 1996.
25 Ibid.
26 Ibid.
27 Éric Rouleau, "America's Unyielding Policy toward Iraq," *Foreign Affairs*, January / February 1995.
28 Barton Gellman, "Allied Air War Struck Broadly in Iraq. Officials Acknowledge Strategy Went beyond Purely Military Targets," *The Washington Post*, June 23, 1991, cited in Larry Everest, *Oil, Power and Empire*, op. cit., p. 145.
29 *The Washington Post*, June 23, 1991, cited in Larry Everest, *Oil, Power and Empire*, op. cit., p. 175.
30 Colin Powell, *My American Journey*, Random House, New York, 1995, p. 516, cited in Larry Everest, *Oil, Power and Empire*, p. 151.
31 Bob Woodward, "Watergate's Shadow on the Bush Presidency," *The Washington Post Magazine*, June 20, 1999.
32 George H. W. Bush and Brent Scowcroft, *A World Transformed*, Vintage, New York, 1999, p. 478.
33 Ibid.
34 Ibid., p. 463.
35 Ibid.
36 Robert Fisk, *La Grande Guerre pour la Civilisation*, op. cit., p. 561.
37 Ibid., p. 580 and 605.
38 Larry Everest, *Oil, Power and Empire*, op. cit., p. 152.
39 Ibid., p. 153.
40 American Gulf War Veterans Association, "Gulf War Vets Question Who Started the Oil Well Fires in Kuwait," press release, February 19, 2003.
41 Sebastião Salgado, *La Main de l'Homme: Une Archéologie de l'ère Industrielle*, La Martinière, Paris, 1993.
42 George H. W. Bush and Brent Scowcroft, *A World Transformed*, op. cit., p. 489.
43 Ibid.
44 Patrick E. Tyler, "Copters a Threat, U.S. Warns Iraqis," *The New York Times*, March 19, 1991.
45 George H. W. Bush and Brent Scowcroft, *A World Transformed*, op. cit., p. 490.
46 Larry Everest, *Oil, Power and Empire*, op. cit., p. 159.
47 Ibid.
48 Ibid., p. 160.
49 *San Francisco Chronicle*, February 16, 2003, cited in ibid., p.161.
50 Larry Everest, *Oil, Power and Empire*, op. cit., p. 161.

51 "No Fly Zone. Why," *The New York Times*, August 28, 1992.
52 George H. W. Bush and Brent Scowcroft, *A World Transformed*, op. cit., p. 383.
53 Ibid., p. 489.
54 Ibid., p. 487.
55 Patrick E. Tyler, "Bush Links End of Trading Ban to Hussein Exit," *The New York Times*, May 21, 1991.
56 Robert Baer, *La Chute de la CIA: Les Mémoires d'un Guerrier de l'Ombre sur les Fronts de L'islamisme*, Folio, Paris, 2003, p. 251–312. First published as *See No Evil*, Random House, New York, 2002.
57 Ibid., p. 262.
58 Martti Ahtisaari, *The Impact of War on Iraq*, United Nations, New York, March 20, 1991.
59 Larry Everest, *Oil, Power and Empire*, op. cit., p. 172.
60 Ibid., p. 174.
61 Ibid., p. 177.
62 Thomas Nagy, "The Secret behind the Sanctions: How the U.S. Intentionally Destroyed Iraq's Water Supply," *The Progressive*, September 2001.
63 Larry Everest, *Oil, Power and Empire*, op. cit., p. 181–182.
64 Iraq Survey Group report, September 30, 2004; and Joy Gordon, "Cool War: Economic Sanctions as a Weapon of Mass Destruction," *Harper's*, November 2002.
65 Iraq Survey Group final report, op. cit.; Judith Miller and Eric Lipton, "Report Cites U.S. Profits in Sale of Iraqi Oil under Hussein," *The New York Times*, October 9, 2004; Jess Bravin, John D. McKinnon, and Russell Gold, "Iraq Oil-for-Food Probe Hits U.S. Exxon, Chevron and El Paso Are Named in CIA Report on Hussein-Era Program," *The Wall Street Journal*, October 11, 2004.
66 Peter Baker, "Iraq's Shortage of Medicine May Grow More Severe," *The Washington Post*, December 19, 2002.
67 Ravij Chandrasekaran, "Iraq's System, Deemed 'Most Efficient in the World,'" Benefits Even Rebel Kurds," *The Washington Post*, February 3, 2003.
68 Naomi Klein, "Why Is War-Torn Iraq Giving $190,000 to Toys R Us ?," *The Guardian*, October 16, 2004.
69 Christopher Marquis, "Over the Years, Cheney Opposed U.S. Sanctions," *The New York Times*, July 27, 2000; and Colum Lynch, "Firm's Iraq Deals Greater Than Cheney Has Said," *The Washington Post*, June 23, 2001.
70 Anthony H. Cordesman, *The Iraq War: Strategy, Tactics, and Military Lessons*, CSIS, Washington, 2003, p. 550; and Colum Lynch, "Firm's Iraq Deals Greater Than Cheney Has Said," op. cit.
71 "Halliburton Exploited Iraq's Oil-For-Food Program While Cheney Was CEO," The Public Record, July 2, 2008, searchable at pubrecord.org.
72 "Child and Maternal Mortality Survey 1999: Preliminary Report," UNICEF, July 1999; and Joy Gordon, *Invisible War: The United States and the Iraq Sanctions*, Harvard University Press, Cambridge, 2012, note 82, p. 255–257.
73 Peter Baker, "Iraq's Shortage of Medicine May Grow More Severe," op. cit.
74 "Punishing Saddam," *60 Minutes*, CBS, May 12, 1996.

Chapter 25: Planetary Harvest

1 Michael T. Klare, *Resource Wars: The New Landscape of Global Conflict*, Metropolitan Books, New York, 2001.
2 Robert Fisk, *La Grande Guerre pour la Civilisation*, op. cit., p. 580.
3 Samuel P. Huntington, "The Clash of Civilizations?," *Foreign Affairs*, Summer 1993.

4 John Esposito, *Unholy War: Terror in the Name of Islam*, Oxford University Press, Oxford, 2002, p. 12.
5 Lawrence Wright, *La Guerre Cachée*, op. cit., p. 158.
6 Ibid., p. 159.
7 Douglas Jehl, "Holy War Lured Saudis as Rulers Looked Away," *The New York Times*, December 27, 2001.
8 Lawrence Wright, *La Guerre Cachée*, op. cit., p. 160.
9 Robert Fisk, *La Grande Guerre pour la Civilisation*, op. cit., p. 41.
10 Craig Unger, *House of Bush, House of Saud*, op. cit., p. 171.
11 Robert Fisk, *La Grande Guerre pour la Civilisation*, op. cit., p. 764.
12 Kim Murphy, "Yemenis Wear Out Their Welcome in Saudi Arabia," *Los Angeles Times*, October 3, 1990.
13 "Angry Welcome for Palestinian in Kuwait," BBC, May 30, 2001, searchable at news.bbc.co.uk /2/hi/middle_east/1361060.stm.
14 Robert Fisk, *La Grande Guerre pour la Civilisation*, op. cit., p. 668.
15 Ibid., p. 668–669.
16 Ibid., p. 560.
17 Michael T. Klare, *Resource Wars*, op. cit., p. 34.
18 Robert Baer, *La Chute de la CIA*, op. cit., p. 313–384.
19 Dan Morgan and David Ottaway, "Drilling for Influence in Russia's Backyard," *The Washington Post*, September 22, 1997, cited in Michael T. Klare, *Resource Wars*, op. cit., p. 89.
20 Robert Baer, *La Chute de la CIA*, op. cit., p. 359.
21 Ibid., p. 345–346.
22 Ibid., p. 347.
23 Ibid., p. 347.
24 Niel A. Lewis, "Questions Arise over Stock Portfolio of Candidate to Head CIA," *The New York Times*, December 13, 1996; David Stout, "CIA Nominee to Pay $5,000 to Settle Inquiry," *The New York Times*, February 8, 1997; and "National Security Adviser Sandy Berger Agrees to Pay $23,000 to Settle Conflict of Interest," Associated Press, November 10, 1997.
25 Robert Baer, *La Chute de la CIA*, op. cit., p. 361.
26 Tim Shorrock, "Crony Capitalism Goes Global," *The Nation*, March 14, 2002.
27 Michael Lewis, "The Access Capitalists," The New Republic, October 18, 1993.
28 Michael Moore, *Fahrenheit 9/11*, 2004.
29 Kenneth Gilpin, "Little-Known Carlyle Scores Big," *The New York Times*, March 26, 1991.
30 Craig Unger, *House of Bush, House of Saud*, op. cit., p. 163–164.
31 Ibid., p. 167.
32 Ibid., p. 168.
33 Ibid.
34 Pascale Robert-Diard, "Angolagate: La Cour d'Appel de Paris Relaxe Charles Pasqua," *Le Monde*, April 29, 2011; and *Le Canard Enchaîné*, July 18, 2012.
35 François-Xavier Verschave, *La Françafrique: Le Plus Long Scandale de la République*, Stock, Paris, 2003.
36 Patrick Benquet, *La Françafrique*, op. cit.
37 "Elf, l'Empire d'Essence," *Les Dossiers du Canard Enchaîné*, April 1998, p. 21.
38 Loïk Le Floch-Prigent, interview with the author, November 2011.
39 Pascale Robert-Diard, "François Mitterrand a Fait Racheter par Elf la Villa du Dr Raillard, pour Faire Son Golf Tous les Lundis," *Le Monde*, May 30, 2003; and Nicolas Lambert, *Elf, la Pompe Afrique: Lecture d'un Procès*, Tribord, Bruxelles, 2006, p. 57–58.

40 "Selon WikiLeaks, Omar Bongo Aurait Détourné des Fonds au Profit de Partis Français," *Le Monde*, December 30, 2010.
41 Lounis Aggoun and Jean-Baptiste Rivoire, *Françalgérie, Crimes et Mensonges d'États,* op. cit., p. 165–169.
42 Loïk Le Floch-Prigent, interview with the author, op. cit.
43 Catherine Simon, "Gabon: Urnes Bourrées, Manque de Bulletins . . . Le Fiasco des Premières Élections Démocratiques," *Le Monde*, September 20, 1990.
44 Loïk Le Floch-Prigent, interview with the author, op. cit.
45 Ibid.
46 Ibid.
47 Patrick Benquet, *La Françafrique*, op. cit.
48 Loïk Le Floch-Prigent, interview with the author, op. cit.
49 Ibid.
50 "Congo: Des Français et des Israéliens pour la Formation de l'Armée et des Forces Spéciales," *Le Monde,* AFP, February 18, 1994.
51 "Congo-Pétrole: Condamnation Lissouba, Exemple Moralisation Publique pour Brazzaville," AFP, December 30, 2001.
52 Alban Monday Kouango, *Cabinda Un Koweït Africain*, L'Harmattan, Paris, 2003, p. 109.
53 "Des Combats avec l'Angola Donnent une Nouvelle Dimension à la Guerre Civile Congolaise," *Le Monde,* AFP, Reuters, October 4, 1997.
54 "Le PDG d'Elf, Philippe Jaffré, a Rencontré le Président Congolais Denis Sassou-Nguesso," *Le Monde*, Reuters, October 29, 1997; Frédéric Fritscher, "Le Centre de Brazzaville, Totalement Dévasté, Est aux Mains des Pillards," *Le Monde*, October 8, 1997; and Claude Angeli, "Elf Dégaine Plus Vite que l'Elysée au Congo," *Le Canard Enchaîné*, November 12, 1997.
55 "Chirac se Met en Quatre pour un Policier Congolais," *Le Canard Enchaîné*, April 7, 2004.
56 Xavier Harel, *Afrique, Pillage à Huis Clos: Comment une Poignée d'Initiés Siphonne le Pétrole Africain*, Fayard, Paris, 2007, chapter 6: "Les Disparus du Beach," p. 89–110.
57 Loïk Le Floch-Prigent, interview with the author, op. cit.

Chapter 26: Grandeur and Decadence

1 International Energy Agency, searchable at iea.org.
2 "Coal: Resources and Future Production, Energy Watch Group," 2007, searchable at energywatchgroup.org.
3 Nicholas Georgescu-Roegen, "The Entropy Law and the Economic Problem," Distinguished Lecture Series no. 1, Department of Economics, University of Alabama 1971, *The Ecologist*, July 1972; and Nicholas Georgescu-Roegen, *La Décroissance: Entropie, Écologie, Économie*, Sang de la Terre, Paris, 1995, p. 43.
4 Nicholas Georgescu-Roegen, *La Décroissance*, p. 62.
5 Ibid., p. 55.
6 Ibid., p. 65.
7 Sadi Carnot, *Réflexions sur la Puissance Motrice du Feu et sur les Machines Propres à Développer Cette Puissance*, Paris, 1824.
8 Nicholas Georgescu-Roegen, *Energy and Economic Myths: Institutional and Analytical Economic Essays*, Pergamon Press, 1976, p. 59.
9 Ibid., p. 62.
10 Ibid., p. 61.
11 Donella H. Meadows, Dennis L. Meadows, Jorgen Randers, and William W. Behrens, *The Limits to Growth*, op. cit.

12 Laure Noualhat, "Le Scénario de l'Effondrement l'Emporte" (interview with Dennis Meadows), *Libération*, June 17, 2012.

13 Charles A. S. Hall and John W. Day Junior, "Revisiting the Limits to Growth after Peak Oil," *American Scientist*, vol. 97, May/June 2009; and Graham Turner, "A Comparison of the Limits to Growth with Thirty Years of Reality," CSIRO *Working Papers*, June 2008.

14 Josette Alia, "Le Chemin du Bonheur" (interview with Sicco Mansholt), *Le Nouvel Observateur*, June 12, 1972.

15 Ibid.

16 Ibid.

17 Ibid.

18 Alexandre Soljenitsyne, *Lettre aux Dirigeants de l'Union Soviétique et Autres Textes*, Seuil, Paris, 1972, cited in François Jarrige, "Soljenitsyne, Dissident Décroissant," *La Décroissance*, June 2012.

19 Laurence Reboul, Alber Te Pass, and Jean-Claude Thill, *La Lettre Mansholt: Réactions et Commentaires*, Éditions Pauvert, Paris, 1972.

20 Friedrich von Hayek, "The Pretense of Knowledge," Stockholm, December 11, 1974, nobelprize.org/nobel_prizes/economic-sciences/laureates/1974/hayek-lecture.html, cited in Thimothée Duverger, "De Meadows à Mansholt," op. cit.

21 "Implications of Worldwide Population Growth for U.S. Security and Overseas Interests," op. cit., p. 41.

22 Ibid., p. 42–43.

23 Cutler J. Cleveland, Robert Costanza, Charles A. S. Hall, and Robert Kaufmann, "Energy and the U.S. Economy: A Biophysical Perspective," *Science*, vol. 225, no. 4665, August 31, 1984, p. 890–897.

24 Energy return on investment, or EROI.

25 Robert Ayres, Leslie Ayres, and Benjamin Warr, "Energy, Power and Work in the US Economy, 1900–1998," op. cit., p. 78.

26 Silicon Valley Toxic Coalition, svtc.org.

27 "Transport Energy Futures: Long-Term Oil Supply Trends and Projections," Bureau of Infrastructure, Transport and Regional Economics, Australian government, 2009, cited in Jean-Marc Jancovici, "Utiliser du Pétrole? Mais pour Quoi Faire," June 2011, searchable at manicore.com.

28 Jean Laherrère, "US Per Capita GDP, Income and Energy Consumption," October 2011; and Jean Laherrère, "Production Mondiale de Combustibles Fossiles par Habitant," July 2013, searchable at aspofrance.viabloga.com.

29 Gail Tverberg, "Ten Reasons Why High Oil Prices Are a Problem," fig. 5, January 17, 2013, searchable at ourfiniteworld.com/2013/01/17/ten-reasons-why-high-oil-prices-are-a-problem.

30 Jean-Marc Jancovici, "La Dématérialisation de l'Économie: Mythe ou Réalité?" *La Jaune et la Rouge* (l'École Polytechnique), September 2007.

31 Philippe Bihouix and Benoît de Guillebon, *Quel Futur pour les Métaux? Raréfaction des Métaux: Un Nouveau Défi pour la Société*, EDP Sciences, Les Ulis, 2010, p. 24–25.

32 Ibid.

33 Energy Information Administration, searchable at eia.gov.

34 Philippe Bihouix and Benoît de Guillebon, *Quel Futur pour les Métaux?*, op. cit., p. 56.

35 OCDE. One exception: Norway.

36 Mary Meeker, "USA Inc.," KPCG, 2011 (sources: US Treasury Department, Federal Reserve).

37 Mike Davis, *Le Pire des Mondes Possibles*, op. cit., p. 157.

38 Ibid., p. 8.

39 Matthieu Auzanneau, "La Banque Mondiale en Passe de Réhabiliter le Rôle de l'État," *Le Monde,* April 20, 2007, searchable at le-monde.fr.
40 *World Urbanization Prospects, the 2011 Revision*: searchable at un.org/development/desa/publications/world-urbanization-prospects-the-2011-revision.html.
41 Mike Davis, *Le Pire des Mondes Possibles*, op. cit., p. 27.
42 Ibid., p. 41
43 Terry Lynn Karl, *The Paradox of Plenty*, op. cit., p. 4 (interview with Pérez Alfonso).
44 Mike Davis, *Le Pire des Mondes Possibles*, op. cit., p. 19.
45 Ibid., p. 19.
46 Ibid., p. 27 and 71.
47 Majid Rahnema, *Quand la Misère Chasse la Pauvreté: Essai,* Actes Sud, Arles, 2003.
48 Ibid., p. 14.
49 According to the Food and Agriculture Organization (FAO).
50 Eleanor Coppola, *Hearts of Darkness: A Filmmaker's Apocalypse* (documentary), 1991.
51 Hazel Smith, *Hungry for Peace: International Security, Humanitarian Assistance, and Social Change in North Korea,* United States Institute of Peace, Washington, 2005, p. 66; and Christine Ahn, *Famine and the Future of Food Security in North Korea*, Institute for Food and Development Policy, 2005, p. 7, searchable at foodfirst.org/publication/famine-and-the-future-of-food-security-in-north-korea.
52 Loïc Wacquant, *Corps et Âme: Carnets Ethnographiques d'un Apprenti Boxeur*, Agone, Marseille, 2002, p. 23.
53 Serge Latouche, "Schmidheiny ou la Farce du Développement Durable," *Politis*, March 29, 2012.
54 Philip Shabecoff, "Global Warming Has Begun, Expert Tells Senate," *The New York Times*, June 24, 1988.
55 Jeremy Leggett, *The Carbon War: Global Warming and the End of the Oil Era*, Routledge, Oxford, 2001, p. 2–3.
56 Ibid., p. 1.
57 Ibid., p. 4.
58 Ibid.
59 Ibid., p. 10–11 and 200–201.
60 "Hohepriester im Kohlenstoff-Klub," *Der Spiegel*, April 3, 1995.
61 Jeremy Leggett, *The Carbon War*, op. cit., p. 12–13.
62 Jack Beatty, "A Forecast of the 2000 Election Predicts Squalls and Continued Global Warming," *The Atlantic*, April 14, 1999; and John Drexhage and Deborah Murphy, *Sustainable Development: From Brundtland to Rio 2012*, United Nations, New York, September 2010, searchable at un.org.
63 Al Gore, *Earth in the Balance: Ecology and the Human Spirit*, Houghton Mifflin, Boston, 1992.
64 Jeremy Leggett, *The Carbon War*, op. cit., p. 80.
65 Ibid., p. 128–129.
66 John H. Cushman, "Public Backs Tough Steps for a Treaty on Warming," *The New York Times,* November 28, 1997.

Chapter 27: Oil's Future Decline Is Announced

1 "The Next Shock," *The Economist*, May 4, 1999.
2 Steve Coll, *Private Empire: ExxonMobil and American Power*, Penguin Books, New York, 2013, p. 59.
3 Ibid., p. 54–57.
4 Ibid., p. 55.

5 Ibid., p. 55–57.
6 Ibid., p. 66.
7 Interview with the author, 2013.
8 Colin Campbell, *Peak Oil Personalities*, Inspire Books, Skibbereen, 2012, p. 77.
9 Colin J. Campbell and Jean H. Laherrère, "The End of Cheap Oil," op. cit., p. 78–83.
10 International Energy Agency, *World Energy Outlook*, 1998, p. 95–98.
11 "PNAC letters sent to President Bill Clinton," searchable at newamericancentury.org.
12 Ibid.
13 "Excerpts from Pentagon's Plan: 'Prevent the Re-emergence of a New Rival,'" *The New York Times*, March 8, 1992.
14 Elizabeth Becker, "Details Given on Contract Halliburton Was Awarded," *The New York Times*, April 11, 2003.
15 "Full Text of Dick Cheney's Speech at the IP Autumn Lunch," searchable at resilience.org /stories/2004-06-08/full-text-dick-cheneys-speech-institute-petroleum-autumn-lunch-1999.
16 Carla Marinucci, "Chevron Redubs Ship Named for Bush Aide," *The San Francisco Chronicle*, May 5, 2001.
17 Steve Coll, *Private Empire*, op. cit., p. 67–68.
18 Ibid., p. 68.
19 "Cheney Energy Task Force Documents Feature Map of Iraqi Oilfields," July 17, 2003, searchable at judicialwatch.org/press-room/press-releases/cheney-energy-task-force -documents-feature-map-of-iraqi-oilfields.
20 Dana Milbank and Justin Blum, "Document Says Oil Chiefs Met with Cheney Task Force," *The Washington Post*, November 16, 2005.
21 Peter Behr and Alan Sipress, "Cheney Panel Seeks Review of Sanctions," *The Washington Post*, April 19, 2001.

Chapter 28: September 11, 2001

1 Craig Unger, *House of Bush, House of Saud*, op. cit., p. 224.
2 Jane Perlez, "Bush Senior, on His Son's Behalf, Reassures Saudi Leader," *The New York Times*, July 15, 2001; and Craig Unger, *House of Bush, House of Saud*, op. cit., p. 234–235.
3 Craig Unger, *House of Bush, House of Saud*, op. cit., p. 249.
4 Eric Lichtblau, "White House Approved Departure of Saudis after Sept. 11, Ex-Aide Says," *The New York Times*, September 4, 2003.
5 Eric Lichtblau, "New Details on FBI Aid for Saudis after 9/11," *The New York Times*, March 27, 2005.
6 Ibid.
7 Craig Unger, *House of Bush, House of Saud*, op. cit., p. 258.
8 Gerald Posner, *Why America Slept: The Failure to Prevent 9/11*, Random House, New York, 2003, p. 188; and Eric Lichtblau, "Conspiracy of Silence," *The New York Times*, October 12, 2003.
9 Simon Wardell, "Three Royal Princes Die within a Week," World Markets Analysis, July 30, 2002, cited in Craig Unger, *House of Bush, House of Saud*, op. cit., note 70, p. 347.
10 Scott Shane, "Waterboarding Used 266 Times on 2 Suspects," *The New York Times*, April 19, 2009.
11 Philip Shenon, *The Commission: The Uncensored History of the 9/11 Investigation*, Twelve, New York, 2008, p. 9–15.
12 Ibid., p. 18.
13 Philip Zelikow and Condoleezza Rice, *Germany Unified and Europe Transformed: A Study in Statecraft*, Harvard University Press, Cambridge, 1995.

14 Philip Shenon, *The Commission*, op. cit., p. 63–64.
15 *The 9/11 Commission Report*, Norton, New York, 2004, p. 254–277.
16 George W. Bush, *The National Security Strategy of the United States of America*, September 17, 2002, searchable at state.gov/documents/organization/63562.pdf; James Mann, *Rise of the Vulcans: The History of Bush's War Cabinet*, Viking, New York, 2003, p. 317; and Philip Shenon, *The Commission*, op. cit., p. 128.
17 Craig Unger, *House of Bush, House of Saud*, op. cit., p. 261.
18 Kurt Eichenwald, "The Deafness before the Storm," *The New York Times*, September 10, 2012.
19 Craig Unger, *House of Bush, House of Saud*, op. cit., p. 231.
20 Joel Mowbray, "Open Door for Saudi Terrorists. The Visa Express Scandal," *National Review*, June 14, 2002.
21 Craig Unger, *House of Bush, House of Saud*, op. cit., p. 262.
22 Ibid., p. 263.
23 Françoise Chipaux, "Hamid Karzaï, une Large Connaissance du Monde Occidental," *Le Monde,* December 6, 2001.
24 Sumana Chatterjee and David Goldstein, "Analyzing *Fahrenheit 9/11*: It's Accurate to a Degree," *The Seattle Times*, July 5, 2004.
25 Daniel Yergin, *The Quest: Energy, Security, and the Remaking of the Modern World,* Penguin Books, New York, p. 146–147.
26 Iraq Survey Group: "CIA's Final Report. No WMD Found in Iraq," Associated Press, April 25, 2005.
27 "Slick Deals Bush Advisers Cashed In on Saudi Gravy Train," *Boston Herald*, December 11, 2001, cited in Craig Unger, *House of Bush, House of Saud*, op. cit., p. 226.
28 "Colin Powell on Iraq, Race, and Hurricane Relief," ABC News, September 8, 2005, searchable at abcnews.go.com/2020/Politics/story?id=1105979.
29 Tim Weiner, "Plot by Baghdad to Assassinate Bush Is Questioned," *The New York Times*, October 25, 1993; and Jim Lobe, "So, Did Saddam Hussein Try to Kill Bush's Dad?," IPS, October 18, 2004.
30 Thomas Donnelly, *Rebuilding America's Defenses: Strategy, Forces and Resources for a New Century: A Report the Project for the New American Century*, September 2000, p. 26.
31 Ibid., p. 50.
32 Ibid., p. 51.
33 Bob Woodward, *Bush at War*, Simon & Schuster, New York, 2002, p. 37.
34 Ron Suskind, *The Price of Loyalty: George W. Bush, the White House, and the Education of Paul O'Neill*, Simon & Schuster, New York, 2004.
35 Edward L. Morse and Amy Myers Jaffe, *Strategic Energy Policy, Challenges for the 21st Century*, James A. Baker III Institute, Council on Foreign Relations, April 2001, p. 46, searchable at cfr.org/report/strategic-energy-policy-challenges-21st-century.
36 Ibid., p. 16.
37 Ibid., p. 2.
38 Ibid., p. 46.
39 William Dowell, "Foreign Exchange. Saddam Turns His Back on Greenbacks," *Time,* November 13, 2000.
40 Edward L. Morse and Amy Myers Jaffe, *Strategic Energy Policy*, op. cit., p. 13.
41 Ibid., p. 46.
42 Ibid., p. 47 and 84.
43 Jane Mayer, "Contract Sport," *The New Yorker*, February 16, 2004.
44 Laurent Muraviec, *La Guerre au XXIe Siècle*, Odile Jacob, Paris, 2000.

45 Jack Shafer, "The PowerPoint That Rocked the Pentagon," Slate, August 7, 2002, searchable at slate.com/articles/news_and_politics/press_box/2002/08/the_powerpoint_that_rocked_the_pentagon.html.
46 "Poland Seeks Iraqi Oil Stake," *BBC*, July 3, 2003.
47 Ibid.
48 Éric Leser, "Halliburton Perd la Distribution d'Essence en Irak," *Le Monde*, January 2, 2004.
49 Elizabeth Becker, "Details Given on Contract Halliburton Was Awarded," op. cit.
50 Eric Schmitt, "Halliburton Stops Billing U.S. for Meals Served to Troops," *The New York Times*, February 17, 2004.
51 "Italy Sent Troops to Iraq to Secure Oil Deal: Report," *Khaleej Times*, May 14, 2005.
52 Alan Greenspan, *The Age of Turbulence*, Penguin Books, New York, 2007, p. 463.
53 Ibid.
54 Ibid., p. 456.
55 Matt Purple, "Hagel Skewers Iraq War, Defends Greenspan's Oil Comments," *CNS News*, July 7, 2008, searchable at cnsnews.com/news/article/hagel-skewers-iraq-war-defends-greenspans-oil-comments.
56 "Abizaid: 'Of Course It's about Oil, We Can't Really Deny That'," *The Huffington Post*, March 28, 2008, searchable at huffingtonpost.com/2007/10/15/abizaid-of-course-its-abo_n_68568.html; and "Courting Disaster: Fight for Oil, Water and a Healthy Planet," youtube.com/watch?v=9sd2JseupXQ, 21'45''.
57 "John Bolton Admits All of These Wars Are for Oil," youtube.com /watch?v=rAgv6HaOHzM, 1'28''.
58 Greg Muttitt, *Fuel on the Fire: Oil and Politics in Occupied Iraq*, Vintage, New York, 2011.
59 Paul Bignell, "Secret Memos Expose Link between Oil Firms and Invasion of Iraq," *The Independent*, April 19, 2011.
60 Ibid.
61 Ibid.
62 Michael Theodoulou and Roland Watson, "West Sees Glittering Prizes Ahead in Giant Oilfield," *The Times*, July 11, 2002.
63 Dan Morgan and David B. Ottaway, "In Iraqi War Scenario, Oil Is Key Issue; U.S. Drillers Eye Huge Petroleum Pool," *The Washington Post*, September 15, 2002.
64 Tom Cholmondeley, "Over a Barrel," *The Guardian*, November 22, 2002.
65 Jeff Gerth, "U.S. Is Banking on Iraq Oil to Finance Reconstruction," *The New York Times*, April 10, 2003.
66 Chip Cummins, Susan Warren, and Bhushan Bahree, "Iraq: New Drill: Inside Giant Oil Industry, a Maze of Management Tensions," *The Wall Street Journal*, April 30, 2003, cited in Larry Everest, *Oil, Power and Empire*, op. cit., p. 263.
67 Chalmers Johnson, *Nemesis*, op. cit., p. 50.

Chapter 29: Shocks and Ruptures

1 Energy Information Administration, "Production of Crude Oil Including Lease Condensate," searchable at eia.gov.
2 Juan Forero, "Documents Show CIA Knew of a Coup Plot in Venezuela," *The New York Times*, December 3, 2004.
3 Eric Schmitt, "U.S. to Withdraw All Combat Forces from Saudi Arabia," *The New York Times*, April 29, 2003.
4 Bob Woodward, *Plan of Attack*, Simon & Schuster, New York, 2004, cited in Craig Unger, *House of Bush, House of Saud*, op. cit., p. 285.

5 Chalmers Johnson, *Nemesis*, op. cit., p. 51–52.

6 Richard Oppel and Diana Henriques, "Bechtel Has Ties in Washington, and to Iraq," *The New York Times*, April 18, 2003.

7 Sherie Winston, "Bechtel Seeks to Register Potential Iraq Rebuilding Subcontractors," Engineering News-Record, April 24, 2003, searchable at enr.com/articles/30373-bechtel-seeks-to-register-potential-iraq-rebuilding-subcontractors; and Stuart Millar, "Pledge To Share Out Bulk of Iraq Pie," *The Guardian*, May 24, 2003.

8 Warren Vieth, "Iraqi Exiles Say They're Excluded from Rebuilding," *Los Angeles Times*, August 10, 2003, cited in Larry Everest, *Oil, Power and Empire*, op. cit., p. 224.

9 "Iraqi Leadership Opens All Industries Except Oil to Foreign Investors," Associated Press, September 22, 2003.

10 Naomi Klein, *La Stratégie du Choc: La Montée d'un Capitalisme du Désastre*, Babel/Actes Sud, Arles, 2010, p. 533.

11 Eric Leser, "Halliburton Perd la Distribution d'Essence en Irak," *Le Monde*, January 2, 2004.

12 Donald L. Barlett and James B. Steele, "Iraq's Crude Awakening," *Time*, May 10, 2003.

13 BP, Statistical Review of World Energy, searchable at bp.com.

14 Mike Davis, *Petite Histoire de la Voiture Piégée*, La Découverte, Paris, 2012, chapter 9.

15 Jennifer Loven, "Bush Gives New Reason for Iraq War," *The Boston Globe*, Associated Press, August 31, 2005.

16 Stephen Daggett, *Costs of Major U.S. Wars*, Congressional Research Service, Washington, June 29, 2010.

17 Sohbet Karbuz "The US Military Oil Consumption," February 26, 2006, searchable at resilience.org/stories/2006-02-26/us-military-oil-consumption.

18 Steve Fainaru, *Big Boys: les Mercenaires d'Irak*, Nomade Store Europe, collection Original Poche, La Rochelle, 2011, p. 205.

19 Ibid., p. 53 and 214.

20 Ibid., p. 55.

21 Ibid., p. 53.

22 Ibid., p. 47–48.

23 Kit Chellel, "Gulf Keystone Wins Lawsuit over Ex-Soldier's Iraqi-Oil Claim," *Bloomberg*, September 10, 2013.

24 Steve Fainaru, *Big Boys*, op. cit., p. 122. (Original quote from Steve Fainaru, *Big Boy Rules: America's Mercenaries Fighting in Iraq*, Da Capo Press, 2009, p. 73.)

25 Patrick Cockburn, "Revealed: Secret Plan to Keep Iraq under US Control," *The Independent*, June 5, 2008.

26 Dafna Linzer, "Past Arguments Don't Square with Current Iran Policy," *The Washington Post*, March 27, 2005.

27 Gareth Porter, "Cheney Tried to Stifle Dissent in Iran NIE," Inter Press Service, November 8, 2007.

28 Seymour Hersh, "Preparing the Battlefield," *The New Yorker*, July 7, 2008.

29 Faiz Shakir, "Exclusive: To Provoke War, Cheney Considered Proposal to Dress Up Navy Seals As Iranians and Shoot at Them," July 31, 2008, searchable at thinkprogress.org/exclusive-to-provoke-war-cheney-considered-proposal-to-dress-up-navy-seals-as-iranians-and-shoot-at-dfde0aaa2c4c.

30 Steve Coll, *Private Empire*, op. cit., p. 560.

31 Ibid., p. 561.

32 Jean-Michel Bezat, "Pétrole Irakien: Le Patron de Total Explique ses Choix et Reconnaît une Certaine Déception," *Le Monde*, December 15, 2009.

33 Sarah Lyall, "In BP's Record, a History of Boldness and Costly Blunders," *The New York Times*, July 12, 2010.
34 Joseph Stiglitz and Linda Bilmes, *The Three Trillion Dollar War: The True Cost of the Iraq Conflict*, Norton, New York, 2008, p. 116.
35 Matthieu Auzanneau and Jean-Michel Bezat, "La Production Pétrolière des Pays non OPEP Décroîtra 'Juste après 2010,' Prévient l'AIE," *Le Monde*, September 20, 2005; and Matthieu Auzanneau, "Sans l'Or Noir Irakien, le Marché Pétrolier Fera Face à un 'Mur' d'Ici à 2015," *Le Monde*, June 27, 2007.
36 Serge Enderlin, Serge Michel, and Paolo Woods, *Un Monde de Brut: Sur les Routes de l'Or Noir*, Seuil, Paris, 2003.
37 Gerald Nadler, "Expert: No Guarantees in Caspian Oil," Associated Press, November 1, 2002.
38 Neela Banerjee, "For Exxon-Mobil, Size Is a Strength and a Weakness," *New York Times*, March 4, 2003.
39 Kjell Aleklett, *Peeking at Peak Oil*, op. cit., p. 262–263.
40 "Ex-Saudi Oil Exec Criticizes US Govt Oil Supply Forecast," Dow Jones Newswires, October 27, 2004.
41 "World Oil Supplies Dangerously Overestimated and Running Out Fast," *The Insider Mailing List*, October 26, 2004.
42 "Ex-Saudi Oil Exec Criticizes US Govt Oil Supply Forecast," op. cit.
43 Interview with the author, see Matthieu Auzanneau, "Pétrole, la Panne Sèche?," *Le Monde 2*, October 1, 2005.
44 Richard W. Stevenson, "Bush and Saudi Prince Discuss High Oil Prices in Ranch Meeting," *The New York Times*, April 26, 2005.
45 Matthew R. Simmons, *Twilight in the Desert*, Wiley, Hoboken, 2005.
46 Matthieu Auzanneau, "Pétrole, la Panne Sèche?," op. cit.
47 Judy Clark, "Saudi Arabia Seeks $623 Billion in Foreign Investment," *Oil and Gas Journal*, May 17, 2005.
48 Robert L. Hirsch, Roger Bezdek, and Robert Wendling, *Peaking of World Oil Production: Impacts, Mitigation, and Risk Management*, February 2005, searchable at netl.doe.gov /publications/others/pdf/Oil_Peaking_NETL.pdf.
49 Ibid., p. 4 and 64.
50 Interview with the author, Matthieu Auzanneau, "Peak Oil: A Conspiracy to Keep It Quiet," *Le Monde*, Oil Man blog, September 16, 2010, searchable at petrole.blog .lemonde.fr/2010/09/16/interview-with-robert-l-hirsch-22.
51 John Mintz, "Outcome Grim at Oil War Game," *The Washington Post*, June 24, 2005.
52 Robert Wisner, "Ethanol Usage Projections and Corn Balance Sheet (mil. bu.)," searchable at extension.iastate.edu/agdm/crops/outlook/cornbalancesheet.pdf.
53 "The Halliburton Loophole" (editorial), *The New York Times*, November 2, 2009.
54 Izabella Kaminska, "Just Like a Giant Secured Loan to Commodity Producers . . .," *The Financial Times*, FT Alphaville, October 19, 2010, searchable at ftalphaville.ft.com/2010 /10/19/374861/just-like-a-giant-secured-loan-to-commodity-producers.
55 See "About Bilderberg Meetings," bilderbergmeetings.org.
56 "Production of Crude Oil Including Lease Condensate," op. cit.
57 Simon Pary, "Revealed: The Ghost Fleet of the Recession Anchored Just East of Singapore," *The Daily Mail*, September 8, 2009.
58 Gail E. Tverberg, "Oil Supply Limits and the Continuing Financial Crisis," *Energy*, vol. 37, no. 1, January 2012, p. 27–34; and Steve Ludlum, "Further Evidence of the Influence of Energy on the U.S. Economy, part 2," April 23, 2009, searchable at theoildrum.com.

59 James D. Hamilton, "Historical Oil Shocks," *Routledge Handbook of Major Events in Economic History,* Routledge, New York, 2013, p. 239–265.
60 International Energy Agency, *World Energy Outlook 2008*, p. 240.
61 Ibid., p. 37 and 41.
62 Ibid., p. 251.
63 Ali Frick, "Report: McCain Received $881,450 from Big Oil Since He Announced Support for Offshore Drilling," July 31, 2008, searchable at thinkprogress.org/report-mccain-received-881-450-from-big-oil-since-he-announced-support-for-offshore-drilling-7fc9bdf2726a.
64 Jane Mayer, "Covert Operation, the Billionaire Brothers Who Are Waging a War against Obama," August 10, 2010.
65 Aaron Mehta, "Haphazard Firefighting Might Have Sunk BP Oil Rig," July 28, 2010, searchable at publicintegrity.org/2010/07/28/2600/haphazard-firefighting-might-have-sunk-bp-oil-rig.

Chapter 30: Winter, Tomorrow?

1 Carl Gustav Jung, "Au Solstice de la Vie" (1930), *La Réalité de l'Âme, 1. Structure et Dynamique de l'Inconscient*, op. cit., p. 398.
2 Richard Heinberg, "Museletter #241: End of Growth Update Part 1," June 11, 2012, searchable at richardheinberg.com/museletter-241-end-of-growth-update-part-1.
3 International Energy Agency, *World Energy Outlook 2014*, p. 78.
4 Interview with the author; see Matthieu Auzanneau, "Washington Considers a Decline of World Oil Production as of 2011," *Le Monde,* Oil Man blog, March 25, 2010, searchable at petrole.blog.lemonde.fr/2010/03/25/washington-considers-a-decline-of-world-oil-production-as-of-2011.
5 "Meeting the World's Demand for Liquid Fuels," April 7, 2009, searchable at eia.gov.
6 Scott McKeown, "Cheer Up, It's Going to Get Worse," June 17, 2009, searchable at bohemian.com/northbay/cheer-up-its-going-to-get-worse/Content?oid=2173388; "Quand le Secrétaire à l'Énergie US Parlait du 'Peak Oil,'" April 18, 2010, searchable at petrole.blog.lemonde.fr/2010/04/18/quand-le-secretaire-a-lenergie-us-parlait-du-peak-oil.
7 US Joint Forces Command, *Joint Operating Environment 2008*, p. 17; and US Joint Forces Command, *Joint Operating Environment 2010,* p. 29.
8 US Joint Forces Command, *Joint Operating Environment 2010*, op. cit., p. 28.
9 Antony Froggatt and Glada Lahn, *Sustainable Energy Security: Strategic Risks and Opportunities for Business,* Lloyd's & Chatham House, London, 2010, searchable at chathamhouse.org.
10 "Peak Oil," Sicherheitspolitische Implikationen Knapper Ressourcen, July 2010, searchable at bundeswehr.de.
11 HSBC Global Research, "Energy in 2050: Will Fuel Constraints Thwart Our Growth Projections?," March 2011, p. 2.
12 Gail Tverberg, "Dr. James Schlesinger: 'The Peak Oil Debate Is Over' at ASPO-USA Conference," searchable at theoildrum.com.
13 "Obama Signals Tough Stance, Drilling to Resume," Reuters, June 8, 2010.
14 Shakuntala Makhijani, "Cashing In on All of the Above: U.S. Fossil Fuel Production Subsidies under Obama," Oil Change International, July 2014, searchable at priceofoil.org.
15 "Pines, Beetles and Bears," *The New York Times*, July 26, 2010.
16 Bill Graveland, "Athabasca Glacier Melting at 'Astonishing' Rate of More Than Five Meters a Year," *The Globe and Mail*, May 24, 2014.
17 Sylvain Lapoix and Daniel Blancou, "Les Pionniers des Gaz de Schiste," *La Revue Dessinée,* no. 1, fall 2013; and Gregory Zuckerman, "Breakthrough: The Accidental Discovery That Revolutionized American Energy," *The Atlantic*, November 6, 2013.

18 Pierre-René Bauquis, *Parlons Gaz de Schiste en Trente Questions*, La Documentation Française, Paris, 2014.
19 Steve Coll, *Private Empire*, op. cit., p. 201.
20 Jerry Dicolo and Tom Fowler, "Exxon: 'Losing Our Shirts' on Natural Gas," *The Wall Street Journal*, June 27, 2012.
21 Ed Crooks, "Shell Chief Warns of Era of Energy Volatility," *The Financial Times*, January 21, 2011.
22 According to Nelson Mandela, the ability to import crude oil was vital for the survival of apartheid; and that the oil embargo was one of the most effective sanctions against the apartheid regime. Cited in Richard Hengeveld and Jaap Rodenburg, *Embargo: Apartheid's Oil Secrets Revealed*, Amsterdam University Press, Amsterdam, 1994, p. 10.
23 Carrie Tait, "Total Shelves $11 Billion Alberta Oil Sands Mine," *The Globe and Mail*, May 29, 2014.
24 Written exchange with the author, January 2014; see "Pourquoi Total s'Intéresse au Pétrole de Schiste Russe," March 28, 2014, searchable at petrole.blog.lemonde.fr/2014/03/28/total-en-discussion-pour-le-petrole-de-schiste-russe.
25 Nafeez Ahmed, "Former BP Geologist: Peak Oil Is Here and It Will "Break Economies," *The Guardian*, December 23, 2013.
26 US Energy Information Administration, *Annual Energy Outlook 2018*, February 6, 2018, p. 45.
27 Pierre-René Bauquis, "Les Nouvelles Découvertes de Gaz de Schiste Retarderont à Peine le Pic Pétrolier," *Le Monde*, May 9, 2012.
28 Katie Hunt, "China Faces Steep Climb to Exploit Its Shale Riches," *The New York Times*, September 30, 2013; and Chen Aizhu, "China's Ragtag Shale Army a Long Way from Revolution," Reuters, March 11, 2013.
29 Andrew Kramer, "Memo to Exxon: Business with Russia Might Involve Guns and Balaclavas," *The New York Times*, August 31, 2011.
30 *World Energy Outlook 2012*, op. cit., p. 110.
31 Interview with the author, April 2012.
32 *World Energy Outlook 2017*, International Energy Agency, November 2017, pp. 185-186.
33 "Nouvelle Chute en 2013 de la Production de Brut des 'Majors,' Désormais Contraintes à Désinvestir," *Le Monde*, March 17, 2014, searchable at petrole.blog.lemonde.fr/2014/03/17/nouvelle-chute-en-2013-de-la-production-de-brut-des-majors-desormais-contraintes-a-desinvestir.
34 Daniel Gilbert and Justin Scheck, "Big Oil Companies Struggle to Justify Soaring Project Costs," *The Wall Street Journal*, January 28, 2014.
35 Yazid Taleb, "Au Rythme Actuel de Consommation d'Énergie, le Potentiel Algérien Est de Vint Ans et Non Cinquante—Tewfik Hasni (audio)," November 12, 2013, searchable at maghrebemergent.info.
36 Melissa Roumadi, "La Rente Pétrolière en Péril," *El Watan*, February 23, 2014.
37 "Did BP Ask for Lockerbie Bomber's Release? U.S. Senators Seek Probe of Al-Megrahi Emancipation," Associated Press, July 13, 2010, searchable at huffingtonpost.com.
38 "La France Aurait Conclu un Accord sur le Pétrole Libyen avec le CNT," AFP, September 1, 2011.
39 Anene Ejikeme, "The Oil Spills We Don't Hear About," *The New York Times*, June 4, 2010.
40 Chineme Okafor, "Nigeria's Crude Oil Production Decline Rates Pegged at 20 Per Cent," March 23, 2014, searchable at thisdaylive.com.
41 Serge Enderlin, Serge Michel, and Paolo Woods, *Un Monde de Brut*, op. cit., p. 224.
42 "Oil Market Weekly," JP Morgan, May 23, 2014, p. 2.
43 "Congo: Nette Baisse de la Production Pétrolière," AFP, August 12, 2013.

44 Guy Chazan, "Kashagan Delays Mar Outlook for ENI, Total and Shell," *The Financial Times*, July 20, 2014.
45 Juliet Samuel, "'Harsh Errors' in Azerbaijan Land BP in More Hot Water," *The Times*, October 12, 2012.
46 *World Energy Outlook 2012*, op. cit., p. 116.
47 Ibid. Also, International Energy Agency, World Energy Outlook 2017, p. 188.
48 Kate Mackenzie, "Saudi Arabia's Real Energy Problem(s)," *The Financial Times*, July 5, 2010, searchable at ft.com.
49 Fatih Birol, "World Energy Outlook 2014, London Presentation," November 12, 2014, p. 5, searchable at iea.org/publications/freepublications/publication/WEO2014.pdf.
50 Ben Hubbard, Clifford Krauss, and Eric Schmittjan, "Rebels in Syria Claim Control of Resources," *The New York Times*, January 28, 2014.
51 On this point, see a number of articles published between 2012 and 2014 by Claude Angeli, "Le Mea Culpa des Financiers Dudjihad," *Le Canard Enchaîné*, September 10, 2014.
52 Peter M. Jackson and Leta K. Smith, "Exploring the Undulating Plateau: The Future of Global Oil Supply," *Philosophical Transactions of the Royal Society A*, vol. 372, January 13, 2014.
53 Marc Préel, "Oil Industry Says Crude Far from Depleted," AFP, December 8, 2011.
54 Bernard Rogeaux and Yves Bamberger, "Quelles Solutions des Industriels Peuvent-ils Apporter aux Problèmes Énergétiques?," *Revue de l'Énergie*, no. 575, January/February 2007.
55 Robert Ayres and Benjamin Warr, "Accounting for Growth: The Role of Physical Work," *Structural Change and Economic Dynamics*, vol. 16, no. 2, June 2005, p. 181–209; Robert Ayres and Benjamin Warr, *The Economic Growth Engine: How Energy and Work Drive Material Prosperity*, Edward Elgar Publishing, Cheltenham, 2009; Gaël Giraud and Zeynep Kahraman, "How Dependent Is Output Growth from Primary Energy?" A Shift Project presentation in Paris on March 6, 2014; and Matthieu Auzanneau, "Gaël Giraud, CNRS: 'Le Vrai Rôle de l'Énergie Va Obliger les Économistes à Changer de Dogme,'" *Le Monde*, April 19, 2014, searchable at petrole.blog.lemonde.fr/2014/04/19/gael-giraud-du-cnrs-le-vrai-role-de-lenergie-va-obliger-les-economistes-a-changer-de-dogme.
56 Jean-Marc Jancovici, "Vous Avez Dit Normal?," *Les Échos*, June 3, 2014.
57 Collectif, *Le Monde en 2030 Vu par la CIA*, Éditions des Équateurs, Paris, 2013, p. 271.
58 Alfred James Lotka, "Contribution to the Energetics of Evolution" and "Natural Selection as a Physical Principle," *Proceeding of the National Academy of Sciences*, 1922, vol. 8, no. 6, p. 147 and 151, cited in François Roddier, "Thermodynamique et Économie, des Sciences Exactes aux Sciences Humaines," exposé presented at CNAM, in Paris on December 2, 2013.
59 Cited in François Roddier, *Thermodynamique de l'Évolution*, op. cit., p. 65.
60 Ibid., p. 31–32; and Isabelle Stengers, "Structure Dissipative," *Encyclopaedia Universalis*, searchable at universalis.fr/encyclopedie/structure-dissipative.
61 François Roddier, *Thermodynamique de l'Évolution*, op. cit., p. 50.
62 Ibid., p. 164.
63 "L'Empreinte Carbone de la Consommation des Français: Évolution de 1990 à 2007," *Observation et Statistiques*, no. 114, March 2012, searchable at statistiques.developpement-durable.gouv.fr /publications/p/1939/1178/lempreinte-carbone-consommation-francais-evolution-1990.html.
64 John M. Broder, "Climate Change Doubt Is Tea Party Article of Faith," *The New York Times*, October 20, 2010.
65 Olivier Razemon, "En France, 8 Millions de Pauvres Peinent à se Déplacer Quotidiennement," *Le Monde*, July 8, 2014, searchable at transports.blog.lemonde.fr/2014/07/08/en-france-8-millions-de-pauvres-peinent-a-se-deplacer-quotidiennement.

66 "'Le Super à 2 i le Litre Est Inéluctable' pour le PDG de Total," *Le Parisien*, April 12, 2011.
67 Nafeez Ahmed, "Peak Oil, Climate Change and Pipeline Geopolitics Driving Syria Conflict," *The Guardian,* May 13, 2013.
68 Olivier Rech, "Croissance = Dette = Facture Énergétique?," *Le Monde,* April 11, 2013, searchable at petrole.blog.lemonde.fr/2013/04/11/croissance-dette-facture-energetique.
69 Safa Motesharrei, Jorge Rivas, and Eugenia Kalnay, "Human and Nature Dynamics (HANDY): Modeling Inequality and Use of Resources in the Collapse or Sustainability of Societies," *Ecological Economics*, vol. 101, May 2014, p. 90–102.
70 François Grosse, "Is Recycling 'Part of the Solution'? The Role of Recycling in an Expanding Society and a World of Finite Resources," *SAPIENS,* vol. 3, no. 1, 2010, searchable at journals.openedition.org/sapiens/906.
71 François Roddier, *Thermodynamique de l'Évolution*, op. cit.

Afterword

1 This hypothesis was developed with talent and steadfastness by Gail Tverberg. See ourfiniteworld.com.
2 Tim McDonnell, "Forget the Paris Agreement: The Real Solution to Climate Change Is in the U.S. Tax Code," Wonkblog, *The Washington Post*, October 2, 2017. See, for instance: Peter Erickson, Adrian Down, Michael Lazarus, and Doug Koplow, "Effect of Subsidies to Fossil Fuel Companies on United States Crude Oil Production," *Nature Energy*, vol. 2, November 2017, p. 891–898.
3 Christopher Hamilton, "The Mysterious Movements of U.S. Oil Production, Demand, Price, and Interest Rates," April 25, 2017, searchable at seekingalpha.com/article/4064796-mysterious-movements-u-s-oil-production-demand-price-interest-rates.
4 Jian-Liang Wang, Jiang-Xuan Feng, Yongmei Bentley, Lian-Yong Feng, and Hui Qu, "A Review of Physical Supply and EROI of Fossil Fuels in China," *Petroleum Science*, vol. 14, no. 4, November 2017, p. 806–821.

Index

Note: Page numbers preceded by *ci* refer to images in the color insert. Page numbers followed by "n" refer to page notes. Page numbers in italic refer to maps, charts, and graphs in the appendix.

post carbon **institute**

Post Carbon Institute provided support for this English edition of *Oil, Power, and War*, including the foreword by Richard Heinberg and technical review of the final three chapters.

Post Carbon Institute envisions a transition to a more resilient, equitable, and sustainable world. It provides individuals and communities with the resources needed to understand and respond to the interrelated ecological, economic, energy, and equity crises of the twenty-first century.

Visit postcarbon.org for a full list of its fellows, publications, and other educational products.

About the Author

Camille Poul

Matthieu Auzanneau is the director of The Shift Project, a European think tank focusing on energy transition and the resources required to make the shift to an economy free from fossil fuel dependence, and also from greenhouse gas emissions. Previously, he was a journalist, based in France, and mostly writing for *Le Monde*. He continues to write his *Le Monde* blog, "Oil Man," which he describes as "a chronicle of the beginning of the end of petroleum." His original edition of this book, the French *Or Noir: La Grande Histoire du Pétrole* (La Découverte, 2015) was awarded the Special Prize of the French Association of Energy Economists in 2016.

About the Foreword Author

Richard Heinberg, senior fellow at the Post Carbon Institute, is the author of thirteen books including *Our Renewable Future* (2016), *The End of Growth* (2011), and *Peak Everything* (2007). He is regarded as one of the world's foremost advocates for a shift away from our current reliance on fossil fuels. His essays and articles have appeared in numerous publications, including *Nature*, *Reuters*, *Wall Street Journal*, *The American Prospect*, *Public Policy Research*, *Quarterly Review*, *Yes!*, and *The Sun*; and on web sites such as Resilience.org, TheOilDrum.com, Alternet.org, ProjectCensored.com, and Counterpunch .org. He lectures widely on energy and climate issues to audiences in fourteen countries, addressing policy makers at many levels, and has appeared in many film and television documentaries, including Leonardo DiCaprio's *11th Hour*. He is a recipient of the M. King Hubbert Award for Excellence in Energy Education, and in 2012 was appointed to His Majesty the King of Bhutan's International Expert Working Group for the New Development Paradigm initiative.

Chelsea Green Publishing is committed to preserving ancient forests and natural resources. We elected to print this title on 30-percent postconsumer recycled paper, processed chlorine-free. As a result, for this printing, we have saved:

35.1 Trees (40' tall and 6-8" diameter)
15 Million BTUs of Total Energy
15,200 Pounds of Greenhouse Gases
3,000 Gallons of Wastewater
120 Pounds of Solid Waste

Chelsea Green Publishing made this paper choice because we and our printer, Thomson-Shore, Inc., are members of the Green Press Initiative, a nonprofit program dedicated to supporting authors, publishers, and suppliers in their efforts to reduce their use of fiber obtained from endangered forests. For more information, visit: www.greenpressinitiative.org.

Environmental impact estimates were made using the Environmental Defense Paper Calculator. For more information visit: www.papercalculator.org.

the politics and practice of sustainable living

CHELSEA GREEN PUBLISHING

Chelsea Green Publishing sees books as tools for effecting cultural change and seeks to empower citizens to participate in reclaiming our global commons and become its impassioned stewards. If you enjoyed *Oil, Power, and War*, please consider these other great books related to energy, politics, and public policy.

LIMITS TO GROWTH
The 30-Year Update
DONELLA MEADOWS, JORGEN RANDERS, and DENNIS MEADOWS
9781931498586
Paperback • $22.50

2052
A Global Forecast for the Next Forty Years
JORGEN RANDERS
9781603584210
Paperback • $24.95

REINVENTING FIRE
Bold Business Solutions for the New Energy Era
AMORY B. LOVINS
9781603585385
Paperback • $29.95

SURVIVING THE FUTURE
Culture, Carnival, and Capital in the Aftermath of the Market Economy
DAVID FLEMING
9781603586467
Paperback • $20.00

For more information or to request a catalog, visit **www.chelseagreen.com** or call toll-free **(800) 639-4099**.